Quantitative Methods
for Business Decisions

Fifth edition

Jon Curwin and Roger Slater

THOMSON

Australia · Canada · Mexico · Singapore · Spain · United Kingdom · United States

THOMSON

Quantitative Methods for Business Decisions – 5th edition

Copyright © **Jon Curwin and Roger Slater**

The Thomson logo is a registered trademark used herein under licence.

For more information, contact Thomson Learning, High Holborn House, 50-51 Bedford Row, London WC1R 4LR or visit us on the World Wide Web at:
http://www.thomsonlearning.co.uk

British Library Cataloguing-in-Publication Data
A catalogue record for this book is available from the British Library

ISBN-13: 978-1-86152-531-4
ISBN-10: 1-86152-531-1

First edition published 1985 Chapman & Hall, reprinted 1986
Second edition published 1988, reprinted 1989
Third edition published 1991, reprinted 1993, 1994
Fourth edition published 1996 International Thomson Press
This edition published 2002 by Thomson Learning
Reprinted 2002, 2003, 2004 and 2005 by Thomson Learning

Designed and produced by Gray Publishing, Tunbridge Wells
Printed in Singapore by Seng Lee Press

Contents

Part 7	Mathematical background	547

Chapter 23	Mathematical relationships	549

Chapter 24	Matrices	576

Chapter 25	Use of calculus	596

Appendices   629

Preface

Welcome to the fifth edition of *Quantitative Methods for Business Decisions*. This new edition, like the previous editions, is intended for students from a variety of backgrounds who find that their courses require a good understanding of mathematics and statistics. We make few assumptions about what the student can remember from previous studies and concentrate on the development of an approach, which will enhance a range of skills with numbers. Each topic is developed using examples that should illustrate the formula or technique being considered, and give some indication of practical application.

All kinds of activities require the use of numbers and all kinds of courses expect students to work with numerical representation. Very few substantial problems will have no numerical content. **We want you to be effective when working with numbers.** An understanding of the quantitative approach should add to the ways you can manage problems and give you reason to feel more confident in the work you do.

Many of the chapters have been substantially revised to reflect subject and course developments. Three new chapters, 'The Quantitative Approach' (the new Chapter 1), 'Managing Data' (the new Chapter 2) and 'Survey Methods' (the new Chapter 3) replace the original first chapter on 'Data Collection'. We stress the importance of an effective approach to problem solving and the importance of methodology. We have continued to incorporate recent developments in personal computing and have increased the use of Excel spreadsheets. We have decided to include Excel output in the form of screenshots so that the reader can become familiar with the way results are typically seen.

Additional material is also included on the new **website**. You will see numerous references to this website throughout the book – indicated by a web icon. This supportive web material includes sets of data, spreadsheet models and further examples. PowerPoint presentations have been added for the lecturer.

The book is still divided into seven parts: Quantitative information, Descriptive statistics, Measuring uncertainty, Statistical inference, Relating variables and predicting outcomes, Modelling, and Mathematical background. Each part has an introduction and an illustrative case study to provide a context for the chapters that follow. A case icon is used in the text to highlight references to case material. The first five parts also include a **Quick Start** so that students can have a concise summary of what follows, and have a quick source of reference.

We would like to acknowledge the help of our colleague, John Alexander, for his work on the use of computer software in project management.

Jon Curwin
Roger Slater

Part 1   Quantitative information

Introduction

In this first part of the book, we are concerned with the ways of approaching problems and managing data. Working with numbers will be an important part of specifying and solving a wide range of problems. We are increasingly able to get data quickly and manipulate that data into forms that are useful to us. However, we need to be sure that we are working on the right problem and that the **data is valid** for our purposes. Managers often complain that:

- too much time was lost trying to solve the wrong problem, or
- the solution was made to fit the data available, or
- the solution found lacked imagination and creativity.

As we will see, it is important to clarify, and if necessary redefine, the problem we are working on. The approach to problem solving needs to use new ideas, take advantage of new opportunities and generate new opportunities. One only needs to look at the history of business decline to see that the same old answers to the same old problems is a recipe for failure. We will encourage you to search for new sources of data and explore that data in a variety of ways.

'Data' is a very general term and can be taken to mean a few recordings or the outcome of an extensive national or international survey. Whether the data is useful or not is another matter! **An item of data becomes information when it informs the user.** We will look at ways to clarify the purpose of data and the ways in which data can be managed. In all aspects of business or organizational life we are likely to encounter increasing quantities of data. New technologies literally put data at our fingertips – for example, share prices in New York, or stock levels in a warehouse some distance away can be known in minutes. The **Internet** has transformed the flow and the availability of data. The ability to manage data, produce information and work with problems are all seen as an important business competencies.

When managing business problems, and indeed problems in general, we need to consider whether to use only the information already available or whether to collect our own additional information. One of the frustrating aspects of research is that time can be spent generating data that is then found already to exist in some other form. It is worth checking what work has already been done. This type of **desk research** will provide figures of a general kind, for example, the numbers of people by sex, region and income, but is unlikely to provide very detailed information required for a specific task. Desk research may also provide information or identify techniques that you had not thought about. It is always helpful, for example, to find a questionnaire that has been used in a previous study and may only require some modification. We are less likely to find pertinent data on attitudes or opinions and will therefore need to develop data collection methods.

Quantitative methods involve more than obtaining a few numbers and working out a few statistics. A **statistic** is merely a descriptive number. To be of any value, statistics need to 'paint a picture' in an acceptable and valid way. Quantitative methods provide a framework for working with statistics. In addition to providing description, quantitative methods also include a number of ways of testing ideas and modelling problems. It can be argued that to better understand what you see, you need to compare your observations with models and representations. It is this ability to compare and contrast, and to break down a problem, that is known as **analysis**.

Chapter 1 considers a creative approach to problem solving that involves examining the definition of the problem and allowing time to generate a number of creative ideas. As an effective problem solver, you will need to be able to justify the methods (methodology) used, apply models when appropriate and be aware of the measurement being achieved. Chapter 2 is about managing data from a variety of sources, including the internet. The use of survey methods is considered in Chapter 3. As you will see, there is no one best approach. The approach used will depend on user requirements, time and cost. It is often said that the answers can only be as good as the questions. Questionnaire design will be examined, and we consider the importance of a high response rate.

Chapter 4 considers the presentation of data using charts and diagrams. Particular reference will be made to the use of software. The personal computer is of increasing importance and now has the capacity to allow for the easy management of large data sets. Computers also provide the opportunity to experiment with and explore data in ways that would not otherwise be possible. Data should not just reveal the obvious but be a source of new ideas and understandings.

Case 1: Shopping Developments Limited – part 1

Shopping Developments Limited have a number of interests in general purpose, smaller retailing units. They have observed the growth of the nearby New Havens Shopping Precinct, and the Hamblug Peoples Store in particular. They would like to know more about the usage of the shopping precinct and plan to review the information they already have.

As part of a previous project on traffic flow, they had recorded the number of cars entering the New Havens Shopping Precinct car park during 10-minute intervals. This data was collated into the form of a table.

In addition to the traffic flow information, the company also has the outcome of a market research survey conducted by a student who had recently completed a placement year in the marketing department. The questionnaire was written with the requirement of the placement project in mind and only part of the questionnaire and coding sheet still remain available to the company. It is known that a quota sampling method was used but that few other details remain. The relevant parts of the questionnaire and coding sheet are as follows (on the next page).

Table of case 1 data. The number of cars entering the New Havens Shopping Precinct car park during 10-minute recording intervals

10	22	31	9	24	27	29	9	23	12	33	29	14	37	28	29	39	25	34	35
30	40	26	30	23	31	24	27	43	33	36	9	43	36	29	10	30	35	29	21
20	22	35	19	32	26	33	32	26	44	25	35	28	23	39	27	37	41	31	36
41	24	29	34	11	29	12	36	28	32	32	37	26	33	13	28	28	9	39	17
27	11	31	10	38	11	28	27	31	10	11	24	28	32	44	36	26	34	23	35
28	43	24	38	22	34	7	12	32	20	21	35	25	31	17	30	33	30	42	30
12	36	36	13	27	28	13	35	25	34	29	48	11	9	11	41	29	37	25	29
38	30	21	29	32	39	29	18	34	22	18	27	30	44	28	27	31	36	21	27
32	25	31	30	16	11	24	27	31	28	46	34	26	35	34	11	38	10	33	23
23	37	12	12	39	34	26	39	10	29	10	33	39	28	23	42	12	31	28	37
31	32	7	10	28	23	33	11	30	30	30	30	13	12	27	19	45	25	43	30
10	28	33	31	11	37	30	38	21	20	38	24	29	35	41	29	29	32	28	25
40	20	22	13	32	14	24	26	36	42	5	16	38	34	25	36	9	24	38	32
28	23	27	30	26	31	30	39	19	22	35	11	31	30	12	21	31	35	29	21
37	35	11	29	37	9	36	8	32	26	33	42	34	40	15	32	26	15	8	41
11	34	35	24	23	27	27	31	30	27	30	19	21	33	26	34	22	43	31	24
33	29	19	28	30	35	11	35	31	34	13	32	24	11	38	28	32	12	30	38
28	9	25	40	12	29	29	23	12	37	9	27	35	31	10	25	30	26	32	17
37	22	41	10	33	26	28	28	22	29	31	24	29	32	27	29	20	27	7	32
34	26	34	22	25	19	36	12	33	24	27	28	10	26	15	33	33	28	27	29

The questionnaire   *Shopping Survey: New Havens Shopping Precinct*

Questionnaire number: _____

1 Is this your first visit to the New Havens Shopping Precinct?
YES/NO
If YES, go to Q.3

2 How often do you visit the New Havens Shopping Precinct? (TICK ONE)
Less than once a week   ☐
Once a week   ☐
Twice a week   ☐
Three times a week   ☐
More often than this   ☐

3 How long did it take you to travel here? (in minutes) _____

4 How did you travel here? (TICK ONLY MAIN MODE OF TRANSPORT)
Bus   ☐
Car   ☐
Walk   ☐
Other, please specify _____

5 Have you ever visited the Hamblug's shop? YES/NO
If NO, go to Q.10

6 Have you been there today? YES/NO
    If NO, go to Q.10

7 Did you buy any of the following items :
    Fresh vegetables   YES/NO
    Fresh fruit   YES/NO
    Fresh meat   YES/NO
    Preprepared meals   YES/NO
    Frozen food   YES/NO
    If NO to all items, go to Q.10

8 Approximately, how much did you spend on food at Hamblug's today? (TICK
    ONE)
    under £5
    £5 but under £10
    £10 but under £15
    £15 but under £20
    £20 but under £30
    £30 but under £40
    £40 or more

9 How many people do you buy food for?
                    (WRITE IN NUMBER) _____

10 Which area do you live in? (TICK ONE)
    Astrag   ☐   Baldon   ☐
    Cleardon   ☐   Other   ☐

The coding sheet

Column number	Description	Coding used
1	Respondent number	as questionnaire
2	Q.1	1 = yes, 2 = no, 9 = not asked
3	Q.2	1 = less than once a week
		2 = once a week
		3 = twice a week
		4 = three times a week
		5 = more often than this
		9 = no usable answer
4	Q.3	as given in minutes
5	Q.4	1 = bus, 2 = car, 3 = walk, 4 = other
6	Q.5	1 = yes, 2 = no, 9 = not asked
7	Q.6	1 = yes, 2 = no, 9 = not asked
8	Q.7 (fresh vegetables)	1 = yes, 2 = no, 9 = not asked
9	Q.7 (fresh fruit)	1 = yes, 2 = no, 9 = not asked
10	Q.7 (fresh meat)	1 = yes, 2 = no, 9 = not asked
11	Q.7 (prepared food)	1 = yes, 2 = no, 9 = not asked

12	Q.7 (frozen food)	1 = yes, 2 = no, 9 = not asked
13	Q.8	2.5 = under £5
		7.5 = £5 but under £10
		12.5 = £10 but under £15
		17.5 = £15 but under £20
		25 = £20 but under £30
		35 = £30 but under £40
		45 = £40 or more
14	Q.9	number given
15	Q.10	1 = Astrag
		2 = Baldon
		3 = Cleardon
		4 = Other

Shopping Developments Limited are expected to plan and undertake more extensive market research but believe the existing information will be useful. It is thought that a better understanding of the shopping precinct and the Hamblug Peoples Store within it will help develop market research and market development strategies.

It is known that Shopping Developments Limited wish to develop the concept of the general purpose outlet within the shopping precinct environment. It is also known that the closeness of a ferry port to the New Havens Shopping Precinct could be an important factor.

Visit the website

The data on the number of cars entering the New Havens shopping Precinct car park in 10-minute intervals is given in the Excel file SDL1.XLS and the data from the questionnaire in the Excel file SDL2.XLS. You should try to find this data and print out a copy. It will be useful to check the number of entries and the correctness of the coding.

Quantitative information

Managing numbers is an important part of understanding and solving problems. The collecting together of numbers, and other facts and opinions provides **data**. This data only becomes **information** when it informs the user.

The quantitative approach is more than just 'doing sums'. It is about making sense of numbers within a context. To understand problems within a context, it can be useful to work through a number of stages: defining (and redefining) the problem, searching for information, problem description (and again redefinition if necessary), idea generation, solution finding and finally, acceptance and implementation. The use of new ideas being particularly important. One of the aims of the book is to make **you** a better problem solver by using techniques and ideas more effectively.

The development of mathematical models can provide a better understanding of the way things work. Relationships are established using **variables** (a quantity or characteristic or interest that can vary within the problem context), **parameters** (values fixed for a given problem) and by making **assumptions** (things accepted as true).

Data can come from existing sources (**secondary data**) or may need to be collected for the purposes of our research (**primary data**). Government publications like *The Annual Abstract of Statistics*, *The Employment Gazette* and *Social Trends* are all good sources of information. Of increasing importance is the **Internet** which can provide a fast link to traditional sources of information but also access to a wide range of data world-wide. We need to ask whether the data found is appropriate, adequate and without bias. Data can come from a **census** (a complete enumeration of all those people or items of interest) or from a **survey** (a selection from the population of interest).

A survey can be based on methods that give every person a known chance of inclusion (probability sampling) or methods that rely on the judgement of the interviewer (non-probability sampling). A survey should be designed and administered in such a way as to minimize the chance of **bias** (an outcome which does not represent the population of interest). If the survey is likely to miss certain people out or there is a relatively high level of non-response, we would be concerned that the results were not representative.

Even a relatively modest survey is likely to generate several hundred pieces of data. More detailed research may give us several thousand values. To understand and develop solutions to a problem we will to need summarize the data. The use of frequency tables, bar charts, pie charts and histograms provide effective ways of summarizing data and communicating findings to others.

1 The quantitative approach

Numbers provide a universal language that can be easily understood and a description of some aspect of most problems. In fact, some problems can be described almost entirely in numerical terms. The development of a departmental budget will involve to a large extent the modelling of cash inputs and outputs (typically on a spreadsheet), and the manipulation of numbers. In contrast, the recruitment of staff or the management of redundancy may require only a limited use of numbers and a great deal of sensitivity to the people involved.

In making business decisions, we need to recognize the importance of the range of information available and to what extent the problem is numerical by nature or non-numerical. It is useful to distinguish between the **quantitative** and **qualitative** approaches to problem solving. Essentially, the quantitative approach will describe and resolve problems using numbers. Emphasis will be given to the collection of numerical data, the summary of that data and the drawing of conclusions from the data. Measurement is seen as important and factors that cannot be easily measured, such as attitudes and perception, are generally difficult to include in the analysis. Qualitative approaches describe the behaviour of people individually, in groups or in organizations. Description is difficult in numerical terms and is likely to use illustrative examples, generalization and case studies. The qualitative approach can use a variety of methods such as observation and the written response to unstructured questions. Data may come in the form of script, for example, transcripts of interviews or observations such as video recordings.

The **quantitative approach** is about using numbers to help define, describe and resolve a wide range of problems. **However, it is about more than just 'doing sums'.** Numbers will need to make sense within a context. It is the context that will give meaning to those numbers and the relative importance of numerical and non-numerical information. The number 5 for example, can mean the age of a child, the number of days in a typical working week or the number of seconds of oxygen left in a diving cylinder (again the significance of the 5 seconds will depend on the context – whether you are on the surface or not). However, the choice is rarely a simple one between a quantitative and qualitative approach, and your research is likely to have some element of both.

Objectives

By the end of this chapter you should be able to:

- contrast quantitative and qualitative approaches
- identify some of the key elements of problem solving
- understand the importance of methodology
- appreciate the use of models
- distinguish between data and information
- identify different types of data

- understand the importance of the level of measurement achieved
- appreciate the importance of quantitative methods in business decision making.

This chapter will be structured as shown in Figure 1.1.

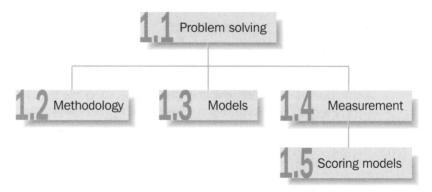

Figure 1.1   Structure of Chapter 1.

1.1 Problem solving

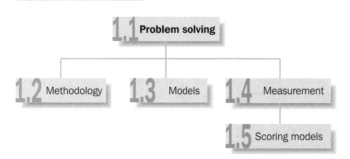

Quantitative methods can be applied in a wide range of problems. The results of a lengthy and complex survey may be summarized by a variety of tables, charts and calculated numbers. These summary numbers are called **statistics**. Trends in the economy, industry sector or market can be identified and compared numerically. Again, a summary number or numbers providing sufficient description to the user. We only need to listen to political debate to become aware of the importance of measures, such as the **Retail Price Index** or the **seasonally adjusted level of unemployment**. The use of summary numbers will allow us to develop concepts like **probability** and the **value of money over time**. The use of quantitative methods will also allow us to build a **model** or **models** that capture our thinking about a particular problem. A model being one or more relationships that make sense of a problem situation of interest. A breakeven model, for example, will describe the costs and revenue of a business (two relationships). When costs exceed revenue the business will make a loss, when revenue exceeds costs the business will make a profit and when these are equal, the business will breakeven.

Numbers can be used at three levels to help us solve problems:

- To describe a wide variety of situations, particularly when large quantities of data are involved. At this level, we would be thinking about statistics such as the mean and standard deviation which would allow us to identify typical values and the spread of values.

- To allow the use of theory. Theory is there to help us. It is useful to assert that measurements follow a normal distribution or that significant in statistical terms.
- To develop models (representations) of real problems and use these mode. to look for improved solutions. A model will illustrate how we think things work, for example, an engineered product such as a car, or a system such as the recording of accounting information.

The use of quantitative methods is a skill that you can develop. It can make you more effective in managing and solving a range of business and non-business problems. We now work in an age of information. It is not unusual for a simple web search to offer 1000 or more sites. Many of the sites will offer only partial information but perhaps a number of other leads. If we are not careful we could drown in a sea of information. We need to be able to focus on the **information required** (rather than all search and no solution!) and focus on the problem itself.

We see value in the old adage that knowing the problem is part of the solution. Business literature is scattered with examples of how individuals and organizations have failed to solve problems, often with very serious consequences. Problems present themselves in very different ways and can often be misleading. It is tempting just to collect some data, undertake some analysis, draw some conclusions and then take action. We advocate such a structured approach to problem solving but we also see problem solving as a process that needs to be managed, that is reflective and is based on an understood methodology.

You will only get the right answer when you have got the right problem!

You will find many examples of how to approach problem solving. The structured approach given in Figure 1.2 is typical.

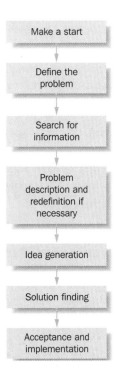

Figure 1.2

1.1.1 Making a start

To illustrate this structured approach to problem solving let's consider a problem that you have (or we think you have). For you to have read this far would suggest to us that you have just started a new quantitative methods course and want to make a success of it. 'Making a start' is about problem sensitivity and being prepared to do something about it. Sensing that there is a problem does not necessarily mean that we know exactly what it is or even how to go about solving it. It does mean that we think a problem exists and we think we need to take some action. What we may observe at this stage are symptoms of the problem. As an analogy, a patient may report to a doctor a tingling feeling in his or her legs which would suggest the source of the problem is also in the legs but these could be the symptoms of a more distant cause, perhaps a back problem. You may at this stage be concerned about the quantitative methods course because it has a reputation for being difficult or you have found mathematics difficult in the past. You may have bought this book because this was the advice given by your lecturer and now feel that you should use it. At this stage awareness is important.

1.1.2 Define the problem

Many regard problem definition as the most important stage of problem solving. Problem definition should **clarify the problem** you are working on for all those involved (often referred to as stakeholders or actors) and be worded in such a way as to give insight into the problem. Two important concepts used in problem definition are **gap** and **problem owner**. A problem can be seen as a gap between what we have and what we want. This also involves the ideas of perception and expectation. It you want a mountain bike and you have a mountain bike, in terms of this definition you don't have a problem. If Roger has a mountain bike but wants a Porsche, then Roger has a problem!

In terms of the new quantitative methods course, the gap could be expressed in terms of the knowledge required to pass the assessment:

in what ways might I acquire the knowledge required to successfully complete this quantitative methods course.

The format 'In what ways might ... ?' (IWWM ... ?) is seen as a useful way to start problem formulation. It is important to note that a number of useful problem definitions are likely to exist. A good problem definition is likely to be thought provoking and likely to have a specific focus (what should I do to pass exams would be seen as rather vague and lacking problem awareness). In this case, the owner is clear (you) and the gap given in terms of knowledge. If you have a good subject knowledge but have difficulties with examinations, you might express the problem in the following form:

in what ways might I use my subject knowledge to improve my examinations performance.

1.1.3 Search for information

A major theme of this book will be turning data into information. In this case, the problem definition, could be used to produce a checklist of the information required. The technique known as the 'Five W's and H', raises the questions of who, what, where, when, why and how. You should be able to develop a range of questions, such as:

who owns this problem?
who can help solve the problem?

what information have I been given on this quantitative methods course?
what subject knowledge do I need?

where can I find this information?
where can I find examples of past examinations papers?

when do the revision classes take place?
when will we be given the assessment dates?

why do I need to attend lectures?
why do I need to buy the text book?

how can I acquire this knowledge?
how can I improve my revision tecniques?

The answers to the questions become information when they can inform your thinking.

1.1.4 Problem description (and redefinition if necessary)

Having considered the problem in more detail and, being more fully informed as a result of fact finding, you can often describe the problem or indeed **redefine** the problem in terms that clarify the possibilities of solution. If for example, you now feel better informed about course content and course support, you may wish to change the problem statement:

in what ways could I organize my time and other resources to successfully complete this quantitative methods course?

1.1.5 Idea generation

It is often suggested that one of the major skills that managers of the future will need is the ability to generate new answers rather than rely on a limited range of current solutions. If mathematics has been a problem to you in the past, perhaps the answer is not to approach your study in the same way. You may find that by taking time to 'brainstorm' your problem, a number of interesting ideas will emerge:

- a study group with other interested students
- using computer-based teaching material
- focusing on what you know rather than what you don't know
- think about problems outside of your course mathematically
- plot progress on a wall chart.

At this stage, it is seen as important to **defer judgement** on the ideas produced even if they seem silly or unrealistic. The concepts of walking around listening to music (the Walkman) or a boat that could fly (the Hovercraft) were seen as ridiculous. The new products and services are based on the ideas and adapted ideas that we are currently prepared to work with. At a later stage we can decide which ideas are worth keeping and working with but at this stage the flow of ideas and the quantity of ideas are of more importance. Ideas are the basis of creativity and it could be the case that an unlikely thought or insight could offer the best future solution. Ideas then are a raw material to work with. To reject ideas at this stage is to limit the number of ways you could consider solving the problem.

1.1.6 Solution finding

Ideas generated are likely to range from rather simple incremental changes to the bizarre. This is the data we now work with. This is also the time to make judgement and develop options that are realistic. Ideas can be clustered together and combined. As a result of fact finding and idea review, a solution or set of solutions may emerge. In many cases the option of 'doing nothing' may also exist. Solutions can often be usefully expressed in terms of what the problem owner can do to close the problem gap. It could be that no single solution will solve the problem and that what is needed is a management of the problem-solving process. As an example, consider a company that is seen as having a quality problem. A number of ideas might emerge that could improve quality, but a gap between what the company produces and what the customer expects could always open at some future point in time. What is important to develop is a problem-solving capability. This capability is enhanced by a fluency in both quantitative and qualitative approaches, a flexibility to manage a diversity of ideas and an ability to challenge barriers to change in an acceptable way.

1.1.7 Acceptance and implementation

Having developed one or more ideas in such a way that they could address aspects of the problem, the next important step is acceptance finding. The ideas may be good, but they also need to be acceptable to the problem owner. We could suggest that you allocate one hour every evening to the study of quantitative methods. You may accept that this has some benefits, but you may find a clash with a part-time job or with other plans. You could, however, take this concept of regular supportive study using a recommended textbook and adapt it to your needs.

Case 1: Shopping Developments Limited – part 1

The case study company, Shopping Developments Limited have an interest in the New Havens Shopping Precinct and have been considering market research and market development strategies.

Suggest ways in which:

(a) the generation and the management of data could help the company, and
(b) decision making (and problem solving) could be approached so as to benefit from the collection of data and the generation of creative ideas.

1.2 Methodology

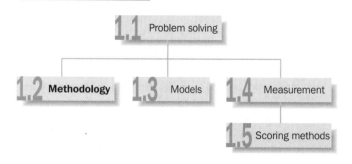

To use any 'old methods' to generate a bit of data is not going to help you become an effective problem solver, or indeed a trusted problem solver within an organizational context. You may well be asked to justify your findings and conclusions against criteria such as **reliability** and **validity**. Methodology is about devising an approach to your research that will work. If the research is repeated then you should expect results to be consistent unless other factors are making a difference. It is often a matter of debate as to whether the differences in new findings are due to the methodology or due to the relationship being observed. If the latest political opinion poll shows a change in voting intentions, we do need to consider whether this is due to a shift in political view or whether it could be an outcome of the sampling process. We know that results are likely to vary a little from sample to sample, but this variation should not overwhelm what you are trying to measure. Your research should also measure what you say it's going to measure. It has been reported, for example, that residents have approved of traffic-calming measures, not because of the impact on traffic flow, but because they are pleased that money is being spent on their area.

In any problem situation there is unlikely to be one obvious best method for getting research results. You will have a choice of approaches. In addition to giving reliable and valid results, your research will also need to be **feasible**. Research is generally carried out within agreed time and cost constraints. You would not want your in-depth, scientifically predictive study of voting to report the day after the election of interest over budget and late!

In the longer-term, it is the ability to manage the problem-solving process in a range of situations that will be of most importance. The skill is knowing what methodology to use. Typically, our approach will be structured and sequential. As a consequence of careful problem definition, a population of interest could be identified. Given the purpose of the research, the type of information required could be clarified and the means of collection devised. Checks could be made on the quality of information and the findings could be reported using various forms of summary.

Suppose, for example, we were asked to assess the views of young people regarding the level of university fees. It would be tempting to list a few questions, ask a few students on the local campus whether they would answer 'yes' or 'no' and then work out the corresponding percentages. **But what does it mean?** We would suggest that mostly such exercises are worthless. They are neither reliable (could be repeated with confidence that the findings would remain the same), valid (is telling you what it should be telling you) or develop your skills in research. It is important to assess research findings and the approach to research critically before deciding whether the outcomes are useful or not. A few questions regarding our example will make the point.

Was the purpose of the research clear?

The answer has got to be <u>no</u>. We would want to know more about the range of views sought and whether we were required to relate these views to other factors such as gender or preferred political party. We would also want to know what was meant by 'young' and the importance placed on other factors, such as interest rates and time allowed for repayment.

Was this research necessary?

Once the purpose of the research is clarified then it is possible to search for existing information. It is likely that someone, somewhere has published findings from similar research. One only has to look at the range of available government publications, studies by the media or the records kept by business on personnel or production to appreciate how easily data accumulate and how extensive data collections are. This kind of enquiry, whether looking through dusty old pamphlets or by giving key words to a search engine on the Internet, is known as '**desk research**'. It is often found that many of the answers already exist and that you get answers to questions you have yet to ask. The use of primary and secondary data will again be considered in Chapter 2.

Was the means of data collection appropriate?

However we collect data, we need to ensure that they are representative and that they are sufficient. It is unlikely that a few students on a local campus will be representative of young people. If our point of reference is the UK, then we need to consider the geographical spread in addition to a fair representation of different socio-economic groups. If we were to ask the students on the campus about student fees, it is likely that most would disagree with the policy (we would need a representative survey of students to really know). However, if we were to ask the same question of a more representative sample of young people (including the correct percentage of those not continuing to higher education) the outcome could be different.

What can we infer?

The purpose of research is generally to provide a better understanding of a phenomenon or situation through description and the development of theory. We use our snapshot of information as a means of extending knowledge in a meaningful way. In this case, the lack of definition and poor methodology give no real basis for the generalization of our results.

The quantitative approach to problem solving and business decisions has developed within a scientific tradition. Science has clearly helped us understand the world we live in and the scientific methods have informed approaches to research and management. As the name implies, the scientific approach is about being systematic in the collection of data, developing law-like relationships and inference. At this point it is useful to distinguish between the **inductive approach** and the **deductive approach**. The inductive approach is based on the collection of empirical evidence in a specific situation and then making a general statement to cover all situations. Having observed the impact on employee motivation of imposed performance related pay schemes in a selected number of companies, for example, we might try to generalize these findings as lessons for all companies. The direction of the inference is from the specific to the general. The deductive approach is concerned with logic and mathematics. A theory or proposition can be built-up from accepted truths. If for example, we where told that Jeff's earnings were greater than Roger's and Roger's were greater than Jon's we could infer than Jeff's were greater than Jon's. That is, if $A > B$ and $B > C$ we can infer that $A > C$ (a deductive relationship). If we accept that a company can have a fixed element and a variable element to cost, then deductively we can infer that total cost is the sum of fixed and variable cost. You will see both the inductive and the deductive approaches used in this book.

Another useful way of understanding research methods is to contrast the **positivist** and **phenomenologist** approaches. The positivist approach comes from the scientific tradition and is essentially deductive. Ideas (hypotheses) are proposed based on reason and logic. Relationships are suggested based on proposed or accepted theory. These relationships are examined using experiments or other empirical data. The approach is structured and can be repeated in a range of ways. Many of the statistical procedures in this book will follow this. The phenomenological approach is based on the way people experience and understand the world. It is more concerned with an understanding, in depth, of the social processes. The researcher is likely to use observation and small-scale stud' the outcome often taking the form of case-histories or case-studies.

The value of your research will be judged partly on methodology. If the m. dology is convincing then your outcomes are more likely to be accepted. The choice of methodology is not necessarily a simple one. You can produce reliable and valid results in a number of different ways. In more complex studies, you may choose to combine several approaches to provide an overall description (probably using quantitative methods) and illustrative examples (qualitative methods).

1.3 Models

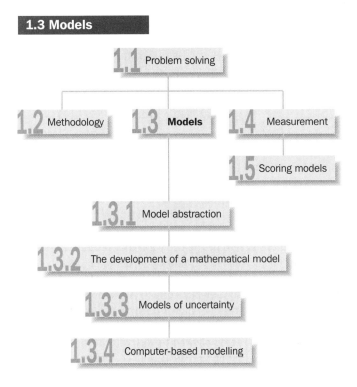

A model is a representation of real objects or situations. Modelling provides a way of expressing our understanding of such objects and situations through the use of simplified constructions, the use of language, the use of diagrams or the use of mathematics. Modelling involves a transformational process where outcomes are explained by a range of inputs and assumptions. If we think about a scaled model of a dam, then the available flow of water (an outcome) will depend on how we manage those inputs we control (variables such as the incoming flow of water during previous time periods) and the assumptions being made (such as loss due to weather conditions and seepage). The transformation process is the modification of flow due to storage. A model of modelling is shown as Figure 1.3.

Models can include the use of descriptive statistics and the use of statistical theory (two major reasons for working with numbers), but, in addition, attempt to show how we believe things work. The benefits of attempting to construct models to explain what we see or understand include:

■ The necessity to have a good understanding of the object or situation. If we do not fully understand the problem, or have missed some important aspect, then the chances are that the model cannot be completed, will not

Figure 1.3 Modelling.

work or will not produce the results expected. If, for example, when modelling a dam, we do not adequately allow for the permeability of surrounding rock, then we may never adequately predict the outgoing water flows. There is a recently constructed dam in Spain that has yet to fill with water and may never do so because of this problem.

- The **recognition of all relevant variables**. To be useful, a model should be a simplification of a problem of interest. To include all possible factors would just make the model more and more complex, and not allow us as users to look at the key issues. We need to strike a balance between those inputs that are significant and those that may have some minor effect but do not impact on the problem characteristics. There have been many attempts to model the way sales respond to changes in the level of supportive advertising. However, to simply plot the level of sales against the level of advertising ignores the fact that sales will also depend on price, and may also depend on price last quarter and advertising last quarter (evidence does suggest that many of these influences are time-lagged). If you were to brainstorm this problem with colleagues you would soon obtain a long list that would include factors such as quality, price of competitors and advertising by competitors. The real issue to be addressed is what you want to achieve with the model. In most cases, models attempt to show how a few key variables impact on the problem and effects of managed changes to these variables.

- The **understanding of relationships**. To determine the outcomes from a model it will be necessary to relate these to the inputs that we can vary and the assumptions being made. A simple breakeven model will determine the level of sales at which profit is zero by the comparison of revenue and costs. Therefore this model requires an understanding of both the revenue and cost functions. These relationships may need to be specified in terms of being 'fit for purpose' rather than perfectly correct. A revenue function, for example, may be specified in terms of a known price multiplied by quantity, which is adequate in terms of a simple breakeven model. However, in other models, we would expect price to be allowed to vary with quantity (like the simple models used to describe supply and demand in economics). In specifying a relationship, the choices need to be fully understood and the assumptions clarified.

- The ability to **undertake analysis**. The process of defining or describing a problem can only be steps in problem solving. The development of a model should allow us to better understand how things work and to better understand the effects of any possible changes. A model should encourage the use of 'what-if' type questions. In looking at the breakeven possibilities for a company, we should be asking 'what if the fixed cost increases' or 'what if the price increases'. As you will see, spreadsheet models in particular, allow the user to easily change values so that the consequences can be followed through. Part of the job of management is the evaluation of alternatives and, as we intend to show, this can be improved by the modelling of a range of activities, such as the control of stock or the characteristics of queues.

Models or modelling is particularly useful when we cannot work directly with the real objects or situations. It is an expensive business to test a real aircraft to destruction or create the conditions to examine company failure. Modelling is seen as a timely and cost-effective way of examining problems that can include both complexity and uncertainty.

1.3.1
Model abstraction

Modelling allows us all the advantages of not working with the real thing. W can consider, for example, the impact of reduced revenue flow or changed check out systems without a business consequence. In addition, modelling should allow us to think more **conceptually** and **imaginatively** about the problems we need to deal with. Models can be classified in terms of their level of abstraction as shown in Figure 1.4.

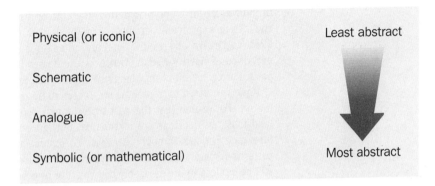

Figure 1.4 Model abstraction.

A **physical model** or **iconic model** generally involves a scaled or simplified version of the real thing. Cardboard cut-outs can be used to represent office furniture or materials-handling equipment. These models find particular application within engineering, operations management and the sciences, but only very limited use (generally for presentational purposes) in quantitative methods.

Schematic models are a more abstract representation of reality and include all forms of graphs and diagrams. Organization charts showing job roles and authority are frequently used to describe how a business works. Flowcharts are used to show how computer software works. As we will see, network diagrams can show the various steps in project management. Schematic models give a visual picture and it is often said 'that a picture is worth a 1000 words'.

Analogue modelling is where one factor, with different properties, is used to describe another. Speed can be represented by a needle on a dial (a speedometer) or workflow by a liquid. Colours on a map, for example, can represent water or forests.

Symbolic models or **mathematical models** use a range of numbers, letters, special characters and symbols to represent problem situations. These models have the precision and neatness of mathematics but are the most abstract. A straight line is defined and can be understood by an equation of the form: $Y = a + bX$. The straight line, itself, may be model of something like 'the steady increase in sales'. These models may be further categorized as **deterministic** or **probabilistic**. A deterministic model will give a certain outcome or outcomes once the inputs have been set. Once costs and revenues are known, a business breaking-even can be modelled deterministically. In contrast, a probabilistic model will need to attach measures of uncertainty to outcomes. The modelling of traffic flows or telephone calls, for example, would need to allow for the natural variation and the uncertainty of events over time.

1.3.2 The development of a mathematical model

A mathematical model will attempt to describe a problem of interest by a number of equations or mathematical procedures. Relationships are established using variables and parameters, and by making assumptions.

A **variable** is a quantity or characteristic of interest that is allowed to change within a particular problem. The marks achieved by students in a mathematics test or the travel time to work would typify the measurement of a variable.

A **parameter** is fixed for a particular problem. When evaluating the cost of running a car for a one-year period, the cost of insurance would be seen as fixed and hence a problem parameter. In the longer term, the insurance costs are likely to change and would be represented by a variable. Parameters are often described as fixed variables, which as the names implies, are given a value for a particular problem but may be given another value for another problem.

An **assumption** is something we accept to be true for the model we are working on. To assume that the cost of insurance is fixed for the year is reasonable (and most assumptions are reasonable) but ignores the fact that we can change insurance policies through the year. To assume travel patterns or expenditure patterns remain stable during the period of a survey may be reasonable but can still be subject to unseasonably bad weather or other unpredictable factors.

Suppose the cost of vehicle hire involves a fixed cost of £100, a daily charge of £25 and a mileage charge of 5 pence per mile. In this case there are two variables, the number of days of car hire (which we will call x) and the number of miles travelled (which we will call y). The fixed cost of £100 will not vary in this problem and is recognized as a problem parameter (which is likely to be different for different hire arrangements). The total cost (c in £'s) can be expressed in the following algebraic form

$$c = 100 + 25x + 0.05y$$

It is assumed that no other costs, such as fuel or parking, need to be included. Typically assumptions are not stated but are important when an interpretation of the results is required.

Mathematical notation (such as the use of $=$, $+$, x and y) allows a simply summary of the way things are related. Once in mathematical form, we can then manipulate the expression to consider a wider range of ideas.

If we are told that the vehicle hire was for five days and that the distance travelled was 723 miles, then by **substitution**, the total cost can be calculated

$$c = 100 + 25 \times 5 + 0.05 \times 723$$
$$= 100 + 125 + 36.15$$
$$= £261.15$$

Suppose now that we are told to work within a budget of £225 for a journey that will take four days. By manipulating the equation, we can calculate the number of miles we can travel before exceeding the budget. By letting $c = 225$, we get

$$255 = 100 + 25 \times 4 + 0.05y$$
$$225 = 100 + 100 + 0.05y$$

We can now rearrange to get the equation in terms of y:

$$0.05y = 225 - 100 - 100$$
$$= 25$$

Dividing both sides by 0.05 gives

$$y = 500$$

This result does need interpretation. If we travel exactly 500 miles then our costs will match the budget; if we travel more we will exceed the amount allowed in our budget and if we travel less then money will be left over.

If you feel concerned about this use of mathematics, then perhaps this is a good time to refer to *Improve Your Maths: A Refresher Course*.

1.3.3 Models of uncertainty

To be able to calculate that a budget of £225 will fund a travel distance of 500 miles is the type of result you would expect from a **deterministic** model. However, many problems have some element of uncertainty and require an understanding of probability. Models that include uncertainty are referred to as **probabilistic** or **stochastic**.

Typically, travel plans can include some uncertainty. Suppose you know that you will need to travel at least 400 miles, but may need to travel further if important customers are prepared to see you. Customer X would add 100 miles to your journey and customers Y and Z would both add 50 miles. However, you do not know if these customers will see you at this stage but you can make a good guess at what the chances are (in Part 3 we will fully develop the concept of probability). Suppose the chances are seen as being 50/50. In the notation of probability, a 50% chance would be written as $\frac{1}{2}$ or 0.5.

We can construct a table to show all the different possibilities – see Table 1.1. Given that all the outcomes are equally likely, we could work out the average.

$$\bar{x} = \frac{400 + 500 + 450 + 450 + 550 + 550 + 500 + 600}{8} = \frac{4000}{8} = 500 \text{ miles}$$

As we shall see in the coming chapters the mean, $\bar{x}$, is important as a descriptive statistic and also as an **expected value**. If we faced this situation a number

Table 1.1 This table shows the number of possible outcomes to your travel plans given that you do not know at this stage whether three important customer (X, Y and Z) are willing (YES) or are not willing (NO) to see you

Customer X (+100)	Customer Y (+50)	Customer Z (+50)	Additional mileage	Total mileage (additional + 400)
No	No	No	0	400
Yes	No	No	100	500
No	Yes	No	50	450
No	No	Yes	50	450
Yes	Yes	No	150	550
Yes	No	Yes	150	550
No	Yes	Yes	100	500
Yes	Yes	Yes	200	600

of times, then **on average**, we would expect to travel 500 miles. It could be that this time all three customers are unwilling to see you and as a result you only need to travel the minimum distance of 400 miles or that all three customers want to see you and you will need to travel the maximum distance of 600 miles. In response to this uncertainty, we are developing a probability model. Given that all eight listed outcomes are equally likely we can talk about a 1 in 8 chance of having to travel 400 miles or a 1 in 8 chance of having to travel 600 miles. We can also talk about this 1 in 8 chance as being $12\frac{1}{2}\%$ ($1/8 \times 100$) or a probability of 0.125. The chance of travelling 500 miles is 2 in 8, or 25%, as this can happen in two ways; either only customer X is willing to see you **or** only customers Y and Z.

Do not feel concerned if this seems a little baffling at this stage; the ideas of descriptive statistics, expected values and probability will be fully developed in later chapters. The above calculation of the mean and the chance of each outcome has been simplified because the chance of a customer being willing to see you has been a uniform 50%. If this chance or probability changes then we would need to develop our approach to manage this. In most problems we will find that the outcomes are not equally likely (e.g. 1 in 8) and we will need to work with a notation and theory.

1.3.4 Computer-based modelling

Computers continue to change our lives, whether in the workplace or in the home. Access to the Internet is both a tool of business and a means of entertainment. Computers bring all the benefits of increased computational power and linkages to a mass of data, but also the threats of information that lacks reliability and validity. Computers can take you on a voyage of discovery or leave you fishing in the sewers!

In this book we are concerned with the ways computing can support your use of quantitative methods. The chances are (a probability statement!) that you will have access to a spreadsheet as part of your course and that in future employment, computer competence will be expected.

The level of abstraction, is a useful way to think about modelling that is computer-based (see Figure 1.5). It is true that we could work through any model without the use of a computer and in that sense a computer is not magical. What we cannot do is match the speed of computers. A computer can do in minutes what would take us many hours with a simple calculator.

Figure 1.5

Computers do make calculations simple. If you model multi-channel queues, you are likely to come across formula like:

$$P_0 = \cfrac{1}{\sum_{i=0}^{S-1} \cfrac{(\lambda/\mu)^i}{i!} + \cfrac{(\lambda/\mu)^S \times \mu}{(S-1)! \times (S \times \mu - \lambda)}}$$

We could do this calculation and the others that go with it by hand, but we think you will quickly agree that this is a rather tedious way to proceed. A range of software is available that will do the calculations for us, including spreadsheets. Having specified the model, in this case a queuing model, a computer is used for computational purposes only.

The ease of calculation and recalculation offered by computer software does facilitate analysis. Once we can easily use formula like the above, we can begin to ask questions like:

- what if the value of S were to change from ... to ...
- what if the value of λ were to change from ... to ...
- what if the value of μ were to change from ... to ...

Spreadsheets are a particularly effective way of developing computational models. They are structured in such a way, that if a critical value, say for example the interest rate, is changed then all the subsequent calculations are updated. If you are not familiar with the use of spreadsheets you could refer to Chapter 14 in *Improve Your Maths: A Refresher Course*.

Those problems that can be solved by the use of mathematical techniques and manipulation are **analytical**. The development of a model showing how a company can make a loss, a profit or breakeven is a good example of the analytical approach. Once the revenue and cost functions of a company are defined, then the difference will measure profitability. The breakeven point is just a special case where profit is equal to zero. However, there are a number of problems that cannot be modelled in this way. There are times when the mathematics is just too difficult or the problem situation is not sufficiently understood. The approaches to modelling are developed in Part 6.

Simulation models (see Chapter 22) are not solved by mathematical manipulation, although they are likely to use equations and distributions. What simulation models attempt to do is replicate the characteristics of the problem situation and then, by experimentation, examine the outcomes of varying inputs. A typical application would be the examination of queues in a supermarket where demand patterns were complex. We could model the flow of customers using existing data and then consider the impact on waiting time and queue length of adjusting the number of checkout points, for example. It would be unnecessarily expensive for the supermarket to make structural alterations just to evaluate whether 4 or 5 or 6 or any other number of checkouts where most appropriate. Even if the most suitable number of checkouts were known for a typical day, the simulation model could be used when events were not typical, for example, the last day of trading before Christmas or the day of local football derby.

Expert systems are concerned not only with the analysis required for problem solution as specified, but also advising on solution. Expert systems attempt to capture the 'best thinking' on problems from a range of sources and produce approaches or ideas that offer solutions. Expert systems are often most effective when relationships are logical rather than mathematical or where the problem is only semi-structured. Expert systems tend to work well when information is incomplete or is not clearly understood. An expert system has three major components:

- a user interface
- a knowledge base, and
- an inference engine.

We expect to see the use of expert systems continue to develop. As the flow of data increases and the means of analysis becomes forever more sophisticated, we would expect improved user support in the form of analysis and advice from the desktop PC.

1.4 Measurement

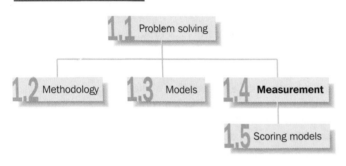

Measurement is about assigning a value or a score to an observation. To label the respondent to a survey as a smoker or non-smoker, or record precisely the dimension of a car component involves measurement. Measurement is the representation of type, size or quantity by numbers. It is this collecting of facts and opinions, often in numerical terms, that we refer to as **data**. Data become **information** when they are organized in such a way as to inform the user.

The properties of the numbers assigned depend on what we wish to measure. Coding a male respondent '0' and a female respondent '1' provides a very different type of measurement to recording their finishing position in a cross-country race (first, second, third, etc.) or recording their income last year. How we work with data will depend on the level of measurement achieved. Measurement can be categorized as **nominal, ordinal, interval** and **ratio**.

If responses are merely classified into a number of distinct categories, where no order or value is implied, only a **nominal** or **categorical** level of measurement is being achieved. The classification of survey respondents on the basis of religious affinity, voting behaviour or car ownership are all examples of nominal measurement. The numbers assigned give no measure of amount or importance. For data processing convenience, we may code respondents 0 or 1 (e.g. YES or NO) or 1, 2, 3 (Party X, Party Y, Party Z), but these numbers do not relate to a meaningful origin or to a meaningful distance. We cannot calculate statistics like the mean and the standard deviation which require measurement made on scales with order and distance. We **can** make percentage comparisons (e.g. 30% will vote for party X), present data using bar charts (see Chapter 4) or use more advanced statistical methods (see Chapter 13).

An **ordinal** level of measurement has been achieved when it is possible to rank order all the categories according to some criteria. The preferences indicated on a rating scale ranging from 'strongly agree' to 'strongly disagree' or the classification of respondents by social class (occupational groupings A, B, C1, C2, D, E) are both common examples where ranking is implied. Individuals are often ranked as a result of performance in sporting events or business appraisal. In these examples we can position a response or a respondent but **cannot weight to numerical differences**. It is as meaningful to code a five point r̶ scale 7, 8, 12, 17, 21 as 1, 2, 3, 4, 5 though the latter is generally expe̶ Only statistics based on order really apply. You will, however, find in market research and other business applications that the obvious codings are made (e.g. 1 to 5) and then a host of computer-derived statistics calculated. Many of these statistics can be useful for descriptive purposes, but you must always be sure about the type of measurement achieved and its statistical limitations.

An **interval** scale is an ordered scale where the differences between numerical values are meaningful. Temperature is a classic example of an interval scale, the increase on the centigrade scale between 30 and 40 is the same as the increase between 70 and 80. However, heat cannot be measured in absolute terms (0°C does not mean no heat) and it is not possible to say that 40°C is twice as hot as 20°C, but we can say it is hotter. In practice, there are few business-related measurements where the subtlety of the interval scale is of consequence.

The highest level of measurement is the **ratio** scale which has all the distance properties of the interval scale and in addition, zero represents the absence of the characteristic being measured. Distance and time are good examples of measurement on a ratio scale. It is meaningful, for example, to refer to 0 time and 0 distance and refer to one journey taking twice as long as another journey or one distance as being twice as long as another distance.

A comparison of the different types of measurement is shown in Figure 1.6.

In summary, it is considered more powerful to achieve measurement at a higher level as this will contain more discriminating information; it is more useful to know how many cigarettes a respondent smokes on average (0 or more) than just whether they smoke or not. The measurement sought will depend on the purpose of the research. If we are only concerned with whether respondents vote or not, then nominal measurement is sufficient. If we want a comparison of types of chocolate bar by sweetness or appearance then we will need, at least, an

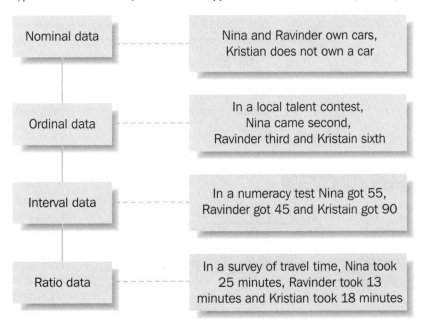

Figure 1.6 Types of data.

ordinal level of measurement. If we want to measure time or distance then we will want all the benefits of ratio measurement.

Another useful system of classification is whether measurement is **discrete** or **continuous**. Measurement is discrete if the numerical value is the consequence of counting. The number of respondents who vote for Party X or the number of respondents who own a car are examples of discrete data (on the nominal scale). Continuous measurement can take any value within a continuum, limited only by the precision of the measurement instrument. Time taken to complete a task could be quoted as 5 seconds or 5.17 seconds or 5.16892 seconds. Time in this case is being measured as continuous on a ratio scale.

When using continuous measurement, you will have to decide the accuracy with which you report your results. To report an average weekly expenditure of £5.917284 would be seen as having too much detail for most practical purposes. Given that such a result would vary from sample to sample and from week to week, there is a mathematical precision that outweighs its business significance. Typically we would quote this result to two decimal places (2dp): £5.92. The process of choosing the number of acceptable (and sensible) decimal places is called **rounding**. If we were to consider annual salary figures we may choose to work with no decimal places and quote the results to the nearest £100, for example, £17 500 and £27 600.

To round decimal places (see *Improve Your Maths: A Refresher Course*), we first decide on the number of decimal places and then consider the next decimal place along. If this figure is **4 or less** we just ignore the remaining decimal places. If the figure is **5 or more** we increase the final digit to be included by 1 and then ignore the remaining decimal places.

Example

304.569432 to three decimal places is 304.569

and

304.569432 to two decimal places is 304.57

The difference between types of data and the importance of these differences will become more apparent as you continue your course in quantitative methods. It is worth noting, however, that the statistics developed for one level of measurement can always be used at the higher level, **but not** (with validity) at a **lower level**.

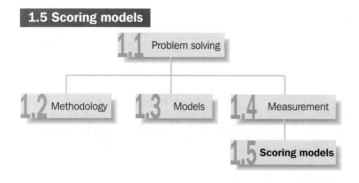

1.5 Scoring models

1.1 Problem solving

1.2 Methodology   1.3 Models   1.4 Measurement

1.5 **Scoring models**

The process of making decisions is likely to be easier if we only need to make calculations on a single factor, such as cost in £'s or travel time in hours. It is also easier to justify our decision if we can offer explanations like 'it was the lowest cost solution' or 'it would minimize overall travel time'.

However, a number of problems require a judgement on a range of information and this information may be based on relatively limited measurement (see Section 1.4, Measurement). **Scoring models** provide a way of combining such information and informing decision-making. The outcome of a model cannot be a substitute for considered decision-making but can provide a useful basis for thinking about the problem.

Example

Suppose a company needs to select a new location and that the choice has been narrowed down to three possible sites. Each of the sites is seen as having some advantages but also some drawbacks. Managers have been consulted, but cannot agree on the best possible choice. A number of meetings had taken place, but these were not always regarded as being productive. One of the problems the managers faced was the number of factors to be considered, and the fact that these factors could not easily be quantified. On one occasion, the availability of skilled labour was accepted as being most critical, while on another occasion it was agreed that the quality of transport links was more important.

The example given is typical of scoring model problems; there are no simple, single criteria to work with and a number of factors need to be considered.

To construct a scoring model all the factors considered important should be listed and a weight assigned indicating their relative importance (in the judgement of the managers). Then, on each of the factors, each site should be given an agreed score out of this 'weight' or an agreed ranking (e.g. site A was given 2 out of 6 for amenities and 2 out of 8 for distance).

In our example, a total for each site is found (by adding the scores in each column), and the site with the largest total is seen as the winner. Table 1.2 shows the use of this simple, additive scoring model.

Table 1.2 An additive scoring model

Factors	Weight	Site A	Site B	Site C
Amenities	6	2	5	3
Distance	8	2	6	3
Housing	10	3	6	3
Safety	0	0	0	0
Services	14	10	8	9
Skilled labour	20	12	8	16
Transport	20	18	10	14
Total		47	43	48

It should be noted that safety has been given a weight of 0, since the same safety standards apply regardless of site (and is therefore not a factor in our decision making). Skilled labour and transport were regarded as equally important and given the highest weight in this case, of 20. In this case, site C achieved the highest score and would be regarded as the most-favoured choice. Given that scoring models are only really seen as bringing together ideas and providing a basis for progressing discussions, the difference between the 48 for site C and the 47 for site A would not be regarded as significant.

The additive scoring model is used when higher levels of measurement, such as interval and ratio scales cannot be achieved. It provides an easy summary of a range of factors that cannot be easily combined using a common unit, such as £'s. All that happens is that once weights and scores have been agreed, a summing takes place and the option, a possible site in this case, with the largest total is seen as the best choice. As a model, it does help focus debate, and can help avoid the problem of 'going round in circles'. It is not seen as giving the answer, but just informing the decision makers of how options 'weigh-up'. The main criticism is that the whole model is subjective. The results will depend on what factors you choose to include, the weights that are chosen and the assignment of scores.

If we change the weighting given to skilled labour and transport (perhaps as part of a 'what-if' question), the scoring model would produce the results shown in Table 1.3.

Table 1.3 The additive scoring model (with changed weights)

Factors	Weight	Site A	Site B	Site C
Amenities	6	2	5	3
Distance	8	2	6	3
Housing	10	3	6	3
Safety	0	0	0	0
Services	14	10	8	9
Skilled labour	10	6	4	8
Transport	10	9	5	7
Total		32	34	33

The outcome of a scoring model depends on the weights and how the scoring has been done. Sites A and C both score strongly on the factors 'skilled labour' and 'transport'; if these factors are reduced in importance (the value of the weighting reduced) we would expect to see a corresponding effect on their overall totals. These models are based on judgement and agreement, and if this cannot be obtained then the value of the modelling is reduced. The models do lend themselves to **'what-if'** type analysis and it can be an interesting exercise to change values and monitor the consequences. It is generally accepted that managers should be encouraged to ask questions like 'what if skilled labour becomes less important?' or 'what difference would this investment make if we became less dependent on good transport links?'.

In many circumstances, it is easy and more realistic to rank options than score them. Typically, in market research, respondents would be asked which 'brand do you prefer' or 'rank the following in order of preference'. In Table 1.4 preferences have been ranked.

We need to be very careful working with ranked data. It is tempting to give the best or the first number one (like the music charts), but if the option with the largest total is to be chosen then the first should have the highest value and the second the next highest value and so on. In this case we rank 3, 2, 1 and not 1, 2, 3. We are looking at a **multiplicative ranking model** (often referred to as a multiplicative scoring model though technically we are working with ordered data). The weight is multiplied by the rank and then the outcomes are added.

Table 1.4 A multiplicative scoring model

Factors	Weight	Site A		Site B		Site C	
Amenities	6	1	6	3	18	2	12
Distance	8	1	8	3	24	2	16
Housing	10	1.5	15	3	30	1.5	15
Safety	0						
Services	14	3	42	1	14	2	28
Skilled labour	20	2	40	1	20	3	60
Transport	20	3	60	1	20	2	40
Total			171		126		171

Sites A and C share the highest score, and in terms of this analysis we would say that we were indifferent between sites A and C.

As you will have seen the answers achieved depend on the model used. In this particular case we would want more detailed information, including financial information, and would want to use more sophisticated modelling.

However, we can use scoring models to make the more general point that analysis is no substitute for the human element in decision making which also includes effective problem formulation, valid and reliable data, appropriate analysis, the generation of creative ideas, effective communication and managerial judgement.

1.6 Conclusions

The use of numbers is likely to be part of most organizational decision making. Numeracy is considered to be a key managerial skill – managers need to understand and work effectively with numbers. In this chapter we have considered the quantitative approach to problem solving.

The real skill of problem solving is not about accepting problems as given, throwing data at them and making decisions quickly. Problems need to be worked on and as managers, you will need to manage the problem solving process. Problem identification and definition are critical stages. There is considerable evidence from organizational studies that problem solving is rushed or that time is spent solving the wrong problem. The example often used to illustrate this latter point is of customers arriving at a busy hotel complaining about queuing for the lifts. Attempts at improving the capacity of the lifts and the speed of the lifts are typically expensive and often make little difference to the level of complaints. If mirrors or other decorative changes are made complaints often stop. Why? The problem was being specified in terms of the speed and capacity of the lifts but, in fact, what customers were unhappy with was the wait for a lift. Once waiting for a lift becomes more interesting (you can look at other people in the mirrors!) the reason for complaint disappears.

We need to build our approach to problem solving on a 'sound foundation'. It is methodology that will make our results acceptable. The objectives of our work should be clear. If we are collecting our own data, then we need to clearly define those people of interest to us (like all those that listen to Radio 1 at least once a week or those entitled to vote at the next election) and ensure that they are fairly represented in any sample selection. We should be able to justify our approach in term of **reliability** (that what we do can be repeated and give similar results) and **validity** (we measure what we say we measure).

In the quantitative approach we want to do more than just describe situations. To undertake analysis we need to understand relationships (between variables), develop concepts and test theories. A model provides a working representation of our problem. We can test whether the ideas we have used to develop the model are adequate. We can change the model and see whether this gives an improved representation. We can change inputs into the model and study any difference this makes. As an analogy, to merely produce performance statistics would not tell us how a car engine works and certainly would not help us if it did not start. Working with data is like working with a car engine, it helps if you know how all the bits fit together.

1.7 Problems

1. A medium-sized business operates a small fleet of vehicles. The existing practice is to allocate to each driver a particular area. This area then becomes their 'patch'. The drivers undertake all the deliveries in their area and only help out in other areas in exceptional circumstances. However, demand has becomes less predictable and some drivers complain that they have an unfair workload. The business would like to see a more flexible use of vehicles and drivers. What aspects of this problem would you regard as quantitative and what aspects would you regard as qualitative?

2. You are a member of student group that has the task of making a presentation on the employment trends within the UK in the next 2 weeks. The group is having difficulty working together and a number of the group members have complained that communications are poor. Define the problem and suggest ways that you could make progress with this problem.

3. You would like to assess the impact on your course of recent staff sickness. How could you ensure that your approach was feasible, and produced results that were reliable and valid.

4. What is the difference between data and information?

5. What characteristics would a model of traffic flow at traffic lights have in common with a model of customer queues at a checkout point?

6. You are planning to buy a new personal computer. What would you see as the fixed and the variable costs?

7. You are given the following information on two types of photocopying machines. The Exone has a fixed cost of £400 per annum and a running cost of £20 per 1000 copies. The Fastrack has a fixed cost of £300 per annum and a running cost of £25 per 1000 copies. Express the cost for each machine by means of an equation and calculate the cost corresponding to a requirement for 10 000, 20 000, 30 000, 40 000 and 50 000 copies per annum. Comment.

8. On what basis could we talk about the chance of success being 50/50?

9. What benefits can we expect from the use of computers?

10. What type of data would you expect to get from the following questions:
    (a) What do you dislike about your travel to work?
    (b) Do you usually travel to work (give main mode of travel)

		Yes	No
(i)	on foot	Yes	No
(ii)	by bus	Yes	No
(iii)	by train	Yes	No
(iv)	in your own car	Yes	No
(v)	in someone else's car	Yes	No
(vi)	by other (please specify) ...		

(c) How long does your journey to work take door to door?

(d) What was your age last birthday?

(e) 'No further motorway construction should take place'

strongly agree	1
agree	2
neither agree or disagree	3
disagree	4
strongly disagree	5

11 Round 234.56397 to

(a) two decimal places

(b) one decimal place.

12 Round £1 285 890 to the

(a) nearest £100

(b) nearest £1000

13 You have been given the following extract from a scoring model used to compare three products, A, B and C in terms of marketing:

	Maximum points	Product A	Product B	Product C
Existing demand	20	12	20	15
Marketing effort	10	4	4	5
Fit with other products	15	10	6	4
Packaging	10	5	6	6
Market trends	5	3	3	4

(a) Using an appropriate model, evaluate the above and comment on your results.

(b) If the weighting for 'market trends' were to double (i.e. maximum points increased to 10), what impact would this have on your results?

(c) Suggest ways this model could be improved.

14 A company has to decide whether to develop a new product, prototype A, or the existing product B. After discussions, it was agreed that the following scoring model would be helpful:

Factor	Prototype A	Existing product B	Score
Time to develop:			
over 6 months	✓		1
3 months but under 6 months			2
under 3 months		✓	3
Research requirements:			
high	✓		1
medium			2
low		✓	3
Changes to production methods:			
high	✓		1
medium			2
low		✓	3
Need for staff development:			
high			1
medium	✓		2
low		✓	3

Product life expectancy:

long	✓		3
medium			2
short		✓	1

Expected returns:

high	✓		3
medium		✓	2
low			1

(a) Using an appropriate method, evaluate the model and indicate whether your results support the case for developing a new product prototype A or developing the existing product B.

(b) It has been agreed that the last two factors are of more importance. Evaluate the model again, but this time give 'product life expectancy' a weighting of 3 and 'expected returns' a weighting of 2. Comment on your results.

15 The following scoring model has been developed to inform managers on the choice of new location. A number of factors have been identified as important and these have been weighted. The locations have been given a rank (highest meaning best) for each of the factors. Evaluate using a multiplicative model.

			Possible locations		
Factor	Weight	Site A	Site B	Site C	Site D
Labour	8	2	4	3	1
Cost	10	2	3	4	1
Transport	8	3	4	2	1
Services	4	1	3	2	4
Housing	2	3	2	1	4
Materials	4	4	3	2	1
Expansion	6	3	4	1	2

2 **Managing data**

The truth is out there somewhere

The collecting together of facts and opinions, typically in numerical form, provides **data**. Whether this data is useful or not depends on what purpose it is required to serve, the method of collection and its collation. The data will become **information** (good or bad) when it informs the decision processes of the user.

The management of data is an important skill to develop. In some situations, the data requirement is established, such as the need to measure the pressure in a gas pipeline at regular intervals of time or sample for defective items in a production process. In other situations the data required is less clear. What data would you need to explain the changing rates of crime or trends in smoking? It is likely that available data on crime or smoking or other topics of interest will only be informative on certain aspects. Mostly you will find some and **need to add some**.

Many business, economic and social questions are not amenable to a simple 'yes' or 'no' answer, they need clarification and discussion. Solutions are likely to be part of a problem solving process (see Section 1.1, Problem solving) and it is important to get this process right. Having clarified the purpose of any research and the resources available, we should be in a better position to identify the data required. The shortage of data is rarely an issue now. A few minutes spent searching on the Internet is likely to produce several hundred leads on most common topics of interest. In terms of data selection, we should be asking whether the data is:

- appropriate
- adequate
- without bias.

Effective problem solving is not just about coming up with answers, but establishing a sustainable process that can be trusted and can give the necessary new insights. It is unlikely that we will have just one problem to research. It can be useful to think about managing a portfolio of problems. As we make progress with one problem we should be developing the skills to solve other problems. As we discover new sources of data, software or techniques applicable to our immediate problems, we should also be developing the knowledge we may need to solve problems in the future.

A typical dictionary definition of data is 'things known and from which inferences may be deduced'. This book is concerned essentially with the use of numerical data. Numbers can be used to describe all kinds of phenomena. In answering a question like 'what sort of business is it' we will soon begin to talk numerically (if we have the information) about size, profitability, product range, the market place, the characteristics of the workforce and a host of other factors.

To be of value, the questions and the responses need to be **appropriate**. If your budget for car purchase is no more than £2000, it is not particularly useful being given a range of prices for Porsche cars between 2 and 4 years old. **Data must serve its purpose.** We could, for example, be given sales figures on an annual, quarterly, monthly, weekly, daily, hourly or minute-by-minute basis. If all we want to know is what the general trend is over time, then quarterly or monthly data might be sufficient. If we want to predict demand for Thursday and Friday of next week, recent sales on a daily or even hourly basis are likely to be required. Working with the numbers given should be informative, but we should also be prepared to take account of other factors. Making sense of the numbers for travel to work time may mean that we take account of local road works or a major sporting or music event. Numbers alone are unlikely to give us an **adequate** understanding of any business problem. We also need to take account of the people involved, the culture of the enterprise, the legal and economic environment.

Knowing whether the information is **adequate** is a problem for the problem solver. It is always possible to collect more and more data. So where do we stop?

You will:

- need to be clear about problem boundaries
- need to know what the problem owner or client expects from you
- need to know if any data is missing (there are many examples of computer files being lost or wiped)
- be expected to work within time and resource constraints
- need to decide whether the current data is sufficient for the purpose (as defined by agreed objectives) or whether additional data should be acquired.

It is important to establish the level of detail required. A survey that is planning to contrast smoking behaviour by gender, age, area and occupation with reasons why will require adequate representation in all these categories, and will therefore require a more sophisticated design and larger numbers than one intended to establish only the percentage that smoke.

The data we have will be the basis of any **inferences** we make. If the data, in some way misrepresents those of interest we have the problem of **bias**. The results of a survey of married women could not, for example, be taken to represent the views of all women. If this survey did not give a fair chance of inclusion to younger married women, then a further source of bias would exist. The underlying principle is that of **fair representation**. We need to be clear about who or what we want to talk about (make inference to) and how the sources of information can fairly represent them.

Objectives

By the end of this chapter, you should be able to:

- discuss the issues of data collection
- demonstrate a knowledge of data sources including those that are internet based
- identify the importance of primary and secondary data
- explain the difference between a census and a survey.

This chapter will be structured as show in Figure 2.1.

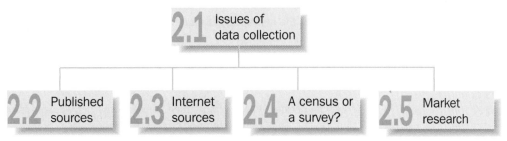

Figure 2.1 Structure of Chapter 2.

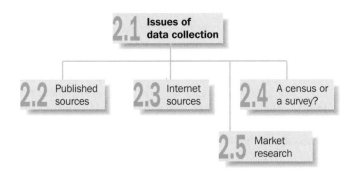

2.1 Issues of data collection

The **five W's and H technique** is frequently used in problem or issue clarification. By asking the questions: *who?*, *what?*, *where?*, *when?*, *why?* and *how?* we should be able to establish more clearly what the problem actually is. We shall use this technique to both illustrate the use of the technique and consider the issues of data collection.

Who? is an important first question in any problem. Data will always relate to a particular group of people or set of items in time and we use this concept to define the population we will be working with.

Definition

The **population** is defined as those people or items of interest.

Given limited resources, including time, the identification of the relevant **population** is essential. If, for example, you were concerned with the acceptability to women of a new contraceptive pill it would be pointless contacting a group of people to find that half were men. A similar problem can arise if the group you have identified as the relevant population does not include everyone for whom the survey is relevant, since a range of views or information will be totally missed. If you were interested in why some people bought foreign-built cars, but failed to contact purchasers of certain Ford models, for example, then you might fail to identify the fact that some buyers do not realize that their car is foreign built. The term 'population' is also be used to describe all the items or organizations of interest. Typically, an audit is concerned with the correctness of financial statements (although we increasingly see references to other types of audit, e.g. environmental audits or creativity audits). The population of interest to an auditor could be all the accounting records, invoices or wage sheets. If we were concerned with job opportunities, the population could be all the local businesses or organizations employing one or more persons.

Exercises

> 1 What is the relevant population to contact regarding a new magazine for
>   women?
> 2 What is the relevant population to contact regarding a new 'numerically
>   controlled' machine tool?

Having decided who, we must then consider whether we need information on
all of them or just a selection.

Definition

A **census** is a complete enumeration of all those people or items of interest
(whereas a **sample** is just a selection from all those people or items of interest).

What? data will depend on the purpose of the research. The more we seek
detail and description, the more likely we are to restrict the size of any survey
we are working with. If our research is concerned with the long-term impact of
smoking on the individual and their family units, the more likely we are to use
illustrative case studies or case histories. However, if we are more concerned
with the purchase of cigarettes by brand for promotional purposes, then we are
likely to choose a design that has good coverage by region, age, gender and other
factors, and is up to date. The data required will also inform decisions on the
method of data collection. If we are interested in the use of car seat-belts, then
we might consider observational methods as most effective. Experience suggests
that respondents do not always accurately report not wearing seat-belts or their
frequency of exceeding the speed limit! Research concerned with opinions and
attitude usually involves asking questions directly to the respondent or the
respondent ticking boxes.

A statistical enquiry may require the collection of new data, referred to as **primary data**, or be able to use existing data, referred to as **secondary data**. Most,
however, require some combination of both sources. Sources of primary data
include observation, group discussions and the use of questionnaires. The distinguishing feature of primary data is its **collection for a specific project**. As a result,
primary data can take a long time to collect and be expensive. Secondary data, in
contrast, has been collected for some other purpose. It is usually available at low
cost but may be inadequate for the purposes of the enquiry. Where the data
requirements are fairly complex, it is normally seen as good practice to first collect
the lower cost secondary data, usually rather more general but of high quality (it
has been published) and add to it as necessary. If, for example, we were considering
the impact of a new shopping centre on the local community, from available statistics, we should be able to describe the demographic characteristics of the surrounding area and we may need this to define the population of interest. Such data may
be descriptive of an area in social and economic terms but would say nothing about
the attitudes, view and opinions of the local residents.

Where? to find the right kind of data **when** you need it or where to find the
people of interest **when** you need them is an important skill. The chances are that
someone somewhere will already have done some research on your topic of interest. You only need to look to discover the number of train passenger miles travelled each year, gross domestic product or the number of diving fatalities in the
previous year. In organizational research it is often useful to distinguish between

internal and **externally** generated data. Recent sales volume, sales value, number of employees, expenditure on advertising and expenditure on research are all examples of internal information that may be available within the organization, but may be difficult to obtain as an outsider. External information would include all the data generated by national governments, local government, chambers of commerce, other commercial sources and the Internet. External information can often be expensive, particularly if generated by the market research industry for commercial purposes. What will typify all data is that it comes in all shapes and sizes, and you still need to **question its validity and reliability**.

This stage of searching for data is often referred to as **desk research** and can add significantly to the information you have and lead to new problem insights. Any data search is likely to leave some gaps. If additional information is required, we again need to ask the question where from and indeed the question why.

Exercises

> 1 What is the retirement age of women in the UK?
> 2 When will compulsory military service end in France?

Why? is seen as part of a questioning approach that should lead to greater clarification and a justification of approach. In fact there is a useful technique called the **why technique**. By probing problems and solutions with the question why, a better understanding of the causes and effects can be achieved. The technique involves repeatedly asking the question why perhaps with probing statements like 'why did you say that?' or 'why should that be the case?' or 'why use that data?'. We should in general be interested in why we need particular data and whether more appropriate data could be obtained by alternative means.

Chapters 2 and 3 are particularly concerned with the **how?**. Having defined the population of interest and the purpose of the research, a number of issues will need to be addressed:

- whether existing published sources provide sufficient information (see Section 2.2)
- whether useful information can be found through an Internet search (see Section 2.3)
- what type of sampling should be used, if any (see Sections 2.4, 3.1 and 3.2)
- how data should be collected (see Section 3.3)
- how questions should be designed, if required (see Section 3.4).

2.2 Published sources

Arguably the most important source of external, secondary UK data are the official statistics supplied by the Office for National Statistics (ONS) and other government departments. The ONS was formed in April 1996 by the merger of the Central Statistical Office (CSO) and the Office of Population Censuses and Surveys (OPCS). Further change is planned following the White Paper 'Building Trust in Statistics' and the launch of National Statistics in April 2000. A useful introduction to the vast range of official statistics is the *Guide to Official Statistics*. The publi-

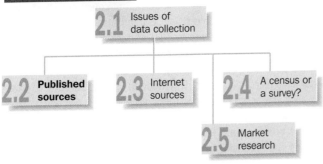

cations listed in Table 2.1 below give some indication of the official statistics readily available.

Table 2.1 Some sources of government statistics

The Annual Abstract of Statistics is regarded as the 'most comprehensive source of statistics in the UK'. It brings together statistics on almost every aspect of economic, social and industrial life. It includes statistical information, generally for the past 10 years, on parliamentary elections, overseas aid, population, labour markets, health, social protection, crime, lifestyles, environment, transport, research and a host of other topics.

The Monthly Digest of Statistics, published monthly (as one would expect!), provides statistical information on some 20 subjects including national accounts, population, labour markets, social services, law enforcement, agriculture, production energy, chemicals, metals, textiles, construction, transport, retailing, external trade, overseas finance, home finance, prices and wages, leisure and weather. The statistics are presented mostly as runs of monthly and quarterly estimates for the past 2 years. Annual figures go back several more years.

Financial Statistics is published monthly and gives a useful range of financial and monetary statistics including public sector finance, money supply, capital issues, exchange rates, interest rates and retail price index. An explanatory handbook, published annually, describes the national accounting framework and the financial instruments used as the basic building blocks.

Economic Trends, also published monthly, presents a range of economic measures including selected monthly economic indicators, income, capital formation, production and prices. Articles relating to economic performance are also included. An annual supplement gives data over longer time periods.

Social Trends provides an annual picture of UK society showing how we live today and how things have changed in recent years. It draws together a range of statistics on topics such as families, education, income and expenditure, health, the labour market, crime and lifestyles. Each year it includes an article that is interpretative of available statistics, examples include 'French and British Societies: a comparison' (1998) and 'A Hundred Years of Change' (2000).

We do recommend that you go online to check the availability of government statistics: http://www.ons.gov.uk. The ONS produce a quarterly newsletter *Horizons* that you can download free directly from the website and is informative on a range of developments. You can also download *Census News* which provides the most recent information on the census.

European statistical sources include *European Official Statistics: A Guide to Databases*, *European Official Statistics: Sources of Information*, *Europe in Figures*, and the *Eurostat Handbook*. Useful non-governmental sources of data include: *Minitel Market Intelligence Reports* (monthly reports on consumer products), *ADMAP* (monthly publication giving advertising expenditure by product categories), *Retail Business* (monthly), *Marketing in Europe* (monthly), *Kompass Directory* (provides information on a variety of companies) and NEILSEN reports on markets and shopping behaviour.

The usefulness of such secondary data will depend on the nature of your research. It is rarely a choice between secondary or primary data. Secondary data will often provide a useful overall description (e.g. economic or social trends) and inform the collection of primary data. Primary data will add specific detail, particularly current attitudes and opinions.

Exercise

> Obtain the most recent figures showing the change over a 5-year period in:
> 1 the numbers smoking by age
> 2 the numbers entering higher education by gender
> 3 participation in the most popular sports
> 4 sales of CDs and cassettes.

Case 1: the use of secondary data

Secondary data collection that could be undertaken by Shopping Developments Limited would include background demographic data on the area, including such factors as car ownership and travel distances. They could also obtain data on retail shops in general (e.g. consumer expenditure figures), profit figures for the sector (e.g. from DataStream), and the number of people using the nearby ferry port.

2.3 Internet sources

The Internet is an <u>international</u> <u>network</u> of computers. Essentially it is a collection of networks which allows the exchange of information and communications.

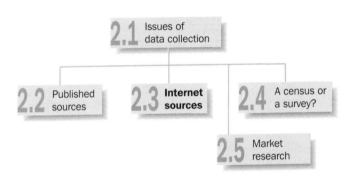

No one actually owns the Internet, though important stakeholders exist. Governments do legislate to restrict the storage and exchange of certain material, for example. Commercial organizations will charge for access to certain information. It is the constituent networks that are owned and although regulation does not effectively exist at a global level, self-regulation does exist at these local levels. It is often said that the Internet offers: '100 million consultants, day or night, on tap – free of charge'.

But what advice do they give? Essentially, we need to be able to **evaluate such information**, and be ready to reject any that is suspect.

The big advantage of the Internet is the scale of the information, which cannot be matched by any imaginable traditional library. The disadvantages include the need to search through sources that are not catalogued in any consistent way and the **lack of any quality control**.

Exercise

> Try the following web addresses and explore possible links:
> ■ http://www.roughguides.com/net/
> ■ http://www.dis.strath.ac.uk/business/
> ■ http://www.ons.gov.uk

Having established access to the Internet and decided that potentially useful information might exist, the next task is to manage a search of the web. The World Wide Web (or just the web or WWW) is the user-friendly face of the Internet and the fastest growing area of the Internet. It can be thought of as pages of information carrying text, sound, pictures and video. It has information on more than one million companies world-wide and many more interest groups and individuals. A good website is likely to link to other similar sites (like the strands of a web) leaving a trail of information. Generally, the quickest way to find information is to use a **search engine** or **directory**. A search engine will have a user-friendly front page with a space for you to enter a key word or phase. The search engine will then search through a database of millions of web pages for suitable matches. A summary is then provided of the number of matches. If the key words are too general then hundreds of possible web sites will be identified and if too specific very few. A balance needs to be achieved between an overwhelming number of sites, often not related closely to the problem of interest, and obtaining sufficiency of sites with potentially valuable information and useful links. A directory will give you a menu of choices and you can gradually focus your information search.

Exercise

Try some of the following:
- Altavista   http://altavista.com
- DejaNews   http://www.dejanews.com
- Excite   http://www.excite.com
- InfoSeek   http://www.infoseek.com
- Lycos   http://www.lycos.com
- Webcrawler   http://webcrawler.com
- Yahoo   http://www.yahoo.com

It is particularly important to check that data from the web is appropriate, is complete and is without bias. This may be particularly difficult to do when you first begin to research a topic. Many pressure groups and organizations do not declare their aims and objectives, or purport to be independent, unbiased researchers. For example, you might find that certain data showing how good chocolate is for you, was in fact, data from a confectionery firm. As well as 'common sense', a useful device is to look for corroboration from two or more Internet sites of particular data or assertions, and then to search for counter assertions. Some data on the net may, in fact, be completely fictitious!

2.4 A census or a sample?

Having identified those people or those items of interest we need to decide whether to include all of them (undertake a **census**) or take a selection (**sample**). Including all has the major advantage that all sections of the population

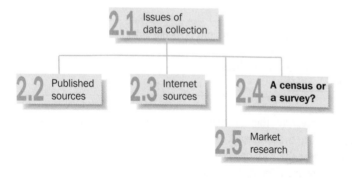

can be fairly represented even if we don't know the characteristics of the population. In addition, we can feel sure that any descriptive numbers calculated, such as the mean and standard deviation, will also be fully representative. However, a census can be prohibitively **expensive**, prohibitively **time consuming** and limit the possible depth of the enquiry. A carefully designed survey can, in many situations, provide the acceptable quality of results at lower cost and greater speed.

A **census** does have a number of reassuring qualities. A census will attempt to include everyone, providing maximum numbers for analysis and avoiding concerns about sampling or selection bias. A survey by its very nature will provide results that can vary from sample to sample (it is for this reason that we talk about sample statistics). A census will often provide a benchmark for a range of research activities by providing localized information on such characteristics as age, gender and ethnic origin, for example. A good example of this type of enquiry is the **population census**, which has been carried out in the UK once every 10 years since 1801 (with the exception of 1941 when rather more urgent matters were being dealt with). At the time of writing the most recent census will have taken place on Sunday 29th April 2001. While this type of enquiry should give highly detailed information and reflect data from all parts of the relevant population, it does take time and can be very costly. The term 'census' is usually associated with a government count of the population, but a census can be any complete count. A census can be taken of all the suppliers to a particular company, all the schools in the UK or all the sports shops in Leeds. A census is of limited use for most business, social or economic applications, unless the identified population is small. A census of all homes, for example, would be an expensive way of estimating the proportion with television sets. In contrast, if you were representing a manufacturer who sold only to a small number of wholesalers and you wanted their views on a new credit-ordering system, then a census would be a suitable method to use and would send the powerful message of inclusion. A census is likely to be the preferred method of enquiry when there are legal requirements for inclusion, inclusion is important and detailed information is required even at a local or sub-group level.

A **survey** is likely to be preferred when:

- it is known as a methodology that works
- cost constraints exist
- time constraints exist.

Most commercial and most governmental research will be based on survey methodology. If the selected sample is **representative** and **sufficiently large**, then the results will be good enough for purpose. The concept of being 'fit for purpose' is important. If you merely need a low-cost commuting car then a small car such as the Ka may meet your needs, if you spend many hours on the road representing your company you may need the larger Mondeo. If you need to demonstrate your business success, perhaps you will need a Jaguar. The right car will depend on your needs.

The procedure used to select the sample is particularly important and this is described by the **sample design**. The sample needs to represent the population in such a way that results from the survey can be used to make generalizations about the population. We talk about making an **inference** from the sample to the population (see Chapters 11 and 12). The concepts of inclusion and exclusion are also important in sample design. If certain people, or items, are excluded because of the sampling procedure we are not able to assess their possible impact on results. If you were to ask the next five people you see how they are likely to vote at the next general election, it is very unlikely that the answers given would be a guide to a general election result. Even if we where to ask the next 10 people or 100 people how they would vote, they are unlikely to represent the electorate as a whole and we would still be concerned whether the numbers where sufficient for analysis. You only need to think about the forma-tion of any queue to realize that you could select only Aston Villa supporters, bus users or dog owners. And who would they represent!

2.4.1 How should we decide sample size?

The size of the sample required will depend on the following factors:

- the accuracy required
- the variability of the population
- the detail required in analysis.

If an accuracy of $\pm1\%$ is required rather than $\pm5\%$ for example, then a larger sample will be necessary. If the average weekly household expenditure on a parti-cular item is only required to an accuracy of $\pm£5.00$ rather than $\pm£0.50$ then a smaller sample should be sufficient. The important point here is that the user or client needs to be able to specify such levels of accuracy. The variability of the population will also be a determining factor in the sample size required. In the extreme case where everyone held exactly the same opinion (no variability existed) we would only need to ask one person to make an inference to the population as a whole. Suppose, for example, we were told that a bag contained 100 coins of the same domination, we would only need a sample of one coin to calculate the value of the bag (the domination of the coin $\times$ 100). As views become variable, larger samples are required. If the mixture of coins in the bag became more variable, a larger sample would be required to achieve the desired level of accuracy. If accuracy is also required by subgroup, e.g. female smokers under 25 years of age, then we would need to ensure that the sample was suffi-ciently large to provide the necessary number in each of the subgroups.

Since most surveys are not designed to find out a single piece of information, but the answers to a whole range of questions, the determination of sample size can become extremely complex. It has been found that samples of about 1000 give results that are acceptable when sampling the general population. (See Chap-ter 11 for the calculation of sample sizes.)

'Gallup and other major organizations use sample sizes of between 1000 and 1500 because they provide a solid balance of accuracy against the increased economic cost of larger and larger samples. If Gallup were to – quite expensively – use a sample of 4000 randomly selected adults each time it did its poll, the increase in accuracy over and beyond a well-done sample of 1000 would be minimal, and generally speaking, would not justify the increase in cost.'

Source: http://www.gallup.com/

2.5 Market research

Market research is seen as a major industry. Data is collected on behalf of a range of organizations, much of it for business use. Data can have considerable commercial value and access can be limited. A variety of methods are used to collect data including face-to-face inter-viewing, telephone interviewing, and group discussions. Market research pro-vides information on people's prefer-ences, attitudes, likes and dislikes, and can help companies understand what

2.1 Issues of data collection

2.2 Published sources

2.3 Internet sources

2.4 A census or a survey?

2.5 Market research

consumers want. National and local government use market research to provide the data to inform policies on everything from planning local transport to the provision of efficient health and social services

Market research can be directly concerned with a market (which will need definition) and can provide information on market size, market trends, market share by brand, customer characteristics and other factors. Aspects of market research include advertising and promotional research, product research and distribution. Market research companies also sell a range of services, and will frequently undertake research for government, both national and local, academic projects and not-for-profit organizations. The Market Research Society provide a range of useful information on their website: http://www.mrs.org.uk (see Figure 2.2).

Figure 2.2

2.6 Conclusions

Obtaining and using data as information is an important part of understanding and solving any problem. There is little doubt about the volume of data now available, and any search of the Internet can easily produce reams of computer print-out. As with all problem solving we need to work within boundaries that ensure the problem remains manageable and yet does not exclude new avenues of enquiry. Given the diversity of possible data sources we need to check that data is appropriate, adequate and without bias.

As discussed, the choice is rarely between secondary data (existing data) or primary data (new data that needs to be collected for the specific purpose). Secondary data will help describe and define the existing problem. The examination of secondary data can also provide guidance on what research methods work and which don't. Primary data will generally be needed to add specific detail.

The purpose of any statistical investigation needs to be clear. A statement that we wish to investigate the management of change within the organization will mean different things to different people. In this case, we need to be clear about our meaning of change or changes, 'management' and the general context. Decisions will need to be made on who to include and who to exclude. In all statistical work the definition of population (all those people or items of interest) is particularly important. If we refer to the workforce, for example, do we mean only full-time employees, those at a particular location or those doing a particular job?

It is a frequently reported experience that 'desk research' yields some of the information required but also yields other data of interest and a wealth of new ideas. It is also worth considering how much research is genuinely original! If the purpose of the statistical investigation requires the collection of original data, then the sample survey is probably the most widely used method in business and economics (see Chapter 3).

Once collected, data needs to be collated and presented (see Chapter 4). Available computer hardware and software now allows data to be stored, manipulated and analysed with relative ease. Many types of computer software are available for dealing with survey data. You could use a standard spreadsheet, such as, Excel or Lotus-123 to record the answers (in a coded form), or you could use more specialized software such as SNAP or SPSS. The choice that you make will depend on the size of the survey, the resources available and the sophistication of the analysis necessary.

2.7 Problems

1   What is the difference between data and information? Provide examples to illustrate your answer.
2   What do we mean by bias? Give examples of how bias might occur.
3   Why is it essential to define clearly the population when undertaking research? Illustrate your answer with reference to a survey on driving.
4   Define the population for:
    (a)   a survey on the attitudes to smoking in the workplace
    (b)   a survey on the attitudes to parking in residential areas near a new leisure centre
    (c)   sampling public vehicles to check maintenance standards
    (d)   sampling components manufactured using new equipment.
5   Search out information (secondary data) on:
    (a)   the number of marriages annually over the last 10 years
    (b)   the annual numbers of medical discharges from United Kingdom service personnel
    (c)   the number of petrol-filling stations
    (d)   the quarterly output from the 'clothing and footwear' industries
    (e)   the quarterly numbers of road casualties
    (f)   the notes and coins in circulation with the public
    (g)   the index of retail prices
    (h)   consumer expenditure on beer
    (i)   the number of new dwellings completed by region
    (j)   employment in manufacturing by region
    (k)   the number of AIDS cases and deaths
    (l)   air pollution
    (m) money donated to charities.
6   Use the Internet to search for information on:
    (a)   graduate vacancies in major companies
    (b)   the cost of travel to holiday destinations

(c)   share prices

(d)   exchange rates

(e)   developments in the leisure industry

(f)   the cost of marketing

(g)   online shopping.

7   Comment on the following: 'The process of polling is often mysterious, particularly to those who don't see how the views of 1000 people can represent those of hundreds of millions.'

3 Survey methods

As described in Chapter 2, there are many useful sources of secondary data. However, in many cases this data (collected for other purposes) will not adequately meet the needs of our particular enquiry. In this chapter we are concerned with the ways survey methods can provide the necessary, **primary data**. The available secondary data is likely to give us some useful overall figures, but not the detail we may require in terms of products, issues or opinions. Secondary data may tell us how many people smoke and how many cigarettes they smoke on average each day, but it is unlikely, for example, to tell us much about brand preference or perceived impact on health. Secondary data, for example, may tell us how many people use local sports facilities, but will not tell us if people are prepared to pay more for an improved services and what they would see an 'improved service' as being.

The chances are that we will need to use **both** secondary and primary data when undertaking research. Generally we have to accept the secondary data as given – with all its limitations. The challenge is collecting primary data on time, within budget and with a quality that meets the needs of our research. It is all too easy to move from a poorly thought-through research idea to a simplistic questionnaire to a reporting of the obvious! What we argue is that for research to be worthwhile (as research) it does need to be **purposeful**, it does need **mechanisms** like carefully designed questionnaires and it does need insightful **analysis**.

Some kind of problem statement or statement of objectives can give purpose to the survey. We need to be clear whether the aim is broad generalization, such as product awareness, or whether we are looking for a small number of illustrative case-histories, as often seen in medical research. It is important to define the people of interest, the **population**. The population can be **all** those people or **all** those households or **all** those items of interest. It should identify those who can be included and those that must be excluded. It is this definitional stage that should clarify what we mean by a survey of home-owners or drivers or customers. We cannot wait to decide whether someone that owns a caravan, or rides a motorbike or buys for someone else should be included.

Having identified those of interest we still need to ensure that the data collected (see Chapter 2) is:

- appropriate
- adequate
- without bias.

Decisions need to be made on method of selection, method of contact and method of data collection. The validity of your work will depend on the **methodology**.

Objectives   This chapter will be structured as shown in Figure 3.1.

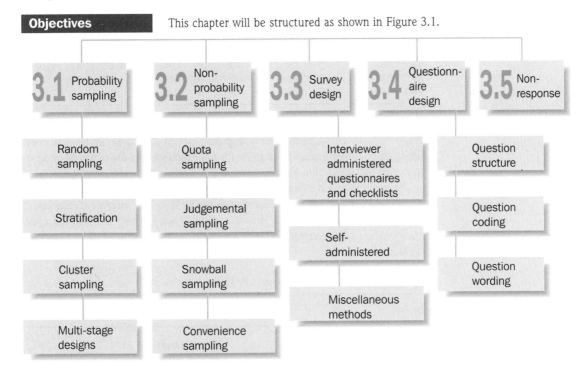

Figure 3.1   Structure of Chapter 3.

3.1 Probability sampling

The essential characteristic of **probability sampling** is that a procedure is devised where each person or item is given a known chance of inclusion and the **procedure** is used for the selection of individuals. We, as researchers, do not influence the actual identification of individuals.

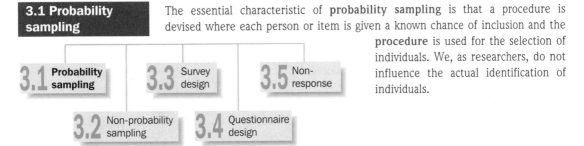

3.1.1 Random sampling

Random does not mean haphazard selection. What it does mean is that each member of the population has some calculable chance of being selected – not always an equal chance as we shall see. It also means the converse that there is no one in the identified population who could not be selected when the sample is set up. A **simple random sample** gives every individual an **equal** chance of selection. To select a random sample a list or **sampling frame** is required. A list of all retail outlets in Greater London or a listing of all students at a particular university or the electoral register are all examples of possible sampling frames. A sampling frame is simply a listing of the population of interest. Typically, each entry on the sampling frame is given a number and a series of random numbers (usually generated on a computer) are used to select the individuals to take part in the survey. Table 3.1 shows an extract from a set of random numbers. There is no (human) interference in the selection of the sample, and samples selected in this way will, in the long run, be representative of the population.

Table 3.1 A typical extract from random number tables

22	17	68	65	84	68	95	23	92	35	87	02	22
57	55	61	09	43	95	06	58	24	82	03	47	10
27	53	96	23	71	50	54	36	23	54	31	04	82
98	04	14	12	15	09	26	78	25	47	47		

It can be noted that not all of those selected will participate in the survey, and this is seen as the problem of **non-response** (see Section 3.5). Typically, we will select more than we require to allow for non-response and other wastage factors (e.g. unreadable questionnaires).

The electoral register is regarded as the most effective sampling frame for individuals and households (we often choose to work with addresses) in the UK. It does, of course, exclude many individuals, those not entitled to vote, and many households. As an example, Table 3.2 provides a small extract from an electoral register.

Table 3.2 An extract from an electoral register

Number	Name	Address
00	J. Hilton	3, York Street
01	A. Mandel	4, York Street
02	J. Johnson	4, York Street
03	U. Auden	5, York Street
04	H. Willis	9, York Street
05	L. Willis	9, York Street
06	Y. Willis	9, York Street
07	A. Patel	11, York Street
08	N. Patel	11, York Street
09	L. Biswas	15, York Street
10	J. Wilson	16, York Street
11	G. Wilson	16, York Street
12	P. McCloud	17, York Street
13	F. Price	17, York Street
14	P. Cross	18, York Street
15	P. Hilton	23, York Street

As you can see, each of the members of the population in Table 3.2 has been given a two-digit number, and we will use this to identify individuals for this example. To select a simple random sample of individuals (entitled to vote), we need to use a procedure to give each an equal chance of inclusion. Working along the list of random numbers given in Table 3.1 from the top left, we can see that the first number 22 does not correspond with anyone in our section of the electoral register (Table 3.2). The first number that does match, working across the table, is the number 02 – J. Johnson. We are using the random numbers in pairs in this case to match the population listing but sets of three and four digits are more usual with a large population. The second acceptable number is 09 – L. Biswas. By continuing this process, the following selection of eight people is obtained (as shown in Table 3.3).

Table 3.3   The selection of eight individuals by simple random sampling

Number	Name	Address
02	J. Johnson	4, York Street
09	L. Biswas	15, York Street
06	Y. Willis	9, York Street
03	U. Auden	5, York Street
10	J. Wilson	16, York Street
04	H. Willis	9, York Street
04	H. Willis	9, York Street
14	P. Cross	18, York Street

No doubt you will see a small problem with this sample (i.e. the same person has been selected twice). How would you overcome this? Essentially, there are two sets of decisions to be made. **First**, you need to decide whether it is more appropriate to sample individuals or households. If you do decide to sample households, you will need to allow for the fact that the number of times an address appears depends on the number of people at that address entitled to vote (i.e. only include the address if it corresponds to the first person listed). The detail of the design will, of course, depend on the purpose of the survey. **Secondly**, you will need to decide whether to **sample with replacement** or **sample without replacement**. In market research, we typically sample without replacement, which does mean that once an individual has been selected they are effectively excluded from further selection.

The simple system outlined above for a simple random sample would work reasonably well for a relatively small population that was concentrated geographically, but would become impractical for any national study. A national study would require the complete electoral register for the whole of the UK and, by chance, we could end up visiting one person on Sark, another in the Shetlands and none in the south-east. By chance, representation might be poor and travel costs might be prohibitive. It would not be impossible, although it would be unlikely, that the whole sample might consist of people living in Wales. The issue of representation would not be too important if the people of Wales were wholly representative of all UK citizens, but on certain issues their views will tend to differ from those of, say England or Scotland, for example. To overcome this type of problem, various other sampling schemes have been developed, but they still retain the basic element of random sampling: that each member of the population has some **calculable** chance of being selected.

Exercise

Check the future availability of the electoral register for sample selection purposes.
Try http://www.open.gov.uk

3.1.2 Stratification

If there are distinct groups or strata within the population that can be identified **before** sample selection takes place, it will be desirable to use this additional knowledge to ensure that each of these groups is represented in the final sample. The final sample is therefore composed of samples selected from each group. The numbers from each group or strata may be proportional to the size of the strata, but if there is a small group, it is often wise to select a rather larger proportion of this group to make sure that the variety of their views is represented. In the latter case it will be necessary to **weight** the results as one group is 'over-represented'. It is often the case, that either by design (numbers not selected in proportion to strata size) or by varying response rates, that results need to be **weighted** (see Section 5.2.3). It can be proven (mathematically) that the use of appropriate strata, for example, the same proportion selected from each region, will improve the accuracy of results.

When it is known that there are appropriate subgroups in the population, but it is not possible to identify them before sample selection, it is usual to ask a question which helps to categorize the respondent. Answers to questions such as:

- 'At the last general election which party did you vote for?' or
- 'Do you regularly smoke cigarettes?' or
- 'Do you own a motor vehicle?'

provide useful ways of partitioning a sample. Again, how you proceed will depend on the purpose of the survey.

This retrospective use of information is known as **post-stratification**. The results from these constructed strata can be weighted to provide more accurate results for the population as a whole.

3.1.3 Cluster sampling

Some populations have groups or **clusters** which adequately represent the population as a whole for the purposes of the survey. It can be argued that pupils from a particular school would have many experiences in common with pupils from similar schools or that the errors in one set of files may be very similar to errors in other sets of files. If this is the case, it will be much more convenient, and much more cost-effective, to select one or more of these clusters at random and then to select a sample or, carry out a census within the selected clusters.

One interesting variant of cluster sampling is the **random walk**. Interviewers are given one or more starting addresses, for example, and then given a procedure like take every fifth house thereafter. In this case, clusters are selected at random and each cluster has a calculable chance of inclusion.

Another interesting variant of cluster sampling is **systematic sampling**. Suppose we wanted to select 10 people from a sampling frame numbered 00 to 199. In this case, we want to select 1 in 20 people. In practice what we would do is start from a random number between 00 and 19 and take every 20th person thereafter. If the random number used as a starting point was 17 then the individuals selected would correspond to 17, 37, 57, 77, 97, 117, 137, 157, 177 and 197. The danger with this design is that many lists (sampling frames) are ordered in some way and you get systematic errors, e.g. one American study found that the interval between the numbers corresponded with the number of apartments on a floor and that those selected all lived in the same relative position of each floor. If all the apartments selected, for example, were next to the lift or railway line this would strongly influence the results.

3.1.4 Multi-stage designs

Even when the designs outlined above are used, there may well be issues over representation and costs. To overcome this, many national samples use a series of sampling stages. What is important is how partitioning takes place at each stage. Administrative regions for gas, electricity, civil defence and television, for example, can be used to partition the UK. The first stage is usually regional and typically all regions are included. Each region consists of a number of parliamentary constituencies, which can usually be classified on an urban–rural scale. A random sample of such constituencies may be selected for each region. This selection may use a systematic sampling procedure and constituencies may be selected with a probability in proportion to their size. Parliamentary constituencies are split into wards, and the wards into polling districts, for which the electoral register is available. Again selection may be with a probability in proportion to size. Typically, the larger the size of constituency, ward or polling district the greater the chance of inclusion. The electoral register is then used (as we have seen in Section 3.1.1) to select individuals or addresses. This type of selection procedure will mean that all regions are represented and yet the travelling costs will be kept to a minimum, since interviewing will be concentrated in a few, specific, polling districts. An example of a possible design is given in Table 3.4.

Table 3.4 Possible design of a national survey

Stage	Sampling unit	Number of units selected
1	Region	all (e.g. 12)
2	Constituency	4 for example
3	Ward	3 for example
4	Polling district	2 for example
5	Individuals or addresses	10 for example

In this case the sample size would be $12 \times 4 \times 3 \times 2 \times 10 = 2880$.

To try to ensure that the resultant sample was more fully representative, further stages could be added, or further stratification (e.g. by social/economic measures) could take place at some or all of the stages.

Exercises

> 1 What form of stratification would you use for a survey on radio listening?
> 2 For what type of information would a class within a school be seen as a suitable cluster?

3.2 Non-probability sampling

In a number of surveys, respondents are selected in such a way that a calculable chance of inclusion cannot be determined. Typically, there is also some element of judgement in the selection. These surveys cannot claim the characteristics of random or simple random sampling (statistical representation). Again, a procedure is devised to justify the sampling method and limit any possible selection bias. However, at some point, selection is not the outcome of predetermined chance, but rather a conscious decision to include, or indeed exclude. As an extreme example, suppose an interviewer is asked to select individuals to take part in the survey and has a particular aversion to say, tall people, then this group may be excluded. If tall people, then, have different views on the subject of the survey from everyone else, this view will not be represented in the results of the survey.

However, a well-conducted non-random survey can produce acceptable results more **quickly**, and at a lower cost, than a random sample; for this reason it is often preferred for market research surveys and political opinion polls.

3.2.1 Quota sampling

The most usual form of non-random sampling is the selection of a **quota sample**. In this case various characteristics of the population are identified as important for the purpose of the survey, for example gender, age and occupation and the proportion of each in the population determined from secondary data. The sample is then designed to achieve similar proportions.

This does suggest that if people are representative in terms of known, identifiable characteristics they will also be representative in terms of the information being sought by the survey. This might be seen like a very big assumption on the part of researchers, but the evidence from many years, from many countries, is that, in general, it is an assumption that does work. Having identified the proportions of each type to be included in the sample, each interviewer is then given a set number, or quota, of people with these characteristics to contact. The **final selection** of the individuals is left up to the interviewer. The interviewers you may have seen or met in shopping precincts are usually working to a quota.

Exercise

> Apart from groups specifically avoided by a poor interviewer, which groups are easily excluded or can easily be under-represented in a quota survey of the general population?

Setting up a quota survey with a few quotas is relatively simple. The results from the *Census of Population* will give the proportions of men and women in the population, and also their age distribution.

Suppose we need to work with the information given in Tables 3.5 and 3.6.

Table 3.5   The distribution by gender of the population aged over 15 years

Gender	Percentage (%)
Male	46
Female	54

Table 3.6   The distribution by age of the population aged over 15 years

Age (in years)	Percentage (%)
15 but under 20	19
20 but under 30	25
30 but under 50	26
50 or more	30

We could devise quotas in such a way that the interviewers would select samples that did reflect the distributions shown in the tables. However, this can lead to two problems. First, we may not achieve the correct age distribution by gender; the interviewer could, for example, select mostly men under 20 and women over 50. Secondly, can we assume the same gender and age distribution at each of the locations used for the survey; we only need to look at occupational and retirement patterns to recognise the weakness of this assumption. Table 3.7 shows jointly the distribution by gender and age.

Table 3.7 The distribution by gender and age of the population aged over 15 years

Age (in years)	Male percentage (%)	Female percentage (%)
15 but under 20	10	9
20 but under 30	12	13
30 but under 50	12	13
50 or more	12	18

This type of table is referred to as a cross-tabulation (see Chapters 4 and 13). We can observe from this table the typical human population characteristic of longer life expectancy for females.

Table 3.7 can be used to determine the quotas for a survey of 1000, say, as shown in Table 3.8.

Table 3.8 The quota required for a sample of 1000 using the distribution given in Table 3.7

Age (in years)	Male	Female
15 but under 20	100	90
20 but under 30	120	130
30 but under 50	120	130
50 or more	120	180

In this case we have just imposed a joint quota by and age and gender, but could add others that are related to the survey topic. We could, for example, add smoking/non-smoking or car ownership, but we would also need to decide whether the quota only needed to be achieved on an overall basis or correctly by subgroup. As further controls are imposed, the implementation of the survey becomes more complex and it is necessary to question whether any additional benefit is worth the additional cost.

Generally, interviewers will work with the same quota (e.g. so many in each of the defined categories) and only exceptionally will these be varied to reflect the characteristics of the location. Quota sampling is regarded as a **method that works**, particularly by the market research industry, and does offer a cost- and time-effective solution for questionnaire-based research. It is important that the characteristics on which the quotas are based are easily identified (or at least estimated) by the interviewer, or else valuable time will be wasted trying to identify the people who are eligible to take part in the survey. If the number of quotas is large, some of the subgroups will be very small, even with an overall sample size of 1000. Such small quotas may lead to problems when these sections of the population are analysed and it may not be possible to generalise the results to the whole population. In this type of situation, it may be necessary to 'over-sample' from these small subgroups in order to get sufficient data for analysis, and then weight results if necessary.

3.2.2 Judgmental sampling

In most non-probability methods there is still an element of chance in the selection of individuals although it may not be calculable. In **judgmental** or **purposive** sampling there is no element of chance and judgement is used to select participants. This approach is typically used when sample sizes are small and the researcher wants to use local knowledge. A teacher, for example, may select certain students to represent the class. A housing department may select a sample of buildings because they have characteristic structural problems. In these cases, the sample is being used for **illustrative purposes** rather than statistical inference to the general population. When testing a new computer system, the sampling of output may be judgmental because the discovery of just one error will indicate a problem with the system. In the same way, the discovery of one child being bullied at school would indicate the need for action.

3.2.3 Snowball sampling

As the name suggests, **snowball sampling**, moves on from the initial starting place (snowballs) to identify possible participants. It is used when possible respondents are difficult to identify and often, relatively rare. Suppose we wanted to interview those that had been homeless in the past 10 years but now had a permanent place of residence and employment, or those people that look after a stray cat. What we try to do is get individuals that seem to fit our description or individuals that have the **right contacts**. Once we have this starting point, we try to establish whether they are eligible and whether they can lead us to others that are eligible. In this type of sampling, typically respondents are asked whether they fit a particular description, for example, own a classic sports car, and if not pass on the enquiry to someone they know that does.

3.2.4 Convenience sampling

As the name suggests, a sample is selected on the basis that it is easy to obtain and does the job. Convenience sampling offers a quick, low-cost solution, but is particularly prone to bias. It may be convenient to select our friends for a particular enquiry, but we are unlikely to get the full range of views. In cases where bias is not regarded as a problem, convenience sampling is an attractive option. If we want to test (pilot) a questionnaire, to take a few different people through the questions can be helpful. If we want to gauge opinion quickly before doing further work, then again to pick a few people that are 'convenient' to work with can be helpful. If we want to illustrate a few of the problems encountered by National Health Service patients or those that shop on the Internet, then a few people can be conveniently selected. Although we could not talk about the population in general, we could identify some real problem, such as waiting times and the concerns about the security of Internet transactions.

Case 1: the use of research methodologies

Looking at the data available to Shopping Developments Limited, we can see that there are two different methodologies in use. The traffic flow data has been collected for 10-minute blocks of time, presumably spread through the day. This could have been collected by electronic means. To be useful it would also need to be spread through the week and have several observations for each time slot. We would also want the additional information to relate this flow to time (produce a time-series). The data collected by questionnaire has used a quota design (probably for both speed and cheapness from the student's point of view) with some attempt to match the characteristics of the sample to those of the local population. We would want to know how the interviewing was spread through the week and what effort was made to ensure that a fair spread of people were included.

3.3 Survey design

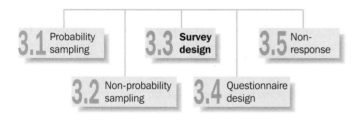

Having considered the relative merits of the various approaches to selecting a sample, in particular whether to use a probability-based or non-probability-based method, it is necessary to consider the method for collecting the data and the design of any checklist or questionnaire. The role of the interviewer is critical in any survey and the level of contact is an important design factor, see Figure 3.2. Other important factors include the purpose of the survey, the resources available (time and cost), the nature of the questions and likely response rate.

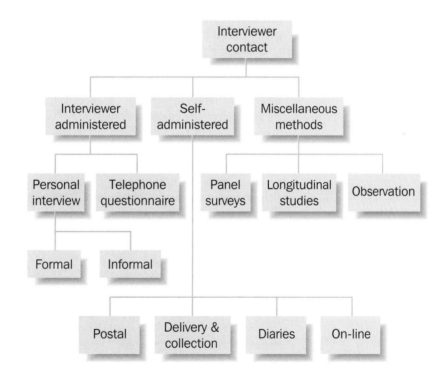

Figure 3.2 Survey methodology,

3.3.1 Interviewer-administered questionnaires or checklists

The important characteristic of interviewer-administered surveys is the **person-to-person contact**. The interviewer can check the details of the respondent, go through a questionnaire at an appropriate pace, 'probe' for more information in a specified way and 'prompt' as a matter of judgement or 'prompt' using aids' such as prompt cards or other illustrative materials. The interviewer may need to locate possible respondents given their names and addresses (as a result of a probability sampling from a sampling frame) or select respondents given certain criteria (quota sampling). Where fair representation is particularly important an interviewer will be given a list of people to interview and will be required to make one or more visits to achieve an interview. Care is taken not to exclude people simply because of unusual working hours or different lifestyle characteristics. An arranged interview in the home has the additional advantages that more detailed and perhaps longer questions can be asked, respondents can check details if they want and more sensitive topics can be included. This form of interview is typically more expensive, can limit the sample design geographically and is still subject to interviewer **bias**. Such bias can result from a range of influences including how the interviewer asks questions, the age and appearance of the interviewer and lack of experience with the type of interviewing. When the interviewer needs to locate the respondent only a third of his or her time is likely to be spent interviewing; the rest being used for travel and locating respondents (40%), editing and clerical work (15%), and preparatory and administrative work (10%).

A more cost-effective approach is the selection of respondents by the interviewer. Probably the most popular form of market research is quota sampling (see Section 3.2.1), where the interviewer is required to select respondents subject to certain conditions such as age and gender. This type of sampling is likely to achieve the target number within time and budget constraints. However, we cannot be sure that the sample is representative; the locations used may not be typical in some way (are those using shopping areas typical of the population as a whole?), the refusal rate is high among certain groups and there can be a bias in the selection by the interviewer.

It is unlikely that a person could just go out and conduct successful interviews; a certain amount of **training** is necessary to help in recording answers correctly and, in the case of open questions, succinctly. An interviewer's attitude is also important, since if it is not neutral or unbiased, it may influence the respondent. Unbiased questions can be turned into biased ones when a bad interviewer lays stress on one of the alternatives, or 'explains' to the respondent what the question really wants to find out. This explanation, or probing, can be turned into an advantage if the interviewer is fully aware of the aims of the survey and can probe without biasing the response.

The aims of the survey are important in the design of the interview. If detailed factual information is required then questions are likely to structured and direct, and the interview more formal, for example, 'how many years have you held a full UK driving licence'. If the subject matter is less clear, say for example the impact of ill health on other family members, the interview is likely to be less formal and more likely to include questions like 'in what ways has this made a difference to other family members'. Formal methods are particularly good for gathering consistent product and service information, both factual and attitudinal, from a wide range of respondents. Informal methods are more effective when a more in-depth understanding is required of relationships that have yet to be fully established. Formal interviews will use a structured questionnaire (see Section 3.4) whereas a more informal interview is likely to use just a checklist of questions, welcome more descriptive response and may well be tape-recorded.

Telephone interviews offer many advantages. The cost per interview is low and a broad spread of the population, whether national or international, can be achieved. The timing of the call can be planned to achieve high response rates, for example, during quieter office hours for workplace research and during the early evenings for household research. Telephone interviews can also be used in conjunction with other forms of sample research to check the correctness of the information given and to seek additional information. At one time telephone interviewing was seen as unfairly excluding certain portions of the population, but now most households (over 95%) and virtually all businesses have a telephone. Generally, response rates are high, but concerns still exist about the representative nature of telephone number listings. As people choose from a wider range of phone providers, including mobile phones, then any lists will need to be reviewed for completeness. Increasingly, questionnaires are being administered by telephone with the interviewer recording responses, often directly on to a computer. A software package, such as MARQUIS, will print the questions on to the screen and the interviewer enters the data directly. The data is then automatically stored and can be used for subsequent analysis. Use of direct data entry will reduce errors introduced in transferring responses from written questionnaires, but errors made by the interviewer cannot be checked. Telephone interviewing is often quicker to organize and complete, but removes any possibility of observing the respondent, or even checking who is being interviewed.

3.3.2 Self-administered

An alternative to interviewing the respondents directly is to have some form of self-completion. These methods have major cost advantages and avoid the problem of interviewer bias. However, bias can still be present in the language used, type of delivery and general presentation. Low response rates can also cause a problem. **Postal surveys** yield a considerable saving in time and cost over an interviewer survey, and will allow time for the replies to be considered, documents consulted or a discussion of the answers with other members of the household (this may be an advantage or disadvantage depending on the type of survey being conducted). Since the interviewer is not present, there is no possibility of observing the respondent or probing for more depth in the answers. This method is more suitable for surveys looking for mostly factual answers given on a yes/no basis or opinions given on relatively simple scales (e.g. from strongly agree to strongly disagree). In general, the questionnaire should be relatively short to maintain interest and encourage response. Postal questionnaires tend to discriminate against the less literate members of society, and are known to have a higher response rate from the middle classes.

Many introductory texts suggest that postal questionnaires inevitably have a low response rate, usually for some of the reasons given above. The most celebrated case is that of the *Literary Digest* survey of 1936 which posted 10 000 000 questionnaires asking how people would vote in a forthcoming US presidential election; they received only a 20% response rate, and also made an incorrect prediction of the result of the election. More recently, surveys have achieved response rates of more than 90% (comparable to interviewer surveys). A low response rate can be avoided if the questionnaires are posted to a **relevant population** (e.g. there is little point in sending a questionnaire on current nursery provision for children under 5 to those over 85!), have a relatively **small number of questions**, are pre-coded, and deal with mostly **factual** issues. The inclusion of a reply-paid envelope (with a stamp, not a pre-paid label) and a sponsoring letter from a well-known organisation are also seen as necessary.

Inducements (such as a food hamper or at least a pen) are also sometimes used to try to boost response rates. Most organizations involved in postal surveys use some form of **follow-up** on initial non-response, usually a week or two after the first letter has been sent out. In a survey of 14–20 year olds this process of reminders helped increase the response rate from 70% after 3 weeks to a final figure of 93.3%.

A variant of postal questionnaires is **delivery and collection** where briefed staff deliver the questionnaire, often directly to the potential respondent, and collect the completed questionnaire at a specified later time. Additional information or materials can be given at the time of delivery (e.g. product samples) and additional information sought when the questionnaire is collected. The use of a cover letter and incentives is again important, and can significantly improve the response rate. The use of **diaries** is particularly good at collecting information over time or picking-up less common events such as the purchase of electrical goods. Diaries are issued for a specified period of time, usually several weeks, and instructions given in their completion. The Family Expenditure Survey, for example, asks respondents to keep a diary of all purchases over a 2-week period.

On-line questionnaires are likely to become more popular as the number of potential respondents who can be contacted by e-mail or the Internet increases. Typically a questionnaire can either be sent by e-mail or potential respondents are asked to complete questions as they access a website. In the case of an e-mail survey, obtaining a suitable list of e-mail addresses remains a problem and questionnaires that arrive by e-mail are often thought of as unwanted junk mail. To be successful, respondents must see the enquiry as legitimate and earlier contact advising them of the purpose of the survey and the questionnaire to come, is helpful. A covering letter can further justify the enquiry. The design of the questionnaire should facilitate easy completion by a keyboard. It is often difficult to show that such a sample can be representative, but it can be an effective way of contacting certain groups of individuals.

3.3.3 Miscellaneous methods

Panel surveys are generally concerned with changes over time. The same respondents are asked a series of questions on different occasions. These questions may be concerned with the individual, the household or an organization. If the same group of respondents is maintained, a panel will have cost advantages and will not have the variation in results that could occur if different samples were selected each time (we can talk about minimizing between sample variation). This method is particularly good at monitoring 'before/after' changes. A panel can be used to assess the effectiveness of advertising, such as the Christmas drink-drive campaign, by comparing before and after responses to a series of attitudinal questions. There are two major problems with panel research. Firstly, respondents can become involved in the nature of the enquiry and as a result change their behaviour (known as **panel conditioning**). Secondly, as panel members leave the panel (known as **panel mortality**), those remaining become less representative of the population of interest. Eventually a decision needs to be taken as to whether to recruit new members to an existing panel or start a new panel. Some panel designs include a gradual replacement of panel members in a phased way to address both the issue of panel conditioning and panel mortality.

A panel may be formed and provide information in a variety of ways including telephone, postal questionnaire, personal interview and electronic means, for example, television tuning (which is different from viewing), may be recorded by an electronic device. Some panels are formed for a specific purpose (known as custom studies), whereas others provide information to a range of possible customers (known as syndicated). The Nielsen homescan grocery panel, for example, can provide brand information (e.g. market share, price) and relate this to other demographic factors (household size, social class classification).

Longitudinal studies follow a group of people, or cohort, over a long period of time. This method tends to require a large initial group and the resources to sustain such a study. It has been used effectively to investigate sociological issues and physical development. The television series 'Seven Up' has followed the progress of a group of people and reported every 7 years on their lives. This type of research can relate adult and childhood experience, for example. Health issues can also be explored by following such a cohort over a long period of time.

Observation can be more effective than questioning. The problem with interviewer-administered or self-administered questionnaires is that we need to work with what people say they do or did. The evidence on a range of products and services is that some aspects of recall are quite good, like whether a particular item was bought or not, but can be poor on other aspects like the date of purchase or the frequency of purchase. The reporting of alcohol consumed is often regarded as unreliable as both under-reporting and over-reporting can both take place for a variety of reasons. Questionnaires are also less likely to reveal the sequence of events that lead to particular decisions or behaviours. If the purchase of an item is being examined, say a car, it can be observed whether people enter a showroom as a group or shop alone, the way they look at display material, how time is spent with the sales assistant and how the visit was concluded. Observation has provided insightful information on a range of topics including the wearing of seat-belts and behaviour at football matches.

Observation can take place under normal conditions or when conditions are controlled in some ways. The way observational methods are managed will depend on the circumstances and the purpose of the enquiry. We could, for example, observe shop assistants working as 'normal' at particular times or observe them working in simulated conditions. Observations can be **structured** or **unstructured**. If the observations are structured then the recording will be done in a standardized way; the observer may be expected to record the frequency of certain movements or expressions. If the observations are unstructured, then an effort is made to capture everything of possible interest. However, if the observer needs to make a judgement or becomes involved with the events, the results can become highly subjective. Is it possible to investigate the safety record of a stretch of motorway objectively by going to the scene of every crash?

As we move more towards the sociological use of survey methods, we can identify two distinct types of observation. **Non-participant observation** is where the researcher merely watches the people involved, such as in work-study, and notes down what is happening. Such non-participant observation may be open, i.e. the people know that they are being watched, or it might be hidden, such as when interview candidates, who are in a waiting room, are watched from behind a two-way mirror. **Participant observation** necessitates the researcher becoming involved in the situation, for example actually going to work on the shop floor, and noting events as a shop-floor worker. There are obvious dangers in this, since the observer may only see part of what is happening, or might become absorbed in the culture of those being observed. There are numerous examples of how the observer or the researcher can become embroiled in the cause of the underdog. In these cases it is important to consider ways of maintaining objectivity.

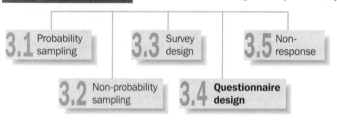

3.4 Questionnaire design

Having identified the relevant population for a survey, and used an appropriate method of selecting a sample of respondents, we now need to decide exactly what questions will be used, and how these questions will be administered. It does not matter how well the earlier stages of the investigation were conducted, if biased questions are used, or an interviewer incorrectly records a series of answers, then results of the survey will loose their value. To be successful, a questionnaire needs both a **logical structure** and **well-thought-out questions**. The structure of the questionnaire should ensure that there is a flow from question to question and from topic to topic, as would usually occur in a conversation. Any radical jumps between topics will tend to disorientate the respondent, and will influence the answers given. It is often suggested that a useful technique is to move from general to specific questions on any particular issue.

It is unlikely that the questions will be right first time. Once a questionnaire is drafted, typically it is tried and tested on a small number of respondents. It is important to know that the language used is accepted and understood by those being interviewed. It is often a last chance to check questionnaire structure and completeness. This stage is often referred to as **piloting the questionnaire** or a **pilot survey**. A pilot survey is generally a small-scale run through of the survey and can also be used to check questionnaire coding and methods of analysis.

3.4.1 Question structure

The Gallup organization has suggested that there are five possible objectives for a question:

1 To find if the respondent is aware of the issue, for example:

Do you know of any plans to build a motorway between Cambridge and Norwich? *YES/NO*

The answers that can be expected from a respondent will depend on the information already available and the source of that information (information available can vary from source to source). If the answer to the above question were *YES* we would then need to ask further questions to ascertain the extent of the respondent's knowledge.

2 To get general feelings on an issue, for example:

Do you think a motorway should be built? *YES/NO*

It is one thing to know whether respondents are informed about plans to build a motorway or indeed the merits of a new product but it is another to know whether they agree or disagree. In constructing such a question, the respondent can be asked to provide an answer on a rating scale such as:

Strongly agree	Agree	Neither agree or disagree	Disagree	Strongly disagree

A scale of this kind is less restrictive than a *YES/NO* response and does provide rather more information.

3 **To get answers on specific parts of the issue,** for example:

Do you think a motorway will affect the local environment?   *YES/NO*

In designing a questionnaire we need to decide exactly what issues are to be included; a simple checklist can be used for this purpose. If the environment is an issue we need then to decide whether it is the environment in general or a number of factors that make up the environment, such as noise levels and scenic beauty.

4 **To get reasons for a respondent's views,** for example:

If against, are you against the building of this motorway because:

(a) there is an adequate main road already;
(b) there is insufficient traffic between Cambridge and Norwich;
(c) the motorway would spoil beautiful countryside;
(d) the route would mean demolishing a house of national interest;
(e) other, please specify ....

The conditional statement '*if against*' is referred to as a **filter**. We need to use filters to ensure that the question asked is meaningful to the respondent. We would not wish to ask a vegetarian, for example, which kind of meat they prefer.

To find the reasons for a respondent's views generally requires questions of a more complex nature. You will first need to know what these views are and then provide the respondent with an opportunity to give reasons why. The above question is **precoded** and does limit the information a respondent can provide. To provide the respondent with the opportunity to give a more complete and detailed answer an **open-ended** question could be used:

Why are you against the motorway being built?

5 **To find how strongly these views are held,** for example:

Which of the following would you be prepared to do to support your view?

(a) write to your local councillor
(b) write to your MP
(c) sign a petition
(d) speak at a public enquiry
(e) go on a demonstration
(f) actively disrupt the work of construction.

To assess the strength of feeling we could use a numerically based rating scale:

How important is the Hall that would be demolished if the motorway is built? (circle answer)

Of great importance						Of no importance
1	2	3	4	5	6	7

The position on a rating scale provides some measure of attitude. The number of points used will depend on the context of the question and method of analysis. Generally, four- and five-point scales are far more common than the seven-point scale shown above. If we want to force respondent away from the neutral middle position (off the fence!) then an even number of points is used.

3.4.2 Question coding

As we have seen, question structure (Section 3.4.1) will allow questions to be answered in a simple yes/no manner, as a rating on a rating scale or with a more open-ended format. Most analysis will be computer based and this will mostly require the conversion of responses to numerical codes. This coding can be part of the question and questionnaire design or can be done at a later stage. Generally, we try to add as much coding to the questionnaire as possible, referred to as *precoding,* leaving only the answers to more complex questions for further analysis or additional coding at a later stage.

Precoded questions give the respondent a series of possible answers, from which one may be chosen, or an alternative specified. These are particularly useful for factual questions, for example:

How many children to do you have?

0 1 2 3 4 5 6 more   (circle answer)

When a limited choice is offered in questions involving opinions or attitudes, some respondents will want to give a conditional response. For example:

Do you agree with the deployment of nuclear weapons in Britain?
Agree   ☐
Disagree   ☐
Don't know   ☐ *(tick box)*

Some respondents might want to say:

'Yes, but only of a certain type' or
'No, but there is no alternative' or
'Yes, provided there is dual control of there operation'.

To improve a question of this kind the range of precoded answers given can be expanded. Alternatively the question could be left open ended. In the example given above, it may be better to ask a series of questions, building up through the objectives suggested by Gallup.

An **open-ended question** will allow the respondent to say whatever he or she wishes:

Why do you choose to live in Kensington?

This type of question will tend to favour the confident, articulate and educated sections of the community, as they are more likely to organize and express their thoughts and ideas quickly. If a respondent is finding difficulty in answering, an interviewer may be tempted to help, but unless this is done carefully, the survey may just reflect the interviewer's views. To **probe** or to **prompt** are seen as useful in some interview situations but generally interviewers need guidance on the use of such techniques. A further problem with open questions is that, since few interviews are tape recorded, the response that is recorded is that written by the interviewer. Given the speed of the spoken word, an interviewer may be forced to **edit** and abbreviate what is said and this can lead to bias. Open questions often help to put people at their ease and help ensure that it is their exact view which is reported, rather than a coding on some precoded list. Open questions can also be used at an early stage of development, perhaps as part of a piloting the questionnaire, to identify common responses for precoding of questions

3.4.3 Question wording

Question wording is critical in eliciting representative responses, as a biased or leading question will bias the answers given. Sources of bias in question design identified by the Survey Research Centre are given below:

1 **Two or more questions presented as one,** for example:

Do you use self-service garages because they are easy to use and clean? YES/NO

Here the respondent may use the garages because they are easy to use, but feel that they are dirty and disorganized, or may find them clean but have difficulty in using the petrol pumps.

2 **Questions that contain difficult or unfamiliar words,** for example:

Where do you usually shop?

The difficult word here is *'usually'* since there is no clarification of its meaning. An immediate response could be *'usually shop for what?'* or *'How often is usually?'* Shopping habits vary with the type of item being purchased, the day of the week the shopping is being done, and often the time of year as well. Technical terms can also present problems:

Did you suffer from rubella as a child?   *YES/NO*

Many people will not know what rubella is, unless the questionnaire is aimed purely at members of the medical profession; it would be much better to ask if the respondent suffered from German measles as a child. This problem will also be apparent if jargon phrases are used in questions.

3 **Questions which start with words meant to soften hardness or directness,** for example:

I hope you don't mind me asking this, but are you a virgin?   *YES/NO*

In this case, the respondent is put on their guard immediately, and may want to use the opening phrase as an excuse for not answering. It is also important to avoid value or judgement loading:

Do you, like most people, feel that Britain should be represented in NATO?   *YES/NO*

There are two possible reactions to this type of leading question:

(a) to tend to agree with the statement in order to appear normal, the same as most people; or, in a few cases

(b) to disagree purely for the sake of disagreeing.

In either case, the response does not necessarily reflect the views held by the respondent.

4 Questions which contain conditional or hypothetical clauses, for example:

How do you think your life would change if you had nine children?

This is a situation that few people will have considered, and would therefore have given little thought to the ways in which various aspects of their life would change.

5 Questions which contain one or more instructions to respondents, for example:

If you take your weekly income, after tax, and when you have made allowances for all of the regular bills, how much do you have left to spend or save?

This question is fairly long and this may serve to confuse the respondent, but there is also a series of instructions to follow before an answer can be given. There is also the problem of complexity of information which could include what we mean by income and how we allocate allowances for the many bills, such as gas, electricity and the telephone, which may be paid monthly or quarterly. In addition, individuals will vary on their level of recall and the way they manage such issues. We need to be careful not to force a respondent or a particular group of respondents into non-response.

The completed questionnaire needs to follow a logical flow and often a flowchart will be used to develop the routes through a questionnaire. Where questions are used to filter respondents, for example *if YES go to question 10 and if NO go to question 18*, then all routes through the questionnaire must be consistent with these instructions. Computer software, such as MARQUIS, allows you to type in the questions, specifying the flow from one to another and then checks for flow and consistency. Using such a package, it is possible to develop a questionnaire and then print copies directly from the program.

Exercise

> Write a few questions that could become part of a (quantitative methods) course evaluation. Try these questions (pilot) on other students. Identify what improvements could be made.

Case 1: the use of the questionnaire

The questionnaire used in this case was written to meet the needs of a student project. We would be concerned whether it was appropriate for business use. We have not been given the objectives of the student project and the company objectives would need further clarification. The content of the questionnaire is very limited and gives the respondent no opportunity to say to what extent they are satisfied or dissatisfied.

We think you will soon identify a number of weaknesses in the questionnaire. The difficult bit is writing a questionnaire that people will want to respond to and will give us the quality and quantity of information that we want.

3.5 Non-response

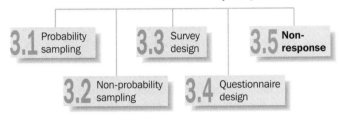

It is almost inevitable that when surveying a human population there will be some non-response, but the researcher's approach should aim at reducing this non-response to a minimum and to find at least some information about those who do not respond. The type of non-response and its recognition will depend on the type of survey being conducted.

For a preselected (random) sample, some of the individuals or addresses that were selected from the sampling frame may no longer exist, for example, demolished houses, since few sampling frames are completely up to date. Once the individuals are identified there may be no response for one or more of the reasons given below.

Unsuitable for interview

The individual may be infirm or inarticulate in English, and while he or she could be interviewed if special arrangements were made, this is rarely done in general surveys.

Those who have moved

These could be traced to their new address, but this adds extra time and expense to the survey; the problem does not exist if addresses rather than names were selected from the sampling frame.

Those out at the time of call

This will often happen but can be minimized by careful consideration of the timing of the call. Further calls can be made, at different times, to try to elicit a response, but the number of recommended recalls varies from one survey organization to another. (The government social survey recommends up to six recalls.)

Those away for the period of the survey

In this case, recalling will not elicit a response, but it is often difficult at first to tell if someone is just out at the time of the call. A shortened form of the questionnaire could be put through the letterbox, to be posted when the respondent returns. Avoiding the summer months will tend to reduce this category of non-response.

Those who refuse to co-operate

There is little that can be done with this group (about 5% of the population) since they will often refuse to co-operate with mandatory surveys such as the Population Census, but the attitude of the interviewer may help to minimize the refusal rate.

Many surveys, particularly the national surveys of complex design, will report the number of non-respondents. In addition, non-respondents may be categorized by reason or cause to indicate whether they differ in any important way from other respondents. In a quota sample, there is rarely any recording of non-response, since if one person refuses to answer the questions someone else can be selected almost immediately.

Exercise

> Why is it desirable to know something about the characteristics of non-respondents?

3.6 Conclusions

Typically, secondary data can only provide so much information. If we want more, then we need to collect it. Essentially our approach can be qualitative or quantitative or aspects of both. This book is concerned with quantitative methods and surveys are one of the most important ways of collecting numerical data. A carefully designed survey will allow us to collect data directly from the people of interest. To avoid problem of adequacy and bias, the sample should be representative and the questions have a probing honesty to achieve meaningful responses.

The use of surveys is an important research tool and the methodology is used extensively by government, business and other organizations. Survey research can become complex and can be expensive. It is important to establish the purpose (the objectives) of the research and what time and cost constraints exist. It is important to be able to justify the chosen methodology. If you are not sure about the purpose of the research, and you are not selective about the data collected, what can be the value of any subsequent analysis? We need to avoid the problem of 'garbage in – garbage out'.

Case 1: the use of methodology

Look again at the case provided and suggest a list of improvements that you would make to both the methodology and the questionnaire.

3.7 Problems

1   Obtain examples of recent surveys from one of the following:
    http://www.gallup.com
    http://www.mori.com
    http://www.nop.com

2   What steps can you take to ensure that all possible respondents to a survey have an equal chance of selection?

3   In a survey of students at a local college it was decided to give part-time students twice the chance of selection because of the smaller numbers involved and the diversity of their study patterns. How would you double the chance of selection of part-time students and how would you allow for this in any subsequent analysis.

4   Why is the electoral register one of the most widely used sampling frames in the UK? What are the potential problems with using this sampling frame?

5   What are the advantages and disadvantages of using postcodes as a sampling frame?

6   How would you construct a sampling frame for a survey of:
    (a)   clothes shops in the West Midlands;
    (b)   people who regularly eat chocolate;
    (c)   customers of a local bakery;
    (d)   students on degree courses in economics.

7   What are the differences between a stratified sample and a quota sample? Illustrate your answer by describing the selection of a national sample of the general population.

8   Non-response rates of 5–15% are often quoted for random sample interviewer surveys, why are figures not quoted for quota surveys? Which of the five types of non-response are relevant for quota surveys?

9   Surveys of political opinion typically use a quota sample of 1000 voters with a new sample being selected each time. In the period before a general election, some organisations use panels. Suggest reasons for using a panel in

preference to the more usual practice.

10 Describe how you would design a survey of
  (a) ethnic small businesses in the south-east
  (b) residents near a new shopping complex
  (c) the use of a sports centre
  (d) the demand for adult education courses.

11 What are the sources of bias in questionnaire design? Write ten biased questions to illustrate your answer.

12 Construct a questionnaire to find the reasons why other students on your course selected your particular college.

13 If you were responsible for briefing interviewers about to conduct a survey on road safety, which points would you stress?

14 Obtain a copy of a recent Family Expenditure Survey report from your library or use the Internet (http://www.statistics.gov.uk) to find out how the sampling is done. Write a brief report on the sample selection methods used.

15 What sort of data collection problems would you expect, if your investigation involved observing people at work?

16 Comment critically on the layout, question ordering and wording of the following questionnaire:

QUESTIONNAIRE ON LIESURE

1 How old are you?

2 What is your martial status?
Single / Marred / Divorced

3 How often do you go to the pub?
less than once a week ☐
once a week ☐
every day ☐
twice a week ☐ (please tick)

3 Do you take partin any sports? YES/NO

  IF YES, How often do you do it?

4 Do you watch any sports? YES/NO

  IF YES, How often do you go?

5 Do you watch television? YES/NO

6 If you had sufficient income and did not have to work and were fit and healthy and young, which sport would you like to take up?

7 How many hours of TV do you watch per week?
<5 ☐
<10 ☐
10 < 20 ☐
20 < 40 ☐
40 or more ☐ (please Tick)

8 Do you watch television to relax in front of interesting
   programmes?   YES/NO

9 What else do you do in your leisure time?

   ....

   ....

   ....

10 How many children do you have?

    1 2 3 4 5 6 more (please circle)

11 Are you employed?   YES/NO

12 Which socio-economic class do you belong to?
    A   ☐
    B   ☐
    C1   ☐
    C2   ☐
    D   ☐
    E   ☐   (Please Tick)

    Ta for you help. Bye.

4

Presentation of data

Once data has been collected, either by you (primary) or by someone else (secondary), then an initial task is to obtain some overall impression of the findings. This is most conveniently done by using **diagrams**. In fact, for some purposes, these diagrams may be all that is required. The mechanics of producing diagrams are made much easier, and the range of diagrams available increased considerably by using spreadsheets. In this chapter we will look at both the construction of diagrams by hand, and by the use of computer packages.

The management of data is a major challenge to organizations of all kinds, and to individuals within organizations. This chapter is concerned with managing data that comes in **numeric form**. Numbers are likely to be generated whenever attempts are made to describe complex business activities. The process of doing business will lead to a numeric description of sales, revenue, costs and other measures of performance. An examination of the business environment may involve an analysis of market trends, disposable income, the effects of pollution or other factors that can be monitored by numerical measurement. The ability to measure and monitor performance in numerical terms has also become increasingly important in 'not-for-profit' organizations, such as hospitals and schools.

The aim of this chapter is to show you some of the main charts and diagrams that are used for presenting data, and to give you a critical awareness of when they might be useful.

Objectives

After working through this chapter, you will be able to:

- ■ construct appropriate tables for different types of data
- ■ present data in a variety of diagrammatic forms by hand
- ■ present data in a variety of diagrammatic forms by computer
- ■ use graphical representations for a range of problems
- ■ choose between different presentations of data.

See Figure 4.1 overleaf for the chapter overview.

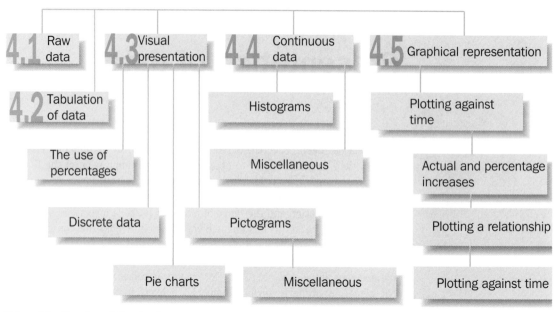

Figure 4.1   Structure of Chapter 4.

4.1 Raw data

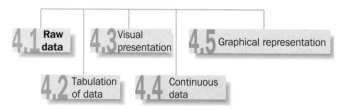

When presenting data we are concerned with the overall picture, rather than a large collection of individual bits. We need to put all the parts together, like a jigsaw, if we want to see a general picture emerge. In attempting to describe a particular market, it is not the single purchase made that is of particular importance, but rather the pattern of purchases being made by a range of possible customers. We may then need to know if this pattern is temporary (for example, due to severe weather), or more permanent (for example, due to changes in tastes and preferences). To examine each of these situations will require different diagrams.

The presentation of data is more than an issue of technical competence in producing the right results; it is a **means of communication**. It is important to know who is going to use your statistical work and what their requirements are. The data you have may have a number of limitations because of collection methods used or the complexity of the topic, and the user needs to be fully aware of these limitations. It is important that the data can fully support any inferences made. The numbers and selected diagrams should tell a story and give an **insight** into the business or organization; it is this that makes such a statistical investigation worthwhile. It is all too easy to create the wrong impression by using computer packages for graphical effect rather than content. Conversely, if you treat such diagrams dismissively, then it is very easy to be fooled into drawing the 'wrong' conclusions. Statistical exploration (to use another metaphor) should be a source of discovery, in addition to a means of reporting results.

Secondary data often comes in a summarized form, such as the tables of information produced by government departments or survey reports. Primary data, inevitably, comes in the form of 'raw' data, such as a list of numbers. (See Section 2.1 on issues of data collection.)

Case study

In the Shopping Developments Limited case, further information has been collected on the number of items bought by shoppers per visit to Hamblug's shop during a week. To try to show the variability and patterns in the data, recordings were made of 420 shoppers. This data is shown in Table 4.1 below. Reading along the first row of data, we can see that the first shopper purchased 12 items, the second shopper purchased 23 items, and so on. Given such data, the challenge is to make sense of it and share the understanding with others. This data is available as an Excel file SDL3.XLS on the website.

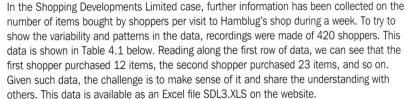

4.2 Tabulation of data

The 420 recordings given in the case would generally be regarded as a relatively small data set; often we need to deal with several thousand values and we will need to develop approaches that will always work regardless of the volume of data. To improve our understanding of this data we could produce a simple table showing how many times a particular number of items were recorded as purchased at Hamblug's shop, and this is shown in Table 4.2.

Table 4.1   Number of items purchased by 420 shoppers at Hamblugs

12	23	25	26	54	92	27	54	36	64	21	88	47	61	52	82	41	61	55	99
14	85	52	61	42	59	37	83	61	147	134	128	93	68	67	18	24	56	81	45
64	68	95	32	18	124	81	61	35	32	104	73	61	38	16	55	72	134	67	49
24	48	95	42	57	48	68	57	19	43	72	58	65	39	57	46	72	68	76	82
35	48	94	55	76	82	46	18	64	53	81	38	57	64	58	42	61	38	51	27
62	48	54	67	48	37	45	51	57	29	48	24	66	83	43	47	72	81	64	67
83	42	43	28	57	55	46	65	58	74	44	38	45	29	67	57	52	48	67	51
79	63	78	95	46	105	46	107	64	53	68	37	83	69	61	53	48	42	51	67
84	68	98	85	45	64	124	56	84	53	56	91	75	64	61	85	82	63	71	45
54	64	49	77	97	50	75	60	77	50	115	51	50	60	52	53	66	73	52	31
29	25	33	41	51	78	56	70	69	92	40	89	60	58	89	65	124	56	40	82
61	38	49	91	65	52	53	55	62	74	62	52	31	17	61	53	20	53	33	95
57	91	55	89	49	65	54	60	90	55	20	91	49	60	90	54	93	70	52	107
84	56	50	64	108	90	71	63	66	58	58	66	69	59	69	22	66	59	92	134
51	49	23	47	36	55	37	56	70	50	39	71	65	65	57	70	49	58	41	120
55	68	34	55	65	33	56	24	63	59	62	56	34	70	65	52	36	62	58	51
64	54	50	63	90	50	31	65	50	60	50	89	22	58	89	65	60	57	90	64
49	27	63	47	20	64	63	69	23	47	59	70	65	55	59	69	47	59	44	72
67	35	62	64	66	78	80	62	66	90	80	59	64	60	65	22	69	60	57	24
48	19	72	40	73	63	60	63	62	39	55	55	49	70	62	55	39	23	35	36
37	56	48	62	36	80	40	58	71	64	60	71	60	59	90	22	70	63	56	34

Definition

Frequency is the number of times a particular value or characteristic occurs.

It is useful to summarize Table 4.1 by looking at the frequency with which a particular number of items was recorded. In this data, for example, there was one occasion (a frequency of 1) when 12 items were recorded, and 14 occasions

Table 4.2 A frequency table showing the number of items purchased by shoppers at Hamblugs (where x is number of items purchased)

x	f	x	f	x	f	x	f	x	f	x	f	x	f	x	f
1	0	21	1	41	3	61	11	81	4	101	0	121	0	141	0
2	0	22	4	42	5	62	10	82	5	102	0	122	0	142	0
3	0	23	4	43	3	63	10	83	4	103	0	123	0	143	0
4	0	24	5	44	2	64	16	84	3	104	1	124	3	144	0
5	0	25	2	45	5	65	13	85	3	105	1	125	0	145	0
6	0	26	1	46	5	66	7	86	0	106	0	126	0	146	0
7	0	27	3	47	6	67	8	87	0	107	2	127	0	147	1
8	0	28	1	48	10	68	7	88	1	108	1	128	1	148	0
9	0	29	3	49	9	69	7	89	5	109	0	129	0	149	0
10	0	30	0	50	9	70	8	90	7	110	0	130	0	150	0
11	0	31	3	51	8	71	5	91	4	111	0	131	0		
12	1	32	2	52	9	72	6	92	3	112	0	132	0		
13	0	33	3	53	8	73	3	93	2	113	0	133	0		
14	1	34	3	54	7	74	2	94	1	114	0	134	3		
15	0	35	4	55	14	75	2	95	4	115	1	135	0		
16	1	36	5	56	11	76	2	96	0	116	0	136	0		
17	1	37	5	57	11	77	2	97	1	117	0	137	0		
18	3	38	5	58	10	78	3	98	1	118	0	138	0		
19	2	39	4	59	9	79	1	99	1	119	0	139	0		
20	3	40	4	60	12	80	3	100	0	120	1	140	0		

when 55 items were recorded. Obtaining these figures directly from Table 4.1 would be difficult and time consuming, and therefore it is usual to produce at least a simple frequency table, as shown in Table 4.2.

This table could be produced using a computer-based method or using the 'tally count' (often called the five-bar gate) method. Essentially, the manual method requires a listing of the number of items purchased (or any other characteristic of interest) and then working systematically through the raw data, adding a '1' or 'down slash' to the appropriate count. In the case of the five-bar gate method, partly shown in Table 4.3, four bars are marked and the fifth is used to 'cross' the other four to complete a count of five. The groupings of five allow an easier summing up at the end (e.g. 5, 10, 15, 20), and are of such a scale that mistakes are considered less likely. Spreadsheets usually require you to list all of the possible values, here 12 to 147, in a blank area of the worksheet, and then use the analysis add-in to perform the frequency count. (Note that in Excel 2000 this add-in is not loaded by default when you load Office 2000, and you will need to install it from inside Excel on the first occasion that you wish to use it.)

Table 4.3 An illustration of the five-bar gate method

Number of items	Tally count
21	|
22	||||
23	||||
24	卅
25	||
Etc.	

We can use Table 4.2 to improve our understanding of the data. We can see that the lowest number of items purchased was 12 and the highest number was 147. We can also see that the most frequent number of items was 64 (recorded on 16 occasions).

If the range of observed values being considered is relatively small, say under 10, then this approach leaves the data both manageable and readable. In the case of pre-coded questions on a questionnaire, the range is likely to be small (often less than five), and even attitude scales normally only range from 1 to 5 or 1 to 7. In these cases we are likely to want to retain the detail of the numbers given for each possible response. However, if the range of observed values is relatively large (e.g. 12–147 or larger), then we can, and generally do, **amalgamate adjacent values to form groups**, as shown in Tables 4.4–4.6.

Each of the tabulations is 'correct' but each conveys a different level of information. In every case **the detail of individual values is lost**. Table 4.4 retains much of the information contained in the original table. We no longer know the

Table 4.4

Number of items (x)	f
Under 10	0
10 but under 20	9
20 but under 30	27
30 but under 40	34
40 but under 50	52
50 but under 60	96
60 but under 70	101
70 but under 80	34
80 but under 90	28
90 but under 100	24
100 but under 110	5
110 but under 120	1
120 but under 130	5
130 but under 140	3
140 but under 150	1

Table 4.5

Number of items (x)	f
Under 20	9
20 but under 40	61
40 but under 60	148
60 but under 80	135
80 but under 100	52
100 but under 120	6
120 and more	9

Table 4.6

Number of items (x)	f
Under 60	218
60 and more	202

lowest value (except that it is under 20) nor the highest value (except that it is between 140 and 150); we only know how many values lie within a given range. The data is further summarized in Table 4.5. The further reduction of intervals in Table 4.6 (to two intervals) means that most of the original information has been lost. The three tables have been produced to show that a judgement is required when constructing tables between the detail that needs to be retained (but detail that might hide a more general pattern) and the clarity given by a more simple summary. **The most important point to consider is whether the management of the data meets the needs of the user.** As a general guide, we would recommend between four and eight intervals.

Exercise

> Search through business publications for examples of the range of tables produced and identify what you regard as good and bad practice.

Where you have the raw data entered onto a spreadsheet such as Excel, then you can use the **built-in functions** to produce the frequency distributions. (For basic guidance on using spreadsheets see *Improve Your Maths*.) You will still need to decided on the groups to be used. Taking the file SDL3.XLS, the raw data is in cells A2–A421. A quick check allows us to find the minimum and maximum for the data, we use the appropriately named functions (click on the f_x button to get a complete list of functions). Finding a frequency distribution is slightly more complex. You need to set up the groups on a blank part of your spreadsheet, with one column for the lower limits and one for the upper limits. (Here we have discrete data, so our upper limits are 19, 39, 59, etc.), see Figure 4.2. You next highlight the column of cells next to your upper limits (here F12–F18) and press the function button. Select FREQUENCY from the *Statistical* list and put the range of cells containing the data into the first box (here A2:A421).

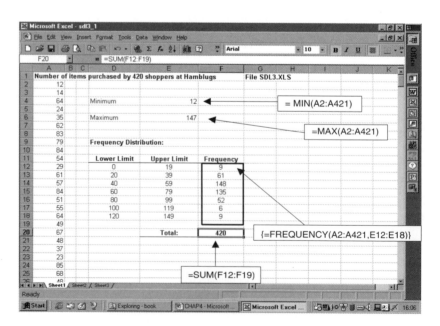

Figure 4.2 A frequency distribution using Excel.

In the second box put in the range of cells containing the **upper limits** (here E12:E18). Then press the three keys Control, Shift and Return together. The frequency distribution will appear in the highlighted cells. (Note that you are using an array, and therefore **cannot change** individual cells in the column you have just created. However, the function is **live** if you change the individual upper limits, the frequencies adjust automatically.)

Case study

Using the original data given in the case on the recorded number of cars (file SDL1.XLS available on the website) find an appropriate frequency distribution. A result is shown in Figure 4.3.

We could also ask the following questions.

■ Are the recorded numbers given in sequential order; if so is it possible to look for a trend or pattern over time? (And if not, why not?)
■ Are there other factors that should be considered; like holiday periods, local road repairs, promotional activities by ferry operators?
■ What is the capacity of the Hamblug's shop; and were people turned away?

It should be checked how representative the timing was of shop opening times or shopping activity.

It is often the case that we have to use data generated by others or that the data is more complex than a simple listing of numbers would suggest. Managing data is rarely as straightforward as it first seems.

The purpose of collected data is to **inform and communicate**. We need to be careful that the method that we choose to aggregate collected data does not mask factors and effects of real interest. If we look at historic data, for example, the number of vegetarians or lager drinkers interviewed was often small, and the trend away from meat eating or the changes in drinking behaviour was in many cases missed.

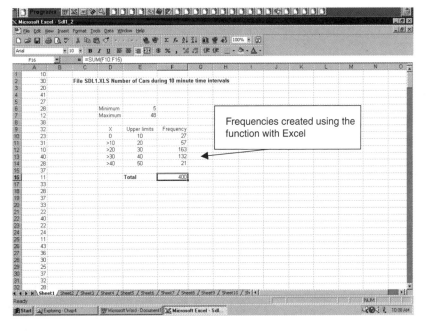

Figure 4.3

Tables that consider only one factor (e.g. the number of items purchased) are likely to limit the analysis that we can do. We are often interested in how one factor relates to another. A table, like Table 4.7, which relates where shoppers reside to the main purpose for shopping could be of particular interest. Note that this table relates to a later survey of 2000 shoppers. A single tabulation would give either the numbers by 'where the shoppers reside' or the numbers by 'main purpose of shopping' but not how the two relate. It can be seen from Table 4.7, that the main purpose of shopping was 'general grocery' (790 from 2000), that most shoppers were from Astrag (860 from 2000) and most shoppers from Astrag were there for 'general grocery' (675 from 860, or about 78.5%).

The cross-referencing of one variable or characteristic against another, as shown in Table 4.7, is referred to as **cross-tabulation**. In general, for large amounts of survey type data, packages such as MICROSTATS, MINITAB, MARQUIS and SPSS are the most effective way of producing the required diagrams and statistics, but for other analysis (for example, breakeven charts), a spreadsheet package like LOTUS or EXCEL is more appropriate.

Table 4.7   The number of shoppers by area of residence and main purpose of shopping

| Main purpose of shopping | Area of residence | | | | |
	Astrag	Baldon	Cleardon	Other	Total
General grocery	675	60	35	20	790
Clothing	30	490	30	20	570
DIY	150	180	235	15	580
Other	5	20	0	35	60
	860	750	300	90	2000

4.3 Visual presentation

4.1 Raw data

4.3 **Visual presentation**

4.5 Graphical representation

4.2 Tabulation of data

4.4 Continuous data

One of the most effective ways of presenting information, particularly numerical information, is to construct a chart or a diagram. This is because many people shy away from tables of numbers, maybe thinking that they will not understand them. Even if this is not the case, it takes more time and effort to elicit information from a table than from a well-constructed diagram.

The choice of diagram depends on the type of data to be presented, the complexity of the data and the requirements of the user. As a guide, we shall make one basic distinction: whether the data is **discrete** or **continuous** (see Section 1.4 on measurement). A set of data is discrete if we only need to make a count, like the number of cars entering a car park or the number of smokers by gender. A set of data is continuous if measurement is made on a continuous scale, such as the time taken to travel to a shopping centre or the yield in kilograms of a manufacturing process. There are, of course, some exceptions. Technically, money is seen as discrete since it changes hands in increments (pence), but is usually treated as continuous because the increments are relatively small. Age is continuous but is often quoted as age last birthday, and therefore becomes discrete.

Tables are generally constructed to show the frequency in each group or category. However, in many cases, it can be more useful to present the percentage. In terms of the language we use, it is often easier to talk about the percentage of shoppers or the percentage of smokers or the percentage that vote for a particular political party. Table 4.8 shows both the frequency and the percentage profile of the number of items purchased.

4.3.1 The use of percentages

As a reminder, to calculate the percentage, you need to take the frequency of interest, say 9, divide by the total and multiply the result by 100: the percentage of occasions when 10 but less than 20 people were recorded

$$= \frac{9}{420} \times 100 = 2.142857 = 2.14\%$$

The value 9/420 gives the fraction when '10 but less than 20 items' were recorded, and 2.14% gives this as a fraction of 100, or a percentage. To maintain the clarity of the table and to work at the level of accuracy usually required of such tables (particularly in market research) the percentages have been rounded to whole numbers. When percentages end with a 0.5%, the usual practice of rounding-up has been followed. The effect of **rounding** has meant the loss of some detail (but how useful is an accuracy of 0.25% on a sample of 400?), but also can lead to the total not necessarily being 100 (in this case the sum is 99). For more details on rounding, see your copy of *Improve Your Maths: A Refresher Course*.

The counts of shoppers by area of residence and 'main purpose of shopping' given in Table 4.7 are discrete data. To allow easy comparisons between areas, we can present the same information as percentages, as shown in Table 4.9.

Table 4.8

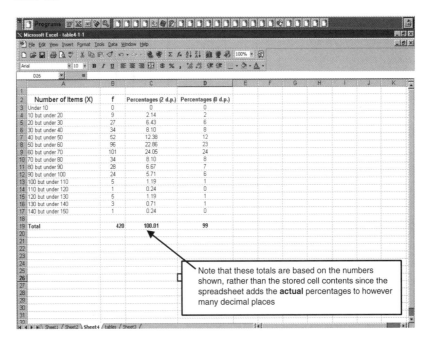

Number of Items (X)	f	Percentages (2 d.p.)	Percentages (0 d.p.)
Under 10	0	0	0
10 but under 20	9	2.14	2
20 but under 30	27	6.43	6
30 but under 40	34	8.10	8
40 but under 50	52	12.38	12
50 but under 60	96	22.86	23
60 but under 70	101	24.05	24
70 but under 80	34	8.10	8
80 but under 90	28	6.67	7
90 but under 100	24	5.71	6
100 but under 110	5	1.19	1
110 but under 120	1	0.24	0
120 but under 130	5	1.19	1
130 but under 140	3	0.71	1
140 but under 150	1	0.24	0
Total	420	100.01	99

Note that these totals are based on the numbers shown, rather than the stored cell contents since the spreadsheet adds the **actual** percentages to however many decimal places

Table 4.9 A percentage breakdown of the 'main purpose of shopping' by area of residence

Main purpose of shopping	Area of residence			
	Astrag	Baldon	Cleardon	Other
General grocery	78.49	8.00	11.67	22.22
Clothing	3.49	65.33	10.00	22.22
D.I.Y.	17.44	24.00	78.33	16.67
Other	0.58	2.67	0.00	38.89
	100.00	100.00	100.00	100.00

It can immediately be seen that 'general grocery' was the 'main purpose of shopping' for 78.49% from Astrag and only 8.00% from Baldon. (To calculate these percentages, we can again find the fractions for each area of residence and then multiply by 100.) Whether a table gives us information as a percentage or a frequency count (discrete data), our understanding can be enhanced by the use of effective diagrams.

4.3.2 The presentation of discrete data

The numbers observed (counts) whether by 'main purpose of shopping', 'area of residence' or some other category can be represented as vertical bars. The height of each bar is drawn in proportion to the number (frequency or percentage) by a vertical ruler scale. Figure 4.4 shows the number of shoppers by area of residence.

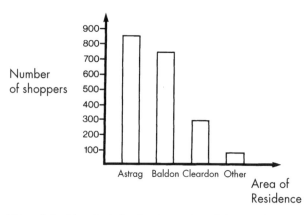

Figure 4.4 A bar chart showing the number of shoppers by area of residence.

Exercise

Construct a bar chart to show the number of shoppers by 'main purpose of shopping'.

We can increase the detail in a bar chart by the use of a key. Where the vertical bars also provide a further breakdown of information, as shown in Figure 4.5, we refer to a component bar chart.

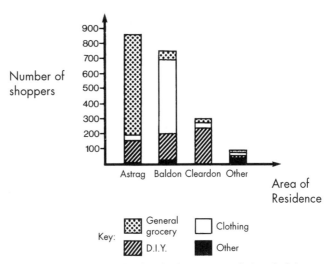

Figure 4.5 A component bar chart showing 'main purpose of shopping' by area of residence.

Exercise

Construct a component bar chart to show how area of residence varies by the 'main purpose of shopping'.

Once we begin to examine the composition of totals, it can become difficult to see the relative size or importance of components if we are working with original frequencies. Often we are interested in measures of share, e.g. percentage market share. In this case, it is often more convenient to construct bar charts on the basis of percentages rather than absolute figures. A percentage component bar chart constructed using a spreadsheet is shown as Figure 4.6 (using figures from Table 4.9).

Constructing these diagrams by hand may occasionally be necessary, but they would normally be produced from a spreadsheet, or other software package. Such packages give much more flexibility in terms of the number of categories used, the labeling attached to the diagram, and the size and scale used. A wide range of alternative representations can be easily drawn. Figures 4.7 and 4.8 are included to show two variants on the 'standard' bar chart and illustrate the choice now available.

4.3.3 Pie charts

In the case of a pie chart, a circle is used as a representation of the total of interest, with **segments being used to represent parts or share**. In almost every case these pie charts are created using a spreadsheet package, but for completeness, we will describe how to construct one from scratch. A circle has 360 degrees (written 360°) and segments are drawn using fractions of this 360°. Table 4.10 shows the determination of degrees required for the construction of a pie chart (you will need to remember how to use a protractor!). In this example, general grocery is the main purpose of shopping for 39.5% of those in the survey and needs to account for that percentage of 360°, i.e. 142.28. The pie chart is shown as Figure 4.9.

Most computer packages will allow you to create pie charts, and will give you the option of presenting as 'exploded slices' (as shown in Figure 4.10) or in

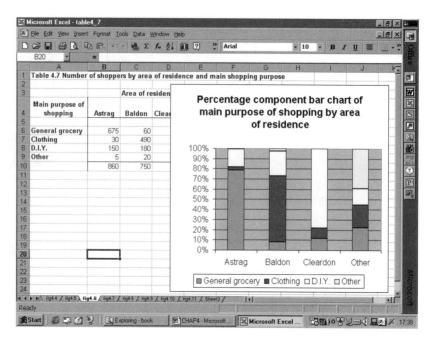

Figure 4.6

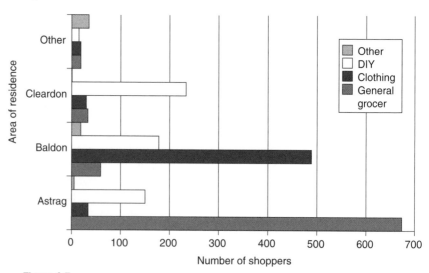

Figure 4.7

Table 4.10 The number of shoppers by main purpose of shopping: the construction of a pie chart

Main purpose of shopping	Number of shoppers	Proportion of total	Proportion of 360°
General grocery	790	0.395	142.2
Clothing	570	0.285	102.6
D.I.Y.	580	0.290	104.4
Other	60	0.030	10.8
	2000	1.000	360.0

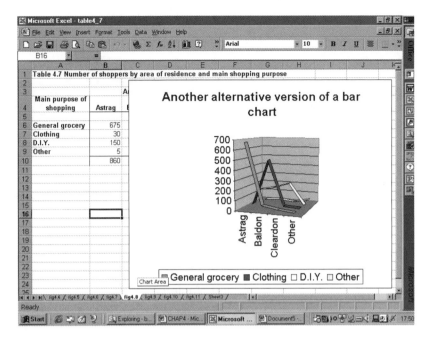

Figure 4.8

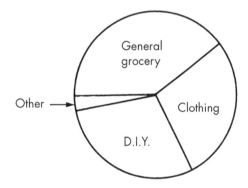

Figure 4.9 A pie chart showing the main purpose of shopping.

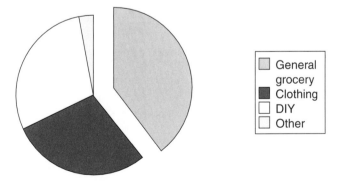

Figure 4.10

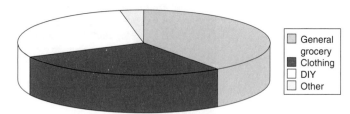

Figure 4.11

three dimensions (as shown in Figure 4.11), which may help the communication process.

It should also be noted that if several pie charts are drawn for the purpose of comparison (not illustrated here), then they should be of the same size for percentage comparisons but their areas should be in proportion to the frequencies involved if frequency comparisons are being made (you will need to remember how to determine the area of a circle to construct these). As a general rule, pie charts are effective for relatively simple representations but become less clear as the number of categories increase and as we attempt to use them for comparative purposes. They are often used in company reports to show how profits have been distributed and by local authorities to explain how the money raised by taxation is spent.

4.3.4 Pictograms

In many types of presentations, it is more important to **attract attention** and **maintain interest** than to given complete statistical accuracy. It may be necessary to make a few important points effectively (think about the methods a politician might employ) and not confuse people with details in the limited time or space available. (More detailed statistical analysis can always be given in briefing papers or be held on a website for easy reference.) A pictogram can be very effective is such circumstances. The bars drawn on a bar chart are replaced by an appropriate picture or pictures, either vertically or horizontally (as shown in Figure 4.12).

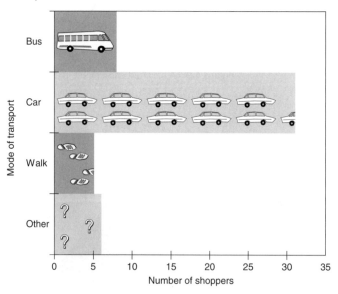

Figure 4.12 A pictogram showing the main mode of transport used by shoppers.

A pictogram can be more eye-catching, but is less accurate than a bar chart (how easily can you tell that eight shoppers used the bus, 31 a car, five walked and six gave a different response?), and, in some circumstances, misleading. It can be particularly confusing if the height and the width of a picture both change as different values are represented. If, for example, we are representing sales growth by a tree, as sales grow we draw taller and taller trees, but unless we increase the width, the tree will look thinner and thinner. The problem is illustrated in Figure 4.13.

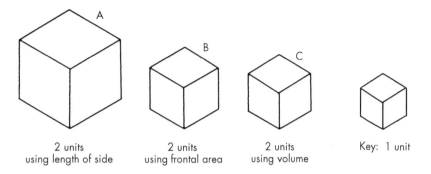

Figure 4.13   Constructing a pictogram to represent a doubling of sales.

In a bar chart, an increase is shown as an increase of height, but in Figure 4.13, the visual impression could be in terms of height, surface area or volume. If we are making a single measurement or count – for example, sales by region, or turnover by company – a one-dimensional representation is generally clearer. We must try to avoid the possible confusion that pictograms of the kind shown in Figure 4.13 can produce; picture A would certainly leave the impression of a more rapid sales increase than picture B or picture C.

4.3.5 Miscellaneous

We have considered bar charts, pie charts and pictograms as typical ways of representing discrete data. The choice depends on the purpose, and we should be prepared to accept variants on the typical representation, if this is likely to be more thought-provoking or effective in other ways.

If we are dealing with a small number of values, a list in ascending or descending order could be sufficient (known as ranking). Table 4.11 shows the number of complaints received each day over a 10-day period and Table 4.12 shows exactly the same data ordered by value.

Table 4.11   The number of complaints received each day over a 10-day period

15	8	14	15	4	15	17	6	18	15

Table 4.12   Number of complaints in rank order

4	6	8	14	15	15	15	15	17	18

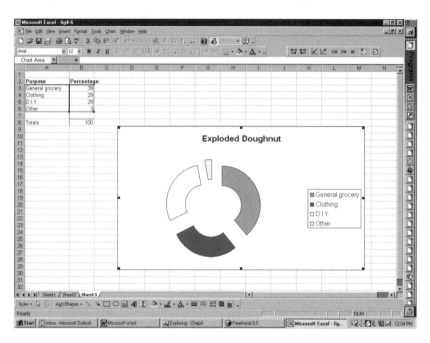

Figure 4.14 A doughnut 'pie chart' showing the main purpose of shopping.

It is far easier to see from Table 4.12 the largest and smallest numbers involved, and that about 15 complaints were received on a 'typical' day, since this could be said to be a 'typical' number. In reality we would need to consider whether the sample size was adequate to represent the number of complaints, what was meant by a 'typical' day and whether, in business terms, any number of complaints is acceptable.

Representations can come in all shapes and sizes, and serve many purposes. A map is a good example of how assumptions allow an effective representation of something as complex as a landscape. The advance of technology has enhanced both choice and quality of representation. In Figure 4.14 a 'doughnut' is used to show in percentage terms the main purpose of shopping. (This is one of the options available in many spreadsheets.)

This form of representation may appeal to some people and not others. Given the choice available it is a matter of judgment of how to most effectively present the data, so that it can be understood (we could loose the meaning of the data in fancy presentation!) and allows a focus on issues of interest.

Small-scale investigations using questionnaires (as described in the case) are likely to produce mostly discrete data by the nature of the questions asked. Questions on gender, occupations, qualifications, car ownership and attitudes will typically produce discrete data. We must be careful not to use the statistics developed for continuous data (e.g. means and standard deviations), but which are easily available, with this type of data without adequate justification.

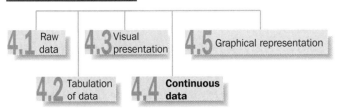

4.4 Presentation of continuous data

Continuous data is the result of making measurements using a measuring device; in this context a ruler or a thermometer would be good examples. The accuracy of the recorded measurement (e.g. 5 or 5.2 or 5.1763) depends on the requirements of the user and the accuracy of the measuring device. Time and length are considered good examples of continuous measurement. However, income is generally treated as continuous because of the larger magnitudes concerned and the fact that we are often working with averaged figures, but money is recorded in discrete chunks, e.g. pence or pounds.

4.4.1 Histograms

The distribution of measurement on a continuous scale is presented by the use of a histogram. As previously discussed, monetary amounts are generally regarded as continuous, and are typically represented by histograms. The amount spent on food in a particular shop by 50 respondents (see question 8 in the Case 1 questionnaire) is shown in Table 4.13.

Table 4.13   The amount spent on food in one particular shop

Expenditure on food	Number of respondents
Under £5	2
£5 but under £10	6
£10 but under £15	8
£15 but under £20	12
£20 but under £30	10
£30 but under £40	4
£40 or more	2
	44

Tables of this kind present a number of problems:

- First, there are two **open-ended groups**; the first and the last. Once data has been collated in this way, it is often difficult to know what the lowest and highest amounts were or are likely to have been. All we can do with groups (or intervals) that are open-ended is to assume *reasonable* lower or upper boundaries on the basis of our knowledge of the data and the apparent distribution of the data. In this case, it may be reasonable to assume a lower boundary of £0 for the first group and an upper boundary of £50 for the last group.
- Secondly, if the numbers are relatively large in the open-ended groups, we could be losing valuable information on those that spend least or most (in this case the numbers are relatively small). This illustrates the point, that **judgement** is needed on how to word questions (the question only asked if spending was £40 or more, and was not more specific) and how to construct tables to best capture and represent information.
- Thirdly, if we were to use bars (a bar chart) to represent the number of respondents in each range, it would appear that there were more in the range '£20 but under £30' (10 respondents) than in the range '£10 but

under £15' (eight respondents). However, the range '£20 but under £30' is twice as wide and we need to take into account this increased chance of inclusion. To fairly represent the distribution, frequencies are plotted in proportion to area.

Histograms are usually constructed with reference to a key as shown in Figure 4.15. Here, expenditure on food is plotted on the horizontal scale. No vertical scale is shown but one could be used for construction purposes and then concealed. As a general rule if you double the interval width then you halve the height (e.g. retain the importance of area). Clearly if the interval is increased by a factor of five, say, then the height of the block would be found by dividing that frequency by five, and so on.

In practice, we choose one of the groups (often the smallest) as the standard and scale the rest, as shown in Table 4.14.

Table 4.14 Method for constructing a histogram

Expenditure on food	Frequency	Width	Scaling factor	Height of block
£0* but under £5	2	5	1	2
£5 but under £10	6	5	1	6
£10 but under £15	8	5	1	8
£15 but under £20	12	5	1	12
£20 but under £30	10	10	½	5
£30 but under £40	4	10	½	2
£40 but under £50*	2	10	½	1

*Assumed boundary.

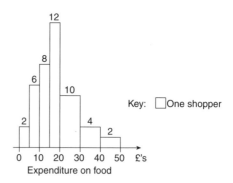

Figure 4.15 A histogram showing the distribution of expenditure on food.

4.4.2 Miscellaneous

Any representation of continuous data needs to ensure that the accuracy of the measurement and the method of recording are adequately captured; there is a difference between exact age and age last birthday, for example. Essentially, a histogram shows the distribution of measurement, and it is this idea of a representative spread that is particularly important. An alternative representation that can achieve the same type of effect is a stem-and-leaf diagram. Suppose we are recording the number of seconds it is taking to complete a cash transaction and the first three recordings were 43, 66 and 32. The most significant digit is the first (4, 6 or 3) and this would be referred to as the **stem**; in this case, the second digit is referred to as the **leaf** (3, 6 or 2, but the decision can be more complex than this). A stem-and-leaf diagram is constructed by placing the 'stem' value to the left of a vertical line and the 'leaf' value to the right of this line as shown in Table 4.15.

Table 4.15 The beginning of a stem-and-leaf diagram showing the first three recorded values

```
3 | 2
4 | 3
6 | 6
```

In Table 4.16 the recordings have been added to and we can read these as 28, 32, 33 and so on. The importance of this diagram is again the representation of distribution.

Table 4.16 A stem-and-leaf diagram based on 30 recorded values

```
2 | 8
3 | 23625
4 | 366418
5 | 5377190674
6 | 682105
7 | 42
```

In general survey work, continuous data is most likely to be generated by questions about time (e.g. travel time to work, age or time taken to complete a task), distance (e.g. distance to work or the dimensions of a component) or value (e.g. weight of a gold bar or income – which is seen as continuous for practical purposes).

4.5 Graphical representation

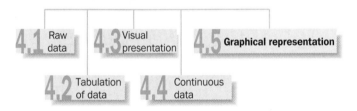

Graphical representation is essentially used for two purposes: either to show changes over time or to explore the relationship between variables. Plotting individual points on a graph is discussed in detail in *Improve your Maths: A Refresher Course* and most people will have drawn a graph at some stage. Here we are concerned with illustrating a particular situation that has been observed rather than a mathematical relationship (this is revisited in Part 7).

4.5.1 Plotting against time

Most problems of substance will have a historic dimension. Governments monitor changes in employment or the balance of trade over time, there is public interest in how birth rates or crime rates are changing over time and businesses look at how a number of measures of performance change over time. Plotting and being able to interpret data recorded over time (see Chapter 14) is a major element of many research projects. A company (like Shopping Developments Limited) might be interested, for instance, in the number of business enquiries related to precinct retail units which have been received over the past 3 years. This is shown in Table 4.17.

Tables of this kind are very common and there are important points to note:

Table 4.17 Number of business enquiries received by Shopping Developments Limited

Year	Quarter	Number of enquiries
1	1	20
	2	33
	3	27
	4	14
2	1	18
	2	29
	3	25
	4	12
3	1	17
	2	27
	3	18
	4	8

- In this case the recording started three years ago, in year 1 (year 3 would be taken as referring to the most recent year) and quarter 1 refers to the months January, February and March.
- Time is always plotted on the *x*-axis.

However plotted, the resultant graph will need interpretation. Figures 4.16 and 4.17 show how the impression given can depend on the scales chosen and how difficult it is to talk about change and rates of change without supportive calculations.

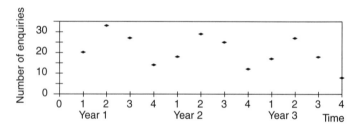

Figure 4.16

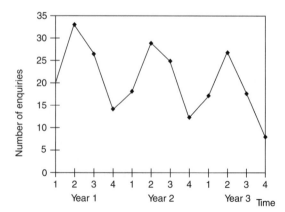

Figure 4.17

Both graphs show a downward trend (which may concern Shopping Developments Limited), but with the trend emphasized in different ways. A regular and perhaps predictable quarterly variation can also be observed, and again this is presented with differing emphasis.

4.5.2 Actual and percentage increases

Suppose the unit sales achieved from two products, Product A and Product B, were recorded as shown in Table 4.18.

The unit sales of product A are shown in Figure 4.18 and the unit sales of product B in Figure 4.19.

The graph for product A has produced the (expected) straight line, showing the constant increase of 8000 units each year. The curve produced in the graph for product B suggests a **constant percentage increase**. To study the rate of change over time, we determine the log values for the y-axis, given in Table 4.19, and then plot against time, as in Figure 4.20.

If values are increasing (or decreasing) at a constant percentage rate (30% in this case), then plotting logs of values against time will produce a **straight line**. We can check these figures with the usual calculations, for example

$$\left[\frac{26\,000 - 20\,000}{20\,000}\right] \times 100 = 30\%$$

[Alternatively, we could antilog the increase in log values (0.1140 or 0.1139) to find the multiplicative factor of 1.30.]

To estimate the sales in the next year we **multiply by 1.3**, that is, increase values by 30%.

Table 4.18 Unit sales of two products

Year	Product A	Product B
1	20 000	20 000
2	28 000	26 000
3	36 000	33 800
4	44 000	43 940
5	52 000	57 122

Unit sales (in thousands)

Figure 4.18 Unit sales of product A.

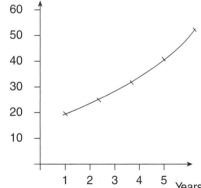

Unit sales (in thousands)

Figure 4.19 Unit sales of product B.

Table 4.19   The log values of sales of product B

Year	Number sold	Log of numbers sold	Increase in log values
1	20 000	4.3010	
2	26 000	4.4150	0.1140
3	33 800	4.5289	0.1139
4	43 940	4.6429	0.1140
5	57 122	4.7568	0.1139

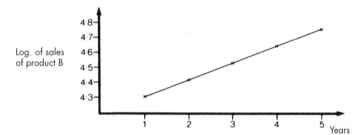

Figure 4.20   The log values of sales of product B plotted against time.

4.5.3 Plotting a relationship

When trying to understand business relationships and other relationships, we need to consider how variables can influence and be influenced by each other. Graphs can often show such relationships very clearly, but we do need to be careful how we assign variables to the x-axis and the y-axis since the direction of the relationship is often implied from this positioning. The variable thought to be responsible for the change is plotted on the x-axis (the horizontal) and is often referred to as the **independent** or **predictor variable**. The variable whose change we are seeking to explain is plotted on the y-axis (the vertical axis) and is referred to as the **dependent variable**. Table 4.20 gives 'travel time in minutes' (see question 3 in the Case 1 questionnaire) and amount spent on food (see question 8) for the first five respondents.

Table 4.20   Data on travel time and expenditure on food

Respondent	Travel time in minutes	Expenditure of food (£)
01	10	17.50
02	15	25.00
03	5	17.50
04	4	2.50
05	25	45.00

In this case, we would probably be trying to explain the differences in the 'expenditure on food' and would be considering a range of factors that could offer some explanation. The dependent variable (y) would be 'expenditure' and the independent variable (x) would be 'travel time'. You could, however, be considering a different scenario where the time that a respondent was prepared to travel did depend on how much they were planning to spend. It is for the researcher to make a judgment on how the analysis should be structured, given the context and the requirements of the analysis. The graph of expenditure against time is shown as Figure 4.21.

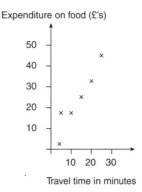

Figure 4.21 Graph of expenditure on food against travel time in minutes.

Such graphs are usually called scatter diagrams. They form a starting point for regression and correlation analysis which is discussed in depth in Part 5 of this book.

4.5.4 The Lorenz curve

One particular application of the graphical method is the Lorenz curve. It is often used with income data or with wealth data to show the distribution or, more specifically, the extent to which the distribution is equal or unequal. This does not imply a value-judgement that there should be equality but only represents what is currently true. To construct a Lorenz curve each distribution needs to be arranged in order of size and then the percentages for each distribution calculated. The percentages then need to be added together to form cumulative distributions which are plotted on the graph.

Let us consider first the information given in Table 4.21. The percentage columns give a direct comparison between population and wealth. It can be seen that the poorest 50% can claim only 10% of total wealth. The cumulative percentage columns allow a continuing comparison between the two. It can also be seen that the poorest 75% of the population can claim 30% of the wealth and the poorest 85% of the population 40%, and so on.

Table 4.21 A percentage comparison of the population and wealth distribution

Group		Percentage of population	Cumulative percentage		Percentage of total wealth	Cumulative percentage	
Poorest	A	50		50	10		10
	B	25	(+50)	75	20	(+10)	30
	C	10	(+75)	85	10	(+30)	40
	D	10	(+85)	95	15	(+40)	55
	E	3	(+95)	98	25	(+55)	80
Richest	F	2	(+98)	100	20	(+80)	100

Note that in Figure 4.22 the point representing zero population and zero wealth (point A) is joined to that representing all of the population and all of the wealth (point B) to show the **line of equality**. If the points were on this line then there would be an equal distribution of wealth: the farther the curve is away from the straight line the less equality there is. The curve can also be used to show how the income distribution changes as a result of taxation. Figure 4.23 shows a progressive tax system where the post-tax income distribution is closer to equality than the pre-tax income distribution.

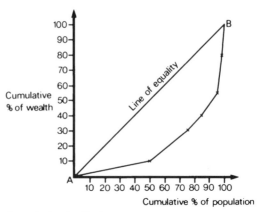

Figure 4.22   A graph showing the Lorenz curve

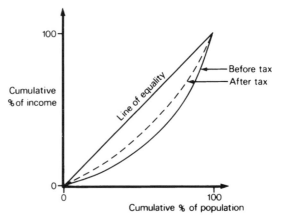

Figure 4.23   The effects of a 'progressive' tax on the distribution of income

4.6 Conclusions

The quantity of data that a business or other kind of organization needs to manage can be immense. There can literally be thousands of figures relating to sales, production and other business activities. Data needs to be summarized and presented so that people, not computers, can understand what is happening. We are not saying that you do not need to use computers to organize and present such data, but, in the final analysis it is essential that what is happening within the organization, or to its environment is **communicated successfully** to those who have to make decisions. Diagrammatic representation offers a quick way of summarizing these large amounts of data and thus getting the general message across. It is not a substitute for statistical analysis, but will often form a starting point. (Senior managers are often most interested in general trends in data, rather than the immense detail contained in the raw data.)

You only need to look at the business press to see the importance of clear, concise presentation.

1 Obtain a number of charts and diagrams used to describe quantitative information. Sources could include, for example, newspaper cuttings, building society pamphlets or textbooks. Classify each as being discrete or continuous data and state reasons why you consider them to be informative or misleading.

2 The sales achieved by telephone over the past three years have been shown as follows:

75 units	100 units	125 units
Year 1	Year 2	Year 3

Prepare an argument for changing this representation and present at least two alternative representations.

3 The following table shows the number of orders received last year by product, carscats and hatueuts, and by country:

	France	Germany	Spain
Carscats	60	40	56
Hatueuts	520	90	120

Produce at least two alternative representations.

4 The number of new orders received by a company over the past 25 working days were recorded as follows:

3	0	1	4	4
4	2	5	3	6
4	5	1	4	2
3	0	2	0	5
4	2	3	3	1

(a) Tabulate the number of new orders in the form of a frequency distribution.
(b) Present this data by means of a bar chart or other appropriate representation.
(c) Comment on the distribution and outline what other information might be of value.

5 The work required on two types of machine, X and Y, has been categorized as routine maintenance, part replacement and specialist repair. Records kept for the past 12 months provide the following information:

	Frequency	
Work required	Type X	Type Y
Routine maintenance	11	15
Part replacement	5	2
Specialist repair	4	3

Present this information using:
(a) pie charts
(b) appropriate bar charts.

6 The mileages recorded for a sample of company vehicles during a given week yielded the following data:

138	164	150	132	144	125	149	157
146	158	140	147	136	148	152	144
168	126	138	176	163	119	154	165
146	173	142	147	135	153	140	135
161	145	135	142	150	156	145	128

(a) Using the data tabulate a grouped frequency distribution starting with '110 but under 120'.
(b) Construct a histogram from your frequency distribution.

7 The average weekly household expenditure on a particular range of products has been recorded from a sample of 20 households as follows:

£8.52	£ 7.49	£ 4.50	£ 9.28	£ 9.98
£9.10	£10.12	£13.12	£ 7.89	£ 7.90
£7.11	£ 5.12	£ 8.62	£10.59	£14.61
£9.63	£11.12	£15.92	£ 5.80	£ 8.31

Tabulate as a frequency distribution and construct a suitable diagram.

8 A company has been able to define the travel necessary for its employees as being 'local', 'commuter' or 'long distance', and has related this to the reported cost of daily travel as shown in the following table:

Reported cost of travel	Type of travel		
	Local	Commuter	Long distance
under £1	60	20	0
£1 but under £5	87	46	17
over £5	12	13	53

Describe this data with appropriate diagrams using the actual figures given and using percentage calculations. What are you able to say about this data?

9 A survey of workers in a particular industrial sector produced the following table:

Weekly income (£)	Number
under £100	170
£100 but under £150	245
£150 but under £200	237
£200 but under £400	167
over £400	124

Describe this data with an appropriate diagram. What difference would it make if you were given the additional information that many of the workers included in the survey were part-time and that a high proportion of the part-time workers were female?

10 Construct a histogram from the data given in the following table:
Journey distance to and from work

Miles	Percentage
under 1	16
1 and under 3	30
3 and under 10	37
10 and under 15	7
15 and over	9

11 Construct a histogram from the information given in the following table:

Error (£)			Frequency
under		−15	20
−15	but less than	−10	38
−10	but less than	−5	178
−5	but less than	0	580
0	but less than	10	360
10	but less than	20	114
20	or more		14

12 On the completion of a survey the following tabulations were presented (with question wording) for questions one and two:

Q1. 'How many years have you been living in this (house/flat)? ...'

Number of years	Frequency
0–1	137
2–4	209
5–9	186
10–19	229
20 +	205

Q2. For each item below ask 'Do you have ...?'
(a) A fixed bath or shower with a hot water supply:

	Frequency
None	67
Shared	27
Exclusive	871
No answer	1

(b) A flush toilet inside the house:

	Frequency
None	83
Shared	30
Exclusive	850
No answer	3

(c) A kitchen separate from living rooms:

	Frequency
None	20
Shared	19
Exclusive	922
No answer	5

Report on the tabulations given using charts and diagrams where appropriate.

13 Use graphical methods to explore the following data for possible causal relationships.

Year	Sales (units)	Research (£000)	Advertising (£000)
1	590	88	142
2	645	99	118
3	495	50	80
4	575	78	42
5	665	97	150
6	810	118	40

14 The sales within an industry have been recorded as follows:

Year	Quarter 1	Quarter 2	Quarter 3	Quarter 4
1	40	60	80	35
2	30	50	60	30
3	35	60	80	40
4	50	70	100	50

Graph this data and discuss the relationship between sales and time.

15 A company's advertising expenditure has been monitored for 3 years, giving the following information:

Year	Quarter 1	Quarter 2	Quarter 3	Quarter 4
1	10	15	18	20
2	14	16	19	23
3	16	18	20	25

Graph this data and write a short report describing the main features.

16 The results of a company were reported as follows:

Year	Turnover (£000)	Pre-tax profit (£000)	Exports (£000)
1	7 572	987	2 900
2	14 651	1 682	6 958
3	17 168	2 229	7 580
4	21 024	3 165	9 306
5	25 718	4 273	10 393
6	37 378	6 247	18 280
7	53 988	9 559	28 229
8	79 258	19 646	48 770
9	122 258	32 714	74 410
10	183 338	49 832	95 029

(a) Graph the three sets of data against time.
(b) Graph the log values for the three sets of data against time.
(c) Comment on your graphs outlining the relative merits of those produced in parts (a) and (b).

17 Construct a Lorenz curve for the following data on income.

Income group	Percentage of people in group	Percentage of income
Poorest paid	10	5
	15	8
	20	17
	20	18
	20	20
	10	15
Highest paid	5	17

18 Construct a Lorenz curve for the following data on wealth-holding in the UK.

Group	No. of wealth holders (thousands)	Amount of wealth (£m)
Poorest	2 099	1 049.5
	3 530	7 060.0
	2 133	8 532.0
	4 414	33 105.0
	2 588	32 350.0
	1 167	20 422.5
	694	15 615.0
	1 018	38 175.0
	320	24 000.0
	94	14 100.0
Wealthiest	31	34 100.0
	18 088	228 509.0

Part 1   Conclusions

Part 1 has been concerned with the collection and presentation of data. The value of such work will depend on the **validity of the data collection methods** used and the **correctness of any representation** through the use of charts and diagrams. In any research it is important to clarify the purpose of the research or objectives and to ensure that the approach, that is the methodology, is capable of delivering meaningful and scientifically acceptable answers. Typically, some information will already exist (as in Case 1) and judgements about its usefulness will need to be made. Data provides a 'snapshot' at a particular point or interval in time, like a balance sheet, and it is always worth questioning its current applicability (data ages). Indeed a PEST (political, economic, social and technological) type analysis could be applied to assess what impact these factors could have had. We would also want to be sure that existing data is representative of the **population** (all the people or items of interest to us) that we are working with. Having identified existing information (from past research, company records, government publication etc.) we can then consider what needs to be added. The chances are that our research will include both **secondary data** (data that already exists and collected for other uses) and **primary data** (data we will need to collect).

It is often the case that the data does exist for broad demographic description, but not for attitudinal and behavioural factors. Judgements need to be made as to whether a **qualitative approach** (descriptive and without numbers, perhaps using illustrative case material) or a **quantitative approach** (using numbers) is most appropriate. The authors believe the approaches to be complementary, and that it is for the researcher to develop the 'craft' skills, based on knowledge, to define effective ways of proceeding. This book is about making numbers work for you, and it is essential that the numbers generated are meaningful within the problem context and that the representations of the numbers provide insightful descriptions.

The Internet is transforming communications in general, and the flows of data in particular. The shortage of secondary data is not likely to be a problem. What we do need to be sure about is the quality of the data. The results you present can only be as meaningful as the methodology that they are built on. **Rubbish data going in is only going to mean rubbish results coming out.** You should always try to ensure that the data you are working with is appropriate, adequate and without bias.

As we have seen already, a range of computer software can support our problem-solving approaches. A package like Excel can make the presentation and analysis of data relatively easy. It is an important skill to be able to interpret the printouts that we can generate and the printouts presented by others. When tackling a business problem, or any other problem of significance, it is unlikely that an obvious, single best solution will immediately emerge. A judgement has to be made on how to collect, collate and present the data. Understanding methods of data collection and presentation can inform this judgement.

Part 2　Descriptive statistics

Introduction

The first part of this book considered ways of collecting and presenting **data**. The quality of the data is fundamental to the work that we do, and yet data collection can be the very stage where time and cost pressures are particularly demanding. It is necessary for the researcher to ensure that data collection methods are adequate and that the data can inform the users in a meaningful way. Fancy analysis, now made easy by computers, cannot compensate for poor data. It is often said (using a machine analogy) that the results coming out can only be as good as the data going in.

However, searching for data (Chapter 2), the collection of data (Chapter 3) and the use of charts and diagrams (Chapter 4) alone are not sufficient for most purposes. Generally, we need to do more than 'paint' an illustrative picture. We need to describe the data with a rigour that summary numbers allow perhaps comparing one set or sub-set of data with another. Users will want numbers to work with, for example market share and rate of increase, to test ideas and to use to identify possible relationships. This process of exploring data using acceptable techniques and theories is termed **analysis**. The calculation of numbers should clarify the data, revealing similarities and differences that were not seen before. The use of numbers can complement the skills of insightful observation and idea generation.

The following three chapters provide ways of summarizing the mass of detail contained in data sets. The use of computer packages, such as MINITAB, SPSS or spreadsheets has made the calculation of descriptive statistics relatively simple. This removal of the burden of calculation or 'number crunching' makes the selection and interpretation of appropriate statistics even more important.

Case 2: Shopping Developments Limited – part 2

Shopping Developments Limited has generally been regarded as an innovative company that has been able to respond quickly to the demands of the market place. Those that can remember the early rapid growth of the company still talk about how one or two conversations could quickly lead to a major business decision being made, even on the same day. However, it is accepted that managers are now seen as being more accountable for the decisions that they make and must justify the risks they are taking.

There has been concern within the company for some time about whether or not managers have all the skills necessary to deal with the new operating environment. The number of business enquiries received and the value of new business (as shown in the following table) has increased this concern.

Table of Case 2 data The number of business enquires received, the value of new business and an index of inflation

Year	Quarter	Number of enquiries	Value of new business (£s)	Inflation rate
1	1	20	32 000	113.8
	2	33	34 000	118.1
	3	27	28 000	122.5
	4	14	17 000	126.8
2	1	18	31 000	131.3
	2	29	33 000	135.9
	3	25	26 500	140.4
	4	12	18 000	145.1
3	1	17	30 600	149.7
	2	27	32 800	154.5
	3	18	26 400	158.8
	4	8	18 000	162.6

The company is particularly concerned about the way market research information has been managed and has been seeking advice on the use of basic statistics. The company is now seen as having the joint challenge of capturing its
early youthful spirit of business enterprise and at the same time making better
use of good business practice.

Quick start	# Descriptive statistics

The presentation of charts and diagrams is not sufficient for most purposes. Most analysis requires a summary of data in the form of descriptive statistics. The mean, median, mode, standard deviation and range can be calculated for different types of data.

Untabulated data (a list of numbers)

Mean: $\bar{x} = \dfrac{\sum x}{n}$, where Σ means 'the summation of'

Median: equals the middle value of an ordered list
Mode: is the most frequent value

Standard deviation: $s = \sqrt{\dfrac{\sum (x - \bar{x})^2}{n}}$

Range: is the difference between the largest and smallest value.

Tabulated (ungrouped) discrete data

Mean: $\bar{x} = \dfrac{\sum fx}{n}$, where f is frequency

Median: find middle value using cumulative frequency
Mode: is the most frequent value

Standard deviation: $s = \sqrt{\dfrac{\sum f(x - \bar{x})^2}{n}}$.

Tabulated (grouped) continuous data

Mean and standard deviation: use mid-points
Median or other order statistics: use Ogive or formula
Mode: use tallest block on the histogram or formula

Spreadsheets

Spreadsheets such as Excel will produce these and other statistics for you. What is important, is to ensure that the statistic is appropriate for the type of data. You can't really talk about the average ethnic origin or mean religious affiliation.

Index numbers

Index numbers are used to describe change over time. A simple index, such as

$$\frac{P_n}{P_0} \times 100$$

will measure the change in price of a single product, from P_0 to P_n, over time. The Laspeyre indices and Paasche indices are important measures of how the cost of a 'basket of goods' would change over time.

5 Measures of location

This chapter refers to 'measures of location' rather than just the average, to emphasize that there is more than one 'typical' summary value (the simple average). What is seen and accepted as a typical value will depend on the data we are considering, for example, the most popular model of car, opinions on service levels as expressed on a questionnaire rating scale, or the weekly cost of groceries. We do not wish to question the use in 'everyday' language of the word 'average' ('she played an average game', 'he has an average job'), but when calculating statistics we need to be more precise. In a shopping survey, for example, we need to know what exactly is meant by the average amount spent or the average number of visits, so that we can use this information more effectively for description or inference. We also need to be aware of all the assumptions being made – so for example, when we talk about the average amount spent on cigarettes, do we mean the average spent by smokers, or the average amount spent by all respondents?

Computer packages make such calculations relatively simple, but they can mask the assumptions being made. We have detailed the manual methods of calculation in this chapter because these are still required by academic courses, are still useful for small data sets, can be necessary for checking results and can enhance your understanding of the particular statistic.

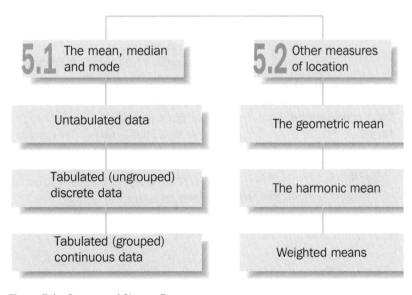

Figure 5.1 Structure of Chapter 5.

Objectives

By the end of this chapter, you should be able to:

■ calculate the mean, median and mode for various types of data
■ understand the relative merits of each measure of location
■ calculate and apply 'weighted' means.

This chapter will be structured as shown in Figure 5.1.

5.1 The mean, median and mode

5.1 The mean, median and mode

5.2 Other measures of location

The **mean** (or arithmetic mean) is the most often used measure of location or average, with the **median** and the **mode** being used for more specific (special case) applications. Each of these statistics has its own characteristics and will generally produce a different result for a given set of data.

For some sets of data it is useful to determine all of these statistics (together they tell us something about the differences in the data), but for other data sets, not all of the calculable statistics may be valid. A major consideration will be the type of data we are dealing with – **categorical**, **ordinal** or **cardinal** (see Chapter 1). As we will see, for example, it is not appropriate to calculate a mean for categorical data. How can you have an average social class or country of origin? Data can be **discrete** or **continuous**. Is giving the average number of children as 1.8 a meaningful answer? Knowing the average number of children could be helpful in a population projection exercise, but less helpful in allocating aircraft seats. We also need to consider the variation in the data (is it all closely bunched together or are there extreme values?). We also need to decide whether we are going to include all values (do we trust all the values we have, or do we wish to exclude some for being unlikely or unrepresentative?).

The **arithmetic mean** (usually just shortened to **mean**) is the name given to the 'simple average' that most people calculate. It is easy to understand and a very effective way of communicating an answer. It does not really apply to categorical data and its interpretation can be difficult when used with ordinal data, but it is used in this way and its use is often justified for practical reasons. The **median** is the middle value of an ordered list of data. It is not as well known as the mean but can be more appropriate for certain types of data. The **mode** is the most frequent value or item, typical examples being the most popular model of car or most common shoe size.

5.1.1 Untabulated data

Untabulated data will usually be presented to us as a list of numbers or rows of numbers (see Case study 1 data on the number of cars entering a car park during 10-minute intervals, file SDL1.XLS). This data can come in any order (unranked rather than ranked) and can range from a few values to several thousand or more. Suppose we consider just the first ten observations from this data set (to make life easy):

$$10 \quad 22 \quad 31 \quad 9 \quad 24 \quad 27 \quad 29 \quad 9 \quad 23 \quad 12$$

The mean

To calculate the mean, the numbers are added together to find the total, and this total is divided by the number of values included. In this case

$$\bar{x} = \frac{10 + 22 + 31 + 9 + 24 + 27 + 29 + 9 + 23 + 12}{10}$$

$$\bar{x} = \frac{196}{10} = 19.6 \text{ cars},$$

where $\bar{x}$ (pronounced x bar) is the symbol used to represent the mean. It is important to clarify the units in use and to give an interpretation to the result. In this example, 0.6 of a car is only meaningful in terms of an average value (unless we are actually cutting up cars). We would also need to decide whether to round our answer to an average of 20 cars every 10 minutes for reporting purposes.

As most statistics require some form of calculation, a shorthand has developed to describe the necessary steps. Using this shorthand, or notation, the calculation of the mean would be written as follows:

$$\bar{x} = \frac{\sum x}{n}$$

where x represents individual values, $\sum$ (sigma) is an instruction to sum values, and n is the number of values.

The median

The median is the value in the middle when numbers have been listed in either **ascending** or **descending** order (typically ascending order). The first step is to rank, or order, the values of interest:

9   9   10   12   22   23   24   27   29   31

The next step is a matter of counting from the left or the right. When working with a listing of values (**but not with continuous data**) the position of the middle value is found using the formula:

$$(n + 1)/2$$

This is easy to use with an **odd** number of values (e.g. given five values the middle one would be given by $(5 + 1)/2$, i.e. the third one). When working with an **even** number of values, such as ten, a $\frac{1}{2}$ emerges and we need to use the two adjacent values. In this case, $(10 + 1)/2$ gives $5\frac{1}{2}$ and we need to consider the fifth value of 22, and the sixth value of 23. Having found the two adjacent values, the practice is to average these to determine a median. In this case the median would be given as 22.5.

This small data set can be used to illustrate two important points. First, the averaging of the two middle values can produce a value **not possible** in the original data (22.5 cars) and secondly, the median is not sensitive to changing values away from the centre. As an extreme example, suppose the largest value of 31 had been wrongly recorded as 310, the mean would change substantially but the median would stay the same. In this sense, the median can be regarded as a more 'robust' statistic.

The mode

The mode is the **most frequently** occurring observation. Given the data:

10   22   31   9   24   27   29   9   23   12

it can easily be seen that 9 occurs twice and is therefore the most frequent value, i.e. it is the mode.

One of the problems with the mode can be illustrated if we consider the first 12 values given in the Case 1 data:

$$10 \quad 22 \quad 31 \quad 9 \quad 24 \quad 27 \quad 29 \quad 9 \quad 23 \quad 12 \quad 33 \quad 29$$

In this case there are two modes, 9 and 29, since both values occur twice. If an extra data value were to become available, say 10, there would then be three modes. If the extra value had been a 9 rather than a 10, there would only be one mode. For some sets of data the mode can be **unstable**. Again, the user should not just accept the statistic at 'face-value', but consider what it really means and what interpretation can be given.

The determination of the mean (using the Excel average function), the median and the mode for this illustrative data is shown in Figure 5.2.

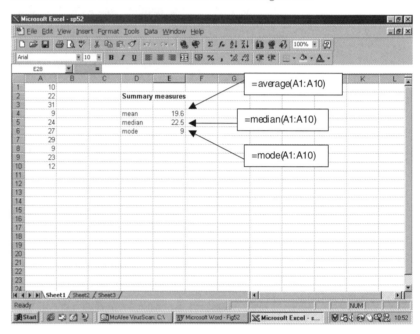

Figure 5.2   The mean, median and mode for illustrative untabulated data.

Case 1: the use of descriptive statistics

It is unlikely that managers in a company like Shopping Developments Limited would be working with data sets as small as 10 values, but it is still worth considering the three descriptive statistics just calculated. The mean was 19.5 cars, the median was 22.5 cars and the mode was nine cars per 10-minute interval. In this case, the statistics are all very different. Most observations were in the range '20 but under 30', but the two values of nine determined the mode and pulled the average downwards. As we shall see, it is important to examine the distribution (spread) of the data, because the distribution will explain the differences in the statistics of location and the differences are important in their own right (see Chapter 6). Often research is more concerned with these differences, and explaining difference (e.g. different buying behaviour) rather than just producing simple summary figures.

Managers need to ensure that the sample size is adequate for the required purpose. A few observations may inform a manager about the magnitude of a problem and allow a few trial calculations, but generally, a rigorous sampling approach is required which produces representative values.

Exercise

Suppose that you have recently joined a company where a substantial part of your earnings will come from commission. You have discovered that the commission earnings of your five colleagues was as follows for the previous year:

£15 000, £15 200, £15 200, £15 700, £18 600

How would you describe typical commission earnings? What commission could you reasonably expect?

You should ensure that you can correctly determine the following summary statistics:

mean = £15 940

median = £15 200

mode = £15 200

highest value = £18 600

lowest value = £15 000

You should note the effect the highest value of £18 600 has had on the mean.

Exercise

The errors in seven invoices were recorded as follows:

−£120,   £30,   £40,   −£8,   −£5,   £20   and   £25

The use of negative and positive signs can be taken to indicate your loss and gain, respectively.

Calculate approximate descriptive statistics. You should get:

mean = −£2.57

median = £20

mode is undefined

lowest value = −£120

highest value = £40

Exercise

Determine the mean, median and mode for the 400 values given as Case 1 data, file SDL1.XLS.

You should get:

mean = 27.045

(you will need to consider rounding when reporting a result like this)

median = 29

mode = 29

You should also add the units, in this case the number of cars in 10-minute recording intervals.

5.1.2 Tabulated (ungrouped) discrete data

It is far easier to manage discrete data in a tabulated form where the highest and lowest values, and the most frequent values, can be quickly identified. The number of working days lost by employees in the last quarter may be of particular interest to a manager. The data could be presented as shown in Table 5.1.

Table 5.1 The number of working days lost by employees in the last quarter

Number of days (x)	Number of employees (f)
0	410
1	430
2	290
3	180
4	110
5	20
	1440

The mean

To calculate the average number of working days lost last quarter, we first need to find the total number of days lost and then divide by the number of employees included. In this example, 410 employees lost no days adding zero to the overall total, 430 employees lost one day adding 430 days to the overall total, 290 employees lost two days adding 580 days to the overall total, and so on. The calculation, as shown in Table 5.2, involves multiplying the number of lost days, x, by the frequency, f, to obtain a column of sub-totals, fx. This new column, fx, is then summed to give a total $\sum fx$ which is then dividing by the number included, n (where $n = $ the sum of frequencies $\sum f$).

Table 5.2 The calculation of the mean from a frequency distribution

x	f	fx
0	410	0
1	430	430
2	290	580
3	180	540
4	110	440
5	20	100
	1440	2090

The mean is $\bar{x} = \dfrac{2090}{1440} = 1.451$ days lost

The formula for this kind of table includes frequency:

$$\bar{x} = \frac{\sum fx}{n}$$

The median

The tabulation of the number of 'lost days' has effectively **ordered** the data (the first 410 employees lost no days, the next 430 lost one day and so on). To find the median, which is a statistic concerned with order, we can either make a 'running' count or use cumulative frequencies. **Cumulative frequency** is the number of items with a given value or less. To calculate cumulative frequency, we just add the next frequency to the running total – see Table 5.3.

Table 5.3 The calculation of cumulative frequency

x	f	Cumulative frequency
0	410	410
1	430	$840 = 410 + 430$
2	290	$1130 = 840 + 290$
3	180	$1310 = 1130 + 180$
4	110	$1420 = 1310 + 110$
5	20	$1440 = 1420 + 20$
	1440	

As the data is discrete, the position of the median is found using the formula $(n + 1)/2$ which gives the $720\frac{1}{2}$th $((1440 + 1)/2)$ ordered observation, i.e. it will lie between the 720th and the 721st 'ordered' employee. It can be seen from the cumulative frequency that 410 employees lost no days and 840 lost one or less days. By deduction, the 720th and the 721st employee both lost one day; the median is therefore one day.

The mode

The mode corresponds to the highest frequency count, which is one day lost (which can be seen easily on the original Table 5.1).

In this case the median and the mode both give the same value of 1 and the mean gives the slightly higher value of 1.451. The effect of a few employees losing a higher number of days is to pull the mean upwards.

Exercise

The following transactions have been recorded on an automatic cash dispenser:

Value of transactions (£s)	Number
10	46
20	57
30	68
40	56
50	47
100	39
200	34

You should get:

$$\text{mean} = £68.56$$
$$\text{median} = £50$$
$$\text{mode} = £30$$

5.1.3 Tabulated (grouped) continuous data

As discussed in Part 1, continuous measurement is the result of using an instrument of measurement and will give values like 5 or 5.2 or 5.1763, depending on the requirements of the user (e.g. to the nearest whole number or ±0.05 or ±0.00005) and the accuracy of the measuring device. Values will either come as a long list (e.g. a data file), or collated in a table where values are grouped by non-overlapping intervals. Table 4.14, reproduced below as Table 5.4, is typical of continuous data.

Table 5.4   The amount spent on food in one particular shop

Expenditure on food	Number of respondents
under £5	2
£5 but under £10	6
£10 but under £15	8
£15 but under £20	12
£20 but under £30	10
£30 but under £40	4
£40 or more	2
	44

If data is given to us in the form of Table 5.4, then we no longer know the exact value of each observation. We only know, for example, that two respondents spent under £5, but not how much under £5. In this type of case, we have to **assume** a value for each group of respondents and **estimate** the descriptive statistic. In practice, we assume that all the values within a group are evenly spread (the larger values tending to cancel the smaller values) and can be reasonably represented by the **mid-point value**. This is the only realistic assumption we can make unless we have additional information (see weighted means – Section 5.2.3). If we were to use the lower limit value, we are likely to under-estimate the mean; if we were to use the upper limit value, we are likely to over-estimate the mean.

Looking at much of the published data, we often find that the first and last groups are left as **open-ended**, for example, 'under £1500' or 'over £100 000'. In these cases it will be necessary to make **assumptions** about the upper or lower limits before we can calculate the mean. There are no specific rules for estimating such end-points, but you should consider the data you are trying to describe. If, for example, we were given data on the 'age of first driving conviction' with a first group labelled 'under 17 years' it would hardly be realistic to use a lower limit of zero! (It is quite difficult to drive at a few months old.) Looking at this data, we might decide to use the minimum age at which a driving licence can normally be obtained, but being caught driving one's parents' Porsche around the M25 at the age of 15 would be likely to lead to some form of conviction. There is no correct answer: it is a question of knowing, or at least thinking about, the data.

The mean

Once we have established the limits for each group we can then find the mid-points and use these as the x values in our calculations.

The formula to use is

$$\bar{x} = \frac{\sum fx}{n}$$

where x now represents the mid-point values.

The procedure is shown in Table 5.5.

Table 5.5 The estimation of the mean using mid-points

Expenditure on food	Mid-point (x)	Number of respondents (f)	fx
£0* but under £5	2.50	2	5.00
£5 but under £10	7.50	6	45.00
£10 but under £15	12.50	8	100.00
£15 but under £20	17.50	12	210.00
£20 but under £30	25.00	10	250.00
£30 but under £40	35.00	4	140.00
£40 but under £50*	45.00	2	90.00
		44	840.00

*Assumed boundary.

$$\text{The mean } \bar{x} = \frac{840.00}{44} = £19.09$$

Care must be taken to clarify the interval range and the mid-points. Generally, for continuous data, the mid-points can be easily found by adding the upper and lower interval boundaries and dividing by two. Discrete data can also be tabulated in a grouped format. If we were considering the number of visitors or enquiries during a given time we might use an interval like '10 but under 20'. This would include 10, 11, 12, 13, 14, 15, 16, 17, 18 and 19 but not 20. The mid-point would be 14.5 and **not** 15. Particular care needs to be taken when working with 'age' data. Age is often given in the form: 10–14 years, 15–19 years. Since age is continuous (and we refer to age last birthday), the mid-points would be 12.5 and 17.5.

The median

We can determine the median either graphically or by calculation. The first step in both cases is to find the **cumulative frequencies**, as shown in Table 5.6.

Table 5.6 The determination of the median

Expenditure on food	Number of respondents (f)	Cumulative frequency (F)
under £5	2	2
£5 but under £10	6	8
£10 but under £15	8	16
£15 but under £20	12	28
£20 but under £30	10	38
£30 but under £40	4	42
£40 or more	2	44

In this example, two respondents spent less than £5, eight respondents spent less than £10 and so on. It should be noted that the cumulative frequency refers to the **upper boundary** of the corresponding interval.

The median – the
graphical method

To find the median graphically, we plot cumulative frequency against the upper boundary of the corresponding interval and join the points with straight lines (this is the graphical representation of the assumption that values are evenly spread within groups). The resultant frequency graph or **ogive** is shown as Figure 5.3.

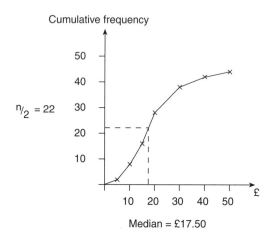

Figure 5.3 The construction of an ogive for the determination of the median.

To identify the median value for continuous data the formula $n/2$ is used (not $(n + 1)/2$). As you will see as you move to more advanced statistics, with continuous data we are dividing a distribution (the area under a curve) in two, and not a list of numbers. In this case, the median is the value of the 22nd observation, which can be read from the ogive as £17.50.

The median – by
calculation

To calculate the median, we must first locate the group that contains the 22nd observation, i.e. the median group. Looking down the cumulative frequency column of Table 5.6, we can see that 16 respondents spend less than £15 and 28 respondents spend less than £20. The 22nd observation (respondent) must lie in the group '£15 but less than £20'. The median must be £15 plus some fraction of the interval of £5. The median observation lies six respondents into this group (22 is the median observation minus the 16 observations that lie below this group). There are 12 respondents in this median group, so the median lies 6/12ths of the way through the interval. The median is equal to:

$$£15 + \frac{6}{12} \times £5 = £17.50$$

In terms of a formula we can write:

$$\text{median} = l + i\left(\frac{n/2 - F}{f}\right)$$

where l is the lower boundary of the median group, i is the width of the median group, F is the cumulative frequency up to the median group and f is the frequency in the median group.

Using the figures from the example above:

$$\text{median} = £15 + £5 \left(\frac{44/2 - 16}{12} \right)$$

$$= £17.50$$

The mode

We can determine the mode either graphically or by calculation. As the data is continuous and given in intervals, the mode can most easily be thought of as the point of greatest density or concentration. To estimate the mode, we need to refer to the histogram, originally drawn as Figure 4.15 and now given as Figure 5.4.

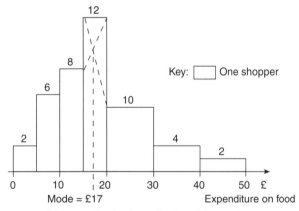

Figure 5.4   The use of a histogram for the determination of the mode.

The graphical method

To estimate the mode, we first identify the tallest block on the histogram (scaling has already taken place) and join the corner points, as shown. The point of intersection locates the mode. In this case the mode is £17 (to the nearest £1).

The mode by calculation

The formula for the mode can be given as

$$\text{mode} = l + \frac{f_m - f_{m-1}}{2f_m - f_{m-1} - f_{m+1}} \times i$$

where l is the lower boundary of the modal group, f_m is the (scaled) frequency of the modal group, f_{m-1} is the (scaled) frequency of the pre-modal group, f_{m+1} is the (scaled) frequency of the post-modal group and i is the width of the modal group.

The frequencies and the scaling effect can be seen in Figure 5.4. The main adjustment to make is the use of five rather than 10 (the interval where the width doubled) for the post-modal frequency. The mode is then:

$$\text{mode} = 15 + \frac{12 - 8}{2 \times 12 - 8 - 5} \times 5$$

$$= 15 + \frac{4}{11} \times 5$$

$$= \pounds 16.82$$

A spreadsheet showing the calculation of the mean, median and mode for tabulated continuous data is given as Figure 5.5. Excel does not provide specific functions for this kind of tabulated data and the spreadsheet needs to be constructed using equations. The cell references have not been shown for the calculations of the median and the mode (which are as above) as the identification of the median and modal intervals are always an important prerequisite and can be ignored if the spreadsheet is modified.

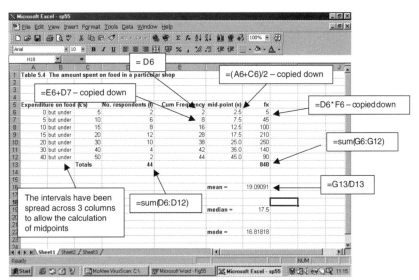

Figure 5.5

Exercise

The results of a travel survey were presented as follows:

Journey distance to and from work	% of journeys
0 but under 3	46
3 but under 10	38
10 but under 20	16

Determine the mean, median and mode.

(Hint: you can use percentages in the same way as frequencies – it is relative magnitude that this important. The same results would be obtained it you used 460, 380 and 160, respectively or 23, 19 and 8, respectively. We often scale frequencies just to make calculations and presentation easier – it is easier to work with 2.8 million than 2 800 000.)

You should get:

mean = 5.56 miles

median = 3.74 miles

mode = 1.82 miles

Cases 1 and 2: understanding data

It is likely that at least some managers in a company like Shopping Developments Limited would need to work with data in the form given in Cases 1 and 2, and in the ways presented this chapter. Numeracy is now accepted as an important management competence and managers do need to work with numbers. The temptation is to get on and calculate a few statistics. However, the more demanding task is to make sense of the data.

We need to question the data. We need to understand whether we are looking at a snapshot in time (like a balance sheet) or whether we are looking at changes over time (like the comparison of profit and loss accounts or cashflow forecasts). In Case 1, we have the results of a traffic flow survey and shopping survey. The charts and diagrams generated, and the statistics calculated should present a picture of the 'life' of the New Havens Shopping Precinct during that period. We need to use judgement to decide whether our results make sense. It is possible that road works or a major sporting event will distort our picture. The data given in Case 2 is about changes over time. Graphs (rather than bar charts) are used to show change over time and allow us to think about the ways this change may continue. It is always tempting to draw a line through graphed data and see the continuation of such a line as sufficient analysis. We may need make simple forecasts on the basis of graphical analysis but will still need to address the basic questions of:

- what information the data carries
- what assumptions can we make and
- what can we infer.

Hopefully, managers will not need to do the calculations shown and will be able to use a friendly PC. However, they will still need to make sure that the answers are correct. You should try to get into the habit of looking at the data and knowing what sort of answers to expect. Given the ranges and frequencies in Table 5.4, we would expect most of the summary location measures to be in the range '£15 but under £20'. Certainly, if our calculations or that of our PC, gave values like −£2.89 or £56.00 (which the authors have seen) we should suspect that something, somewhere is wrong. It is important that managers try to **develop this intuition** of knowing roughly the right values for a set of data.

5.2 Other measures of location

The mean, median and mode as described are not the only measures of location. The geometric mean and the harmonic mean are included here for the completeness of the chapter (and for reference purposes), but their use is limited to particular types of data and is rather specialist. The weighted mean is used if we wish to adjust the representation of the raw data; it is particularly important in market research.

5.1 The mean, median and mode

5.2 Other measures of location

5.2.1 The geometric mean

The geometric mean is defined to be 'the nth root of the product of n numbers' and is particularly useful when we are trying to average percentages. (It is also used with index numbers.)

Given the percentage of time spent on a certain task, we have the following data:

$$30\% \quad 20\% \quad 25\% \quad 31\% \quad 25\%$$

Multiplying the five numbers together gives

$$30 \times 20 \times 25 \times 31 \times 25 = 11\,625\,000$$

and taking the fifth root gives the geometric mean as 25.8868%. A simple arithmetic mean would give the answer 26.2% which is an over-estimate of the amount of time spent on the task.

5.2.2 The harmonic mean

The harmonic mean is used where we are looking at ratio data, for example miles per gallon or output per shift. It is defined as 'the reciprocal of the arithmetic mean of the reciprocals of the data'. For simple data it is not too difficult to calculate.

For example, if we have data on the number of miles per gallon achieved by five company representatives:

$$23 \quad 25 \quad 26 \quad 29 \quad 23$$

then, to calculate the harmonic mean, we find the reciprocals of each number:

$$0.043478 \quad 0.04 \quad 0.038461 \quad 0.034483 \quad 0.043478$$

find their average:

$$\frac{0.043478 + 0.04 + 0.038461 + 0.034483 + 0.043478}{5}$$

$$= 0.03998$$

and then take the reciprocal of the answer: 25.0125.

Thus the average fuel consumption of the five representatives is 25.0125 miles per gallon. (The arithmetic mean would be 25.2 mpg.)

5.2.3 Weighted means

Suppose that we were given a grouped frequency distribution of weekly income for a particular group of workers and in addition the average income within each of these categories. In this case we could use the set of averages rather than mid-points to calculate an overall mean. In terms of our notation we would need to write the formula as:

$$\bar{x} = \frac{\sum \bar{x}_i f_i}{n}$$

where $\bar{x}$ remains the overall mean, $\bar{x}_i$ is the mean in category i and f_i is the frequency in category i.

The calculation using a set of averages is shown in Table 5.7. We would note from the table, for example, that the first 10 workers have an average weekly income of £170, and together earn £1700. The overall mean is

Table 5.7 The weighting of means

Weekly income	Category average ($\bar{x}_i$)	Number of workers (f_i)	$\bar{x}_i f_i$
less than £200	£170	10	1 700
£200 but less than £300	£260	28	7 280
£300 but less than £400	£350	42	14 700
£400 but less than £600	£590	50	29 500
£600 or more	£750	20	15 000
		150	68 180

$$\bar{x} = \frac{68\,180}{150} = £454.53$$

The same result could have been obtained as follows:

$$\bar{x} = \left(170 \times \frac{10}{150}\right) + \left(260 \times \frac{28}{150}\right) + \left(350 \times \frac{42}{150}\right)$$

$$+ \left(590 \times \frac{50}{150}\right) + \left(750 \times \frac{20}{150}\right)$$

$$= £454.53$$

In terms of describing this procedure the formula can be rewritten as:

$$\bar{x} = \sum \left[\bar{x}_i \times \left(\frac{f_i}{n}\right)\right]$$

where f_i/n are the weighting factors.

These weighting factors can be thought of as a measure of size or importance. They can be used to correct inadequacies in data or to collate results from a survey which was not completely representative.

5.3 Conclusions

We seek to understand a range of issues through what is typical or average. Statistics like the mean, median and mode provide easy summary measures for numerical information. As we have seen, they are likely to give different answers and provide different information. The mean can be thought of as giving the 'centre of gravity', the median divides the distribution in two and the mode gives the highest point of the distribution; as shown in Figure 5.6.

The relative positions of the mean, median and mode will also tell us something about the distribution of the data, as shown in Figure 5.7. (See Chapter 6 for measures of skewness.)

Typically, income or wealth data will produce a positive skew, with a few individuals (not the authors!) on very high incomes or very wealthy. These relatively few large values will tend to pull the mean upwards, leaving over 50% of individuals below the mean – typically 55% to 60% in the UK context.

The mean remains the most commonly used statistic, with its advantages of being easily understood and generally accepted. However, it can be misleading if the data is heavily skewed or includes very different sub-groups. It is unlikely, for example, that a single summary statistic could describe the income of a work-

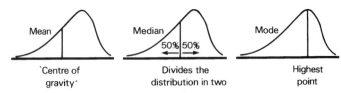

Figure 5.6   Interpreting the mean, median and mode.

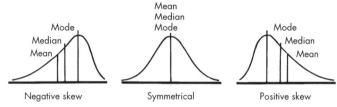

Figure 5.7   The relationship between the mean, median and mode and the shape of the distribution.

force if that workforce included employees doing very different jobs and on very different rates of pay.

Case 2

We would argue that that it is better to have some statistics, even if they are inadequate, than have none. For managers of a company like Shopping Developments Limited, summary statistics can be both **descriptive** and **interpretative**. Managers can use statistics to describe and explain their decisions to others, and they can use the statistics to better understand the world they work in. It may well be the case that there is no substitute for intuition, innovation and creative flair, but this can only be enhanced by an understanding and confidence to work with numbers.

5.4 Problems

1 Which measures of central location would most effectively describe:
   (a) travel distance to work?
   (b) the most popular model of car?
   (c) earnings of manual workers in the UK?
   (d) cost of a typical food item
   (e) holiday destinations?
   (f) working days lost through strikes?

2 'The arithmetic mean is the only average ever required.' Write a brief, critical assessment of this statement.

3 'The median provides a better description of typical earnings than the mean.' Discuss.

4 The number of new orders received by a company over the past 25 working days were recorded as follows:

3	0	1	4	4
4	2	5	3	6
4	5	1	4	2
3	0	2	0	5
4	2	3	3	1

Determine the mean, median and mode.

5 The number of faults in a sample of new cars has been listed as follows:

0	0	1	3	1	0	0	2	3
1	1	0	1	0	0	4	4	1
1	2	3	0	0	1	1	4	3
0	4	2	2	1	0	0	0	3

Determine the mean, median and mode.

6 The mileages recorded for a sample of company vehicles during a given week yielded the following data:

138	164	150	132	144	125	149	157
146	158	140	147	136	148	152	144
168	126	138	176	163	119	154	165
146	173	142	147	135	153	140	135
161	145	135	142	150	156	145	128

Determine the mean, median and mode. What do these descriptive statistics tell you about the distribution of the data?

Data now becomes available on the remaining ten cars owned by the company and is shown below. How does this new data change the measures of location which you have calculated?

234	204	267	198	179	210	260	290	198	199

7 The average weekly household expenditure on a particular range of products has been recorded from a sample of 20 households as follows:

£8.52	£ 7.49	£ 4.50	£ 9.28	£ 9.98
£9.10	£10.12	£13.12	£ 7.89	£ 7.90
£7.11	£ 5.12	£ 8.62	£10.59	£14.61
£9.63	£11.12	£15.92	£ 5.80	£ 8.31

(a) Determine the mean and median directly from the figures given.
(b) Tabulate the data as a frequency distribution and estimate the mean and median from your table.
(c) Explain any differences in your results from part (a) and part (b).

8   A company has produced the following table to describe the travel of its employees:

Reported cost of travel	Local	Type of travel Commuter	Long distance
Under £1	60	20	0
£1 but under £5	87	46	17
Over £5	12	13	53

Using appropriate calculations compare the cost of the different types of travel.

9   A survey of workers in a particular industrial sector produced the following table:

Income (weekly)	Number
under £100	170
£100 but under £150	245
£150 but under £200	237
£200 but under £400	167
over £400	124

Determine the mean, median and mode, and comment on the shape of the distribution.

10 Determine the mean, median and mode from the following information given on journey distance to work:

Miles	Percentage
under 1	16
1 and under 3	30
3 and under 10	37
10 and under 15	7
15 and over	9

11 Determine the mean and median from the information given in the following table:

Error (£)		Frequency
under −15		20
−15	but less than −10	38
−10	but less than −5	178
−5	but less than 0	580
0	but less than 10 .	360
10	but less than 20	114
20	or more	15

12 The number of breakdowns each day on a section of road were recorded for a sample of 250 days as follows:

Number of breakdowns	Number of days
0	100
1	70
2	45
3	20
4	10
5	5
	250

Determine the mean, median and mode. Which statistic do you think best describes this data and explain why.

13 Comparisons are being made between the percentage wage rises given in a group of industries. Each group of workers want to know if they have been given a wage rise which is greater than the average for all of the industries. Given the data below, use both the arithmetic and the geometric means to determine the average rise in wages. What would you report to the various workers?

Industry	Wage rise (%)
A	3.9
B	5.2
C	7.2
D	6.1
E	13.9

14 Included in the following table is the average income for those in each income group:

Income (£)	Average income	Number
under 40	35	175
40 but under 80	64	229
80 but under 120	101	241
120 but under 200	158	269
200 or more	240	86

(a)   Determine the mean using the average income figures given.

(b)   Is the method used in part (a) as precise as working with a listing of original data?

15 A company files its sales vouchers according to their value so that they are effectively in four strata. A sample of 200 is selected and the strata means calculated.

Stratum	Number of vouchers	Sample size	Sample mean (£)
above £1000	100	50	1800
£800 but under £1000	200	60	890
£400 but under £800	500	50	560
less than £400	1000	40	180
		200	

Estimate the mean value and total value of the sales vouchers.

16 Extract the most recent data on personal income from the *Annual Abstract of Statistics* and determine the mean and median personal income levels using a spreadsheet model. Estimate the percentage whose personal income falls below the mean and comment on your findings.

6 Measures of dispersion

In Chapter 5 we considered several measures of the typical, or average value. The mean is widely regarded as the most important descriptive statistic. When references are made to the average time or the average weight or the average cost it is generally the mean that has been calculated. Knowledge of the mean, the median and the mode will increase our understanding of the data but will not provide a sufficient understanding of the differences in the data.

In many applications it is the differences that are of particular interest to us. In market research, for example, we are interested not only in the typical values but also in whether opinions or behaviours are fairly consistent or vary considerably. A niche market is defined by difference. Quality control, whether in the manufacturing or the service sector, is concerned with difference from the expected.

In this chapter we introduce ways of measuring this variability, or **dispersion**, and then consider ways of comparing different distributions. Measures of dispersion can be **absolute** (considering only one set of data at a time and giving an answer in the original units e.g. £'s, minutes, years), or **relative** (giving the answer as a percentage or proportion and allowing direct comparison between distributions).

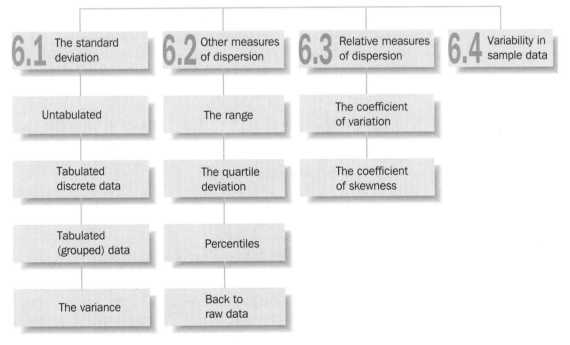

Figure 6.1 Structure of Chapter 6.

By the end of this chapter, you should be able to:

- calculate the standard deviation for various types of data
- determine the range, quartiles and percentiles for various types of data
- understand the relative merits of the different measures of dispersion
- use the concept of variability to better understand survey data.

This chapter will be structured as shown in Figure 6.1.

6.1 The standard deviation

The standard deviation is the most widely used measure of dispersion, since it is directly related to the mean. If you choose the mean as the most appropriate

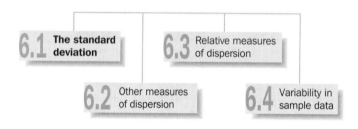

measure of central location, then the standard deviation would be the natural choice for a measure of dispersion. Unlike the mean, the standard deviation is not so well known and **does not have the same intuitive meaning**. The standard deviation measures differences from the mean – a larger value indicating a larger measure of overall variation. The standard devia-

tion will also be in the same units as the mean (£'s, minutes, years) and a change of units (e.g. from £'s to dollars, or metres to centimetres) will change the value.

The application of computer packages will generally make the determination of the standard deviation a relatively straightforward procedure, but it is worth checking what version of the formula is being used (the divisor can be n or $n - 1$). We will continue to follow the practice of showing the calculations by hand, as you may still need to do them. Such calculations do have the additional advantage of showing how the standard deviation is related to the mean.

The standard deviation is particularly important in the development of statistical theory, since most statistical theory is based on distributions described by their mean and standard deviation. We will use the mean and standard deviation extensively in Chapter 10 on the normal distribution, in Chapters 11 and 12 on statistical inference, and in Chapters 15 and 16 on forecasting.

6.1.1 Untabulated data

We have already seen, in Section 5.1.1, how to calculate the mean from simple data. We will need this calculation of the mean before we calculate the standard deviation. We can again use the first 10 observations on the number of cars entering a car park in 10-minute intervals:

$$10 \quad 22 \quad 31 \quad 9 \quad 24 \quad 27 \quad 29 \quad 9 \quad 23 \quad 12$$

The mean of this data is 19.6 cars.

The differences about the mean are shown diagrammatically in Figure 6.2.

To the left of the mean the differences are negative and to the right of the mean the differences are positive. It can be seen, for example, that the observation 9 is 10.6 units below the mean, a deviation of −10.6. The sum of these differences is zero – check this by adding all the deviations. This summing of deviations to zero illustrates the physical interpretation of the mean as being the centre of gravity with the observations as a number of 'weights in balance'. A

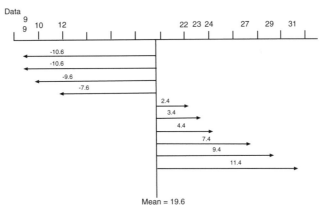

Figure 6.2   The differences about the mean.

proof that deviations from the mean sum to zero is given in the appendix to this chapter.

To calculate the standard deviation we follow six steps:

1   Compute the mean $\bar{x}$.
2   Calculate the differences from the mean $(x - \bar{x})$
3   Square these differences $(x - \bar{x})^2$
4   Sum the squared differences $\sum (x - \bar{x})^2$
5   Average the squared differences to find variance:

$$\frac{\sum (x - \bar{x})^2}{n}$$

6   Square root variance to find standard deviation:

$$\sqrt{\left[\frac{\sum (x - \bar{x})^2}{n}\right]}$$

The calculations are shown in Table 6.1.

Exercise

> The errors in seven invoices were recorded as follows:
>
> $$-£120, \quad £30, \quad £40, \quad -£8, \quad -£5, \quad £20 \quad \text{and} \quad £25$$
>
> Calculate the mean and standard deviation.
>   You should get:
> $$\text{mean} = -£2.57$$
> $$\text{standard deviation} = £50.66$$

Exercise

> To check the working consistency of a new machine, the time taken to complete a specific task was recorded on five occasions. On each occasion the recorded time was 30 seconds. Calculate the mean and standard deviation.
>   You should get:
> $$\text{mean} = 30 \text{ seconds} \qquad \text{standard deviation} = 0 \text{ seconds}$$
>
> This is clearly the result you would expect. If there is no variation, then the measure of difference should be 0.

Table 6.1 The calculation of the standard deviation

x	$(x - \bar{x})$	$(x - \bar{x})^2$
10	−9.6	92.16
22	2.4	5.76
31	11.4	129.96
9	−10.6	112.36
24	4.4	19.36
27	7.4	54.76
29	9.4	88.36
9	−10.6	112.36
23	3.4	11.56
12	−7.6	57.76
196		684.40

where $\bar{x} = \dfrac{\sum x}{n} = \dfrac{196}{10} = 19.6$ cars

and $s = \sqrt{\left[\dfrac{\sum(x - \bar{x})^2}{n}\right]}$

$= \sqrt{\left[\dfrac{684.40}{10}\right]}$

$= \sqrt{68.440}$

$= 8.27$ cars

6.1.2 Tabulated discrete data

Table 6.2, showing the number of working days lost by employees in the last quarter, typifies the tabulation of discrete data. (See Section 5.1.2 for the determination of the mean, median and mode using such data.)

Table 6.2 The number of working days lost by employees in the last quarter

Number of days (x)	Number of employees (f)
0	410
1	430
2	290
3	180
4	110
5	20

We need to allow for the fact that 410 employees lost no days, 430 lost one day and so on by including **frequency** in our calculations. In this example there are 1440 employees in total and we need to include 1440 squared differences. The formula for the standard deviation becomes

$$s = \sqrt{\left[\frac{\sum f(x - \bar{x})^2}{n}\right]}$$

The calculations are shown in Table 6.3.

Table 6.3 Standard deviation from tabulated discrete data

x	f	fx	$(x - \bar{x})$	$(x - \bar{x})^2$	$f(x - \bar{x})^2$
0	410	0	−1.451	2.1054	863.214
1	430	430	−0.451	0.2034	87.462
2	290	580	0.549	0.3014	87.406
3	180	540	1.549	2.3994	431.892
4	110	440	2.549	6.4974	714.714
5	20	100	3.549	12.5954	251.908
	1440	2090			2436.596

$$\bar{x} = \frac{2090}{1440} = 1.451 \text{ days lost}$$

$$s = \sqrt{\left[\frac{\sum f(x - \bar{x})^2}{n}\right]} = \sqrt{\left[\frac{2436.596}{1440}\right]} = 1.301 \text{ days lost}$$

Exercise

The following transactions have been recorded on an automatic cash dispenser:

Value of transaction (£)	Number
10	46
20	57
30	68
50	56
100	47
150	39
200	34

Determine the mean and standard deviation.
You should get:

$$\text{mean} = £68.56$$

$$\text{standard deviation} = £61.33$$

6.1.3 Tabulated (grouped) data

When data is presented as a grouped frequency distribution we must determine whether it is discrete or continuous (as this will affect the way we view the range of values) and determine the mid-points (see Section 5.1.3). Once the mid-points have been determined we proceed as before using mid-point values for x and frequencies, as shown in Table 6.4.

The approach shown clearly illustrates how the standard deviation summarizes differences, but would be extremely tedious to perform by hand. Some algebraic manipulation of the formula given in Section 6.1.2 (see the appendix to this chapter), will provide a simplified formula that is easier to work with for both calculations by hand and the construction of spreadsheets.

The simplified formula is usually presented as follows:

$$s = \sqrt{\left[\frac{\sum fx^2}{\sum f} - \left(\frac{\sum fx}{\sum f}\right)^2\right]}$$

The formula does lose its intuitive appeal but is easier to use. Formula of this

Table 6.4 The estimation of the standard deviation using mid-points

Expenditure on food	Number of respondents (f)	Midpoint (x)	fx	(x − x̄)	(x − x̄)²	f(x − x̄)²
£0* but under £5	2	2.50	5.00	−16.59	275.228	550.456
£5 but under £10	6	7.50	45.00	−11.59	134.328	805.968
£10 but under £15	8	12.50	100.00	−6.59	43.428	347.424
£15 but under £20	12	17.50	210.00	−1.59	2.528	30.336
£20 but under £30	10	25.00	250.00	5.91	34.928	349.280
£30 but under £40	4	35.00	140.00	15.91	253.128	1012.512
£40 but under £50*	2	45.00	90.00	25.91	671.328	1342.656
	44		840.00			4438.632

$$\bar{x} = \frac{\sum fx}{n} = \frac{840}{44} = £19.09$$

$$s = \sqrt{\left[\frac{\sum f(x-x)^2}{n}\right]} = \sqrt{\left[\frac{4438.632}{44}\right]} = £10.04$$

*Assumed boundary.

kind can be presented in a variety of ways. Using a formula presented in different ways should not be a problem. What you do need to be sure about are the stages required in the calculations (e.g. what columns to add) and the assumptions being made (e.g. is n or $(n − 1)$ being used as the divisor?).

The use of this simplified formula is illustrated in Table 6.5.

Table 6.5 The estimation of standard deviation using an alternative formula

x	f	fx	x²	fx²
2.50	2	5.00	6.25	12.50
7.50	6	45.00	56.25	337.50
12.50	8	100.00	156.25	1 250.00
17.50	12	210.00	306.25	3 675.00
25.00	10	250.00	625.00	6 250.00
35.00	4	140.00	1225.00	4 900.00
45.00	2	90.00	2025.00	4 050.00
	44	840.00		20 475.00

$$s = \sqrt{\left[\frac{\sum fx^2}{n} - \left(\frac{\sum fx}{n}\right)^2\right]} = \sqrt{\left[\frac{20475.00}{44} - \left(\frac{840}{44}\right)^2\right]} = £10.04$$

Calculations using a spreadsheet are shown in Figure 6.3.

Exercise

The results of a travel survey were presented as follows:

Journey distance to and from work (miles)	% of journeys
0 but under 3	46
3 but under 10	38
10 but under 20	16

Determine the mean and standard deviation.

You should get:

mean = 5.56 miles

standard deviation = 4.71 miles

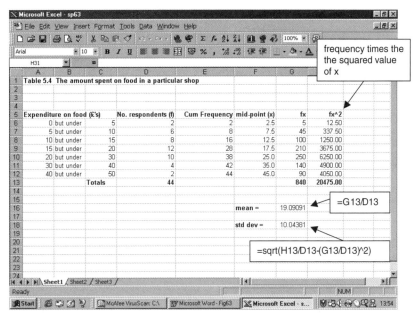

Figure 6.3 The determination of the mean and standard deviation using a spreadhseet.

6.1.4 The variance

The variance is the **squared value of the standard deviation**, and therefore is calculated easily once the standard deviation is known. It is sometimes used as a descriptive measure of dispersion or variability rather than the standard deviation, but its importance lies in more advanced statistical theory. As we will see, you can add variances but you cannot add standard deviations. Variance is mentioned here for completeness.

Case 2: using the standard deviation

We have seen the standard deviation calculated for a list of numbers in Section 6.1.1 (s = 8.27 cars), for tabulated discrete data in Section 6.1.2 (s = 1.301 days lost) and tabulated grouped data in Section 6.1.3 (s = £10.04). Managers are likely to see the standard deviation produced for a range of applications and see the type of results that we have produced. So what does it mean? As presented, the standard deviation is descriptive – the bigger the value, the greater the variation. If, for example, the standard deviation increased for the observed number of cars entering a car park, or the number of working days lost, then we would know that these were becoming more variable. The standard deviation also provides a rough guide (to be described in more detail later) on the range of the data. As a rule of thumb, we could say that most observations lie in the range

the mean $\pm 2 \times$ standard deviation

So, for example, most observations for cars entering a car park in a 10-minute interval would lie in the range

19.6 cars $\pm 2 \times 8.27$

or more conveniently, in the range

3 cars to 36 cars

This provides a quick check on whether our answer is of the right order of magnitude. If the standard deviation of 8.27 cars had been wrongly recorded as 0.827 cars or 82.7 cars, this can be easily spotted.

Managers in a company like Shopping Developments Limited should be looking beyond the calculation of a standard deviation. They should be asking what it means in a particular context and should try to understand the importance of the differences observed.

6.2 Other measures of dispersion

While the standard deviation is the most widely used measure of dispersion, it is **not the only one**. As we saw when looking at measures of location (Chapter 5), different measures (mean, median and mode) are appropriate for different situations and the same is true for measures of dispersion. Furthermore, some of the measures of dispersion are specifically linked to certain measures of location and it would not make sense to mix and match the statistics.

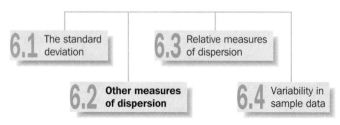

6.2.1 The range

The range is the most easily understood measure of dispersion as it is the **difference between the highest and lowest values**. If we were again concerned with the 10 observations:

10   22   31   9   24   27   29   9   23   12

the range would be 22 cars (31 − 9).

It is, however, a rather crude measure of spread, being dependent on the two most extreme observations. It is also highly unstable as new data is added. If this measure is to be used, it may well be better to **quote** the highest and lowest figure, rather than the difference. The range has, however, found a number of specialist applications, particularly in quality control (range charts). When dealing with data presented as a frequency distribution we will not always know exactly the highest and lowest values, only the group they lie in. If the groups are open-ended (e.g. 60 and more), then any values used will merely be based on assumptions that we have made about the widths of the groups. In such cases there seems little point in quoting either the range or the extreme values.

6.2.2 The quartile deviation

If we are able to quote a half-way value, the median, then we can also quote quarter-way values, the **quartiles**. These are order statistics like the median and can be determined in the same way. With untabulated data or tabulated discrete data it will merely be a case of counting through the ordered data set until we are a quarter of the way through and three quarters of the way through and noting the values; this will give the **first quartile** and **third quartile**, respectively.

When working with tabulated continuous data, further calculations are necessary. Consider for example the data given in Table 6.6 (see Table 5.6 for the determination of the median).

Table 6.6   The determination of the quartiles

Expenditure on food	Number of respondents (f)	Cumulative frequency (F)
under £5	2	2
£5 but under £10	6	8
£10 but under £15	8	16
£15 but under £20	12	28
£20 but under £30	10	38
£30 but under £40	4	42
£40 or more	2	44

The lower quartile (referred to as Q_1), will correspond to the value one-quarter of the way through the data, the 11th ordered value:

$$\frac{n}{4} = \frac{44}{4} = 11$$

and the upper quartile (referred to as Q_3) to the value three-quarters of the way through the data, the 33rd ordered value:

$$\frac{3n}{4} = \frac{3}{4} \times 44 = 33$$

The graphical method

To estimate any of the order statistics graphically, we plot cumulative frequency against the value to which it refers, as shown in Figure 6.4. The value of the lower quartile is £12 and the value of the upper quartile is £25 (to an accuracy of the nearest £1 which the scale of this graph allows).

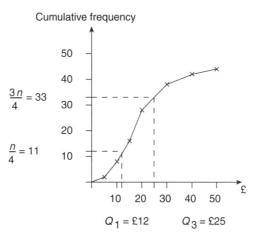

Figure 6.4   The determination of the quartiles.

Calculation of the quartiles

We can adapt the median formula (see Section 5.1.3) as follows:

$$\text{Order value} = l + i\left(\frac{O - F}{f}\right)$$

where O is the order value of interest, l is the lower boundary of corresponding group, i is the width of this group, F is the cumulative frequency up to this group, and f is the frequency in this group.

The lower quartile will lie in the group '£10 but under £15' and can be calculated thus:

$$Q_1 = 10 + 5\left(\frac{11-8}{8}\right) = 10 + 5 \times \frac{3}{8} = £11.88$$

The upper quartile will lie in the group '£20 but under £30' and can be calculated thus:

$$Q_3 = 20 + 10\left(\frac{33-28}{10}\right) = 20 + 10 \times \frac{5}{10} = £25.00$$

The quartile range is the difference between the quartiles:

$$\text{Quartile range} = Q_3 - Q_1$$
$$= £25.00 - £11.88$$
$$= £13.12$$

and the quartile deviation (or semi-interquartile range) is the average difference:

$$\text{Quartile deviation} = \frac{Q_3 - Q_1}{2}$$
$$= \frac{£13.12}{2} = £6.56$$

As with the range, the quartile deviation may be misleading. If the majority of the data is towards the lower end of the range, for example, then the third quartile will be considerably further above the median than the first quartile is below it, and when we average the difference of the two numbers we will disguise this difference. This is likely to be the case with a country's personal income distribution. In such circumstances, it would be preferable to quote the actual values of the two quartiles, rather than the quartile deviation.

6.2.3 Percentiles

The formula given in Section 6.2.2 for an order value, O, can be used to find the value at any position in a grouped frequency distribution of continuous data.

For data sets that are not skewed to one side or the other, the statistics we have calculated so far will usually be sufficient, but heavily skewed data sets will need further statistics to fully describe them. Examples would include some income distributions, wealth distributions and times taken to complete a complex task. In such cases, we may want to use the 95th percentile, i.e. the value below which 95% of the data lies. Any other value between 1 and 99 could also be calculated. An example of such a calculation is shown in Table 6.7.

For this wealth distribution, the first quartile and the median are both zero. The third quartile is £4347.83. None of these statistics adequately describes the distribution.

To calculate the 95th percentile, we find 95% of the total frequency, here

$$0.95 \times 26\,700 = 25\,365$$

and this is the item whose value we require. It will be in the group labelled

Table 6.7 Wealth distribution

Wealth	Number (f)	Cumulative frequency
Zero	15 000	15 000
Under £1 000	3 100	18 100
Under £5 000	2 300	20 400
Under £10 000	2 300	22 700
Under £25 000	1 600	24 300
Under £50 000	1 000	25 300
Under £100 000	800	26 100
Under £250 000	300	26 400
Under £500 000	170	26 570
Under £1 000 000	80	26 650
Over £1 000 000	50	26 700

'under £100 000' which has a frequency of 800 and a width of 50 000 (i.e. 100 000 − 50 000). Using the formula, we have:

$$95\text{th Percentile} = £50\,000 + £50\,000 \left(\frac{25\,365 - 25\,300}{800} \right)$$

$$= £54\,062.50$$

Exercise

Using the same data (Table 6.7), calculate the 90th percentile and the 99th percentile for wealth.
   You should get:
                          £22 468.75 and £298 529.41

6.2.4 Back to raw data

So far this chapter has taken us from individual numbers (raw data) through ordered data to grouped data, looking at the methods used to find the measures of dispersion. The previous chapter did the same for measures of location. However, the idea of grouping the data developed when calculation had to be done by hand, or at least using slide-rules and calculators. It was the only practical method when large amounts of data were being analysed. Now we have computers and suitable software, which can deal with huge amounts of data very quickly and easily, without having to make assumptions about an even spread of data within each group, or guessing what the highest or lowest value was. Add to this that most data starts life as individual bits of raw data, and you can see that most of the descriptive statistics we have been discussing can be found very easily, provided someone has recorded them electronically. An example using Excel is shown as Figure 6.5.

An example of the output from SPSS is shown as Figure 6.6.

If you are trying to describe secondary data for which you only have tabulated data, then, of course, you have to go back to the methods we have been discussing.

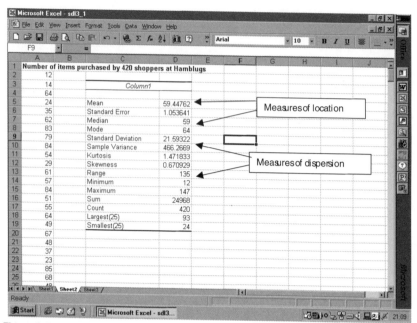

Figure 6.5 Using the descriptives function in Excel.

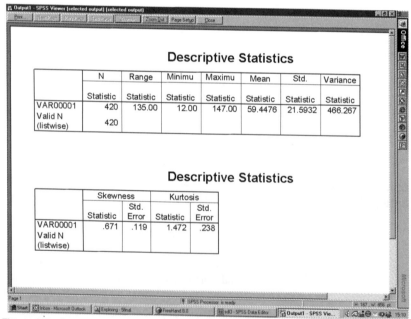

Figure 6.6 Example of the output of SPSS.

6.3 Relative measures of dispersion

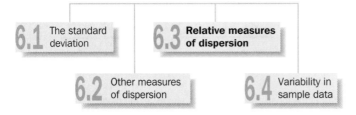

All of the measures of dispersion described earlier in this chapter have dealt with a single set of data. In practice, it is often important to compare two or more sets of data, maybe from different areas, or data collected at different times. In Part 4 we look at formal methods of comparing the difference between sample observations, but the measures described in this section will enable some initial comparisons to be made. The advantage of using relative measures is that they do not depend on the units of measurement of the data.

6.3.1 Coefficient of variation

This measure calculates the standard deviation from a set of observations as a percentage of the arithmetic mean:

$$\text{coefficient of variation} = \frac{s}{\bar{x}} \times 100$$

Thus the higher the result, the more variability there is in the set of observations. If, for example, we collected data on personal incomes for two different years, and the results showed a coefficient of variation of 89.4% for the first year, and 94.2% for the second year, then we could say that the amount of dispersion in personal income data had increased between the two years. Even if there has been a high level of inflation between the two years, this will not affect the coefficient of variation, although it will have meant that the average and standard deviation for the second year are much higher, in absolute terms, than the first year.

6.3.2 Coefficient of skewness

Skewness of a set of data relates to the shape of the histogram which could be drawn from the data. The type of skewness present in the data can be described by just looking at the histogram, but it is also possible to calculate a measure of skewness so that different sets of data can be compared. Three basic histogram shapes are shown in Figure 6.7, and a formula for calculating skewness is shown below.

$$\text{coefficient of skewness} = \frac{3(\text{mean} - \text{median})}{\text{standard deviation}}$$

A typical example of the use of the coefficient of skewness is in the analysis of income data. If the coefficient is calculated for gross income before tax, then the coefficient gives a large positive result since the majority of income earners receive relatively low incomes, while a small proportion of income earners receive high incomes. When the coefficient is calculated for the same group of earners using their after tax income, then, although a positive result is still obtained, its size has decreased. These results are typical of a progressive tax system, such as that in the UK. Using such calculations it is possible to show that the distribution of personal incomes in the UK has changed over time. A discussion of whether or not this change in the distribution of personal incomes is good or bad will depend on your economic and political views; the statistics highlight that the change has occurred.

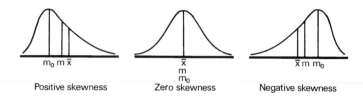

Figure 6.7   Where $\bar{x}$ is the mean, m is the median and m_0 is the mode.

Exercise

Using the data given in Table 6.4 determine the coefficient of variation and the coefficient of skewness.
  You should get:

the coefficient of variation $= 52.59\%$

the coefficient of skewness $= 0.47$

6.4 Variability in sample data

We would expect the results of a survey to identify differences in opinions, income and a range of other factors. The extent of these differences can be summarized by an appropriate measure of dispersion (standard deviation, quartile deviation, range). Market researchers, in particular, seek to explain differences in attitudes and actions of distinct groups within a population. It is known, for example, that the propensity to buy frozen foods varies between different groups of people. As a producer of frozen foods you might be particularly interested in those most likely to buy your products. Supermarkets of the same size can have very different turnover figures and a manager of a supermarket may wish to identify those factors most likely to explain the differences in turnover. A number of clustering algorithms have been developed in recent years that seek to explain differences in sample data.

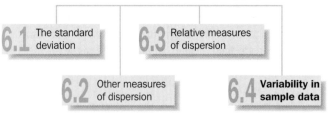

  As an example, consider the following algorithm or procedure that seeks to explain the differences in the selling prices of houses:

1 Calculate the mean and a measure of dispersion for all the observations in your sample. In this example we could calculate the average price and the range of prices (Figure 6.8).

$$\bar{x} = £35\,000$$
$$\text{Range} = £40\,000$$

Complete sample

Figure 6.8

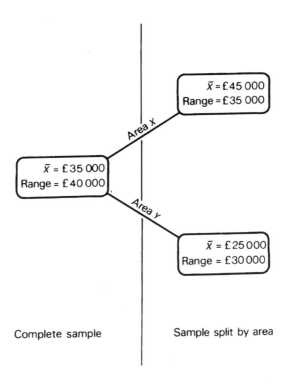

$\bar{x} = £45\,000$
Range $= £35\,000$

Area X

$\bar{x} = £35\,000$
Range $= £40\,000$

Area Y

$\bar{x} = £25\,000$
Range $= £30\,000$

Complete sample

Sample split by area

Figure 6.9

It can be seen from the range that there is considerable variability in price relative to the average price. Usually the standard deviation would be preferred to the range as a measure of dispersion for this type of data.

2   Decide which factors explain most of the difference (range) in price, for example, location, house-type, number of bedrooms. If location is considered particularly important, we can divide the sample on that basis and calculate the chosen descriptive statistics (Figure 6.9).

In this case we have chosen to segment the sample by location, areas X and Y. The smaller range within the two new groups indicates that there is less variability of house prices within areas. We could have divided the sample by some other factor and compared the reduction in the range.

3   Divide the new groups and again calculate the descriptive statistics. We could divide the sample a second time on the basis of house-type (Figure 6.10).

4   The procedure can be continued in many ways with many splitting criteria.

A more sophisticated version of this procedure is known as the automatic interactive detection technique.

Case 2: using measures of difference and performance

Managers are likely to meet a number of measures of difference and increasingly also various measures of performance (benchmarking, for instance, has become an important management tool, where targets are determined using the performance of the 'best' organizations on certain measures). Managers need to be able to respond to this type of information with insight and confidence.

It is important for managers to clarify what these measures mean in business terms and what the underlying assumptions are. In the same way that you don't need to be an accountant to use accounting information, you don't need to be a statistician to use statistical information. Managers should look for a business understanding in the information they are given and develop responses that allow their organization to interpret and apply such information. Knowing the assumptions will reveal some of the thinking of those that devised them. Management is a process that involves a judgement as to what is appropriate and when.

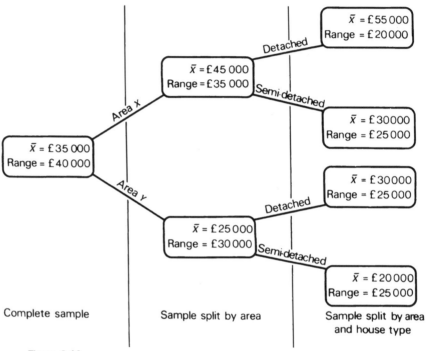

Figure 6.10

6.5 Conclusions

Describing the variability in data is more complex than describing its location since there is less intuitive meaning to the statistics that are used. When observations are close to the average, we will obtain relatively low values for the measure of dispersion and conversely when observations are more widely spread, then larger values will be obtained. We need to be aware of the units being used (a change from £'s to pence will increase the order by 100). We also need to be aware of what is meant by large or small in the context of the problem. A small change in sales (which may look small on every measure of change) could lead to business failure. As we shall see later in the book, there can be a difference between statistical significance and business significance.

When describing situations, measures of location and measures of dispersion are giving us two kinds of information. It is useful to know the average travel distance or the average overtime payment. It is also useful to know whether these averages are increasing or decreasing over time. This type of averaged information will inform decision makers. It is also important to recognize that we are also dealing with individuals and that policy decisions need to accommodate differences. Average travel distance could disguise the fact that some individuals travel very long distances on a regular basis or that average hours of overtime worked could disguise the fact that overtime is only available to a proportion of the workforce. It could also be the case, that the different behaviour or views of a few could be the beginning of a social or economic trend. Only a few years ago, the requirements of the vegetarian customer were generally ignored (and would have accounted for only a small proportion of the responses in any survey), but they now define new market segments. Market research is particularly concerned with how some groups of individuals differ from other groups, and how a market segment could, for example, be defined by those who buy fresh orange juice, own a family car or smoke on a regular basis.

The standard deviation and the other measures of variation will provide measures of difference, the skill is in the **interpretation** of these differences.

6.6 Problems

1 Which descriptive statistics would most effectively describe the differences in:
   (a) travel distance to work?
   (b) the most popular model of car?
   (c) earnings of manual workers in the UK?
   (d) the cost of a typical food item?
   (e) holiday destinations?
   (f) working days lost through strikes?
   (g) defective parts in a production process?
   (h) the wealth of the richest 10%?

2 Write a brief statement (200 words) explaining the meaning of standard deviation to someone who knows nothing about statistics.

3 'The standard deviation provides a better measure of the differences in typical earnings than the range.' Discuss.

4 The number of new orders received by a company over the past 25 working days were recorded as follows:

3	0	1	4	4
4	2	5	3	6
4	5	1	4	2
3	0	2	0	5
4	2	3	3	1

Determine the range, quartile deviation and standard deviation.

5 The number of faults in a sample of new cars has been listed as follows:

0	0	1	3	1	0	0	2	3
1	1	0	1	0	0	4	4	1
1	2	3	0	0	1	1	4	3
0	4	2	2	1	0	0	0	3

Determine the range, quartile deviation and standard deviation. Which of these measures of dispersion would you consider most appropriate for the data given? Explain the reasons for your choice.

6 The mileages recorded for a sample of company vehicles during a given week yielded the following data:

138	164	150	132	144	125	149	157
146	158	140	147	136	148	152	144
168	126	138	176	163	119	154	165
146	173	142	147	135	153	140	135
161	145	135	142	150	156	145	128

Determine the range, quartile deviation and standard deviation from these figures.

The data below now becomes available on the mileages of the other ten cars belonging to the company.

234	204	267	198	179	210	260	290	198	199

Recalculate the range, quartile deviation and standard deviation, and comment on the changes to these statistics.

7 The average weekly household expenditure on a particular range of products has been recorded from a sample of 20 households as follows:

£8.52	£ 7.49	£ 4.50	£ 9.28	£ 9.98
£9.10	£10.12	£13.12	£ 7.89	£ 7.90
£7.11	£ 5.12	£ 8.62	£10.59	£14.61
£9.63	£11.12	£15.92	£ 5.80	£ 8.31

(a) Determine the mean and standard deviation directly from the figures given.
(b) Tabulate the data as a frequency distribution and estimate the mean and standard deviation from your table.
(c) Explain any differences in your results from part (a) and part (b).

8 A company has produced the following table to describe the travel of its employees:

Reported cost of travel	Type of travel		
	Local	Commuter	Long distance
Under £1	60	20	0
£1 but under £5	87	46	17
Over £5	12	13	53

Using measures of location and measures of dispersion, describe and contrast the various types of travel.

9 A survey of workers in a particular industrial sector produced the following table:

Income (weekly)	Number
Under £100	170
£100 but under £150	245
£150 but under £200	237
£200 but under £400	167
Over £400	124

Estimate the standard deviation, range and quartile deviation.

10 Determine the quartile deviation and standard deviation from the data given in the following table:

Journey distance to and from work in miles	Percentage
Under 1	16
1 and under 3	30
3 and under 10	37
10 and under 15	7
15 and over	9

11 Determine the mean and standard deviation from the data given in the following table:

Error (£)	Frequency
Under −15	20
−15 but less than −10	38
−10 but less than −5	178
−5 but less than 0	580
0 but less than 10	360
10 but less than 20	114
20 or more 15	

12 The number of breakdowns each day on a section of road were recorded for a sample of 250 days as follows:

Number of breakdowns	Number of days
0	100
1	70
2	45
3	20
4	10
5	5
	250

Calculate the range, quartile deviation and standard deviation.

13 Given the following distribution of weekly household income:

Household income (£)	% of all households
Under 30	13.7
30 but under 40	7.6
40 but under 60	11.6
60 but under 80	13.4
80 but under 100	14.2
100 but under 120	13.0
120 but under 150	12.3
150 or more	14.2

(a) calculate the mean and standard deviation
(b) estimate the percentage of households with a weekly income below the mean
(c) determine the median and quartile deviation
(d) contrast the values you have determined in parts (a), (b) and (c) and comment on the skewness (if any) of the distribution.

14 The following annual salary data has been collected from two distinct groups of skilled workers within a company:

Annual salary (£)	No. from group A	No. from group B
8 000 but under 10 000	5	0
10 000 but under 12 000	17	19
12 000 but under 14 000	21	25
14 000 but under 16 000	3	4
16 000 but under 18 000	1	0
18 000 but under 20 000	1	0

(a) Determine the mean and standard deviation for each group of skilled workers;

(b) Determine the coefficient of variation and a measure of skew for each group of skilled workers;

(c) Discuss the results obtained in parts (a) and (b).

15 The time taken to complete a particularly complex task has been measured for 250 individuals and the results are shown below:

Time taken	No. of people
Under 5 minutes	2
Under 10 minutes	2
Under 15 minutes	3
Under 20 minutes	5
Under 25 minutes	5
Under 30 minutes	18
Under 40 minutes	85
Under 50 minutes	92
Under 60 minutes	37
Over 60 minutes	1

Estimate the maximum time taken by someone in the quickest:

(a) 1%, (b) 5%, (c) 10%.

16 You have been given the following data from a sample of 20 individuals:

Code number	Sex	Age	Employment	Amount spent weekly on alcoholic drinks (£)
1		20	0	8.83
2	1	33	0	4.90
3	1	50	1	0.71
4	0	48	0	5.70
5	0	47	0	6.20
6	0	19	0	7.40
7	1	21	1	3.58
8	0	64	0	4.80
9	1	32	0	4.50
10	1	57	1	2.80
11	0	49	0	4.60
12	0	18	0	5.30
13	1	39	1	3.42
14	0	28	0	10.15
15	0	51	0	6.20
16	1	43	0	4.80
17	1	40	0	3.82
18	0	22	1	7.70
19	1	30	0	6.20
20	0	60	0	4.45

Age: number of years   Employment: working 0
Sex: male 0   not working 1
    female 1

Measure and explain the variation in the amount spent weekly on alcoholic drinks with reference to the other factors given.

6.7 Appendix

For untabulated data (e.g. number of frequencies) proof that $\sum(x - \bar{x}) = 0$.

$$\sum(x - \bar{x}) = \sum x - \sum \bar{x}$$

$$= \sum x - n\bar{x}$$

$$= \sum x - n \cdot \frac{\sum x}{n}$$

$$= 0$$

For tabulated data

$$\sum f(x - \bar{x}) = \sum fx - \sum f\bar{x}$$

$$= \sum fx - \bar{x} \sum f$$

$$= \sum fx - \bar{x} \cdot n$$

$$= \sum fx - \frac{\sum fx}{n} \cdot n$$

$$= 0$$

Proof that $\sqrt{\left[\dfrac{\sum f(x - \bar{x})^2}{n}\right]} = \sqrt{\left[\dfrac{\sum fx^2}{n} - \left(\dfrac{\sum fx}{n}\right)^2\right]}$

$$\frac{1}{n}\sum f(x - \bar{x})^2 = \frac{1}{n}\sum f(x^2 - 2x\bar{x} + \bar{x}^2)$$

$$= \frac{1}{n}\left(\sum fx^2 - 2\bar{x}\sum fx + \bar{x}^2 \sum f\right)$$

$$= \frac{1}{n}\left(\sum fx^2 - 2n\bar{x}^2 + n\bar{x}^2\right)$$

$$= \frac{1}{n}\left(\sum fx^2 - n\bar{x}^2\right)$$

$$= \frac{1}{n}\left[\sum fx^2 - n\left(\frac{\sum fx}{n}\right)^2\right]$$

$$= \frac{\sum fx^2}{n} - \left(\frac{\sum fx}{n}\right)^2$$

Notes: $\bar{x}$ is a constant as far as $\sum$ is concerned; $\sum f = n$; $\sum fx = n\bar{x}$ since $\bar{x} = \sum fx/n$.

7 Index numbers

It is often necessary to describe and interpret changes in economic, business and social variables **over time**. Information on change may come from different types of data, recorded in different ways. Index numbers can provide a simple summary of change by aggregating the information available and making a comparison to a starting figure of 100. A typical index then could take the form of 100, 105, 107, where 100 is the starting point, 105 shows the relative increase one year later and 107 shows the relative increase two years on. Index numbers, therefore, are not concerned with absolute values but rather the movement of values. The **retail prices index** (RPI) is one the better known indices and is a general measure of how the prices of goods and services change, rather than as an indicator of the absolute amounts we actually spend each week.

By the end of this chapter, you should be able to:

- understand the concept of an index number
- scale number series for comparative purposes
- construct an index
- understand the Retail Prices Index (RPI).

This chapter will be structured as shown in Figure 7.1.

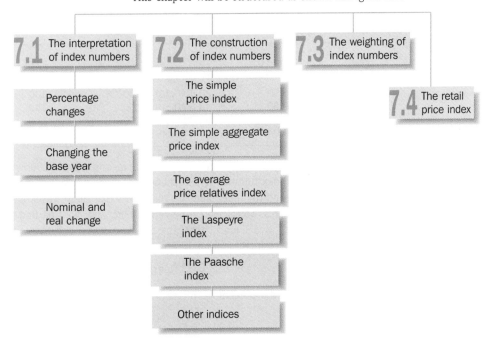

Figure 7.1 Structure of Chapter 7.

7.1 The interpretation of an index number

Indices provide a measure of change over time, making reference to a base year value of 100.

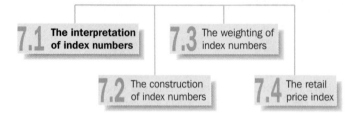

7.1.1 Percentage changes

An index is a **scaling** of numbers so that a start is made from a **base figure of 100**. Suppose, for example, the price for bread over the past 4 years was as shown in Table 7.1.

Table 7.1   The price of bread over a four-year period

Year	Price (£)
1	0.50
2	0.60
3	0.80
4	0.94

We would first need to decide which year should be used for the **base year**, and then scale all figures accordingly. If year 1 was chosen for the base year, we would divide all the prices by 0.50 and multiply by 100, as shown in Table 7.2.

Table 7.2   Scaling to produce an index

Year	Price (£)		Index
1	0.50	(0.50/0.50) × 100	100
2	0.60	(0.60/0.50) × 100	120
3	0.80	(0.80/0.50) × 100	160
4	0.94	(0.94/0.50) × 100	188

The index numbers given for years 2–4 all measure the **change from the base year**. The index number of 120 shows that there was a 20% increase from year 1 to year 2, and the index number 160 shows that there was a 60% increase from year 1 to year 3. To calculate a percentage increase, we first find the difference between the two figures, divide by the base figure and then multiply by 100. The percentage increase from 100 to 188 is

$$\frac{188 - 100}{100} \times 100 = 88\%$$

In the same way, the percentage increase in the price of bread from £0.50 to £0.94 is

$$\frac{0.94 - 0.50}{0.50} \times 100 = 88\%$$

One important feature of index numbers is that by starting from 100, the percentage increase from the base year is found just by subtraction. However, the differences thereafter are referred to as **percentage points**. It can be seen from Table 7.2 that there was a 28 percentage point increase from year 3 to year 4. The percentage increase, however, is

$$\frac{188 - 160}{160} \times 100 = 17.5\%$$

7.1.2 Changing the base year

There are no hard and fast rules for the choice of a base year and, as shown in Table 7.3, any year can be made into the base year from a purely mathematical point of view.

Table 7.3   Indices that all measure the same change

Year	Index 1	Index 2	Index 3	Index 4	Index 5
1	100	90.91	83.33	74.07	66.67
2	110	100.00	91.67	81.48	73.33
3	120	109.09	100.00	88.89	80.00
4	135	122.73	112.50	100.00	90.00
5	150	136.36	125.00	111.11	100.00

Each of the indices measures the same change over time (to two decimal places). The percentage increase from year 1 to year 2 using index 2, for example, is

$$\frac{100.00 - 90.91}{90.91} \times 100 = 10.00\%$$

To change the base year (move the 100) requires only a scaling of the index up or down. If we want index 1 to have year 2 as the base year (construct index 2), we can use the equivalence between 110 and 100 and multiply index 1 by this scaling factor 100/110.

Exercise

Scale index 2 in such a way that the base year becomes year 5 (as index 5).

In practice, there are a number of important considerations in the choice of a base year. As the index gets larger the same percentage change is represented by a larger increase in percentage points. A change from 100 to 120 is the same as a change from 300 to 360 but the impression created can be very different. If, for example, our index were used as a measure of inflation, like the Retail Prices Index, we would not want the index to move very far from 100. We would like the seen change (points) to be close to the actual change (percentages).

An index number is typically a summary of what is happening to a group of items (often referred to as a **basket of goods**). From time to time we may review and change the items to be included and this is often when the index is again started at 100. Footnotes or other forms of referencing may indicate these changes. Suppose a manufacturer constructed a productivity index using as a measure of productivity the times taken to make the most popular products. As new products appear and established products disappear, the manufacturer would need to reconsider the basis of the index. The manufacturer would also need to consider the compatibility of the indices produced as the new products may be adding a different level of value and involve different methods of production. An index can be unadjusted or adjusted. To show the general (underlying) trend in unemployment, the index can be adjusted to allow for predictable changes through the year, like the number of school leavers (see also Chapter 14).

A change in base year is shown in Table 7.4.

Table 7.4 A change of base year

Year	'Old' index	'New' index
3	120	
4	135	
5	150	100
6		115
7		125

We can use the equivalence of 150 in the 'old' index with 100 in the 'new' index at year 5. We either scale down the 'old' index using a multiplication factor of 100/150 as shown in Table 7.5 or scale up the 'new' index using 150/100 as shown in Table 7.6

Table 7.5 Scaling down the 'old' index

Year		'New' index
3	$120 \times 100/150 =$	80
4	$135 \times 100/150 =$	90
5		100
6		115
7		125

Table 7.6 Scaling up the 'new' index

Year		'Old' index
3		120
4		135
5		150
6	$115 \times 150/100 =$	172.5
7	$125 \times 150/100 =$	187.5

Exercise

A company has constructed an efficiency index to monitor the performance of its major production plant. After 3 years it was decided that a new index should be started owing to the major changes in the production process. The indices are given below:

Year	Existing index	New index
1	140	
2	155	
3	185	100
4		105
5		107

1 Calculate the percentage increase in efficiency from year 1 to year 5. You should get 41.4%.
2 Construct another index using year 2 as the base year. You should get 90.3, 100.0, 119.4, 125.3, 127.7.

Note that no allowance has been made for changes in the methods used to construct the indices.

7.1.3 Nominal and real change

Index numbers allow us to distinguish between *nominal* and *real* values. Suppose your annual entertainment allowance had increased from £500 to £510. This £10 increase is referred to as nominal value (and is given in the original units of measurement, in this case £'s). However, you may be more concerned with the purchasing power of the new £510 allowance and how this compares with the £500 allowed in the previous year. Suppose that you are now told that the cost of entertainment has increased by 5%. To maintain your purchasing power you would need £525 (£500 plus the extra 5%, which is £25). We would now say that in real terms your purchasing power has decreased. Indices can measure change in real or nominal terms.

Case 2: the use and construction of indices

Managers are always likely to be concerned with measures of change over time. Case 2 data gives the number of business enquiries received, the value of new business and an index of inflation over a 3-year period. It is relatively clear from the figures that the important business performance measurements of the number of enquiries and the value of new business are declining. It is less clear that the rate of inflation is about 3% and what impact the inflation rate is having on the 'real' value of new business. Table 7.7 shows the construction of indices to illustrate the changes using year 1, quarter 1 for the base.

The columns in Table 7.7 for the number of enquiries and the value of new business both show the downward trend and a quarterly variation. The indices confirm this trend and show the greater variation (in percentage terms) in the number of enquiries.

The spreadsheet shown as Table 7.8 can also be constructed to show how the 'real' value of new business has declined.

It can be seen that if we are working with the purchasing power of the £ in year 1, quarter 1 (real as opposed to nominal pounds), the drop in the value of new business is even greater. The spreadsheet has also been used to show that the rate of inflation and how that has been declining.

Table 7.7

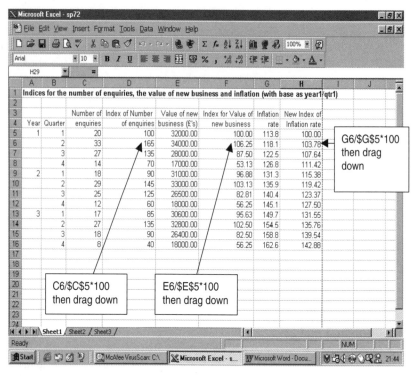

Table 7.8

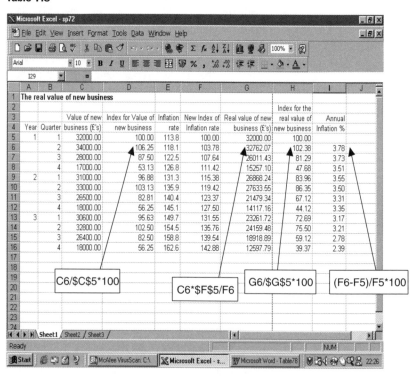

A mechanistic approach to business statistics is unlikely to capture the 'early youthful spirit' of business enterprise, but should better inform managers, and indeed other stakeholders, about the position of the company. The figures 'on the surface' look bad, but need to be interpreted within their business context (business significance rather than statistical significance). Trading conditions might have become particularly difficult and the company may still have done better than other rivals (the business could consider benchmarking against best practice wherever that is to be found). The figures may reflect a change in company strategy where new business of this kind has not been sought and existing business has been consolidated. What is important is that the analysis is able to inform a debate and highlight the realities that the company may face (which is better understood through analysis and debate). Analysis should also inform policy formulation and change. It is important to explore the data for improved insight; you could, for example, calculate the ratio of the value of new business to the number of enquiries and consider what these figures mean.

Exercise

Check the figures given, and the method of calculation used in Tables 7.7 and 7.8. Adapt the spreadsheet to provide the information given in Tables 7.7 and 7.8, and to show the impact of a constant inflation rate of 2%, 3% and 4%.

7.2 The construction of index numbers

Index numbers are perhaps best known for measuring the change of price or prices over time. To illustrate the methods of calculation, we will use the information given in Table 7.9.

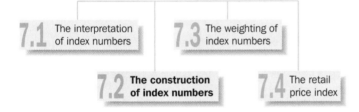

Table 7.9 The prices and consumption of tea, coffee and chocolate drinks by a representative individual in a typical week

Drinks	Year 0		Year 1		Year 3	
	Price	Quantity	Price	Quantity	Price	Quantity
Tea	32	15	48	12	64	10
Coffee	60	3	68	3	72	4
Chocolate	88	1	92	3	96	5

The price can be taken as the average amount paid in pence for a cup and the quantity as the average number of cups drunk per person per week.

7.2.1 The simple price index

If we want to construct an index for the price of one item only we first calculate that ratio of the 'new' price to the base year price, the price relative, and then multiply by 100. In terms of a notation

$$\frac{P_n}{P_0} \times 100$$

where P_0 is the base year price and P_n is the 'new' price.

A simple index for the price of tea, taking year 0 as the base year, can be calculated as in Table 7.10.

Table 7.10   A simple price index

Year	Price	P_n/P_0	Simple price index
0	32	1.0	100
1	48	1.5	150
2	64	2.0	200

The doubling of the price of tea from 32p to 64p over the 3-year period gives a 100% increase in the index, from 100 to 200. The increase from 48p to 64p is a 50 percentage point increase (the index increases from 150 to 200) or a percentage increase of $33\frac{1}{3}\%$.

In reality we are likely to drink more than just tea. When constructing an index of beverage prices, for example, we may wish to include coffee and chocolate drinks.

Exercise

> Calculate a simple price index for coffee with year 0 as the base year.
>   You should get 100, 113, 120

7.2.2 The simple aggregate price index

To include all items, we could sum the prices year by year and construct an index from this sum. If the sum of the prices in the base year is $\sum P_0$ and the sum of the prices in year n is $\sum P_n$, then the simple aggregate price index is

$$\frac{\sum P_n}{\sum P_0} \times 100$$

The calculations are shown in Table 7.11.

Table 7.11   The simple aggregate price index

Drinks	P_0	P_1	P_2
Tea	32	48	64
Coffee	60	68	72
Chocolate	88	92	96
	$\sum P_0 = 180$	$\sum P_1 = 208$	$\sum P_2 = 232$

Year	$\sum P_n / \sum P_0$	Simple aggregate price index
0	$180/180 = 1.00$	100
1	$208/180 = 1.16$	116
2	$232/180 = 1.29$	129

This particular index ignores the amounts consumed of tea, coffee and chocolate drinks. In particular, the construction of this index ignores both consumption patterns and the units to which price refers. If, for example, we were given the price of tea for a pot rather than a cup, the index values would differ.

7.2.3 The average price relatives index

To overcome the problem of units, we could consider price ratios of individual commodities instead of their absolute prices and treat all price movements as equally important. In many cases, the goods we wish to include will be measured in very different units. Breakfast cereal could be in price per packet, potatoes price per kilo and milk price per pint bottle. As an alternative to the simple aggregate price index we can use the average price relatives index:

$$\frac{1}{k}\sum(P_n/P_0) \times 100$$

where k is the number of goods. Here the price relative, $\sum P_n/\sum P_0$, for a stated commodity will have the same value whatever the units. The calculations are shown in Table 7.12.

Table 7.12   The average price relatives index

Drinks	P_0	P_1	P_2	P_1/P_0	P_2/P_0
Tea	32	48	64	1.50	2.00
Coffee	60	68	72	1.13	1.20
Chocolate	88	92	96	1.05	1.09
				$\sum(P_1/P_0) = 3.68$	$\sum(P_2/P_0) = 4.29$

Year	$(1/k)\sum(P_n/P_0)$	Average price relatives index
0	1.00	100
1	$(1/3)(3.68) = 1.23$	123
2	$(1/3)(4.29) = 1.43$	143

Comparing Tables 7.12 and 7.11 we can see that the average price relatives index, in this case, shows larger increases than the simple aggregate price index. To explain this difference we could consider just one of the items: tea. The value of tea is low in comparison to other drinks so it has a smaller impact on the totals in Table 7.11. In contrast, the changes in the price of tea are larger than any of the other drinks and this makes a greater impact on the totals in Table 7.12. To construct a price index for all goods and sections of the community we need to take account of the quantities bought.

It is not just a matter of comparing what is spent year by year on drinks, food, transport or housing. If prices and quantities are both allowed to vary, an index for the amount spent could be constructed but not an index for prices. If we want a price index we need to control quantities. In practice, we consider a typical basket of goods in which the quantity of goods of each kind is fixed and we find how the cost of that basket has changed over time. To construct an index for the price of beverages we need the quantity information for a selected year as given in Table 7.9.

7.2.4 The Laspeyre index

This index uses the quantities bought in the base year to define the typical basket. It is referred to as a base-weighted index and compares the cost of this basket of goods over time. This index is calculated as

$$\frac{\sum P_n Q_0}{\sum P_0 Q_0} \times 100$$

where $\sum P_0 Q_0$ is the cost of the base year basket of goods in the base year and $\sum P_n Q_0$ is the cost of the base year basket of goods in any year (thereafter) n.

It can be seen from Table 7.13 that we only require the quantities from the chosen base year (Q_0 in this case). The index implicitly assumes that whatever the price changes, the quantities purchased will remain the same. In terms of economic theory, no substitution is allowed to take place. Even if goods become relatively more expensive it assumes that the same quantities are bought. As a result, this index tends to overstate inflation.

Table 7.13 The Laspeyre index

Drinks	P_0	Q_0	P_1	P_2
Tea	32	15	48	64
Coffee	60	3	68	72
Chocolate	88	1	92	96

$P_0 Q_0$	$P_1 Q_0$	$P_2 Q_0$
480	720	960
180	204	216
88	92	96
748	1016	1272

Year	$\sum P_n Q_0 / \sum P_0 Q_0$	Laspeyre index
0	1.00	100
1	$1016/748 = 1.36$	136
2	$1272/748 = 1.70$	170

7.2.5 The Paasche index

This index uses the quantities bought in the current year for the typical basket. This current year weighting compares what a basket of goods bought now (in the current year) would cost, with cost of the same basket of goods in the base year. This index is calculated as

$$\frac{\sum P_n Q_n}{\sum P_0 Q_n} \times 100$$

where $\sum P_n Q_n$ is the cost of the basket of goods bought in the year n at year n prices and $\sum P_0 Q_n$ is the cost of the year n basket of goods at base year prices. The calculations are shown in Table 7.14.

Table 7.14   The Paasche index

Drinks	P_0	P_1	Q_1	P_2	Q_2
Tea	32	48	12	64	10
Coffee	60	68	3	72	4
Chocolate	88	92	3	96	5

P_1Q_1	P_0Q_1	P_2Q_2	P_0Q_2
576	384	640	320
204	180	288	240
276	264	480	440
1056	828	1408	1000

Year	$\sum P_nQ_n / \sum P_0Q_n$	Paasche index
0	1.00	100
1	1.28	128
2	1.41	141

As the basket of goods is allowed to change year by year, the Paasche index is not strictly a price index and as such, has a number of disadvantages. Firstly, the effects of substitution would mean that greater importance is placed on goods that are relatively cheaper now than they were in the base year. As a consequence, the Paasche index tends to **understate inflation**. Secondly, the comparison between years is difficult because the index reflects both changes in price and the basket of goods. Finally, the index requires information on the current quantities and this may be difficult or expensive to obtain.

7.2.6 Other indices

The Laspeyre and Paasche methods of index construction can also be used to measure quantity movements with prices as the weights.

Laspeyre quantity index using base year prices as weights:

$$\frac{\sum P_0Q_n}{\sum P_0Q_0} \times 100$$

Paasche quantity index using current year prices as weights:

$$\frac{\sum P_nQ_n}{\sum P_nQ_0} \times 100$$

To measure the change in value the following 'value' index can be used:

$$\frac{\sum P_nQ_n}{\sum P_0Q_0} \times 100$$

These calculations are shown, along with the Laspeyre and Paasche index, in Table 7.15.

Table 7.15

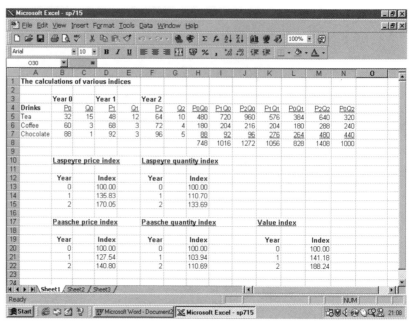

Having constructed a spreadsheet you can experiment by making changes to price or quantity information and observing the overall effect. (This spreadsheet (sp715.XLS) is available on the website).

7.3 The weighting of index numbers

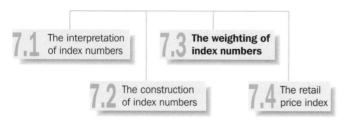

Weights can be considered as a measure of importance. The Laspeyre index and the Paasche index both refer to a typical basket of goods. The prices are weighted by the quantities in these baskets. In measuring a diverse range of items, it is often more convenient to use amount spent as a weight rather than a quantity. If we consider travel, for example, it could be more meaningful to define expenditure on public transport than the number of journeys. In the same way, we would enquire about the expenditure on meals bought and consumed outside the home rather than the number of meals and their price. Expenditure on public transport, meals outside the home and other items are additive since money units are homogeneous; the number of journeys, number of meals and number of shirts are not.

In constructing a base-weighted index we can use

$$\frac{\sum wP_n/P_0}{\sum w} \times 100$$

where P_n/P_0 are the price relatives (see Section 7.2.1) and w are the weights. Each weight is the amount spent on the item in the base year.

Consider again our example from Section 7.2 (Table 7.16).

It is no coincidence that this base-weighted index is identical to the Laspeyre index of Table 7.13. The identity is proven below:

Exercise

site Look it up on the WWW

1 Load the spreadsheet sp715.XLS. Vary the prices and the quantities, and discuss the effects.

2 Using the data given below, set up a spreadsheet to calculate the following all-items index numbers:

 1 Laspeyre price index;
 2 Paasche price index;
 3 Laspeyre quantity index;
 4 Paasche quantity index; and
 5 Value index

for years 1 and 2 with year 0 as the base year.

	Year 0		Year 1		Year 2	
Items	Price	Quantity	Price	Quantity	Price	Quantity
Bread	25	4	27	5	29	4
Potatoes	8	6	9	7	10	9
Carrots	21	2	22	2	22	3
Swede	35	1	35	1	40	1
Cabbage	45	1	46	2	55	1
Soup	18	5	23	4	25	5
Cake	53	2	62	2	78	1
Jam	40	1	45	2	56	1
Tea	23	3	25	3	29	4
Coffee	58	2	60	3	60	3

Answers:

		Price	Quantity
Laspeyre	1	110.85	122.87
	2	124.31	110.56
Paasche	1	109.54	121.41
	2	120.68	107.33
Value	1		134.59
	2		133.43

Check the working of your spreadsheet by seeing the effects of changing the details on potatoes in Year 2 to Price = 15 and Quantity = 6 and for coffee to Price = 70 and Quantity = 1.

$$\text{Laspeyre index} = \frac{\sum P_n Q_0}{\sum P_0 Q_0} \times 100$$

Let the weights, w, equal the amount spent on items in the base year, e.g. $w = P_0 Q_0$:

$$\text{Laspeyre index} = \frac{\sum P_n Q_0}{\sum w} \times 100$$

If we note that $Q_0 = w/P_0$ then

$$\text{Laspeyre index} = \frac{\sum wP_n/P_0}{\sum w} \times 100$$

The weights only need to represent the relative order of magnitude and in practice are scaled to sum to 1000. (If we were to multiply each of the weights in Table 7.16 by 1000/748, the value of the index would not change but the sum of weights would add to 1000.) The items included in the Retail Prices Index are assigned weights in this way.

Table 7.16 Weighted price index

Drinks	P_0	Q_0	w	P_1	P_1/P_0	P_2	P_2/P_0
Tea	32	15	480	48	1.50	64	2.00
Coffee	60	3	180	68	1.13	72	1.20
Chocolate	88	1	88	92	1.05	96	1.09

Drinks	w	$w \times P_1/P_0$	$w \times P_2/P_0$
Tea	480	720.00	960.00
Coffee	180	203.40	216.00
Chocolate	88	92.40	95.92
	748	1015.80	1271.92

Year	$(\sum wP_n/P_0)/\sum w$	Base-weighted index
0	1.00	100
1	1015.80/748 = 1.36	136
2	1271.92/748 = 1.70	170

7.4 The Retail Prices Index

The general Retail Prices Index (RPI) is the main index used to measure of inflation in the UK. It measures the average change on a monthly basis of the prices of goods and services purchased by most households. It is the measure of inflation reported in the media, debated by politicians and used to revise benefits and pensions. Increases in wages are often justified in terms of the RPI, with recent or anticipated changes often forming the basis of a wage claim. In many cases, savings and pensions are index-linked; they increase in line with the index. All forms of economic planning take some account of inflation, and economists will use both real and nominal values in their analysis (see Section 7.1.3).

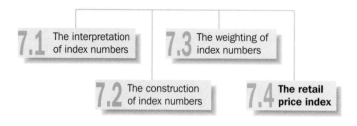

The interpretation of index numbers

The construction of index numbers

The weighting of index numbers

The retail price index

The RPI covers a range of goods and services bought by a typical household. It is useful to think of the RPI as representing the changing cost of a large 'basket of goods and services' reflecting the full range of things that people buy including leisure goods, fuel, food and footwear. In many ways, the RPI is not a 'cost of living' index as it does not attempt to provide a measure of the cost of staying alive. A 'cost of living' index would imply some definition or knowledge of what were essential purchases. Who could make such a judgement? The index reflects what people choose to buy; for example, some people buy cigarettes and alcohol, so these are included in the index. Coverage includes housing and travel but excludes items like savings, investments, charges for credit, betting and cash gifts. The expenditure of certain higher income households and of pensioner households mainly dependent on state benefit is excluded.

The 'basket of goods' is kept fixed for a year at a time, so that only changes in prices are recorded that year. The basket is reviewed each year to keep it as up to date as possible. Changes made in 'year 2000' basket include broccoli, pre-packed salad and takeaway/delivered pizza in the 'food and catering category', and the introduction of PC printers in the 'leisure goods' category. It has also been decided that the RPI should begin to reflect Internet prices, and some books and toys typically bought on the Internet have been included.

The prices of more than 600 separate goods and services are collected each month. The movements in these prices are taken as representative of all price movement in the goods and services covered by the index. There are six price indicators for beef, for example (January 2000), which are combined together to estimate the overall change in beef price. The base period is January of each year and current prices are compared to this base period. The RPI for any month is calculated by weighting (averaged) price relatives. Essentially, the RPI is a Laspeyres base-weighted index. For a more detailed description of the RPI refer to http//www.statistics.gov.uk or http//www.ons.gov.uk (see Figure 7.2).

Figure 7.2   Web-based information on the Retail Prices Index.

Weights are used to allow for the relative importance of the various categories of goods and services. The weights are derived mainly from the **Family Expenditure Survey** (see Figure 7.3). The Family Expenditure Survey is based on a set sample size of about 11 000 households each year (11 400 in 1998/99 – see

Figure 11.3) and uses addresses from the postcode address file. Selected households are asked to keep records of what they spend over a 2-week period and are also asked to give details of their major purchases over a longer period. The response rate given in Figure 7.3 is 59%, with a more recent response rate of 63% being reported. Analysis is based on about 7000 households.

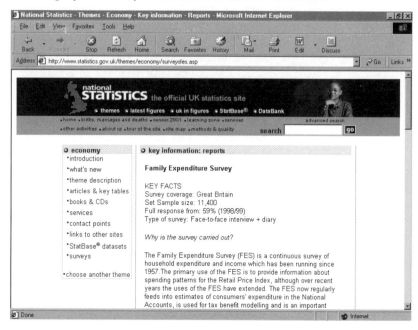

Figure 7.3 Web-based information on the Family Expenditure Survey.

The weights used in the RPI for 1990, 1995 and 2000 are shown in Table 7.17.

Table 7.17 The weights used for the Retail Prices Index

Categories	1990	1995	2000
Food	158	139	118
Catering	47	45	52
Alcoholic drink	77	77	65
Tobacco	34	34	30
Housing	185	187	195
Fuel and light	50	45	32
Household goods	71	77	72
Household services	40	47	56
Clothing and footwear	69	54	58
Personnel goods and services	39	39	43
Motoring expenditure	131	125	146
Fares and travel costs	21	19	21
Leisure goods	48	46	46
Leisure services	30	66	66
Total	1000	1000	1000

As far as the RPI is concerned, using the weights given in Table 7.17, food accounted for 15.8% of the typical basket in 1990, 13.9% in 1995 and 11.8% in

2000. You could also note the recent decrease in the relative expenditure on alcoholic drink (from 7.7% in 1990 to 6.5% in 2000) and the recent increase on the relative expenditure on motoring (from 13.1% in 1990 to 14.6% in 2000). The weightings therefore provide a useful guide to changing patterns of expenditure. The weighting can be used to demonstrate the effect of a price change in one category on the overall RPI. If, for example, the 'price' of fuel and light increased by 10% in 2000, the overall impact would be 0.32% (under $\frac{1}{2}$%), as the category accounts for 3.2% of expenditure (a weight of 32 out of 1000). To show this using weights, the weight for fuel and light would increase from 32 to 35.2 (a 10% increase). The sum of the weight would become 1003.2 and the percentage increase would be:

$$\frac{1003.2 - 1000}{1000} \times 1000 = 3.2\%$$

In practice, calculations can be more complex because a number of changes take place at the same time.

The typical basket of goods and services indicated by the weights shown in Table 7.17 will only reflect completely the expenditure of a proportion of households. Some families will spend more on some items and less on others, particularly on an annual basis. The RPI, like all aggregated statistics, will have an averaging-out effect and will reasonably describe most families most of the time.

Exercise

> Table 7.18 below gives the weights used for the one-person and two-person pensioner household in 1990, 1995 and 2000. Compare and contrast these sets of weights.

Table 7.18 Weights used in pensioner indices

Categories	One person 1990	One person 1995	One person 2000	Two person 1990	Two person 1995	Two person 2000
Food	320	285	275	330	289	267
Catering	31	32	39	26	29	36
Alcoholic drink	28	26	32	38	40	44
Tobacco	28	35	37	36	36	31
Housing	0	0	0	0	0	0
Fuel and light	173	152	115	124	104	84
Household goods	90	105	110	89	105	113
Household services	82	101	95	53	61	56
Clothing and footwear	61	55	49	67	51	45
Personnel goods and services	58	47	65	56	57	58
Motoring expenditure	22	34	51	90	103	129
Fares and travel costs	21	22	21	16	18	18
Leisure goods	49	46	51	49	45	50
Leisure services	37	60	60	26	62	69
Total	1000	1000	1000	1000	1000	1000

7.5 Conclusions

Index numbers play an important role in describing the economy, managing the economy and measuring the performance of business. Percentage increases from the base year can be seen at a glance, and that the numbers provide a manageable and understandable sequence. As we have seen, we are able to aggregate a wide range of different items into a single index series, which will enhance our comprehension of an overall situation, for example, the level of inflation in an economy. In the presentation of accounting information allowance needs to be made for inflation. Historic cost accounting (with no allowance for inflation) only works well in periods of stable prices. In current cost accounting (CCA), adjustments are made in proportion to relevant indices. The Office for National Statistics publishes price index numbers for current cost accounting.

Index numbers can be misleading if care is not taken. When an index is rebased it is important to compare the last value of the previous series to the starting value of the new series and make any necessary adjustments. When items are excluded, or new items included in an index, there may be drastic movements in the series, which do not reflect major changes in prices or quantities, but merely the changed composition of the index. Crime statistics, in particular, are statistically (and politically) very sensitive to changes in definition and reporting.

Case 2: the use of indices

Managers may need to use indices generated externally, like the Retail Prices Index, or construct their own, using internal company information or other data. Indices provide useful measures of change for those factors that are quantifiable (see Tables 7.7 and 7.8).

Changes in prices or costs can be fairly easily described by indices and they are important measures of business and organisational performance. However, other important aspects, such as culture and motivation, cannot be described so easily, and as a consequence can be overlooked. As always, good management is about interpretation and index numbers should make this easier for some types of (quantitative) information.

7.6 Problems

1

Year	Index
1	100
2	115
3	120
4	125
5	130
6	145

(a) Change the base year for this index to year 4.
(b) Find the percentage rises from year 3 to year 4 and year 5 to year 6.

2

Year	'Old' index	'New index'
1	100	
2	120	
3	160	

4	190	100
5		130
6		140
7		150
8		165

(a) Scale down the 'old' index for years 1 to 3.
(b) Scale up the 'new' index for years 5 to 8.
(c) Explain the reasons for being cautious when merging indices of this kind.

3 The following information was recorded for a range of DIY items.

Items	No. of items bought			Price per item (£)		
	Year 0	Year 1	Year 2	Year 0	Year 1	Year 2
W	4	5	7	3	5	10
X	3	3	4	4	6	15
Y	3	2	2	4	7	19
Z	2	2	1	5	9	25

(a) Construct a simple price index for item Y using year 0 as the base year.
(b) Determine the simple aggregate price index using year 0 as the base year.
(c) Determine the price relatives index using year 0 as the base year.
(d) Calculate the Laspeyre price index using year 0 as the base year.
(e) Calculate the Paasche price index using year 0 as the base year.
(f) Calculate the Laspeyre quantity index using year 0 as the base year. Note: here we wish to keep prices fixed, and thus the appropriate formula is

$$\frac{\sum(P_0 Q_n)}{\sum(P_0 Q_0)} \times 100$$

(g) Calculate the Paasche quantity index using year 0 as the base year. Note: here we are using the current year prices as the fixed weights, and thus the appropriate formula is

$$\frac{\sum(P_n Q_n)}{\sum(P_n Q_0)} \times 100$$

(h) Why do the Laspeyre and Paasche indices give such different answers for year 2 in parts (d) to (g)?

4 The following data gives the wages paid to four groups of workers and the number of workers in each group.

Groups	Year 0		Year 1		Year 2	
	Wage	No. of workers	Wage	No. of workers	Wage	No. of workers
Managerial	300	40	330	50	390	70
Skilled	255	60	270	70	270	70

| Semi-skilled | 195 | 60 | 240 | 60 | 270 | 70 |
| Labourer | 90 | 100 | 150 | 100 | 240 | 80 |

(a) Construct a simple index of wages for skilled workers using year 0 as the base year.
(b) Calculate the simple aggregate index of wages using year 0 as the base year.
(c) Calculate the Laspeyre index of wages using year 0 as the base year.
(d) Calculate the Paasche index of wages using year 0 as the base year.

5 The average prices of four commodities are given in the following table:

| Commodity | Average price per unit (£) | | |
	Year 0	Year 1	Year 2
A	101	105	109
B	103	106	107
C	79	93	108
D	83	89	86

The number of units used annually by a certain company is approximately 400, 200, 600 and 100 for the commodities A, B, C and D respectively. Calculate a weighted price index for years 1 and 2 using year 0 as the base year.

6 Extract from published statistics, indices showing monthly changes in prices and average earnings since January 2000.
(a) Rebase these indices so that the current year becomes the base year, stating any assumptions made.
(b) Compare the indices produced (part a).
(c) Construct an index for the value of 'real' earnings.

7 Find the most recent weights for the (all items) Retail Prices index
(a) compare with those given for the year 2000 (Table 7.17).
(b) identify the impact of a 1%, 3% and 5% change in the price of food.

8 Over the past 15 years data has been recorded on prices and wages in a certain country. Index numbers for this data are shown in the table below.

Year	Prices (old)	Prices (new)	Wages (old)	Wages (new)
0	145.3		100.0	
1	148.5		103.2	
2	154.7		107.5	
3	163.6		112.4	
4	172.3		123.4	
5	194.1	100.0	136.3	
6		109.5	145.1	
7		114.9	146.3	
8		119.4	150.5	
9		125.6	158.1	100.0
10		130.6		105.4
11		136.6		111.9

12	144.2	121.1
13	150.3	130.0
14	158.9	141.3
15	171.3	154.6

As can be seen, both index series have been rebased during this time period.

(a) Create a single index series for prices with year 0 as the base year.

(b) Create a single index series for wages with year 0 as the base year.

(c) Construct a graph showing these two index series.

(d) Find the year-by-year percentage change in prices and the year-by-year percentage change in wages and graph your results.

(e) Comment on the implications of your results for the standard of living within the country. What reservations might you have about using this data to infer conclusions about the standard of living? What extra information would you require to make more positive statements about the standard of living?

9 Use a spreadsheet with the data below to find the all-items index numbers listed below for years 1 and 2 with year 0 as the base year.

(a) Laspeyre price index

(b) Laspeyre quantity index

(c) Paasche price index

(d) Paasche quantity index

(e) a value index

| Item | Prices | | | Quantities | | |
	Year 0	Year 1	Year 2	Year 0	Year 1	Year 2
A	3.00	3.25	3.40	198	237	287
B	2.30	2.45	2.55	300	307	296
C	6.10	6.10	6.50	800	755	789
D	4.20	4.33	4.44	200	290	300
E	5.70	5.89	5.99	351	427	389
F	12.50	12.60	12.89	107	110	104
G	0.56	0.76	0.79	1106	1473	1145
H	1.60	1.66	1.89	852	841	773
I	13.60	13.99	14.99	390	409	400
J	29.99	33.99	49.99	17	29	50

Further information now comes to light which changes various price and quantity data in the table. Use your spreadsheet to find the effect on the ten index numbers you have already constructed in each of the following:

(f) Year 2, quantities change to:
   $A = 207$, $B = 400$, $C = 1545$;

(g) Year 2, prices change to:
   $A = 3.70$, $B = 2.75$, $D = 4.54$,
   $F = 12.99$, $G = 0.60$, $J = 59.99$;

(h) Year 2, quantities change to:
   $B = 196$, $C = 710$, $E = 289$,
   $F = 85$, $G = 70$.

Part 2   Conclusions

Part 2 has been concerned with methods of description using a variety of statistics. It is not usually sufficient just to have data or to know where data can be found. We have to make sense of the data. The availability of published statistical information and the accessibility of databases, particularly on the Internet, can provide us with more and more figures but not necessarily the insight that is being sought. It is important to 'paint an illustrative picture' so that numbers can add to the detail and complexity of the landscape.

There are very few problems of any significance that exclude the use of numbers (or exclude the qualitative factors). The calculation of the mean, the standard deviation, indices and other measures should provide a summary at a glance. As a result of studying Parts 1 and 2 of this book you should also be aware of the limitations of calculated figures and, in particular, the importance of methodology and data quality. Given the availability of software that will quickly produce all the requested statistics, it is becoming even more important than before to clarify the purpose of the enquiry and the reasons for requiring the chosen statistics.

Part 3　Measuring uncertainty

Nothing in this world is certain but death and taxes
　　　　　　　　　　　　　　　　　　　　　　　– Benjamin Franklin (1706–1790)

Introduction

This assertion is still true today, since we do not live in a world of certainty. We cannot say with certainty how some individual will react to an advertisement, or which horse will win a race, or which numbers will be drawn in a lottery. While the uncertainty of life is perhaps most graphically illustrated by games and wagers, which we will indeed use to develop the ideas in this part of the book, those in business also have to deal with chance events. Chance plays a part in everyone's life, whether they choose to bet against some chance event by, for example, playing the lottery, backing a horse, or buying life insurance; or even if they eschew such gambles. Most activities require the assessment of risk, for example, when crossing the road you usually assess your chances of getting across safely as 100%, or else you do not set off.

In situations which are deterministic, then data collection alone can provide the basis for decision making. If we know the sales of a product are going to be 100 next month, and we know the price to be £10, then we know the total revenue will be $100 \times £10 = £1000$ (we look at this sort of model in Part 7). Even if we are not certain of the values next month, we may still be able to identify the minimum and maximum values for sales and price. Say, for example, that we know prices can vary from £9 to £11 and we know that sales can vary from 90 to 120. We now have a basis for looking at the worst possible, and best possible cases. In the worst case, we sell 90 at £9 each and get £810. In the best case, we sell 120 at £11 and get £1320. While we have identified the range of possible outcomes, *each is not equally likely to happen*. This concept will provide a vital clue in understanding the treatment of chance (or probability) in a business context.

Most of us think we have some idea about chance (or more formally, risk assessment) because we have played various games involving dice or cards. Even something as simple as tossing a coin, we 'expect' an equal chance of it coming down 'heads' or 'tails'. This intuitive idea of chance will help in developing some of the basic ideas in this part of the book, and in fact, we shall use cards, dice and coins as examples in Chapter 8. However, the intuitive idea of probability is capable of leading us to question some results which 'feel' wrong, – it is not a foolproof guide in this case. (see, for example, the birthdays problem in Chapter 8).

We also need to distinguish between risk and uncertainty. In the former case, we can assess that there are some outcomes that have some chance (probability) of occuring. Such situations can be modelled and developed in such a way that we can build the risk factor into the decision making process. With uncertainty, there are a large number of factors involved in some complex way which means that the probabilities of the occurrence of any event cannot be assessed. It is still possible to build business decision-making models in this situation, but they are by no means as simple as those where probabilities can be calculated.

Organization of Part 3

In this part we will begin by looking at the basic concepts of probability in Chapter 8, especially in relation to discrete events. In Chapter 9 we move on to known probability distributions which will be useful in a business context and will speed calculation. Finally we look at continuous distributions in Chapter 10.

Case 3: Carroll Imitations

Carroll Imitations

Carroll Imitations began in the 1990s in Coventry. It was started by two friends who, after spending some time unemployed, decided to go it alone. They produced copies of pictures, prints or small *object d'art* to order. While these were not exact copies, they bore enough of a passing resemblance to fool some people some of the time. There have been no problems with the copyright laws so far.

Originally these were on a 'one-off' basis to preserve exclusivity, but they quickly found that the market for such items was very small at the prices they were forced to charge, and that the number of pictures which people actually wanted copying was rather limited. This led them to producing small batches of copies which companies used as marketing and promotional aids. (In fact, some of these are now rather sought after and are expected to become 'collector's items' in their own right.) These items take rather longer to produce, since it is necessary to incorporate the sponsoring company's logo or message into the picture in an acceptable way, but the return to Carroll Imitations is relatively higher, given the volume produced.

In the early days, getting commissions from companies or individuals was a major problem. Sometimes they would get no work for several weeks, and then get three or more orders at the same time. This often led to very long working sessions for the partners. Some jobs took a couple of hours, others might take up to a week.

Carroll Imitations are now considering expanding by producing much larger batches as well as keeping their 'exclusive' copies for individual clients or company promotions. As a friend of the partners you have been asked to help them assess what they have achieved so far and understand the uncertainties in their market place.

Quick start	**Measuring uncertainty**

Probability is about assessing the chance of an event or a series of events happening.

The basic relationships of probability are:

- $P(A \text{ or } B) = P(A) + P(B)$ *for mutually exclusive events*
- $P(A \text{ or } B) = P(A) + P(B) - P(A \text{ and } B)$ *for non-mutually exclusive events*
- $P(A \text{ and } B) = P(A).P(B)$ *for independent events*
- $P(A \text{ and } B) = P(A).P(B|A)$ *for non-independent events*

There is no definition of probability as such – none of the various attempts actually work adequately. Instead, the theory can be built up from a series of axioms.

All of the ideas within probability can be built up from these basic relationships, although the more common relationships which relate to combinations of events are summarized into probability distributions as seen below.

Most useful discrete probability distributions:

- Uniform – *all outcomes have same probability*
  – applies to dice or random number tables. Often used as fall back position if you have no idea of what might happen.

- Binomial –

$$P(r) = \binom{n}{r} p^r q^{n-r}$$

  – used where there are two outcomes and the probability of a particular outcome remains constant, no matter how many times you combine the events

- Poisson –

$$P(x) = \frac{\lambda^x e^{-\lambda}}{x!}$$

  – used when the probability of 'success' is very small and there are a large number of events (also a distribution in its own right) (see Chapter 19).

Most useful continuous distribution:
- Normal distribution – *many factors having small effect each, give a symmetrical distribution that occurs in many 'natural' and other situations. A known distribution for which there are tables of probabilities. Strong links to sampling theory.*

8 Probability

Chance and the assessment of risk play a part in everyone's life. It might be something as simple as tossing a coin at the start of a sports match, playing card games, owning Premium Bonds or playing the National Lottery, or it could be in relation to more 'serious' issues such as car insurance, life insurance or being selected for a clinical trial of a new drug. Even events which seem to have sound logical reasons for occurring may have their timing selected by chance. Other events, especially those involving large groups of people or items, often have characteristics that can be represented, or modelled, by some reference to probability.

Probability was first studied in relation to gambling, and many examples including some in this chapter, may still be drawn from the use of cards, dice, roulette wheels, etc., simply because these items are familiar to many people. However, it quickly became apparent to mathematicians working in this area that the ideas being investigated had a much wider application, initially in relation to rates to be charged for insurance of freight carried by sea.

Probability has found a wide range of business applications. In addition to the calculation of risk in the banking and insurance industries, probability provides the basis of many of the sampling procedures used in market research and quality control. Investment appraisal requires an assessment of risk and a measure of expected outcomes. The planning of major projects needs to take account of uncertainties, whether it is the effects of the weather on building site schedules or fashions in the market place.

However, an understanding of probability will give you more than an ability to get the right answers to statistics questions, or a way of incorporating chance into project planning. Probability represents a new set of conceptual tools. Rather than looking at the world as consisting of **deterministic** situations, where everything is known with certainty, we can now consider a range of outcomes to every situation. More than this, by treating the world as **stochastic**, we can assess the chances of particular outcomes happening in a given situation. This, in fact, may lead us to reconsider the number of outcomes from that situation, so that we can acknowledge that certain outcomes are possible, even if, very unlikely. In some circumstances, assessing the probabilities of the various outcomes may be done by using tried and tested probability models, but this is by no means always the case.

Objectives

By the end of this chapter, you should be able to:

- describe the concept of probability
- identify mutually exclusive events
- identify independent events
- solve a range of problems involving probability

■ understand conditional probability
■ construct a probability tree
■ calculate expected values
■ use a simple Expected Monetary Value decision model
■ use a Markov chain.

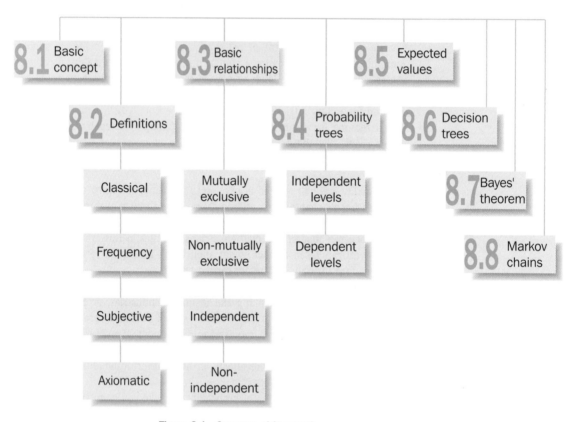

Figure 8.1 Structure of Chapter 8.

8.1 Basic concepts   Before we look at the formal approaches to probability, we will consider a small group of people who have taken part in a survey. On the basis of their answers, we know a little about them, and this will allow us to consider what would be

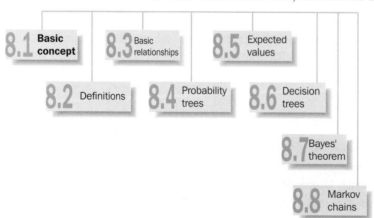

likely to happen if we were to select an individual or subgroup from the respondents.

The sample survey had 500 respondents, of whom 200 were men and 300 were women. In relation to a question about their leisure time activities which specifically asked for their favourite Sunday morning activity, the responses were as shown in Table 8.1.

Table 8.1 Sunday morning activities

Response	Number of respondents
Walking the dog	50
Reading the papers	100
Washing the car	20
Staying in bed	300
Other	30

Imagine for a moment that you were to pick just one person from these respondents. What is the chance of picking a man? Well, there are 200 men, and there are 500 people in total, so the chance (or probability) is:

$$\text{Probability(Man)} = 200/500 = 0.4$$

Similarly, you can find the probability that the person who is selected has walking the dog as their favourite Sunday morning activity:

$$\text{Probability(Walking the dog)} = 50/500 = 0.1$$

What this information doesn't tell you is how many men think walking the dog is their favourite Sunday morning activity. However, we might make a guess from these probabilities. A probability of 0.1 is another way of saying 10%, so we can say that 10% of people in the survey had walking the dog as their favourite Sunday morning activity. Since there were 200 men in the survey, we might legitimately assume that 10% of them chose walking the dog. This would give us 20 (200 × 0.1) men. Therefore the anticipated probability of selecting a man who chose walking the dog would be:

$$20/500 = 0.04$$

Note the assumption that we have to make in order to make this calculation work! It is that men and women acted in the same way in choosing their favourite Sunday morning activity. You might question this assumption in this case! Where we have survey results of this type, then we can actually check on the specific probability for these people, since we can produce a **cross-tabulation** of the answer to the question on favourite Sunday morning activity with the gender of the respondent. This might give the table shown in Table 8.2.

Table 8.2 Sunday morning activities and gender

Response	Men	Women	Total
Walking the dog	30	20	50
Reading the papers	40	60	100
Washing the car	15	5	20
Staying in bed	100	200	300
Other	15	15	30
Total	200	300	500

From the actual data we can see that the chance of selecting a man who chose walking the dog is actually:

$$\text{Actual proportion} = 30/500 = 0.06$$

which is higher than our guess just using probability. Where we do not have access to the actual data, then the probability method is the best we can do. We return to this issue in Chapter 13 when we consider chi-squared tests.

We can draw a few more conclusions about probability from this set of responses. The probability of selecting a man was 0.4, and you can easily work out that the probability of selecting a woman is 0.6. This means that:

$$P(\text{Man}) + P(\text{Woman}) = 0.4 + 0.6 = 1$$

but this can be turn around so that we have:

$$P(\text{Woman}) = 1 - P(\text{Man})$$

and since being a man is 'not being a woman', we could also write this as:

$$P(\text{Woman}) = 1 - P(\text{not a Woman})$$

In this context, the manipulation above seems very obvious, but in many cases it will be easier to find the probability of something not happening than to find the probability that it does occur. For example, if items were packed in boxes of 1000 and we wanted to find the probability that there were two or more defective items in a box, then to calculate this directly, we would need to use the following relationships:

$$P(2 \text{ or more}) = P(2) + P(3) + P(4) + \ldots + P(999) + P(1000)$$

However, if we notice that the only alternatives to '2 or more' are 'no defective items' or 'one defective item', then

$$P(2 \text{ or more}) = 1 - P(\text{not 2 or more}) = 1 - [1 \text{ or less}] = 1 - [P(0) + P(1)]$$

and this will usually be very much easier to evaluate.

Probabilities are often used to suggest what is likely to happen. If a fair coin were tossed 500 times you would expect 250 heads and 250 tails. The number of trials or samples (n) multiplied by a probability (in this case $n = 500$ and $P(\text{head}) = P(\text{tail}) = \frac{1}{2}$) gives the **expected value**. As with all probabilistic measures what is most likely to occur may not occur. You could toss a fair coin 500 times and get 251 heads or 235 heads or 300 heads; all these outcomes are possible even if they are less likely. What expectation tells you is what will happen **on average**. The general formula is given by:

Expected outcome	=	(probability of that particular outcome on a single trial)	×	(total number of trials)	
e.g. expected no. of heads =			0.5	×	500 =	250

With considerably more mathematical calculation, we can also work out the probability of getting exactly the expected number of outcomes, but this will involve us using probability distributions described in later chapters. (Just for interest, the answer here is $P(\text{exactly 250 heads}) = 0.036$, a relatively low chance.)

8.2 Definitions

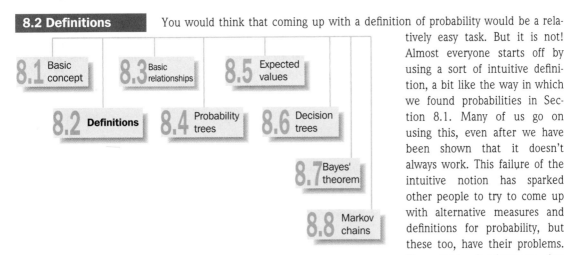

You would think that coming up with a definition of probability would be a relatively easy task. But it is not! Almost everyone starts off by using a sort of intuitive definition, a bit like the way in which we found probabilities in Section 8.1. Many of us go on using this, even after we have been shown that it doesn't always work. This failure of the intuitive notion has sparked other people to try to come up with alternative measures and definitions for probability, but these too, have their problems. Theoretical development has been from an axiomatic starting point. In this section we will look at these various attempts to define probability, since each adds something to our overall understanding of the concept.

8.2.1 The 'classical' definition

This definition was the first to be used, and involves thinking about the problem and applying logic. In the cases we have discussed above, we have effectively used this definition, since we have counted up the number of ways of selecting someone with a particular characteristic (e.g. that of being male), and divided by the total number of possible results (e.g. the total number of people). Generalizing this idea for an event E, we have:

$$P(E) = \frac{\text{no. of ways } E \text{ can occur}}{\text{total number of outcomes}}$$

This definition is widely used when trying to assess a particular situation but it is not complete. If you consider a die, there are six different faces, and this definition would suggest that $P(5) = 1/6$. However, if the die has been weighted so that, say, 6 will appear each time it is thrown, then the probability of a 5 is zero $(P(5) = 0)$ despite the fact that there is one 5, and there are six faces on the die. To complete the definition, we need to add that each outcome is equally likely; equally likely means equally probable, and thus we have a definition of probability which uses probability in that definition, i.e. a tautology. Even if we are sure that the outcomes are equally likely, the definition cannot deal with situations where there are an infinite number of outcomes. Despite these comments, most people will use this definition when considering simple probability situations.

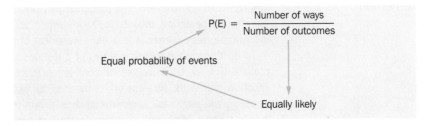

Figure 8.2   Tautology in the classical definition.

Exercise

> Think of a coin and write down the probability of a 'head'. Now think of tossing three coins and, by working out all of the possible outcomes, write down the probability of three 'heads'.
>   You should get 0.5 and 0.125.

8.2.2 The frequency definition

If the 'sit back and think about it' approach to assessing probabilities can lead us into problems, as outlined above, then maybe we should take a more proactive approach to the problem. An alternative way to look at probability would be to carry out an experiment to determine the proportion in the long run, sometimes called the frequency definition of probability.

$$P(E) = \frac{\text{no. of times } E \text{ occurs}}{\text{no. of times the experiment was conducted}}$$

This would certainly overcome the problem involving the biased die given above, since we would have $P(5) = 0$ and $P(6) = 1$. One problem with this definition is assessing how long the long run is. Experiments with a theoretical, unbiased coin do not necessarily conclude that the probability of a 'head' is $\frac{1}{2}$, even after 10 million tosses, and it can be shown that this frequency definition will not necessarily ever stabilize at some particular proportion. A second problem with this definition is that it must be possible to carry out repeated trials, whereas some situations only occur once, for example the chance of a sales person selling more than the target set for next March.

Exercise

> 1 Roll a die ten times and count the number of 6s. What do you estimate the probability of a 6 to be from your experiment?
> 2 Continue your experiment until you have rolled the die 100 times. What is the estimate of $P(6)$ now? Is there a lesson here?

As an example of the frequency definition, if possible, go to the website and download the file c8freqdef.xls. This shows a sample of 20 tosses of a coin and calculates the frequency definition of probability. Press F9 to recalculate the spreadsheet. An example of the screen you are likely to see is shown in Figure 8.3.

8.2.3 The subjective definition

It is often said that there is no substitute for experience, and it is also often observed that someone doing a particular job, say stock control, does not need to refer to mathematical models to decide when and how to act. Maybe we can incorporate this experience into our assessment of probability.

To do this we would ask a series of questions to find out people's subjective probabilities, or their degree of confidence in certain events happening. This situation will work for 'one-off' situations as well as recurrent positions. For example, you might ask about the chance of rain tomorrow for a particular area; it may be that locals will give you 'better' answers than strangers to the area. Similarly, you might ask a series of shopkeepers about the likely success of new

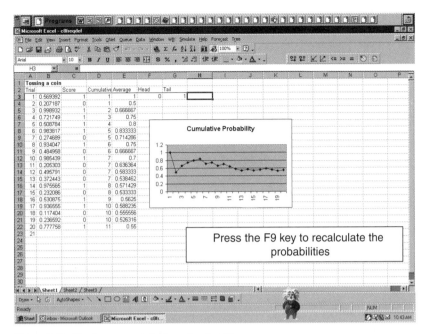

Figure 8.3 The frequency definition of probability.

packaging for your product. As with all sampling, the larger the number of the relevant population who take part in the survey, the better the results.

This idea is similar to the management technique known as Delphi forecasting.

Case study

Who would you ask to assess the likely profitability of Carroll Imitations in the next year?

8.2.4 The 'axiomatic' definition

The logical difficulty of defining probability is recognized by the modern trend that bases the whole of probability theory on a number of axioms from which theorems are deduced. The axioms and theorems are formulated in terms of set theory and, avoiding details, some of them may be paraphrased as follows:

1 The probability of an event lies in the interval $0 < P(E) < 1$ and no other values are possible.
2 If something is **certain** to occur, then it has a **probability** of 1.
3 If two or more different outcomes of a trial or experiment cannot happen at the same time, then the probability of one or other of these outcomes occurring is the **sum of the individual probabilities,** for example: if $P(E_1) = \frac{1}{4}$, $P(E_2) = \frac{1}{2}$, then $P(E_1 \text{ or } E_2) = \frac{1}{4} + \frac{1}{2} = \frac{3}{4}$.

Fortunately, the theory of probability which can be and is built on these axiomatic foundations is supportive of the various distributions and 'laws' which were worked out before the axioms were formally stated.

As was said earlier, you will tend to use the 'classical' definition when you think about probability problems, and this is quite acceptable, provided that you recognize its limitations and do not allow it to blinker your thinking.

8.3 Basic relationships in probability

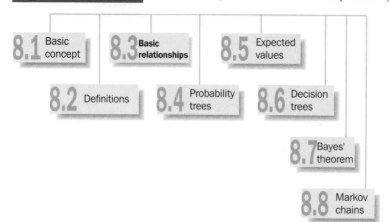

Once we have developed our ideas of what is meant by probability, we can think of the chance of certain events happening, as well as what those events might be. However, the assessment of the probability of a single event is likely to be of limited interest. Particularly in a business context, we are concerned about the chances of two or more events happening at the same time, or maybe, not happening together. In this section we will look at ways of classifying events, and the various approaches to finding probabilities of these combined events.

8.3.1 Mutually exclusive events

A mutually exclusive event is the situation represented in part 3 of the axiomatic definition of probability, and, as we have seen, we add the probabilities together to find the probability that one or other of the events will occur. For example, if a group of people consists of 20 single and 40 married men with 30 single and 10 married women, then, selecting a person at random we have:

$$P(\text{single man}) = 20/100$$
$$P(\text{single woman}) = 30/100$$
$$\text{and} \quad P(\text{single person}) = 20/100 + 30/100 = 50/100 = 0.5$$

It can be seen in Figure 8.4 that 50 out of 100 are single.

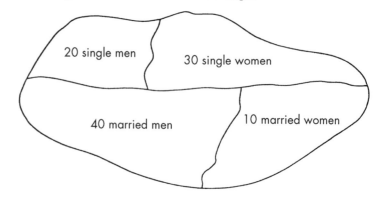

20 single men

30 single women

40 married men

10 married women

Figure 8.4

8.3.2 Non-mutually exclusive events

Where one outcome has, or can have, more than one characteristic, then these outcomes are said to be **non-mutually exclusive**. In this case it will not be possible simply to add the probabilities together as we did above to find the overall probability of one characteristic or another, since this would involve counting some outcomes twice. For example, if a group of people contains both men and women, and these people either agree or disagree with some proposition, then to find the probability of selecting a person who is either a man, or disagrees, will be

$$P(\text{man or disagree}) = P(\text{man}) + P(\text{disagree}) - P(\text{man and disagree})$$

Since the men who disagree appear in each of the first two probabilities, we need to subtract the probability of this group from our required probability. If, for example, there were 30 men, of whom 10 disagree with the proposition, and 70 women, of whom 40 disagree with the proposition then:

$$P(\text{man}) = 30/100$$
$$P(\text{disagree}) = 50/100$$
$$P(\text{man and disagree}) = 10/100$$

Therefore

$$P(\text{man or disagree}) = 30/100 + 50/100 - 10/100 = 70/100 = 0.7$$

It can be seen in Figure 8.5 that there are 30 women who agree with the proposition, leaving the remaining 70 who are women or men who disagree (here we have again used the concept of **not**).

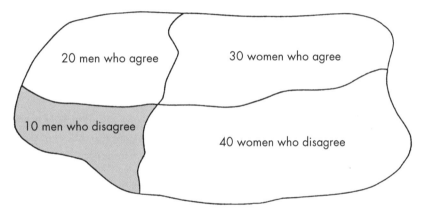

Figure 8.5

8.3.3 Independent events

When one outcome is known to have no effect on another outcome, then the events are said to be **independent**. For example, if the probability of a machine breaking down is 1/10 and the probability of stoppage of raw material supplies is 1/8, then we can find the probability of the two events happening together by multiplying the two probabilities, because the occurrence of one of these events does not affect the probability of the other. Thus

$$P(\text{breakdown and stoppage of supplies}) = 1/10 \times 1/8 = 1/80 = 0.0125$$

8.3.4 Non-independent events

If two events are such that the outcome of one affects the probability of the outcome of the other, then the probability of the second event is said to be **dependent** on the outcome of the first. From a group of ten people, five of whom are men and five women, the probability of selecting a man is $P(\text{man}) = 5/10 = 0.5$. If a second person is now selected from the remaining nine people, the probabilities will depend on the outcome of the first selection.

If a man is selected first, then $P(\text{man}) = 4/9$ and $P(\text{woman}) = 5/9$
If a woman is selected first, then $P(\text{man}) = 5/9$ and $P(\text{woman}) = 4/9$
$P(\text{man}\,|\,\text{man}) = 4/9$, and $P(\text{man}\,|\,\text{woman}) = 5/9$

where the notation $P(\text{man}\,|\,\text{woman})$ is read as 'the probability of selecting a man when a woman has been selected at the first selection'. (*Note that if the second selection had been from the original group of 10 people, then the two probabilities would be independent.*)

8.3.5 Summary

These relationships are fundamental to understanding and working with probabilities. If you are able to identify the type of event or events with which you are dealing, then you are a considerable way into solving a problem. At this stage, practice in answering questions on probability is likely to be the best way of 'growing' this understanding. The different types of event can be summarized as follows:

If events A and B are mutually exclusive, then

$$P(A \text{ or } B) = P(A) + P(B)$$

If events A and B are non-mutually exclusive, then

$$P(A \text{ or } B) = P(A) + P(B) - P(A \text{ and } B)$$

If events A and B are independent, then

$$P(A \text{ and } B) = P(A) \times P(B)$$

If events A and B are dependent, then

$$P(A \text{ and } B) = P(A) \times P(B|A)$$

where $P(B|A)$ is the probability that B occurs given that A has already happened (often referred to as a **conditional probability**).

Probability occasionally gives answers which do not seem to match our own guesses. For instance, if you were asked 'What is the probability of two or more people in a group of 23 having the same birthday in terms of day and month, but not necessarily year?' what would you guess the answer to be? It is in fact more than $\frac{1}{2}$!

To build up to this answer, consider a group size of two. Whenever the first person's birthday, if all birthdays are equally likely, then the probability that they have the same birthday will be 1/365 or 0.0027397 (ignoring leap years), and that they do not 364/365. For a group of three, given the first person's birthday, the probability that the second has a different birthday is 364/365 and that the third has a different birthday is 363/365. Thus

$$P(\text{at least 2 same}) = 1 - P(\text{all different}) = 1 - \frac{364}{365} \times \frac{363}{365} = 0.00802$$

For a group of four, we have:

$$P(\text{at least 2 same}) = 1 - \frac{364}{365} \times \frac{363}{365} \times \frac{362}{365} = 0.01636$$

We can continue this process, to give Table 8.3.

Table 8.3

No. of people	Probability	No. of people	Probability
2	0.002 739 7	26	0.598 24
3	0.008 024 2	27	0.626 859
4	0.016 355 9	28	0.654 46
5	0.027 135 57	29	0.680 968 5
6	0.040 462 48	30	0.706 3
7	0.056 235 7	31	0.730 45
8	0.074 335	32	0.753 347 5
9	0.094 623 8	33	0.774 97
10	0.116 948	34	0.795 3
11	0.141 141 4	35	0.814 38
12	0.167 025	36	0.832 18
13	0.194 4	37	0.848 738
14	0.223	38	0.864 06
15	0.252 9	39	0.878 2
16	0.283 6	40	0.891 23
17	0.315	41	0.903 15
18	0.346 91	42	0.914
19	0.379 1	43	0.923 9
20	0.411 4	44	0.932 885
21	0.443 668	45	0.940 97
22	0.475 695	46	0.948 25
23	0.507 297	47	0.954 77
24	0.538 34	48	0.960 597
25	0.568 699 7	49	0.965 779 6

Exercises

1 When will the probability reach 1?
2 If the group are all from the same country or cultural background, what factors will increase the probabilities proposed in the model given above?

8.4 Probability trees

As we have already seen in Chapter 4, diagrams are a very useful way of representing a situation. In the case of probability a diagram may help you to explain the problem to other people, or may help you to clarify the problem in your own mind. For example, it is a useful way of ensuring that you have taken into account all of the possible outcomes. One method of illustration is the probability tree. It is important to distinguish between trees that illustrate independent events, and those that show dependent events. It may seem very obvious, but in the former case, the events can be shown in any order, and the results will still be the same. In

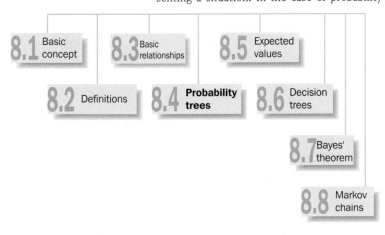

the latter case, the order in which the events are depicted is crucial, since the probabilities change with the order of events.

8.4.1 Independent events

As an illustration of the case of independent events, consider people's preferences in terms of chocolate and the fact of their eye colour. The two things are not related in any way. Suppose we know that 30% of people have brown eyes, 40% have green eyes and 30% have blue eyes. Also, we have asked for preferences in type of chocolate, and found that 20% of people prefer plain, 70% prefer milk and 10% prefer white chocolate. Now, if we wish to know the proportion of people with a certain eye colour who prefer a particular type of chocolate, we can use the rule for independent events, and multiply the probabilities. For example, those who prefer plain chocolate and have brown eyes:

$$P(\text{plain}) \times P(\text{brown}) = 0.2 \times 0.3 = 0.06$$

We could use a diagram such as Figure 8.6 to show all of the possibilities and their probabilities.

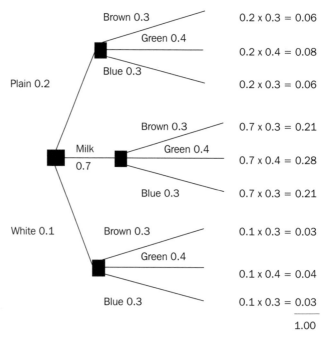

Figure 8.6

8.4.2 Dependent events

If we have three groups of people, a red team of 10 men and 10 women, a blue team of seven men and three women and a yellow team of four men and six women, then using a two-stage selection procedure firstly selecting a team, and then selecting an individual, the probability of selecting a woman will be dependent on which team is selected. In Figure 8.7 the probabilities of selecting a red, blue or yellow team are respectively 0.4, 0.4 and 0.2. The probabilities of being a man or a woman were derived from the numbers in each team see Figure 8.7.

Since individuals can only belong to one team, then we can treat the teams as mutually exclusive. This means that we can add the probability of selecting a

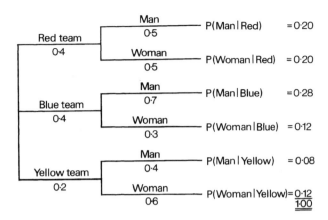

Figure 8.7

woman in the red, blue and yellow teams to get the overall probability of selecting a woman:

$$P(\text{woman}) = P(\text{woman|red}) + P(\text{woman|blue}) + P(\text{woman|yellow})$$
$$= 0.2 + 0.12 + 0.12$$
$$= 0.44$$

8.5 Expected values

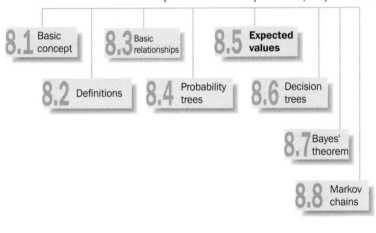

When evaluating problems or situations where uncertainty exists, the concept of **expected values** is particularly important. Consider a simple game of chance. If a fair coin shows a head you win £1 and if it shows a tail you lose £2. If the game were repeated 100 times you would expect to win 50 times, that is £50 and expect to lose 50 times, that is £100. Your overall loss would be £50 or 50 pence per game on average. This average loss per game is referred to as **expected value** (EV) or **expected monetary value** (EMV). Given the probabilistic nature of the game, sometimes the overall loss would be more than £50, sometimes less. Rather than work with frequencies, expected value is usually determined by weighting outcomes by probabilities (see Section 5.2.3). As we have seen, the probabilities may have been derived from relative frequency. In this simple game, the expected value of the winnings is

$$£1 \times \frac{1}{2} + (-£2) \times \frac{1}{2} = -£0.50 \text{ or } 50 \text{ pence}$$

where −£2 represents a negative win, or loss.

You will of course, never lose 50 pence in a single game, you will either win £1 or lose £2. Expected values give a **long-run, average result**. In general

$$E(x) = \Sigma(x \times P(x))$$

where $E(x)$ is the expected value of x.

If we link together the idea of probability trees and the concept of expected values, then we can show what the situation might be if two characteristics or events were independent. Take the results of the survey which we used in Table 8.2 above. There were five different choices for the favourite activity on a Sunday morning. There were two genders. Assuming that the two factors are independent, we can construct the probability tree shown in Figure 8.8 below.

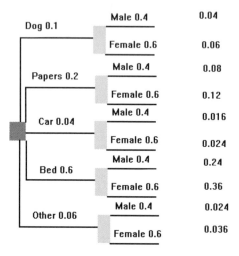

Figure 8.8 Probability tree assuming independence.

Since we know that there were 500 people in the survey, we can take the results of the probability calculations (in Figure 8.8) and multiply each of these by 500 to find the **expected** numbers. This is shown in Table 8.4 below.

Table 8.4 Sunday morning activity and gender

Outcome	Probability	Expectation	Actual
Walking the dog and male	0.04	20	30
Walking the dog and female	0.06	30	20
Reading the papers and male	0.08	40	40
Reading the papers and female	0.12	60	60
Washing the car and male	0.016	8	15
Washing the car and female	0.024	12	5
Staying in bed and male	0.24	120	100
Staying in bed and female	0.36	180	200
Other and male	0.024	12	15
Other and female	0.036	18	15

In Table 8.4 we have included the actual numbers of people who answered the survey. As you can see, in some cases there is a considerable difference between what we might have 'expected' to happen, and what actually occurred. Looking at whether or not this difference is significant will be deferred until Chapter 13.

Exercise

> What reasons would you suggest for the differences between the final two columns of Table 8.4?

8.6 Decision trees

We have now brought together enough ideas to make practical use of probability within business decision making. Managers, as part of their job, have to make choices between various courses of action. Each course of action will have consequences that will depend on other factors. It is quite likely that probabilities can be attached to these other factors, and hence a range of expected consequences found. If we assume a rational decision maker, then the choice will be the course of action which has the most favourable expected consequences.

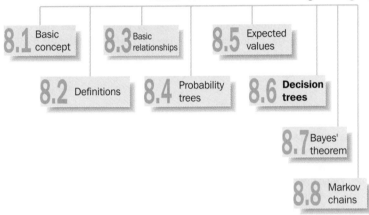

Suppose now you are given two opportunities to invest your savings. The first opportunity, option A, is forecast to give a profit of £1000 with a probability of 0.6 and a loss of £400 with a probability of 0.4. The second opportunity, option B, is forecast to give a profit of £1200 with a probability of 0.5 and a loss of £360 with a probability of 0.5. These two opportunities actually give you three choices, the third choice being not to invest. The possible decisions are represented in the **decision tree** shown in Figure 8.9.

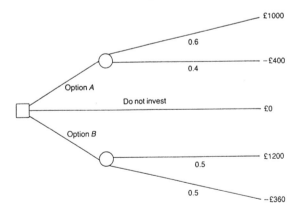

Figure 8.9

A decision tree shows the decisions to be taken (a decision node is denoted by a □) and the possible outcomes (a chance node is denoted by a ○). The expected value associated with each chance node is:

$$E(\text{option } A) = £1000 \times 0.6 + (-£400) \times 0.4 = £440$$
$$E(\text{option } B) = £200 \times 0.5 + (-£360) \times 0.5 = £420$$

On the basis of highest expected value, option A would be chosen. However, once the decision is made and if the forecasts are correct, option A will yield a profit of £1000 or a loss of £400. It is important to recognize that £440 is an average, like 2.5 children, and may never occur as an outcome. Indeed, if you only make this type of decision once, you either win or lose. The highest expected value provides a useful decision criterion if the type of decision is being repeated many times, e.g. car insurance, but may not be appropriate for a one-off decision, e.g. whether to extend an existing factory. There are other decision criteria. A risk taker, for example, would be attracted to the larger possible profits of option B. A risk avoider would be deterred by possible losses and choose not to invest or choose the option with least loss, i.e. option B. Different decision criteria do lead to different choices, so the selection of criteria is important.

As an example of a more complex decision tree, consider the following scenario.

Example

A company has developed a new product. It needs to decide whether or not to product test and market test before launch, and has been advised that, even though these processes do cost money, they increase the likelihood of success for the product. (Note that it has been agreed within the company that you can only market test a product once it has passed product testing. If a product fails either test it is regarded as worthless.) You have been able to obtain details of the costs of these testing processes, together with historical data which suggests how much the likely success of the product is enhanced by successful testing. Launching the product will cost £300 000 and the estimates of profit are as follows:

$$\text{Highly successful} = £2\,000\,000$$
$$\text{Moderately successful} = £1\,000\,000$$
$$\text{Low level success} = £500\,000$$
$$\text{Failure} = £100\,000$$

The historic data that has been collected gives the following results:

Chance of:	No testing	Product testing	Product testing & market testing
High success	0.1	0.1	0.2
Medium success	0.2	0.4	0.45
Low success	0.4	0.4	0.3
Failure	0.3	0.1	0.05

Product testing costs £100 000 and Market testing costs £100 000. Should the product fail either of these tests it is abandoned. The probability of passing product testing is 0.8 and the probability of passing market testing is 0.9
The alternative to this process is to sell the product design for £500 000
What do you advise the company to do?

We can draw a diagram to represent this situation, such as the one below or we could use some software such as INSIGHT which is published by Thomson Learning. At each stage Figure 8.10 shows the expected outcomes and nominates the 'best' one (based on the assumption of rational decision makers who choose the highest expected monetary value (EMV)). As you can see, the diagram suggests that selling the design is the best course of action.

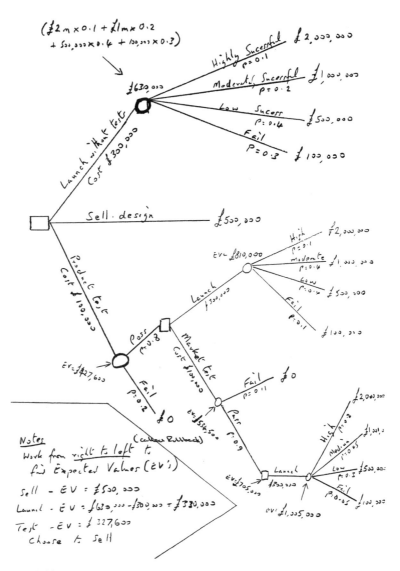

Figure 8.10

Exercise

> What would be the 'best' decision if the profit from the highly successful
> product were estimated to be £3 500 000 rather than £2 000 000?

Selling the design for £0.5m is the only other option that gives an expected
profit. If *we assume rational decision makers*, then they will choose the highest
expected profit, which means that they should product test and market test, and
only if these are successful should they launch the product.

Note that the exact level of profit of £0.505m will not be made.

This type of decision tree and decision criteria is perhaps the simplest. Should
you study management decision making in more detail, you will find many complex decision rules.

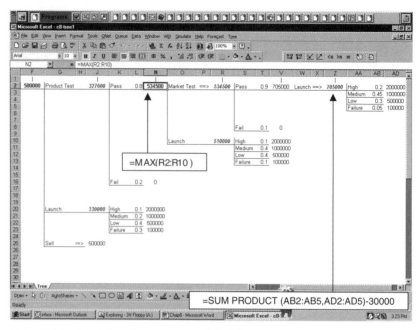

Figure 8.11 Output from the tree program in Insight.

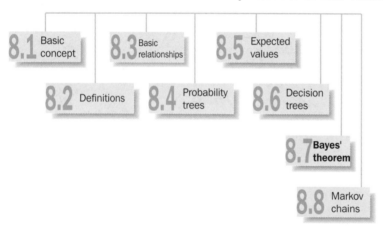

8.7 Bayes' theorem

As we have seen in Section 8.4.2, when events are described as dependent, this means that the probabilities of their occurrence depend on the **outcome of some previous event**. Bayes' theorem is a way of looking at this situation. It proposes that we are at the end of the sequence of events, so that we know the final outcome, and we wish to find the probability that certain events occurred in the sequence. This probability is known as a **posterior probability**, since it is calculated with the benefit of hindsight. This probability will almost always be different from the **prior probabilities**, which were those we knew before the sequence began. This problem can be illustrated by a diagram such as that in Figure 8.12. This figure illustrates the situation where a company uses three machines, labelled A, B, and C, and 40% of production goes to A, 50% to B and the remaining 10% to C. Each machine has a different wastage rate, due to differing ages and manufacturing tolerances. When defective items are found, it is not immediately possible to say from which machine they came. However, we can use Bayes' theorem to identify the machine which is most likely to have produced a faulty item, and check that one first. This will save time and cost, although it, obviously, will not be 'correct' every time. The wastage rates are 6% for Machine A, 1% for Machine B, and 10% for Machine C (which is now very old).

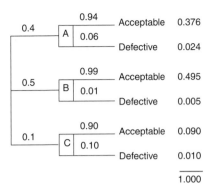

Figure 8.12

Looking at the probabilities given in Figure 8.12, we can see that the probability of getting a defective item is:

$$P(\text{defective}) = 0.024 + 0.005 + 0.010 = 0.039$$

This could also be written as:

$$P(\text{defective}) = P(\text{defective}|\ A \text{ or } B \text{ or } C)$$

To look at a particular outcome, for example for Machine A and a defective item, we could write the probability as:

$$P(A \text{ and defective}) = \frac{P(\text{defective from Machine } A)}{P(\text{defective from any machine})}$$

and here the probability would be:

$$P(A \text{ and defective}) = \frac{0.024}{0.039} = 0.615$$

The more general formula for Bayes' theorem is:

$$P(A_i|X) = \frac{P(X|A_i)P(A_i)}{\displaystyle\sum_{i=1}^{i=n} P(X|A_i)P(A_i)}$$

where X is the known outcome, and A_i is the particular route in which we are interested. This formula could be applied to each possible route to the known outcome, and the sum of the probabilities would be one. Using the information from Figure 8.12, we can construct Table 8.5.

Table 8.5 Prior and posterior probabilities (given defective)

Machine	Prior probability	Posterior probability
A	0.4	0.615
B	0.5	0.128
C	0.1	0.256
Total	1.0	0.999*

*Note that sum does not equal 1 due to rounding.

8.8 Markov chains

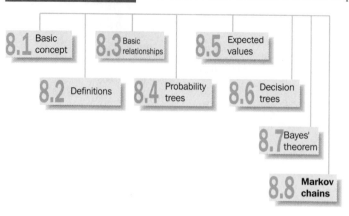

A Markov chain combines the ideas of probability with those of matrix algebra as discussed in Chapter 24. We will present this section as if you are already familiar with matrix algebra in order to keep the continuity of the probability concept, but you should refer to Chapter 24 if that is not so.

The Markov chain concept assumes that probabilities remain fixed over time, but that the system that is being modelled is able to change from one state to another, using these fixed values as transition probabilities. Consider, for example, the following transition matrix:

$$\mathbf{P} = \begin{array}{c} \\ E_1 \\ E_2 \end{array} \begin{array}{c} E_1 \quad E_2 \\ \begin{bmatrix} 0.8 & 0.2 \\ 0.3 & 0.7 \end{bmatrix} \end{array}$$

This means that if the system is in some state labelled E_1, the probability of going to E_2 is 0.2. If the system is at E_2, then the probability of going to E_1 is 0.3, and the probability of remaining at E_2 is 0.7. This transition matrix could be represented by the **directed graph** (Figure 8.13).

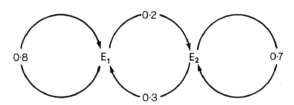

Figure 8.13

If we consider the movement from one state to another to happen at the end of some specific time period, and look at the passage of two of these periods, we have a situation as shown in Figure 8.14.

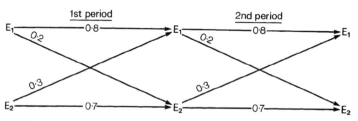

Figure 8.14

The probability of ending at E_1 after two periods if the system started at E_1 will be:

$$P(E_1 \to E_1 \to E_1) + P(E_1 \to E_2 \to E_1) = (0.8)(0.8) + (0.2)(0.3) = 0.70$$

Starting at E_1 and ending at E_2:

$$P(E_1 \rightarrow E_1 \rightarrow E_2) + P(E_1 \rightarrow E_2 \rightarrow E_2) = (0.8)(0.2) + (0.2)(0.7) = 0.30$$

Starting at E_2 and ending at E_1:

$$P(E_2 \rightarrow E_1 \rightarrow E_1) + P(E_2 \rightarrow E_2 \rightarrow E_1) = (0.3)(0.8) + (0.7)(0.3) = 0.45$$

Starting at E_2 and ending at E_2:

$$P(E_2 \rightarrow E_2 \rightarrow E_2) + P(E_2 \rightarrow E_1 \rightarrow E_2) = (0.7)(0.7) + (0.3)(0.2) = 0.55$$

Thus the transition matrix for *two* periods will be:

$$\begin{array}{cc} & \begin{array}{cc} E_1 & E_2 \end{array} \\ \begin{array}{c} E_1 \\ E_2 \end{array} & \begin{bmatrix} 0.70 & 0.30 \\ 0.45 & 0.55 \end{bmatrix} \end{array}$$

but note that this is equal to $\mathbf{P}^2$, i.e. the square of the transition matrix for one period. To find the transition matrix for four periods, we would find $\mathbf{P}^4$ and so on.

The states of the system at a given instant could be an item working or not working, a company being profitable or making a loss, an individual being given a particular promotion or failing at the interview, etc. In all transition matrices, the movement over time is from the state on the left to the state above the particular column, and thus, since something must happen, the sum of any row must be equal to 1.

A state is said to be **absorbent** if it has a probability of 1 of returning to itself each time. In the matrix

$$P = \begin{array}{c} E_1 \\ E_2 \\ E_3 \end{array} \begin{array}{ccc} E_1 & E_2 & E_3 \end{array} \begin{bmatrix} 0.4 & 0.2 & 0.4 \\ 0.2 & 0.7 & 0.1 \\ 0 & 0 & 1 \end{bmatrix}$$

the state E_3 is an absorbent state, since each time the system reaches state E_3 it remains there. This can again be shown by a directed graph (Figure 8.15).

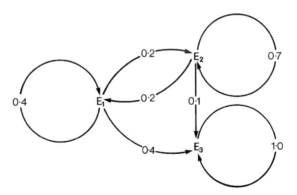

Figure 8.15

To use these transition matrices for predicting a future state, we need to know the initial state, which is written in the form of a vector. For example, if state E_1 in Figure 8.13 were a company being profitable, and E_2 were a company not being profitable, then we would consider what would happen to a group of companies. If the group consists of 150 profitable companies and 50 non-profitable companies, the initial vector will be

$$\begin{matrix} & E_1 & E_2 \\ \mathbf{A}_0 = & [150 & 50] \end{matrix}$$

To find the situation after one time period, say one year, we postmultiply the initial state vector by the transition matrix:

$$[150 \quad 50] \begin{bmatrix} 0.8 & 0.2 \\ 0.3 & 0.7 \end{bmatrix} = [135 \quad 65] = \mathbf{A}_1$$

$\mathbf{A}_1$ now represents the situation after one time period, where we would expect 135 companies to be profitable, and 65 not to be profitable. To find the situation after two time periods, we either multiply the initial state vector by P^2

$$[150 \quad 50] \begin{bmatrix} 0.7 & 0.3 \\ 0.45 & 0.55 \end{bmatrix} = [127.5 \quad 72.5] = \mathbf{A}_2$$

or postmultiply the vector $\mathbf{A}_1$ by P:

$$[135 \quad 60] \begin{bmatrix} 0.8 & 0.2 \\ 0.3 & 0.7 \end{bmatrix} = [127.5 \quad 72.65] = \mathbf{A}_2$$

Both calculations give a row vector labelled A_2, which represents the expected number of companies in each state after 2 years. Note that some of the companies that are currently profitable may not have been profitable after only 1 year.

The process given above can continue with any transition matrix for any number of time periods; however, it is likely that the probabilities within the matrix will become out of date, and thus not fixed, if a prediction into the distant future is made. Markov chains are often used in manpower planning exercises and in market predictions.

Multiplying out matrices can be a very time consuming process and offers many opportunites for mistakes. Fortunately, we can use various programs and packages to do the multiplication for us. One example is INSIGHT^{TM} which can be seen on our website.

To use this package, start Excel in the normal way and then load MARKOV.XLS from the website.(you will have to install INSIGHT the first time you use it) You should see a transition matrix as follows:

	A	B	C
A	65%	10%	0%
B	30%	75%	10%
C	5%	15%	90%

And you should also notice that the matrix is arranged so that the columns add to 100%. To the right of the transition matrix is a table which shows the situation after 1, 2, 3, etc. time periods. Looking across this table, you can see that convergence occurs fairly quickly in this case. (Note that time period 0 is the initial state.) Clicking on the tab at the bottom of the screen will allow you to see a graph of this table.

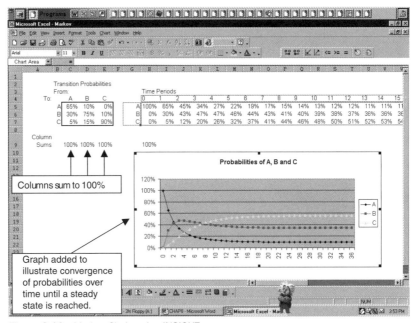

Figure 8.16   Markov Chain using INSIGHT.

This basic spreadsheet will allow you to solve any Markov chain problem. For transition matrices which are 2×2, then make the C row and C column equal to zeros. For larger transition matrices, insert new rows and columns as necessary. (Note: you should not insert over the A column or row.) You will also need to increase the number of rows in the table to the right, and copy the formulae down the columns, and then across the table.

Exercise

> Try solving the problem given in this section using INSIGHT.

8.9 Conclusions

Probability is about a way of looking at the world and we have set out some of the implications of this probabilistic view in this chapter. We all have some intuitive ideas about probability, whether we recognize it or not, and this chapter gives you a framework for organizing your own ideas. Many business decisions involve elements of chance (e.g. the chance that a competitor will match your price reduction), and thus we need to include aspects of probability when we consider these situations. In the absence of concrete information, we may opt to act as if the most likely outcome will actually occur; but then, of course, we need to define and calculate the most likely outcome.

In looking at a situation it is always important to identify which factors are independent and which are dependent. It is also necessary to see if events happen in sequence or in parallel. Both of these concepts can be dealt with using probability, and the consequences of such dependencies can easily be shown. The ideas shown in Sections 8.1 and 8.3, although very simple, are particularly powerful in analysing situations.

Case study

From their experience, the two friends think that the probability of getting a '2-hour' job is 0.5 (because they try to be selective over who they approach) and the probability of getting a '3-day' job 0.1. After a rather slack period, they decide to increase their marketing effort, and approach 30 firms which they suspect can provide work for them. On the basis that the partners work an 8-hour day, how much work would you expect them to generate from these efforts? What factors would you want to take into account to assess how realistic this is likely to be?

(You should get 4.875 days.)

8.10 Problems

Many of the exercises below use coins, dice and cards; this does not imply that this is the only use to which these ideas can be put, but they do provide a fairly simple mechanism for determining if you have absorbed the ideas of the previous section.

1 If you toss two coins, what is the probability of two heads?
2 If you toss two coins, what is the probability of a head followed by a tail?
3 If you toss three coins, what is the probability of a head followed by two tails?
4 If you toss three coins, what is the probability of two tails?
5 How do the probabilities in questions 1–4 change if the coins are biased so that the probability of a tail is 0.2?
6 If the experiment in question 1 were done 100 times, what is the expected number of times that two heads would occur?
7 If the experiment in question 5 were done 1000 times, what is the expected number for each of the outcomes?
8 Construct a grid to show the various possible total scores if two dice are thrown. From this, find the following probabilities:
    (a) a score of 3
    (b) a score of 9
    (c) a score of 7
    (d) a double being thrown.
9 When two dice are thrown, what is the probability of a 3 followed by a 5?
10 When a biased die (with probability of a six $= 0.25$) is thrown twice, what is the probability of not getting a six?
11 A die is thrown and a coin is tossed, what is the probability of a 1 and a tail?
12 In the experiment in question 11, what is the probability of a 1 or a tail or both?
13 If two coins are tossed and two dice are thrown, what is the probability of two heads and a double five?
14 From a normal pack of 52 cards, consisting of four suits each of 13 cards, on taking out one card, find the following probabilities:
    (a) an ace
    (b) a club
    (c) an ace or a club
    (d) the ace of clubs
    (e) a picture card (i.e. a jack, queen or king)
    (f) a red card
    (g) a red king
    (h) a red picture card.

15 What is the probability of a queen on either or both of two selections from a pack:
   (a) with replacement
   (b) without replacement?

16 What is the probability of a queen or a heart on either or both of two cards selected from a pack without replacement?

17 A company has 100 employees, of whom 40 are men. When questioned, 60 people agreed that they were happy in their work, and of these 30 were women. Find the probabilities that:
   (a) a man is unhappy;
   (b) a woman is happy.

18 A company has three offices, A, B and C which have 10, 30 and 40 people in them respectively. Company policy is that three people from each office are selected for promotion interviews. This year there will be three promotions: find the probability for a person in each office of being promoted.

19 A survey of 1000 people were asked which political party they would vote for. The respondents included 600 women. Response was as follows:
   Party A   350
   Party B   320
   Party C   300
   Don't know   30
   If sex and party are independent, find the expected numbers in each of the following categories:
   (a) men supporting Party A
   (d) women supporting Party C
   (c) women supporting either Party A or Party B.
   Would you be surprised to learn that 150 men supported Party C?

20 A company manufactures red and blue plastic pigs; 5% red and 10% blue are misshapen during manufacture. If the company makes equal numbers of each colour, what is the probability of selecting a misshapen pig on a random selection? How would the probability change if 60% of the pigs manufactured were blue? In a sample of three, what is the probability of getting two misshapen red pigs?

21 Two events are independent. There are three different outcomes (A, B, C) of the first event, with probabilities of 0.2, 0.3 and 0.5 respectively. There are two outcomes from the second event (X and Y), with probabilities of 0.1 and 0.9 respectively
   (a) Construct a probability tree to represent this situation.
   (b) What is the probability that the outcome of both events includes outcome Y?

22 A student group contains 40 men and 50 women. Of the men, 60% support longer opening hours for the Union Bar, whilst the corresponding figure is 80% for the women. What is the probability of:
   (a) selecting a woman
   (b) selecting someone against longer opening hours
   (c) selecting someone against longer opening hours, given you have selected a man.

23 If the probability of getting across a set of traffic lights without having to stop is 0.3, what is the probability of getting across a series of four sets of traffic lights without having to stop? What assumption have you made to do this calculation? Is it valid in this case?

24 Five candidates are called for an interview; three are men and two are women. All of the men and one of the women hold PhDs. If a candidate is selected at random, evaluate the following probabilities:
(a)   selecting a particular candidate
(b)   selecting a man
(c)   selecting someone with a PhD
(d)   selecting a man with a PhD
(e)   that the person selected is the woman who has a PhD.

25 A company produces plastic elephants in two colours for the novelty trade market. Production in the factory is on one of three machines; 10% is on machine A, 30% on machine B, and the remainder on machine C. Machine A's production consists of 40% blue elephants and 60% pink elephants. Machine B's production consists of 30% blue elephants and 70% pink elephants. Machine C's production has 80% pink elephants with the remainder being blue.
(a)   What proportion do blue elephants form of total production?
(d)   If a particular elephant is pink, what is the probability it was made by machine B?

26 In a certain year the general public is divided in their support of three political parties in the ratio $35:35:30$ (Con : Lab : Lib). The sex distribution of support for the parties is $60:40$ for Con, $50:50$ for Lab and $45:55$ for Lib, each ratio being male : female. When questioned on a current issue, male-Con were $80:20$ in favour, whilst female-Con were $70:30$ in favour. Male-Lab were $70:30$ in favour and female-Lab were $60:40$ in favour. Male-Lib were $20:80$ in favour and female-Lib were $30:70$ in favour.
(a)   Find the proportion of the sample who were male-Con in favour of the proposal.
(b)   If an individual is in favour of the proposition, what is the probability that they are a Con supporter.
(c)   If an individual is against the proposition, what is the probability the person is male.

27 A switch has a 0.9 probability of working effectively. If it does work, then the probability remains the same on the next occasion that it is used. If, however, it does not work effectively, then the probability it works on the next occasion is 0.1. Use a tree diagram to find the probability:
(a)   it works on three successive occasions
(b)   it fails, but then works on the next two occasions
(c)   on four occasions it works, fails, works and then fails.

28 The forecasted profits from a new project have been allocated the following probabilities:

Forecasted profit/(loss)*	Probability
(2000)	0.2
(1000)	0.3
500	0.1
1600	0.1
3000	0.3

*Note that brackets used in this way indicate negative values.

Calculate the expected profit and explain the meaning of this figure.

29 You have been given the probability distributions of possible profits from two projects, A and B:

Project A	
Probability	Profit (£)
0.6	4 000
0.4	8 000

Project B	
Probability	Profit (£)
0.2	2 000
0.3	2 500
0.3	4 000
0.1	8 000
0.1	12 000

Determine the expected profit from each project and state your project choice. What factors should a decision maker take into account when looking at the possible profits from these projects.

30 A small company has developed a new product for the electronics industry. The company believes that an advertising campaign costing £2000 would give the product a 70% chance of success. It estimates that a product with this advertising support would provide a return of £11 000 if successful and a return of £2000 if not successful. Past experience suggests that without advertising support a new product of this kind would have a 50% chance of success giving a return of £10 000 if successful and a return of £1500 if not successful.

Construct a decision tree and write a report advising the company on its best course of action.

31 In order to be able to meet an anticipated increase in demand for a basic industrial material a business is considering ways of developing the manufacturing process. After meeting current operating costs the business expects to make a net profit of £16 000 from its existing process when running at full capacity. All the data relates to the same period.

The Production Manager has listed the following possible courses of action.
(a) Continue to operate the existing plant and not expand to meet the new level of demand.
(b) Undertake a research programme which would cost £20 000 and has been given a 0.8 chance of success. If successful, a net profit of £60 000 is expected (before charging the research cost). If not successful a net profit of £5000 is expected.
(c) Undertake a less expensive research programme costing £8000 which has been given a 0.5 chance of success. If successful, a net profit of £5000 is expected and if not successful a net profit of £4000 is expected.

Present a decision tree. On the basis of this analysis determine the most profitable course of action. Comment on your findings.

32 A manufacturer has developed a new product and must decide whether to shelve the new product, sell the design for £60 000 or manufacture the new product. If the new product is successful then expected profits are £120 000 and if not successful £20 000.

The manufacturer also needs to decide whether or not to commission a market research survey. The cost of the survey would be £4000.

It is accepted by the manufacturer that new products of this kind generally have a 60% chance of success. However, it is also accepted that if the results of the market research survey are favourable then the chances of success increase to 90% and if not favourable decrease to 30%. In the past, 50% of market research surveys for this type of new product have given favourable results.

Construct a tree diagram to represent the decisions the manufacturer has to make. Using the criteria of expected monetary value, advise the manufacturer as to the best course of action.

33 Draw a directed graph for each of the transition matrices given below:

(a)

$$\begin{array}{cc} & \begin{array}{cc} E_1 & E_2 \end{array} \\ \begin{array}{c} E_1 \\ E_2 \end{array} & \begin{bmatrix} 0.7 & 0.3 \\ 0.5 & 0.5 \end{bmatrix} \end{array}$$

(b)

$$\begin{array}{c} & \begin{array}{ccc} E_1 & E_2 & E_3 \end{array} \\ \begin{array}{c} E_1 \\ E_2 \\ E_3 \end{array} & \begin{bmatrix} 1.0 & 0 & 0 \\ 0.1 & 0.8 & 0.1 \\ 0.1 & 0.8 & 0.1 \end{bmatrix} \end{array}$$

In case (b) what will eventually happen to the system?

34 For the transition matrix $\mathbf{P}$, given below, find $\mathbf{P}^2$, $\mathbf{P}^4$, $\mathbf{P}^8$.

$$\mathbf{P} = \begin{bmatrix} 0.2 & 0.8 \\ 0 & 1.0 \end{bmatrix}$$

35 A particular market has 100 small firms, 50 medium-sized firms and 10 large firms, and it has been noticed that the transition matrix from year to year is represented by the matrix given below:

	Size next period			
Size this period	Small	Medium	Large	Bankrupt
Small	0.7	0.2	0	0.1
Medium	0.3	0.5	0.1	0.1
Large	0	0.3	0.6	0.1
Bankrupt	0.1	0	0	0.9

No firms are bankrupt initially.

Find the expected number of firms in each category at the end of:

(a) 1 year
(b) 2 years
(c) 3 years
(d) 4 years.

36 A firm has five levels of intake of staff, and wishes to predict the way in which staff will progress through the various grades. Data has been collected to allow the construction of a transition matrix, including those who leave the firm. This is shown below:

$$
\begin{array}{c c}
 & \begin{array}{cccccc} 1 & 2 & 3 & 4 & 5 & \text{Left} \end{array} \\
\begin{array}{c} 1 \\ 2 \\ 3 \\ 4 \\ 5 \\ \text{Left} \end{array} &
\begin{bmatrix}
0.3 & 0.3 & 0.1 & 0.1 & 0.1 & 0.1 \\
0 & 0.3 & 0.2 & 0.2 & 0.2 & 0.1 \\
0 & 0 & 0.4 & 0.3 & 0.2 & 0.1 \\
0 & 0 & 0 & 0.5 & 0.4 & 0.1 \\
0 & 0 & 0 & 0 & 0.8 & 0.2 \\
0 & 0 & 0 & 0 & 0 & 1.0
\end{bmatrix}
\end{array}
$$

Six hundred and fifty people join in a particular year, at grades of 1–5 as set out below:

$$[200 \quad 150 \quad 150 \quad 100 \quad 50]$$

Use this information to predict:
(a)   the numbers in each grade after one year
(b)   the numbers in each grade after four years.
      The wages for each grade from 1 to 5 are given below:

$$[50 \quad 60 \quad 80 \quad 100 \quad 120]$$

Find the total wage bill of the company for this cohort:
(c)   as soon as they join
(d)   after 4 years with the company (assuming the same wage levels).

37 Within a company, an individual's probability of being promoted depends on whether or not they were promoted in the previous year. They may also be made redundant. This situation may be modelled by a Markov process, with the following transition matrix:

| | This period | | |
Last period	*Same job*	*Promotion*	*Redundant*
Same job	0.7	0.2	0.1
Promotion	0.9	0	0.1
Redundant	0	0	1.0

If, in the last time period, 100 people retained the same job, 5 were promoted and none were made redundant, find the expected numbers in each category after:
(a)   one period
(b)   two periods
(c)   three periods.

After three periods, the economy becomes more buoyant, and the threat of redundancy is lifted. The transition matrix now becomes:

$$\begin{array}{c@{\qquad}c@{\qquad}c} & \textit{Same job} & \textit{Promotion} \\[4pt] \begin{array}{r} \textit{Same job} \\ \textit{Promotion} \end{array} & \begin{bmatrix} 0.7 & 0.3 \\ 0.8 & 0.2 \end{bmatrix} \end{array}$$

For those still remaining with the company from the original cohort, find the expected numbers in each category after:

(d) the next period, i.e. period 4
(e) the following period, i.e. period 5
(f) period 6.

If you have access to the software, try the following questions:

A1 Use the Excel spreadsheet C8freqdef.xls (on the website) to find the probability of a head after 20 trials. Record your answer. Press F9 and again note the answer. Continue until you have 20 answers, and then find the average of your answers. What does this tell you? (see Section 10.4).

A2 As a comparator, repeat the exercise A1, but note the answer after only five trials. How does your final answer differ from your previous one?

A3 If you have access to INSIGHT, use the TREE.XLS spreadsheet to suggest the rational decision for the following problem. A business believes that the market for its product is very buoyant and will expand. In order to take advantage of the expectation, the company has to decide whether or not to expand is production capabilities and has identified that it could expand by 30% at a cost of £10m, expand by 10% at a cost of £2m or not expand at all (zero cost). Market analysts suggest that the probability of an expansion of 40% is 0.1, of an expansion of 20% is 0.3, an expansion by 10% is 0.4, or else it will remain constant. The following table gives the likely increase in profit to the company for various combinations of company expansion and market growth:

| Market | Company growth | | |
Growth	30%	10%	0%
+40%	+£50m	+£30m	+£4m
+20%	+£30m	+£30m	+£4m
+10%	+£12m	+£12m	+£4m
+0%	+£2m	+£2m	+£0m

A4 A new product can be launched, market tested, the design sold on (for £25m), or the company can delay the decision for a year. Given the following data, suggest what the rational decision would be.

Product is launched at a cost of £4m:

Outcome	Probability	Profit increase
Big success	0.1	+£100m
Success	0.3	+£30m
OK	0.4	+£5m
Fails	0.2	−£5m

Product is market tested at cost of £0.5m:

Outcome	Probability
Pass	0.8
Fail	0.2

The company can then decide whether to launch, wait a year (at a cost of £0.2m) or sell the design for £35m. If they launch (cost = £4m), then the following table applies:

Outcome	Probability	Profit increase
Big success	0.1	+£100m
Success	0.5	+£30m
OK	0.3	+£5m
Fails	0.1	−£5m

If they launch after a year (cost £4m), the following table applies:

Outcome	Probability	Profit increase
Big success	0.15	+£90m
Success	0.55	+£25m
OK	0.25	+£10m
Fails	0.05	−£5m

Were the company to wait for a year (cost = £0.2m), they could then sell the design for £30m or launch at a cost of £4m with the following probabilities and outcomes:

Outcome	Probability	Profit increase
Big success	0.1	+£90m
Success	0.4	+£25m
OK	0.4	+£10m
Fails	0.1	−£5m

A5 If you have access to INSIGHT, use the MARKOV program to solve the following Markov chain:

	From		
To	A	B	C
A	0.1	0.4	0.5
B	0.5	0.1	0.4
C	0.4	0.5	0.1

What is the position after four time periods? What will be the equilibrium position?

A6 If you have access to INSIGHT, use the MARKOV program to solve the following Markov Chain:

	From			
To	A	B	C	D
A	0.9	0	0	0
B	0	0.9	0.1	0
C	0.1	0.1	0.8	0
D	0	0	0.1	1.0

What is the position after four time periods? What will be the equilibrium position?

9

Discrete probability distributions

The previous chapter on basic probability concepts has introduced you to some useful ideas, but the problem about using them to assess business situations is that you have to start from scratch with each new problem. Complex problems will take some time to work through, particularly if there are many ways of achieving a 'desired' outcome, since each one will have to be evaluated (cf. probability trees). One way of summarizing probabilities is by using known probability distributions, and fortunately, a fairly small number of these are able to describe a surprisingly wide range of situations. This means that if you can understand these distributions and when to apply each, then you can deal with the majority of business related probability problems. (There will always, however, be a few situations that have to be analysed from first principles!)

Objectives

After working through this chapter, you will be able to:

- appreciate the importance of probability distributions
- describe a uniform distribution
- recognize a binomial situation
- calculate probabilities for a binomial
- calculate probabilities for a Poisson.

Probability distributions form a frame of reference that summarizes the theoretical position. In fact, for some distributions, there are tables of values, so that all you need to do is look up the required probability, rather than do the calculations. Even though the calculations can be simplified considerably by the use of tables or computers, there is no substitute for a basic understanding of the principles of probability. The sort of experience you have developed if you have worked your way through Chapter 8 will help you in deciding which probability distribution is most appropriate in each situation.

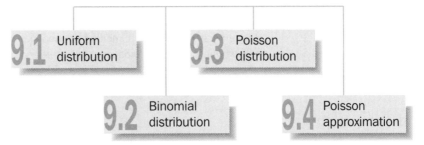

Figure 9.1 Structure of Chapter 9.

We cannot deal with every possible probability distribution, and so have chosen the most frequently used ones. To simplify dealing with these, we will look at discrete distributions in this chapter and postpone consideration of continuous distributions until Chapter 10. (For notes on the distinctions between discrete and continuous data, see Chapter 1.)

9.1 Uniform distribution

The simplest possible probability distribution is one which we have already met. This is the position we assume when tossing a 'fair' coin. There are two possible outcomes, and each has a probability of 0.5 A definition might be:

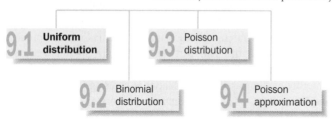

If we generalize this situation we might define a discrete uniform distribution as one where there are a given number of possible outcomes, and each has the same probability. Putting this another way, if there are n outcomes, then the probability of any particular outcome will be $1/n$.

In fact, we have already met the uniform distribution in Chapter 3 where we used random number tables (which are based on the premise that each digit has an equal probability of occurrence) and in Chapter 8 where we used coins and dice (each of which assumed that each outcome was equally likely).

Thus the theoretical distribution for a set of random numbers is such that each of the digits 0, 1, 2, 3, 4, 5, 6, 7, 8 and 9 occurs an equal number of times. This is illustrated in Figure 9.2. If you were to look through a set of random number tables (such as Appendix K) you would see that each digit occurs an approximately equal number of times, provided that you take a big enough sample.

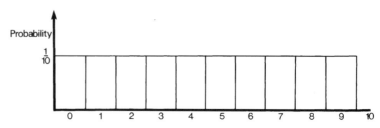

Figure 9.2

Exercise

> Take a page of random numbers from Appendix K and create a frequency distribution. From this produce a bar chart and compare your result with Figure 9.2

Exercise

> Most spreadsheets will allow you to create random numbers, and you can use these both to illustrate the concept of the Uniform Distribution and to

check on the numbers produced if you create a column of random numbers and then draw a bar chart of the results. Try this for a sample of 10, then 20 then 200 random numbers.

You could download the spreadsheet probsim.xls from the website and then use the F9 key to recalculate the answers.

What you should notice is that typically you do not get the theoretical distribution shown in Figure 9.2, but that, as the sample size increases, the histogram looks more and more like this theoretical diagram.

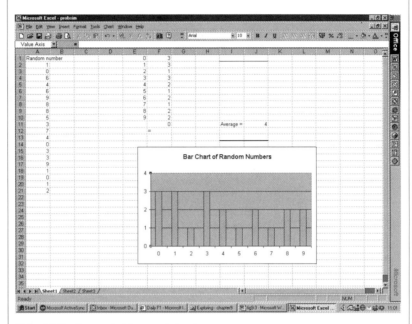

Figure 9.3 Screenshot from probsim.xls.

Add to the spreadsheet so that you have a bar chart of a sample size of 1000 and comment on the shape achieved.

The uniform distribution is useful when we are trying to compare what has actually happened with our preconception that all of the outcomes were equally likely. For example, if you toss a coin, you presume that heads and tails are equally likely (i.e. you have an unbiased coin). We could check a particular coin by tossing it many times and counting the number of heads and tails. If it turned out that we got an equal number of each, then it would be reasonable to conclude that the coin was unbiased. If we did not get exactly equal numbers, then it would be necessary to use some form of significance test to determine whether or not the coin was biased (see Chapter 13).

You might also like to note that if we add together the results from two uniform distributions, then we do not get a uniform distribution. Think back to the case of the two die. Each can take values from 1 to 6, and each has a uniform distribution. If we add the values on the two die together, then the results can vary from 2 to 12, but the distribution is far from uniform; see Figure 9.4.

The uniform distribution is also used within simulation models, and is discussed again in Chapter 22.

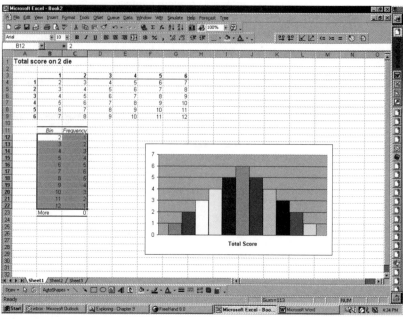

Figure 9.4 Total score on two die.

9.2 Binomial distribution

In many cases, the variable of interest is dichotomous, has two parts or two outcomes. Examples include questions that only allow a 'YES' or 'NO' answer, or a classification such as male or female, or recording a component as defective or not defective. If the outcomes are also independent, e.g. one respondent giving a YES answer does not influence the answer of the next respondent, then the variable is binomial.

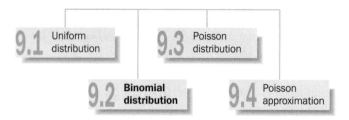

> If we generalize this situation we might define a binomial distribution as one where there are only two defined outcomes of any particular trial *and* the probability of each outcome remains constant from trial to trial (i.e. the events are independent).

Consider the situation of items coming off the end of a production line, some of which are defective. If the proportion of defective items is 10% of the flow of items, we can regard the selection of a small sample as consisting of independent selections, and thus the probability of selecting a defective item will remain constant as $P(\text{defective}) = 0.1$. (Note that if there were a small, fixed number of items in total and selection were without replacement, then we would have a situation of conditional probabilities, and $P(\text{defective})$ would not remain constant.) Samples will be selected from production lines to monitor quality, and thus we will be interested in the number of defective items in our sample. However, the number of defectives may be related to the size of the sample and to whether or not the process is working as expected. This type of system was often used in quality control procedures. Many organizations have attempted to move to quality assurance, where the checking is performed at earlier and earlier stages, and an attempt is made to ensure 100% of production meets the standards (zero defectives).

Exercise

> Which would lead to most concern, one defective in a sample of 10, or 10 defectives in a sample of 100?

We shall return to this exercise to compare your impressions with the theoretical results of the appropriate probability model, but first we will consider much smaller samples.

For a sample of size one, the probability is that the item selected is defective is 0.1 and the probability that it is not defective is 0.9.

For a sample of two, for each item the probabilities are as above, but we are now interested in the sample as a whole. There are four possibilities:

1 both items are defective
2 the first is defective, the second is OK
3 the first is OK, the second is defective
4 both items are OK.

Since the two selections are independent, we can multiply the unconditional probabilities together, thus:

1 $P(\text{both defective}) = (0.1)(0.1) \qquad = 0.01$
2 $P(\text{1st defective, 2nd OK}) = (0.1)(0.9) = 0.09$
3 $P(\text{1st OK, 2nd defective}) = (0.9)(0.1) = 0.09$
4 $P(\text{both OK}) = (0.9)(0.9) \qquad = \underline{0.81}$
$$1.00$$

However, (2) and (3) both represent one defective in the sample of two. If we are not interested in the order in which the events occur, then:

$$P(\text{2 defective}) \quad = 0.01$$
$$P(\text{1 defective}) \quad = 0.18$$
$$P(\text{no defectives}) = \underline{0.81}$$
$$1.00$$

If p = probability of a defective, then $q = (1 - p)$ is the probability of a non-defective item, and we have:

$$P(\text{2 defective}) = p^2$$
$$P(\text{1 defective}) = 2pq$$
$$P(\text{0 defective}) = q^2$$

This situation is illustrated in Figure 9.5 (where S = defective and F = OK)..

For a sample of three, the following possibilities exist (where def. is an abbreviation for defective):

Figure 9.5

1 3 defective	$P(3$ def.$)$ $= (0.1)^3$	$= 0.001$
2 1st, 2nd defective; 3rd OK	$P(1,2$ def.; 3 OK$) = (0.1)^2(0.9)$	$= 0.009$
3 1st, 3rd defective; 2nd OK	$P(1,3$ def.; 2 OK$) = (0.1)(0.9)(0.1)$	$= 0.009$
4 2nd, 3rd defective; 1st OK	$P(2,3$ def.; 1 OK$) = (0.9)(0.1)^2$	$= 0.009$
5 1st defective; 2nd, 3rd OK	$P(1$ def.; 2,3 OK$) = (0.1)(0.9)^2$	$= 0.081$
6 2nd defective; 1st, 3rd OK	$P(2$ def.; 1,3 OK$) = (0.9)(0.1)(0.9)$	$= 0.081$
7 3rd defective; 1st, 2nd OK	$P(3$ def.; 1,2 OK$) = (0.9)^2(0.1)$	$= 0.081$
8 all OK	$P(3$ OK$)$ $= (0.9)^3$	$= 0.729$
		$\overline{1.000}$

Again, we can combine events, since order is not important; (2), (3) and (4) represent two defectives; (5), (6) and (7) represent one defective. Thus:

$$P(3 \text{ defectives}) = 0.001 = p^3$$
$$P(2 \text{ defectives}) = 0.027 = 3p^2q$$
$$P(1 \text{ defective}) = 0.243 = 3pq^2$$
$$P(0 \text{ defectives}) = 0.279 = q^3$$
$$\overline{1.000}$$

This situation is illustrated in Figure 9.6.

We could continue with the procedure, looking at sample sizes of four, five and so on, but a pattern is already emerging from the results given above. At the extreme, the probability that all of the items are defective is p^2 or p^3, and for a sample of size n, this will be p^n. With other outcomes, these consist of a series of possibilities, each with the same probability, and these are then combined. For example, in a sample of 10, the probability of four defectives would consist of a series of outcomes, each of which would have a probability of $(0.1)^4(0.9)^6 = p^4q^6 = 0.0000531$. (Note that four defective items means that there are also six items which are OK, since the sample size is 10.)

Continuing with the diagrammatic approach, consider Figures 9.6 and 9.7. These illustrate the binomial position with four and five trials, respectively. Looking at the shaded squares you should be able to identify a pattern emerging from the series of diagrams from 9.4 to 9.8.

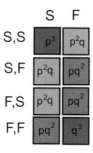

Figure 9.6 Binomial properties with three trials.

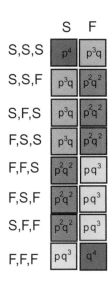

Figure 9.7 Bionomial properties with four trials.

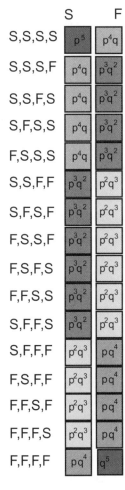

Figure 9.8 Binomial probabilities with five trials.

Where we have large numbers of trials the question that needs to be answered now is:

'How many of the outcomes from a set of trials have the same number of defective items?'

To answer this question we will use the idea of **combinations**. The number of combinations of r defective items in a sample of n items is given by:

$$^nC_r = \binom{n}{r} = \frac{n!}{r!(n-r)!}$$

where ^nC_r and $\binom{n}{r}$ are the two most commonly used notations for combinations, and $n!$ is the factorial of n. This means, n times $(n-1)$ times $(n-2)$, etc., until 1 is reached.

For example:

2! = 2 × 1 = 2
3! = 3 × 2 × 1 = 6
4! = 4 × 3 × 2 × 1 = 24
10! = 10 × 9 × 8 × 7 × 6 × 5 × 4 × 3 × 2 × 1 = 3 628 800

However, 0! = 1.

Note also that these factorials can be written as:

10! = 10 × 9 × 8 × 7!
    = 10 × 9!
    = 10 × 9 × 8 × 7 × 6 × 5 × 4!

since this will help when calculating the number of combinations. Returning now to the sample of 10, the number of ways of getting 4 defective items will be when $n = 10$ and $r = 4$:

$$^{10}C_4 = \binom{10}{4} = \frac{10!}{4!(10-4)!} = \frac{10!}{4!\,6!}$$

If we note the highest factorial in the denominator is 6, we have

$$\binom{10}{4} = \frac{10 \times 9 \times 8 \times 7 \times 6!}{4 \times 3 \times 2 \times 1 \times 6!} = \frac{10 \times 9 \times 8 \times 7}{4 \times 3 \times 2 \times 1} = 210$$

There are 210 different ways of getting four defectives in a sample of 10, and thus

$$P(4 \text{ defective in } 10) = 210(0.1)^4(0.9)^6 = 0.011\,151$$

For a small sample it may be preferable to use Pascal's triangle to find the number of combinations, or the coefficients for each term in the Binomial probability model. This begins with the three 1s arranged thus:

$$1$$
$$1 \quad 1$$

To find the next line, which will have three terms, the first and last will be 1s, the middle term will be *the sum of the two terms just above it*:

$$1$$
$$1 \quad 1$$
$$(1+1)$$
$$1 \quad 2 \quad 1$$

and this will apply to a sample size of two.

This process continues, to give the next line

$$(1+2)(2+1)$$
$$1 \quad 3 \quad 3 \quad 1$$

which will apply for a sample of size three. The process can continue until the desired sample size is reached.

													Sample size:
						1							
					1		1						
				1		2		1					2
			1		3		3		1				3
		1		4		6		4		1			4
	1		5		10		10		5		1		5
1		6		15		20		15		6		1	6
1	7	21	35	35	21	7	1						7
1	8	28	56	70	56	28	8	1					8
1	9	36	84	126	126	84	36	9	1				9
1	10	45	120	210	252	210	120	45	10	1			10
1	11	55	165	330	462	462	330	165	55	11	1		11
1	12	66	220	495	792	924	792	495	220	66	12	1	12

Example

The number of different combinations of four defective items from a sample of 11 is the fifth term from the left in the corresponding row of the triangle, i.e. 330.

Alternatively, the binomial coefficient is $^{11}C_4$ given by

$$\binom{11}{4} = \frac{11!}{4! \, 7!} = 330$$

Looking back to the probabilities of different numbers of defectives in a sample of three, these can now be rewritten as follows:

$$P(3 \text{ defectives}) = p^3$$

$$P(2 \text{ defectives}) = \binom{3}{2} p^2 q$$

$$P(1 \text{ defective}) = \binom{3}{1} pq^2$$

$$P(0 \text{ defectives}) = q^3$$

The general formula for a Binomial probability will be:

$$P(r \text{ items in a sample of } n) = \binom{n}{r} p^r q^{(n-r)}$$

Example

What is the probability of more than three defectives in a sample of 12 items, if the probability of a defective item is 0.2?

The required probability is:

$$P(4) + P(5) + P(6) + P(7) + P(8) + P(9) + P(10) + P(11) + P(12)$$

But this may be written as:

$$1 - [P(0) + P(1) + P(2) + P(3)] \qquad - \textit{from the basic rules in Chapter 8}$$

which will considerable simplify the calculation.

We have: $n = 12; \ p = 0.2;$ and $q = 1 - p = 1 - 0.2 = 0.8;$ so

$P(0)$	$= q^{12}$	$= (0.8)^{12}$	$= 0.0687195$
$P(1)$	$= 12pq^{11}$	$= 12(0.2)(0.8)^{11}$	$= 0.2061584$
$P(2)$	$= 66p^2q^{10}$	$= 66(0.2)^2(0.8)^{10}$	$= 0.2834678$
$P(3)$	$= 220p^3q^9$	$= 220(0.2)^3(0.8)^9$	$= 0.2362232$
			0.7945689

therefore the required probability is:

$$1 - 0.7945689 = 0.2054311$$

or more simply 0.21.

An alternative to this calculation would be to use tables of the cumulative binomial distribution (see Appendix A). For example, if we require the probability of five or more items in a sample of 10, when $p = 0.20$, from the table we find that $P(5 \text{ or more}) = 0.0328$.

For the same sample, if the required probability were for five or fewer, then we would look up $P(6 \text{ or more}) = 0.0064$ and subtract this from 1:

$$P(5 \text{ or less}) = 1 - 0.0064 = 0.9936$$

Returning now to the problem of whether it is a greater matter of concern to find more than one defect in a sample of 10, or more than 10 defectives in a sample of 100, we see that:

for a sample of 10 with $p = 0.1$, $P(2 \text{ or more}) = 0.2639$;
for a sample of 100 with $p = 0.1$, $P(11 \text{ or more }) = 0.4168$.

What conclusion would you draw from these figures?

Exercise

Determine these probabilities.

Thus, if the process is working as was proposed, giving 10% of items defective in some way, then the probability of finding two or more in a sample of 10 is very much lower than the probability of finding 11 or more in a sample of 100, i.e. in both cases finding more than the expected number in a sample. However, from the note on expectations in Chapter 8, we know that we are unlikely always to get the expected number in a particular sample selection. Even so, the small sample result would suggest more strongly that something was wrong with the process and would therefore be cause for more concern.

While it is important to see the development of the binomial probabilities, in practice you would use tables or a computer. We have built a small spreadsheet which will allow you to find probabilities for values of n up to 20 and draw a histogram of the whole distribution.

Use this spreadsheet (labelled binomial.xls at the website) to confirm the results shown in this section.

Try putting in different values of n and p and watching what happens to the bar chart. You should be able to draw some conclusions from this process about the shape of the binomial distribution.

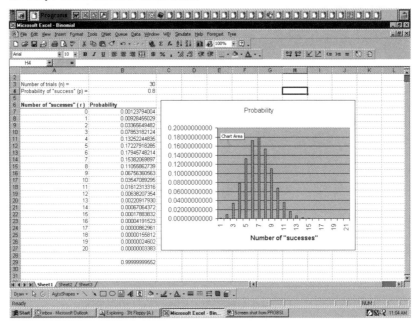

Figure 9.9 Screenshot from binomial.xls.

For a binomial distribution, the **mean** can be shown to be np and the **variance** to be npq. Thus, for a sample of size 10 with a probability $p = 0.3$, the average, or **expected number**, of items per sample with the characteristic will be

$$np = 10 \times 0.3 = 3$$

the variance will be

$$npq = 10 \times 0.3 \times 0.7 = 2.1$$

Case study

The partners at Carroll Imitations have, unknowingly, been sent a faulty batch of paper. The effect of this is that one in every 20 of the sheets (on average) has a fault such that ink will just run off. This, of course, ruins the drawing or illustration which is being prepared. If Paul, one of the partners, starts to prepare a set of 10 illustrations (one per sheet), what is the probability that none of the drawings will be ruined by the faulty paper? What is the probability that more than one drawing will be ruined?

(Answer: $P(0) = 0.5987$; $P(>1) = 0.0861$.)

The binomial distribution is a very powerful tool in looking at probabilistic situations, since, even where there are several outcomes, it is often possible to group these together into 'good' and 'bad' or some other appropriate categorization. Making use of the binomial distribution is limited by the two restrictions set out at the start of this section – there are two outcomes and the probability of each remains constant from trial to trial (i.e. events are independent). We have already suggested a way around the first of these, but the criterion of independence is a necessary requirement, which is often rather more difficult to identify in practice.

9.3 Poisson distribution

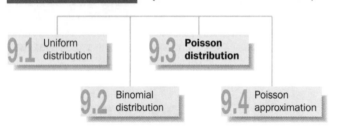

The binomial model is successful at modelling a very wide range of business and production situations. However, where there are very large numbers of trials involved, and the probability of 'success' is very small, then the Poisson distribution will give a better representation of the situation. An example might be where a very large number of people are potential buyers of a product, but the chance of an individual actually purchasing it are extremely small. A second example might be where lots of components are packed into boxes for delivery, and there is a small chance of each being faulty. It seems to work well in situations where we are looking at the time taken to complete a task, the majority of the people completing quite quickly, but a few taking a very long time, and maybe some never competing it.

> If we generalize this situation we might define a Poisson distribution as one in which each trial is independent so that the probability of 'success' remains constant and the number of trials is large. In fact the Poisson distribution is completely defined by its **average**.

The model works with the expected or average number of occurrences; if this is not given it can be found as np.

The probability model is: $P(x) = (\lambda^x e^{-\lambda}/x!)$, where lambda ($\lambda$) is the average number of times a characteristic occurs and x is the number of occurrences (x may be any integer from 0 to infinity). For example, if a company receives an average of three calls per 5-minute period of the working day, then we can calculate the probabilities of receiving a particular number of calls in a randomly selected 5-minute period.

The average number of calls, $\lambda = 3$ and $e^{-3} = 0.0498$, so

$$P(0 \text{ calls}) = \frac{3^0(0.0498)}{0!} = 0.0498$$

$$P(1 \text{ call}) = \frac{3^1(0.0498)}{1!} = 0.1494$$

$$P(2 \text{ calls}) = \frac{3^2(0.0498)}{2!} = 0.2241$$

$$P(3 \text{ calls}) = \frac{3^3(0.0498)}{3!} = 0.2241$$

$$P(4 \text{ calls}) = \frac{3^4(0.0498)}{4!} = 0.168075$$

As you may have noticed, there is a **recursive relationship** between any two consecutive probabilities , such that:

$$P(4 \text{ calls}) = \frac{3}{4} \times P(3 \text{ calls}) = \frac{3}{4} \times 0.2241 = 0.168075$$

Or, more generally

$$P(N \text{ calls}) = (\lambda/N) \times P(N - 1 \text{ calls})$$

If the company discussed above has only four telephone lines, and calls last for at least 5 minutes, then there is a probability of

$$P(\text{no calls}) + P(1 \text{ call}) + P(2 \text{ calls}) + P(3 \text{ calls}) + P(4 \text{ calls})$$
$$= 0.0498 + 0.494 + 0.2241 + 0.2241 + 0.168075$$
$$= 0.815475$$

or approximately 0.815 of the switchboard being able to handle all incoming calls. Put another way, you would expect the switchboard to be sufficient for 81.5% of the time, but for callers to be unable to make the connection during 18.5% of the time. This raises the question of whether another line should be installed.

$$P(5 \text{ calls}) = \frac{3}{5} \times P(4 \text{ calls}) = 0.1008$$

The switchboard would now be in a position to handle all calls for an extra 10% of the time, but whether or not this is worth while would depend on the likely extra profits that this would create, against the cost of installation and running an extra telephone line.

Again, there is an alternative to calculating all of the probabilities each time, by using tables of cumulative Poisson probabilities (see Appendix B). For example, if the average number of faults found on a new car at its pre-delivery inspection is five, then from tables we can find that

(a) P(3 or more) = 0.8753
(b) P(5 or more) = 0.5595
(c) P(10 or more) = 0.0318

and, as before, these can be manipulated. From (a), we see that $1 - 0.8753 = 0.1247$ so that the probability of a car having fewer than three faults is 0.1247; or we would expect only 12.47% of cars that have pre-delivery inspections to have fewer than three faults.

For the Poisson distribution it can be shown that the mean and variance are both equal to λ.

Visit the website

As with the binomial, whilst seeing the development of the ideas is important, if you want to do things with the Poisson distribution you will use tables or a computer. If you click on the second tab of the spreadsheet (labelled binomial.xls at the website) you will find the various Poisson probabilities and a bar chart.

Use this spreadsheet (see Figure 9.10) to investigate the shape of the Poisson distribution as the value of the average changes. Are there any conclusions that you can draw from this?

The Poisson distribution has been successfully used where very small probabilities are encountered in relatively large sets of trials or batches, for example more than one defective item in a batch of 1000, where the probability of a

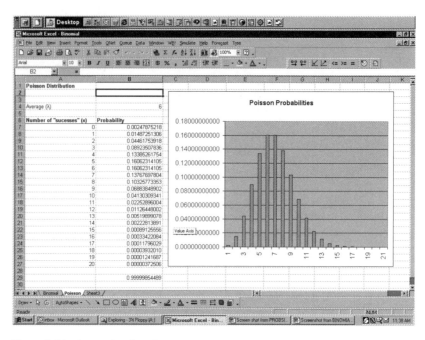

Figure 9.10   Screenshot from Poisson in binomial.xls.

defect is 0.001 – say something like circuit boards. The distribution also plays a very important role in modelling how a queue functions. We will return to this use in Chapter 21.

Case study

Carroll Imitations allow customers a 30-day payment period for outstanding trade credit. However, not all customers abide by this ruling and previous trading experience suggests that the average time over this limit is, in fact, 6 days. What percentage of customers are likely to take over 40 days to pay?

(Answer: from table (Appendix B) $P(>10)=0.0839$, or 8.39%.)

9.4 Poisson approximation to the binomial

Both distributions are discrete probability models, but for many values of $\lambda = np$ the Poisson model is considerably more skewed than the binomial. However, for small values of p (less than 0.1), and large values of n, it may be easier to use a Poisson distribution. (Note that if p is very small, $(1 - p)$ will be close to 1 and hence $np(1 - p) \approx np = \lambda$ which is both the mean and the variance of the Poisson distribution.)

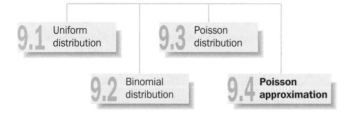

Example

If the probability of a fault in a piece of precision equipment is 0.0001, and each completed machine has 10 000 components, what is the probability of there being two or more faults?

(a) Using Poisson distribution:

$\lambda = np = 10\,000 \times 0.0001 = 1$

$P(0) = e^{-1} = 0.3679$

$P(1) = e^{-1} = 0.3679$

$P(0) + P(1) = 0.7358$

Therefore $P(2 \text{ or more}) = 1 - 0.7358 = 0.2642$

(b) Using binomial distribution:

$P(0) = (0.9999)^{10\,000} = 0.3679$

$P(1) = 10\,000\,(0.00001)(0.9999)^{9999} = 0.3679$

$P(0) + P(1) = 0.7358$

Therefore $P(2 \text{ or more}) = 1 - 0.7358 = 0.2642$

Comparing these two answers it is suggested that method (a) is very much easier to work with than method (b).

9.5 Conclusions

Many business situations involve discrete events which are stochastic in nature. Rather than seeing each of these as unique problems which require time-consuming solutions, we can often use well-known distributions to save considerable time and effort. In this chapter we have looked at three of the most commonly met discrete distributions – uniform, binomial and Poisson – and illustrated the type of situation where they can be applied.

As with many areas of statistics and modelling, even if the theoretical distribution is not an exact match to the reality, it may be close enough to allow us to draw some conclusions, even if we have to impose some restrictions, for example 'this only works with sample sizes below a certain number'.

When looking at a problem for the first time, try to identify the parameters (these values are always given in some form in traditional examination questions) and the assumptions that you may need to make, e.g. events are independent. If your problem does match a known distribution, then clearly you have at least one well-established method of solution which may only require reference to statistical tables.

As has been suggested several times in this chapter, the shape of a distribution can be very helpful in identifying the approach to adopt. Most of the time this sort of pattern recognition will be the best way to proceed, but it is not something which should be done blindly. There is a danger of only ever applying vertical thinking and thus getting trapped into previously known solutions, and sometimes it will be necessary to be more creative and apply lateral thinking to solve a problem.

9.6 Problems

1   Evaluate the following expressions:

(a) $\begin{pmatrix} 3 \\ 1 \end{pmatrix}$ (b) $\begin{pmatrix} 10 \\ 3 \end{pmatrix}$ and $\begin{pmatrix} 10 \\ 7 \end{pmatrix}$ (c) $\begin{pmatrix} 20 \\ 6 \end{pmatrix}$ and $\begin{pmatrix} 20 \\ 0 \end{pmatrix}$

(d) $\begin{pmatrix} 10 \\ 2 \end{pmatrix}$; $\begin{pmatrix} 10 \\ 1 \end{pmatrix}$ and $\begin{pmatrix} 10 \\ 0 \end{pmatrix}$ (e) $\begin{pmatrix} 52 \\ 13 \end{pmatrix}$

Part (e) represents the number of different possible hands of 13 cards that could be dealt with a standard pack of playing cards.

2 Given the answer to question 1(e), what is the probability of getting a complete suit of cards in one of the four hands dealt from a standard pack of cards?

3 A binomial model has $n = 4$ and $p = 0.6$. Find the probabilities of each of the five possible outcomes (i.e. $P(0)$–$P(4)$). Construct a bar chart of this data.

4 Attendance at a cinema has been analysed, and shows that audiences consist of 60% men and 40% women for a particular film. If a random sample of six people were selected from the audience during a performance, find the following probabilities:
(a) all women are selected
(b) three men are selected
(c) fewer than three women are selected.

5 How would the probabilities in question 4 change if the sample size were eight?

6 A quality control system selects a sample of three items from a production line. If one or more is defective, a second sample is taken (also of size three), and if one or more of these is defective then the whole production line is stopped. Given that the probability of a defective item is 0.05, what is the probability that the second sample is taken? What is the probability that the production line is stopped?

7 The probability that an invoice contains a mistake is 0.1. In an audit a sample of 12 invoices are chosen from one department; what is the probability that fewer than two incorrect invoices are found?

8 Find each of the Poisson probabilities from $P(0)$ to $P(5)$ for a distribution with an average of 2. Construct a bar chart of this part of the distribution.

9 For a Poisson distribution with an average of 2, find the probability of $P(x > 4)$ and $P(x > 5)$.

10 Find the probabilities $P(0)$ to $P(4)$ for a Binomial distribution with $n = 10\,000$ and $p = 0.000\,15$.

11 The number of accidents per day on a particular stretch of motorway follows a Poisson distribution with a mean of one. Find the probabilities of 0, 1, 2, 3, 4 or more accidents on this stretch of motorway on a particular day. Find the expected number of days with 0, 1, 2, 3, 4 or more accidents in a one-year period (assuming 365 days per year). If the average cost of policing an accident is £1000, find the expected cost of policing accidents on this stretch of motorway for a year.

12 The number of train passengers who fail to pay for their tickets in a certain region has a Poisson distribution with a mean of four per train. If there are 4800 trains per month in the region, find the expected number of trains with more than three non-fare-paying passengers. Calculate the average cost to the train company during a month if the average cost of a ticket is £7.84.

13 A man has four cars for hire. The average demand on a weekday is for two cars. Assuming 312 weekdays per year, obtain the theoretical frequency distribution of the number of cars demanded during a weekday. Hence estimate to the nearest whole number, the number of days on which demand exceeds supply. (Assume demand does not surpass nine cars per day.) Would you suggest that the man buys another car?

14 (a) Items are packed into boxes of 1000, and each item has a probability of 0.001 of having some type of fault. What is the probability that a box will contain fewer than three defective items?

(b) If the company sells 100 000 boxes per year and guarantees fewer than three defectives per box, what is the expected number of guarantee claims?

(c) Replacement of a box returned under the guarantee costs £150. What is the expected cost of guarantee claims?

(d) Boxes sell at £100 but cost £60 to produce and distribute. What is the company's expected profit for sales of boxes?

15 Twenty per cent of the population are thought to be carriers of a certain disease, although they themselves may show no symptoms. If this is true, evaluate the following probabilities for a sample of five people drawn at random from the population:

(a) that none are carriers

(b) that all five are carriers

(c) that fewer than two are carriers.

10 The Normal distribution

In developing the ideas of probability we have, so far, restricted our analysis to dealing with discrete data only. Such a restriction makes considerable sense when you first approach this subject area, since we all find it rather easier to think in terms of individual events and discrete outcomes. Also, of course, many business problems involve discrete events. However, the concept of probability can also be successfully applied to variables which are continuous, or can be treated as continuous (e.g. money). As with the discrete case, there are a large number of different continuous probability distributions. Several of these have been identified, and will be used in other parts of this book, but for most statistical work there is one particular distribution which stands out as being the most useful in a very wide variety of circumstances. This is the Normal distribution. The importance of this distribution, both to the use of statistics in practical situations, and the development of the theory, is difficult to overstate.

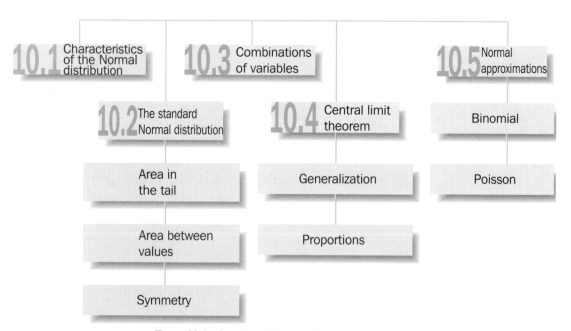

Figure 10.1 Structure of Chapter 10.

Objectives

After working through this chapter, you will be able to:

- describe the Normal distribution graphically and with parameters
- state the conditions which give rise to a Normal distribution
- calculate standard values
- find areas under the normal curve
- apply the normal distribution to discrete data
- understand the central limit theorem.

**10.1 Characteristics
of the Normal
distribution**

Although the normal distribution does occur in many situations and is probably the most widely used statistical distribution, the word 'normal' does not imply any sort of moral meaning or value judgement. What we have here is a distribution that is symmetrical about its mean; see Figure 10.2.

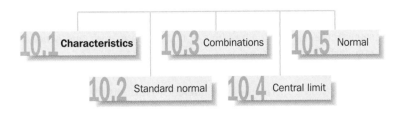

To generalize, when a variable is continuous, and its value is affected by a large number of chance factors, none of which predominates, then it will frequently have a Normal distribution.

An example would be the weights of male adults. Their weight is affected by genetic factors inherited from their parents, their diet, their age, their build, the amount of exercise taken, diseases they may have or have had, where they live, and many other things. Therefore, we would expect the distribution of male adult weights to be approximately a normal distribution. Also, however, within a business context, manufactured items are affected by the quality of the raw materials used, the sources used, the types of machines, their ages, the wear on the tools, and so on; so we might also expect that certain dimensions on these products will also have a normal distribution. Similarly, peoples' opinions reflect their age, culture, education, political affiliations, etc., and *maybe* opinions could be normally distributed unless something has happened to change this situation, for example, a marketing campaign.

Normal distributions come in many shapes and sizes, some will be relatively 'flat', and have a high standard deviation, whilst others will appear 'tall and thin' and have a relatively small standard deviation. (The shape you see, of course, will

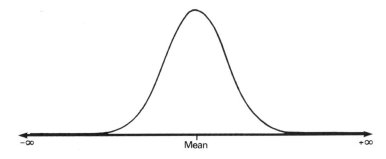

Figure 10.2 A Normal distribution.

also depend on the scale you use to draw the graph!) These distributions are often summarized by their mean and variance (usually labelled μ and σ^2, respectively). If a variable X has a Normal distribution, this may be written as $X \sim N(\mu, \sigma^2)$. Normal distributions are characterized particularly by the areas in various sectors of the distribution. If these areas are considered as a proportion of the total area under the distribution curve, then they may also be considered as the probabilities of obtaining a value from the distribution in that sector.

Theoretically, to find the area under the distribution curve in the sector less than some value x we should need to evaluate the integral

$$\int_{-\infty}^{x} \frac{1}{\sigma\sqrt{2\Pi}} \exp\left[-\frac{(x-\mu)^2}{2\sigma^2} \right] dx$$

which tends to 1 as x tends to infinity.

If it were necessary to perform this bit of mathematics every time that you wanted a probability, then the Normal distribution would not be widely used!

Fortunately there is an **easier method** of finding these areas, and hence the associated probabilities.

10.2 The standard Normal distribution

As we have said earlier, there are many different Normal distributions, but they do all have certain criteria in common, and it is these criteria which form the basis for finding the areas and probabilities quickly and easily. Looking at the horizontal scale of the graph of a Normal distribution (like Figure 10.2), we can see that, at least in theory, the values go off to infinity in both directions. This, at first, doesn't seem very helpful, but it implies that all values of the variable are theoretically possible. If we now subtract the mean of the distribution from every value, all we will be doing is shifting the distribution along the axis so that the mean of the new distribution is zero, but the distribution still goes off to infinity in both directions – see Figure 10.3.

10.1 Characteristics

10.3 Combinations

10.5 Normal

10.2 Standard normal

10.4 Central limit

Looking again at the horizontal axis, we know that it is measured in whatever units X was measured in (e.g. pounds, time, etc.), but the second thing which characterizes a Normal distribution, after the mean, is its standard deviation. If we now divide all of the values on the horizontal axis by the standard deviation, we will have a scale which has a mean of zero and goes off in 'number of standard deviations' in either direction – see the third part of Figure 10.3.

When you do this to any Normal distribution you arrive at something called the standard Normal distribution. It is this which makes the Normal distribution concept so useful, since, no matter what the variable X represents, and no matter what units it is measured in, we can almost immediately reduce it to this standard Normal distribution.

So, if we define a variable, Z, as the standard Normal variable, we can write it as:

$$Z = \frac{X - \mu}{\sigma}$$

This is known as a **transformation** of the original variable, and we now find that

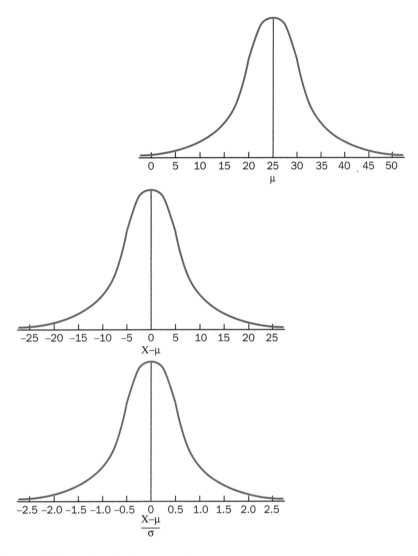

Figure 10.3 Constructing the standard Normal.

the areas under this standard Normal distribution are contained in published tables, such as those in Appendix C.

Example

If a variable X has a Normal distribution with a mean of 250 and a standard deviation of 20, then:

for X = 275 Z = (275 – 250)/20 = 1.25
for X = 200 Z = (200 – 250)/20 = –2.5
for X = 284 Z = (284 – 250)/20 = 1.7

The area excluded in the right-hand tail of the distribution is given in Appendix C and is shown in Figure 10.4.

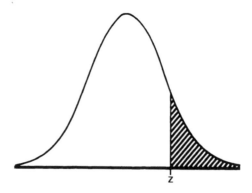

Figure 10.4

10.2.1 Area in a tail

This involves just looking up the value in the tables. For example, the area in the right-hand tail above $Z = 1.03$ is 0.1515.

Since the total under the standard Normal curve is 1, this area is also the probability of obtaining a value from the original distribution more than 1.03 standard deviations above the mean.

Manipulating the value from the table, we see that the probability of obtaining a value below 1.03 standard deviations above the mean is $1 - 0.1515$ (= 0.8485).

10.2.2 Area between two Z values

To find the probability of a value between 1 and 1.1 standard deviations above the mean, we have to subtract one from the other as follows:

$$\text{area above } Z = 1 \text{ is } 0.1587$$
$$\text{area above } Z = 1.1 \text{ is } 0.1357$$

so the area between $Z = 1$ and $Z = 1.1$ is $0.1587 - 0.1357 = 0.0230$.

10.2.3 Symmetry

Since the standard Normal distribution is symmetrical about its mean of 0, an area to the right of a positive value of Z will be identical to the area to the left of the corresponding negative value of Z. (Note that areas cannot be negative.) Thus to find the area between $Z = -1$ and $Z = +1$, we have:

area to the left of $Z = -1$ is 0.1587
area to the right of $Z = +1$ is 0.1587
area outside the range $(-1, +1)$ is $0.1587 + 0.1587 = 0.3174$
so, area between $Z = -1$ and $Z = +1$ is $1 - 0.3174 = 0.6826$.

For any Normal distribution, 68.26% of the values will be within one standard deviation of the mean. (*Hint: it is often useful to draw a sketch of the area required by a problem and compare this with Figure 10.4.*)

Exercise

What percentage of values will be within 1.96 standard deviations of the mean?
(Answer: 95%.)

If a population is known to have a normal distribution, and its mean and variance are known, then we may use the tables together with the z transformation to express facts about this population. Looking at the algebra, we have:

$$Z = \frac{X - \mu}{\sigma}$$

so

$$X = \mu + Z\sigma$$

For example, if $\mu = 200$ and $\sigma = 20$ and we want to know the X value which is two standard deviations above the mean, then:

$$X = 200 + 2 \times 20 = 200 + 40 = 240$$

Case study

Carroll Imitations have been offered the chance to have a small stand at a trade fair being held at the NEC near Birmingham. The cost of the stand will be £5000, the cost of promotional materials will be £500, and the partners include a cost of £200 for their time. The organizers of the fair say that the likely number of enquiries will average out at 3000 and you have decided to assume that these enquiries are normally distributed with a standard deviation of 300. Past experience has given a conversion rate of 5% of enquiries into actual jobs, and the partners work on an average profit of £50 per job. What is the probability that there will be insufficient jobs generated to cover the cost of attending the fair?

(Answer: Total cost £5700, profit per job £50, therefore break-even is at 114 jobs; this implies 2280 enquiries.)

Enquiries distribution is shown in Figure 10.5, with the second horizontal axis being the equivalent Z scores.

Looking in Appendix C, we find the area to the left of $Z = -2.4$ is 0.0082. Therefore, we would predict a probability of 0.0082 (or just under 1%) of not breaking even. It is therefore worthwhile taking the stand at the trade fair.

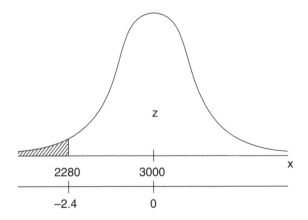

Figure 10.5

10.3 Combinations of variables

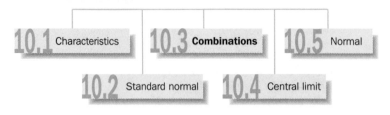

What we have looked at so far will allow us to deal with situations where there is only one value whose probability we need to assess. However, in many cases, we are dealing with two or more values being combined. For example, variations in the dimensions or weights of manufactured items may also be the result of a wide variety of factors and these manufactured items are often brought together as a series of components to produce some good for sale. Each of the items could be described by its mean and standard deviation. To consider the characteristics of this assembled product we will need to combine the means and standard deviations from the constituent parts. Similarly, in assessing the results of a survey (see Sections 13.2.1 and 13.2.2 for examples) we may need to combine results with several sources of variation. We may wish to compare the difference in annual income by region or by sex, for example.

If X and Y are two independent, Normally distributed random variables with means of μ_1 and μ_2 and variances of σ_1^2 and σ_2^2, respectively, then:

$$\text{for } X + Y \quad \text{mean} = \mu_1 + \mu_2$$
$$\text{variance} = \sigma_1^2 + \sigma_2^2$$
$$\text{for } X - Y \quad \text{mean} = \mu_1 - \mu_2$$
$$\text{variance} = \sigma_1^2 + \sigma_2^2 \quad \textbf{(note the plus sign)}$$

If we are adding variables, the mean of the sum is the sum of the means, and the variance of the sum is the sum of the variances. Note that we **add variances**, not standard deviations. The standard deviation is calculated by taking the square root of variance:

$$\text{standard deviation} = \sqrt{(\sigma_1^2 + \sigma_2^2)}$$

Example

An assembled product is made up from two parts. The weight of each part is normally distributed with the following characteristics:

Part 1: mean = 15 g   standard deviation = 4 g
Part 2: mean = 20 g   standard deviation = 2 g

What percentage of these products weighs more than 36 grams?

Consider this problem as a combination of weights. We can then use the formulae above to get:

$$\text{Mean} = \mu = 15 + 20 = 35 \text{ g}$$
$$\text{Standard deviation} = \sigma = \sqrt{(4^2 + 2^2)} = 4.4721$$

Taking 36 g as the X value, we have:

$$Z = \frac{X - \mu}{\sigma} = \frac{36 - 35}{4.4721} = 0.22$$

Using Appendix C we find that the area to the right of $Z = 0.22$ is 0.4129, so the percentage of the finished products that we would expect to over 36 g is 41.29%.

Case study

Carroll Imitations have predicted that their turnover for next year will be £450 000 and that their costs will be £400 000. They thus expect to make a profit of £50 000, and are quite happy. As you talk to the partners, however, you realize that there are a whole host of factors which will affect both the turnover and costs figures. You persuade them to look at the problem as if each figure were a variable and suggest that they treat them as having Normal distributions. Given the nature of the business, you decide to assume a standard deviation of £25 000 for each distribution. What is the probability that the partnership makes a profit?

(Answer: Profit is $(\mu_1 - \mu_2)$ where μ_1 is the mean of turnover, and μ_2 is the mean of costs. We now know that this will have a Normal distribution with a standard deviation of $\sqrt{(\sigma_1^2 + \sigma_2^2)}$, or $\sqrt{(25^2 + 25^2)}$, or 35.355. (Note we are working in thousands to simplify the arithmetic.)

We want the probability that profit will be over zero, so our Z score will be:

$$Z = (0 - 50)/35.355 = -1.414$$

From Appendix C, we get a probability of 0.0787 (approx.) for $Z = 1.414$, so the probability that Z is above -1.414 is $1 - 0.0787 = 0.9213$.

The probability that they make a profit is approximately 92%.

10.4 Central limit theorem

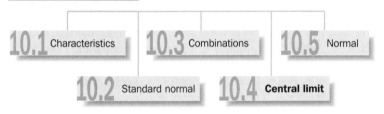

Since the Normal distribution appears in both the 'natural world' and as a result of some manufacturing processes, it has been found to be a particularly useful distribution for modelling behaviour. However, it has been found to have an even wider application in the field of sampling theory. The interpretation of survey results is the subject of Part 4 and will develop the application of the central limit theorem in those chapters. In this section we will look at the basic idea.

The mathematical derivation of why and how the Normal distribution applies to sampling situations is rather beyond a book of this type and involves considerable use of calculus. However, we can look at a specific example of a population and the samples that could be drawn from it to illustrate the concept. The concept can then be applied to a business context. You don't need to know the proof of the central limit theorem, only that it exists.

In order to develop the ideas behind the central limit theorem without using calculus we need to use our imagination. Normally when you take a sample from a population, you take just the one sample, as we discussed in Chapter 2. To develop these ideas, you need to think about what would happen if you took very many samples from the same population: in fact all of the possible different samples. This is not too difficult to think about, but would be very hard to do, even for relatively small samples.

Consider a population that has a Normal distribution with a mean μ and a variance σ^2 as shown in Figure 10.6. If we took every possible sample of one from this distribution and drew a graph of all of the results, then we would just obtain a graph which looked exactly the same as the original population, and the diagram would be exactly as in Figure 10.6.

However, if we increase the sample size to two, and calculate the mean, there will be a change in the distribution obtained. Think about a single sample for a

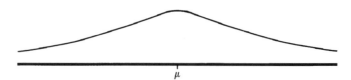

Figure 10.6

moment. When you calculate the average of two numbers, the answer will be a value in-between the original numbers. From the use of the probability tables, we know that we are much more likely to get a sample value from somewhere near the mean than from a point on the distribution that is a long way from the mean. Thus, the probability that both values in our sample of two will be close to the mean will be considerably higher than the probability that both values will be a long way below the mean, or a long way above the mean. This will give us a very high probability of getting a sample mean close to the population mean, and a relatively small probability of getting a sample mean a long way from the population mean. Again, considering every possible sample of two from the distribution, and calculating the mean, there will be more sample means close to the population mean than there were original population values, since one small value and one large value will give a mean close to the centre of the distribution.

This situation is illustrated in Figure 10.7 where we also see that the average of all of the sample averages will be the population mean μ. As we increase the sample size, the probability of getting all of the sample values, and hence the sample average in an extreme tail of the original population distribution, becomes extremely small, while the probability of the sample mean being close to the original population mean increases.

The illustration in Figure 10.8 shows that as the sample size increases, the distribution of sample means remains a Normal distribution with μ as its mean; however, the variance of the distribution decreases as the sample size

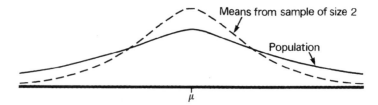

Figure 10.7

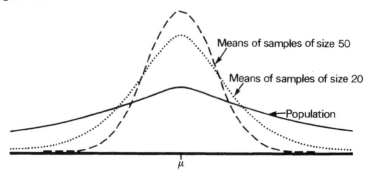

Figure 10.8

increases. It can be shown that the sample mean has the following distribution:

$$\overline{X} \sim N\left(\mu, \frac{\sigma^2}{\sqrt{n}}\right)$$

where n is the sample size and the standard deviation of this sampling distribution or the **standard error** is given by $\sigma^2/\sqrt{n}$.

It is certainly **not** being suggested that, in any particular situation, all possible samples of a certain size would be selected. What we are arguing is that if we know the theoretical distribution of the sample means, then we can compare this to the particular result that we get from our **one** sample.

Example

The time spent queuing in traffic on the way to work by the employees of a large firm has a mean of 60 minutes and a standard deviation of 20 minutes. If a sample of employees is selected at random, what is the probability that the sample average will be over 64 minutes if the sample size is (a) 40 and (b) 100.

To determine the probabilities of sample statistics, such as the mean, we need to establish the distribution concerned, in this case the Normal distribution, and the measure of spread, here the standard error.

(a) with a sample of 40, we have:

$$\text{standard error} = \frac{20}{\sqrt{40}} = 3.162278$$

and

$$Z = \frac{64 - 60}{3.162278} = 1.2649$$

and, using the Normal distribution tables, P(over 64) = 0.1020 (or about 10%)

(b) with a sample of 100, we have:

$$\text{standard error} = \frac{20}{\sqrt{100}} = 2$$

and

$$Z = \frac{64 - 60}{2} = 2$$

and, using the Normal distribution tables, P(over 64) = 0.02275 (or about 2%).

This example illustrates the fact that, as sample size increases, the chances of getting an extreme result diminish.

The Z transformation for sample means is given by:

$$Z = \frac{\overline{x} - \mu}{\sigma/\sqrt{n}}$$

We are now concerned with how many standard errors the sample mean is away from the true mean.

10.4.1 Generalization   The logical argument and the examples given so far should convince you that where a population is normally distributed, then means of samples drawn from that population will also be distributed normally.

The central limit theorem allows even more generalization of the results than we have alluded to, since for **any population distribution**, whether it is discrete or continuous, skewed, rectangular or even multimodal, the distribution of the sample means will – remarkably – be approximately Normal if the sample size is sufficiently large. No matter whether we know the population distribution or not, if we take large enough samples, we will be able to use the Normal distribution to analyse and understand the results we obtain.

The difficulty, as you can no doubt see, is the phrase '*if the sample size is sufficiently large*'. We cannot, at this stage, fully define sufficiently, but see Chapter 11. (A working definition may be that the sample size is over 30.)

10.4.2 Proportions

In the case of a proportion of a sample, we are effectively considering a **Binomial** situation, and, as n (the sample size) becomes large, the Binomial distribution can be approximated by the Normal distribution.

Thus the sampling distribution of a proportion will also be a Normal distribution. For a distribution with a population proportion Π we have the distribution of the sample proportion, P, as:

$$P \sim N\left(\Pi, \frac{\Pi(1-\Pi)}{n}\right)$$

Example

It is known that 60% of a group (0.6 as a proportion) have tried to loose weight in the last year. What are the chances of a random sample showing less than half have tried to loose weight if:
(a) a sample of 60 is chosen
(b) a sample of 200 is chosen.
Again we need to adapt the Z transformation and recognize the measure of spread of the sampling distribution, the standard error, is the square root of the variance given in the formula above.
This is given by:
(a) with $n = 60$:

$$\sqrt{\frac{0.6 \times 0.4}{60}} = \sqrt{0.004} = 0.06325$$

and

$$Z = \frac{0.5 - 0.6}{0.06325} = -1.581$$

using the Normal distribution tables, P(less than 50) = 0.0571 (i.e. 5.7%)
(b) with $n = 200$:

$$\sqrt{\frac{0.6 \times 0.4}{200}} = \sqrt{0.0012} = 0.03464$$

and

$$Z = \frac{0.5 - 0.6}{0.03464} = -2.887$$

using the Normal distribution tables, P(less than 50) = 0.00193 (i.e. 0.2%).

This example illustrates again that the chances of an extreme sample result are reduced if the sample size is increased. It is assumed, of course, that the selection method is random, and the sampling frame and questionnaire are valid.

The Z transformation for a sample proportion p is given by:

$$Z = \frac{P - \Pi}{\sqrt{\dfrac{\Pi(1 - \Pi)}{n}}}$$

where P is the sample proportion. In this case we are concerned with how many standard errors the sample proportion is away from the true proportion. It is worth noting that many problems of this kind are specified in terms of percentages and can be managed in exactly the same way. However, the bottom line of the formula will become:

$$\sqrt{\frac{\Pi(100 - \Pi)}{n}}$$

10.5 Normal approximations

The normal distribution, despite being a continuous distribution, can also be used to approximate to certain other (discrete) distribution under certain circumstances. This is usually where we are dealing with a large value for n (the number of trials). Two examples are given in this section.

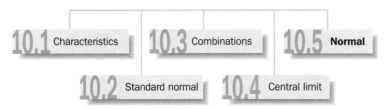

10.5.1 Normal approximation to the binomial

Although the binomial distribution is a **discrete** probability distribution, and the Normal distribution is **continuous**, it will be possible to use the Normal distribution as an approximation to the Binomial if n is large and $p > 0.1$. (As we saw in the last chapter if $p < 0.1$ we would use the Poisson approximation to the binomial.) To see why this will work, consider a binomial distribution with a probability, p, of 0.2. For various values of n, we have distributions as shown in Figure 10.9.

Looking at the various parts of Figure 10.9, we see that with $n = 2$, we have a highly skewed distribution; the mean will be $np = 0.4$. As n increases, the amount of skewness decreases: in Figure 10.9b, the mean is 2 and in Figure 10.9c, the mean is 4, and even at this stage, we are beginning to see the typical 'bell shape' of the Normal distribution curve. In Figure 10.9d, the mean is 20, and although the shape of the histogram is not exactly that of the normal curve, it is very close.

If we wish to use the Normal distribution as an approximation to the binomial distribution, we must develop a method of moving from a discrete distribution to a continuous one. To see how to do this, look at Figure 10.10 where a curve has been superimposed on the histogram.

Here we see that as the curve cuts through the midpoints of the blocks of the histogram, small areas such as B are excluded, while other areas, such as A, are included under the curve but not in the histogram. These areas will tend to cancel each other out. Since each block represents a whole number, often the number of successes, it can be considered as extending from 0.5 below that integer to 0.5 above. Thus in the example above, the block representing 52 successes extends from 51.5 to 52.5.

In order to find the area and hence the probability for a series of outcomes, it will thus be necessary to go from 0.5 below the lowest integer to 0.5 above

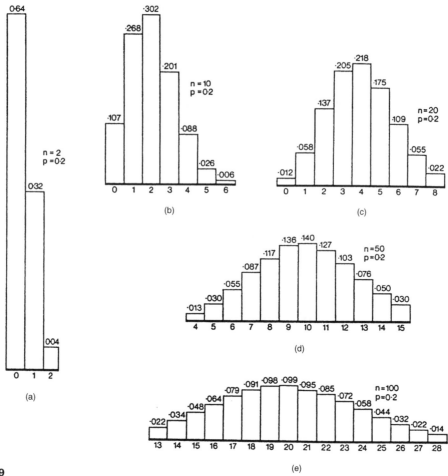

Figure 10.9

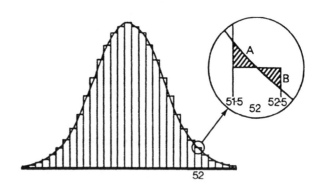

Figure 10.10

the highest integer. If from Figure 10.10 we wanted to find the probability of 49, 50, 51 and 52 successes, then we would need to find the area under the normal curve from $X = 48.5$ to $X = 52.5$, or if we wanted to find the probability of 52 or more successes, we should require the area to the right of $X = 51.5$.

To find areas, we must transform the X values into Z values, on the standard Normal distribution.

From the previous chapter, we know that for a binomial distribution, the mean $= np$ and the standard deviation $= \sqrt{[np(1-p)]}$; and these values can be used to calculate the Z value:

$$Z = \frac{X - np}{\sqrt{[np(1-p)]}}$$

Example

An insurance broker deals with many enquiries during a week, but needs 40 new policies per week to break even in the life assurance department. During a particular week, the broker has 95 enquiries and past experience suggests that the conversion rate of enquiries to policies is 50%. What is the probability that the broker does not at least break even in that week?

This is a binomial situation since p is assumed fixed at 0.5 (or 50%) and each enquiry either leads to a new policy, or it doesn't. Using the ideas above, we have:

$$\text{Mean} = np = 95 \times 0.5 = 47.5$$
$$\text{Standard deviation} = \sqrt{95 \times 0.5 \times 0.5} = 4.8734$$

The broker will at least break even if there are 40 or more new policies, and therefore we need the area under the normal curve to the right of $X = 39.5$. For this value of X

$$Z = \frac{39.5 - 47.5}{4.8734} = -1.6416$$

and the area to the left of this negative value is 0.0505 (from Appendix C). By subtraction, the area to the right of this point is $1 - 0.0505 = 0.9495$. The probability that the broker will at least break even is therefore 0.9495, or about 95%.

If the binomial probability had been computed directly, then we would have to find:

$$P(\text{at least } 40) = P(40) + P(41) + P(42) + \ldots + P(95)$$

A rather more lengthy process!

10.5.2 Normal approximation to the Poisson

In a similar way to the binomial distribution, as the mean, $\lambda = np$, gets larger and larger, the amount of skewness in the Poisson distribution decreases, until it is possible to use the Normal distribution. (From Chapter 9 you may recall that the variance of a Poisson distribution is λ.) To transform values from the original distribution into Z values, we use:

$$Z = \frac{X - \lambda}{\sqrt{\lambda}}$$

(Note that it is usual to use the Normal approximation if $\lambda > 30$. Again, we should allow for the fact that we are going from a discrete to a continuous distribution.)

Example

The average number of broken eggs per lorry load is known to be 50. What is the probability that there will be more than 70 broken eggs on a particular lorry load?

$$\text{Mean} = \lambda = 50$$
$$\text{Standard deviation} = \sqrt{\lambda} = 7.07107$$

The area required is the area above $X = 70.5$, so the value of Z will be:

$$Z = \frac{70.5 - 50}{7.07107} = 2.9$$

From Appendix C we can find the area to the right of this value to be 0.00187, so the probability that there will be more than 70 broken eggs on the lorry will be 0.00187 (or 0.187%).

10.6 Conclusions

Dedicating a whole chapter to a single distribution may seem a little extravagant, but if you have worked through the various examples and seen the implications of the central limit theorem, then we hope that you will agree that it was worthwhile. The Normal distribution is simply the most important continuous probability distribution that there is. Not only is it capable of describing many naturally occurring phenomena, it can also be applied to many other situations which arise in a business context. Furthermore, it provides the basis for moving forward to draw implications from the results of surveys. This is known as statistical inference and is dealt with in detail in the Part 4 of this book. Even where the Normal distribution is not an exact match to the data, we can often use it as a comparator, and thus draw conclusions that it would otherwise be impossible to draw.

Quite simply, the Normal distribution is central to a substantial proportion of statistics as applied to business situations.

10.7 Problems

1   A Normal distribution has a mean of 30 and a standard deviation of 5; find the Z values equivalent to the X values given below:
(a) 35;   (b) 27;   (c) 22.3;   (d) 40.7;   (e) 30.

2   Use the tables of areas under the standard Normal distribution (given in Appendix C) to find the following areas:
(a)   to the right of $Z = 1$
(b)   to the right of $Z = 2.85$
(c)   to the left of $Z = 2$
(d)   to the left of $Z = 0$
(e)   to the left of $Z = -1.7$
(f)   to the left of $Z = -0.3$
(g)   to the right of $Z = -0.85$
(h)   to the right of $Z = -2.58$
(i)   between $Z = 1.55$ and $Z = 2.15$
(j)   between $Z = 0.25$ and $Z = 0.75$
(k)   between $Z = -1$ and $Z = -1.96$
(l)   between $Z = -1.64$ and $Z = -2.58$
(m) between $Z = -2.33$ and $Z = 1.52$
(n)   between $Z = -1.96$ and $Z = +1.96$.

3   Find the Z value, such that the standard Normal curve area:
(a)   to the right of Z is 0.0968
(b)   to the left of Z is 0.3015
(c)   to the right of Z is 0.4920
(d)   to the right of Z is 0.992 66
(e)   to the left of Z is 0.9616

(f) between $-Z$ and $+Z$ is 0.95

(g) between $-Z$ and $+Z$ is 0.9.

4 Invoices at a particular depot have amounts which follow a Normal distribution with a mean of £103.60 and a standard deviation of £8.75.

(a) What percentage of invoices will be over £120.05?

(b) What percentage of invoices will be below £92.75?

(c) What percentage of invoices will be between £83.65 and £117.60?

(d) What will be the invoice amount such that approximately 25% of invoices are for greater amounts?

(e) Above what amount will 90% of invoices lie?

5 Items coming from the end of a production line are measured for their diameter. The measurements have a mean of 1450 mm and a variance of $0.25 \, \text{mm}^2$.

(a) To meet quality control checks, the items must be between 1448.4 mm and 1451.5 mm. What proportion will meet these standards?

(b) At a later stage in the production process, this item has to fit into a hole with a diameter which has a Normal distribution with a mean of 1451 mm and a standard deviation of 1 mm. What proportion of these holes will have diameters over 1448.5 mm?

(c) What proportion of holes will have diameters above 1451.5 mm?

6 Thirty per cent of the general public have bought a certain item in the past month. If a sample of 1000 people is selected at random find the following probabilities:

(a) more than 310 have bought the product

(b) less than 295 have bought the product

(c) more than 285 have bought the product.

7 On average, one in 10 000 people have the skills and talents to become major successes as singers. If a sample of 20 000 people were selected at random from a particular region, calculate the probability that there would be:

(a) no suitable people

(b) more than three suitable people.

8 (a) In a very large population there are 40% of the people who support a change in government policy on regional aid. If a sample of 1000 people is chosen at random, what is the probability that this sample will contain over 425 people who support a change in policy?

(b) What is the probability that fewer than 390 people support a change in policy?

9 The average number of customers in a shop per week is 256. Calculate the probability of there being:

(a) more than 240 customers in a week

(b) less than 280 customers in a week

(c) 234 to 290 customers in a week.

10 A process yields 15% defective items. If 180 items are randomly selected from the process, what is the probability that the number of defectives is 30 or more?

11 If the probability of selecting a smoker is 0.3, determine the probability of selecting more than 20 smokers:

(a) in a randomly selected sample of 80

(b) at a conference of 70 doctors.

State any assumptions made.

12 In a certain manufacturing process, 25% of the items produced are classified as seconds. If 5000 items have been produced, what is the probability that at least 1300 items will be available as seconds for a sale?

13 A switchboard receives 42 calls per minute on average. Estimate the probability that there will be at least:
(a) 40 calls in the next minute
(b) 50 calls in the next minute.

14 The average demand for a particular item from stock is found to be 45 per week. Estimate the probability of a demand for 50 or more:
(a) in the next week
(b) in the next 2 weeks.

15 The time taken, in minutes, to complete three tasks, machining, assembly and packaging, have been recorded as follows:

	Mean	Standard deviation
Machining	25	7.5
Assembly	15	5.5
Packaging	15	4.5

If a particular job requires all three tasks, what is the probability that the job will take more than an hour?

16 It has been estimated that the average weekly wage in a particular industry is £172 and that the standard deviation is £9.
(a) What is the probability that a random sample of 10 employees will have an average weekly wage of £180 or more?
(b) What is the probability that an individual will have an average weekly wage of £180 or more if it can be assumed that wages follow a normal distribution?

17 An industrial chemical that comes in granular form is packaged and sold to the trade in 20 kg plastic sacks. To minimize the number of underweight sacks, the filling process has been adjusted so that the mean is 21 kg. The weight of the sacks varies because of the nature of the product and this has been measured by a standard deviation of 2.5 kg.
(a) Determine the probability of an underweight sack.
(b) If the sacks can also be bought in batch quantities of 30, determine the probability that the average weight per sack in the batch is less than 20 kg.
(c) Comment on your results.

18 It has been claimed that only 45% of customers find changes to an invoicing system an improvement. Assuming this is the case, determine the probability that a market research survey of 100 customers will show that 50% or more report an improvement. What are the implications for the design of the survey?

Part 3   Conclusions

This part has introduced you to one of the key concepts in statistics, the idea of probability. We have shown that certain basic relationships exist and that it is possible to look at combinations of probabilities. You should be convinced that probability has an impact on everyone. You should now try to list some other examples of probability in your everyday life.

Try the following exercise to consolidate you understanding of probability.

Exercise

Select a situation where a queue forms awaiting service, and arrange to observe the behaviour of the queue over 10 periods of 10 minutes. Note the arrival times of customers.

When you have collected this data, draw up a table of the amount of time between arrivals in each of the 10-minute periods, constructing histograms to illustrate the data.

Pool all of the data which you have collected, and calculate the probabilities of new customers arriving in under one minute, under 2 minutes, and so on. Does this distribution match any of the distributions discussed in this part?

From your probability distribution, construct a table of expected percentages of customers arriving in each of the various periods. Observe the same queue on one further occasion, calculating the percentage of customers arriving for various time periods and compare your results with the expected percentages.

What can you conclude from this exercise?

You could also try putting 'Probability' into one of the search engines on the internet, and you will discover some interesting sites, although you should always be slightly wary of information gathered in this way.

Part 4 Statistical inference

So far, most of this book has been about describing situations. This is useful and helps in communication, but would not justify doing a whole book or course. In this part we are going to the next step and looking at ways of extending or generalizing our results so that they not only apply to the group of people or set of objects which we have measured, but also to the whole population. As we saw in Part 1, most of the time we are only actually examining a sample, and not the whole population. Although we will take as much care as possible to ensure that this sample is representative of the population, there may be times when it cannot represent everything about the whole group. The exact results which we get from a sample will depend on chance, since the actual individuals chosen to take part in a survey may well be chosen by random sampling (where every person has a particular probability of being selected).

We need to distinguish between values obtained from a sample, and thus subject to chance, and those calculated from the whole population, which will not be subject to this effect. We will need to also distinguish between those true population values that we have calculated, and those that we can estimate from our sample results. Some samples may be 'better' than others and we need some method of determining this. Some problems may need results that we can be very sure about, others may just want a general idea of which direction things are moving. We need to begin to say how good our results are.

Sample values are no more than estimates of the true population values (or parameters or population parameters). To know these values with certainty, your sample would have to be 100%, or a census. In practice, we use samples that are only a tiny fraction of the population for reasons of cost, time and because they are adequate for the purpose. How close the estimates are to the population parameters will depend upon the size of the sample, the sample design (e.g. stratification can improve the representativeness of the sample), and the variability in the population. It is also necessary to decide how certain we want to be about the results; if, for example, we want a very small margin of sampling error, then we will need to incur the cost of a larger sample design. The relationship between sample size, variability of the population and the degree of confidence required in the results is the key to understanding the chapters in this part of the book.

The approach in Chapter 13 is different, as it is concerned with data that cannot easily or effectively be described by parameters (e.g. the mean and standard deviation). If we are interested in characteristics (e.g. smoking/non-smoking), ranking (e.g. ranking chocolate products in terms of appearance) or scoring (e.g. giving a score between 1 and 5 to describe whether you agree or disagree with a certain statement), a number of tests have been developed that do not require description by the use of parameters.

After working through these chapters you should be able to say how good your data is, and test propositions in a variety of ways.

Case study

The Arbour Housing Trust was founded some ten years ago in response to the general level of dereliction and decay in Tonnelle, an outer-city, run-down area. There have been a number of changes in recent times and there are signs of the long-awaited economic improvement. The local population has continued to decline with a movement away from the locality by younger people. The proportion of elderly has increased and some of the Victorian housing is again attracting a more affluent group of residents, many of whom are professional, and commute to the city centre.

The Arbour Housing Trust recently completed a survey of 300 representative households within the locality as part of a review of local housing conditions. A summary of the work done so far on the responses to some of the questions is given below.

Q2. How long have you been resident in Tonnelle?

Number of years	Frequency
Under 1	21
1 but under 5	66
5 but under 10	69
10 but under 20	84
20 or more	60

Q4. How would your property be best described?

Type	Frequency
House	150
Flat	100
Bedsit	45
Other	5

Q5. How long have you been living in this property?

Number of years	Frequency
Under 1	28
1 but under 5	78
5 but under 10	81
10 but under 20	71
20 or more	42

Q6. For each item below, ask 'Do you have ...?'

(a) A fixed bath or shower with a hot water supply:

	Frequency
None	20
Shared	34
Exclusive	246
No answer	0

(b) A flush toilet inside the house:

	Frequency
None	8
Shared	58
Exclusive	234
No answer	0

(c) A kitchen separate from the living room:

	Frequency
None	2
Shared	28
Exclusive	269
No answer	1

Q10. How often do you use the local post office?

	Frequency
Once a month	40
Once a week	200
Twice a week	50
More often	10

Q15–Q17 were concerned with mortgage payment.

Analysis already undertaken on the survey data revealed 100 respondents having a mortgage which was costing, on average, £253 a month. Additional information suggests that the standard deviation will be about £70.

Q18–Q20 were concerned with rent.

The following table has already been produced to summarize monthly rent:

Rent	Frequency
Under £50	7
£50 but under £100	12
£100 but under £150	15
£150 but under £200	30
£200 but under £250	53
£250 but under £300	38
£300 but under £400	20
£400 or more	5

It is known that a similar trust, the Pelouse Housing Trust, has also completed a similar survey in Sauterelle. Sauterelle shares many of the same problems as Tonnelle, but has not shown any signs of economic recovery. The Pelouse Housing Trust survey was of a similar size and conducted about the same time and is thought of as particularly useful for comparative purposes.

Inference quick start

Inference is about generalizing your sample results to the whole population.

The basic elements of inference are:

- confidence intervals
- parametric significance tests
- non-parametric significance tests.

The aim is to reduce the time and cost of data collection while enabling us to generalize the results to the whole population. It allows us to place a level of confidence on our results which indicates how sure we are of the assertions we are making. Results follow from the central limit theorem and the characteristics of the Normal distribution for parametric tests.

Key relationships are:
- Ninety-five percent confidence interval for a mean:

$$\mu \text{ is between } \overline{X} \pm 1.96\sigma\sqrt{n}$$

- Ninety-five percent confidence interval for a percentage:

$$\Pi \text{ is between } P \pm 1.96\sqrt{\left[\frac{\Pi(100-\Pi)}{n}\right]}$$

- Significance tests take seven steps:
  1. hypotheses
  2. significance level
  3. critical value(s)
  4. calculation, e.g. $z = \dfrac{\overline{X} - \mu}{\sigma/\sqrt{n}}$
  5. comparison
  6. decision
  7. Interpretation and 'business' significance.

- Where there is no cardinal data, then we can use non-parametric tests such as chi-squared.

11 Confidence intervals

This chapter allows us to begin to answer the question:

'What can we do with the sample results we obtain, and how do we relate them to the original population?'

Sampling, as we have seen in Chapter 3, is concerned with the collection of data from a (usually small) group selected from a defined, relevant population. Various methods are used to select the sample from this population, the main distinction being between those methods based on random sampling and those which are not. In the development of statistical sampling theory it is assumed that the samples used are selected by simple random sampling, although the methods developed in this and subsequent chapters are often applied to other sampling designs. Sampling theory applies whether the data is collected by interview, postal questionnaire or observation. However, as you will be aware, there are ample opportunities for bias to arise in the methods of extracting data from a sample, including the percentage of non-respondents. These aspects must be considered in interpreting the results together with the statistics derived from sampling theory.

The only circumstance in which we could be absolutely certain about our results is in the unlikely case of having a census with a 100% response rate, where everyone gave the correct information. Even then, we could only be certain at that particular point in time. Mostly, we have to work with the sample information available. It is important that the sample is adequate for the intended purpose and provides neither too little nor too much detail. It is important for the user to define their requirements; the user could require just a broad 'picture' or a more detailed analysis. A sample that was inadequate could provide results that were too vague or misleading, whereas a sample that was overspecified could prove too time-consuming and costly.

Objectives

After working through this chapter, you should be able to:

- understand and apply the concept of inference
- determine a confidence interval for a sample mean and percentage
- use the concept of a confidence interval to determine sample size
- determine confidence intervals for the difference between sample means and sample percentages
- apply the finite population correction factor
- apply the *t*-distribution
- determine confidence intervals for the median (large sample approximation).

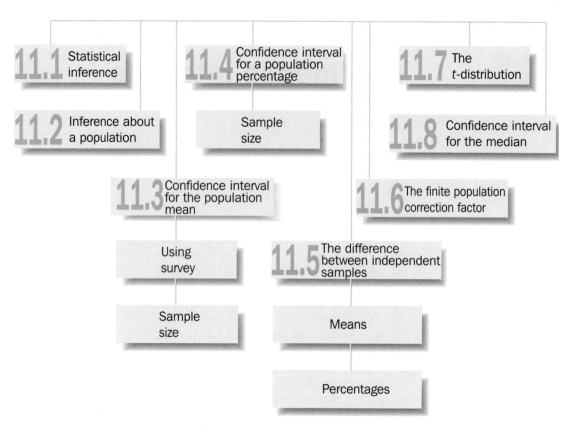

Figure 11.1 Structure of Chapter 11.

11.1 Statistical inference

The central limit theorem (see Section 10.4) provides a basis for understanding how the results from a sample may be interpreted in relation to the parent population; in other words, what conclusions can be drawn about the population on the basis of the sample results obtained. This result is crucial, and if you cannot accept the relationship between samples and the population, then you can draw no conclusions about a population from your sample. All you can say is that you know something about the people involved in the survey. For example, if a company conducted a market research survey in Buxton and found that 50% of their customers would like to try a new flavour of their sweets, what useful conclusions could be

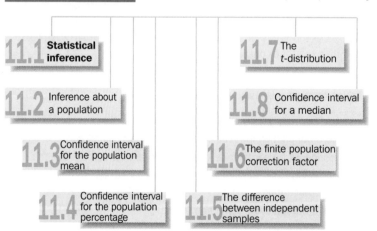

drawn about all existing customers in Buxton? What conclusions could be drawn about existing customers elsewhere? What conclusions could be drawn about potential customers? It is important to clarify the link being made between the selected sample and a larger group of interest. It is this link that is referred to as inference. To make an inference the sample has got to be sufficiently representative of the larger group, the population. It is for the researcher to justify that the inference is valid on the basis of problem definition, population definition and sample design.

Often results are required quickly, for example the prediction of election results, or the prediction of the number of defectives in a production process may not allow sufficient time to conduct a census. Fortunately a census is rarely needed since a body of theory has grown up which will allow us to draw conclusions about a population from the results of a sample survey. This is statistical inference or sampling theory. Taking the sample results back to the problem is often referred to as business significance. It is possible, as we shall see, to have results that are of statistical significance but not of business significance, e.g. a clear increase in sales of 0.001%.

Statistical inference draws upon the probability results as developed in Part 3, especially from the Normal distribution. It can be shown that, given a few basic conditions, the statistics derived from a sample will follow a Normal distribution. To understand statistical inference it is necessary to recognize that three basic factors will affect our results; these are:

- the size of the sample
- the variability in the relevant population
- the level of confidence we wish to have in the results.

As illustrated in Figure 11.2, these three factors tend to pull in opposite directions and the final sample may well be a compromise between the factors.

Increases in sample size will generally make the results more accurate (i.e. closer to the results which would be obtained from a census), but this is not a simple linear relationship so that doubling the sample size does not double the

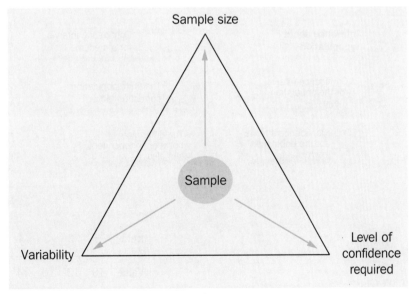

Figure 11.2 Factors affecting the result.

level of accuracy. Very small samples, for example under 30, tend to behave in a slightly different way from larger samples and we will look at this when we consider the use of the *t*-distribution. In practice, sample sizes can range from about 30 to 3000. Many national samples for market research or political opinion polling require a sample size of about 1000. Increasing sample size, also increases cost.

If there was no variation in the original population, then it would only be necessary to take a sample of one; for example, if everyone in the country had the same opinion about a certain government policy, then knowing the opinion of one individual would be enough. However, we do not live in such a homogeneous (*boring*) world, and there are likely to be a wide range of opinions on such issues as government policy. The design of the sample will need to ensure that the full range of opinions is represented. Even items which are supposed to be exactly alike turn out not to be so, for example, items coming off the end of a production line should be identical but there will be slight variations due to machine wear, temperature variation, quality of raw materials, skill of the operators, etc.

Since we cannot be 100% certain of our results, there will always be a risk that we will be wrong; we therefore need to specify how big this risk will be. Do you want to be 99% certain you have the right answer, or would 95% certain be sufficient? How about 90% certain? As we will see in this chapter, the higher the risk you are willing to accept of being wrong, the less exact the answer is going to be, and the lower the sample size needs to be.

11.2 Inference about a population

Calculations based on a sample are referred to as sample statistics. The mean and standard deviation, for example, calculated from sample information, will often be referred to as the sample mean and the sample standard deviation, but if not, should be understood from their context. The values calculated from population or census information are often referred to as population parameters. If all persons or items are included, there should be no doubt about these values (no sampling variation) and these values (population statistics) can be regarded as fixed within the particular problem context. (This may not mean that they are 'correct' since asking everyone is no guarantee that they will all tell the truth!)

If you have access to the web, try looking at the spreadsheet sampling.xls which takes a very small population (of size 10) and shows every possible sample of size 2, 3 or 4. The basic population data is as follows:

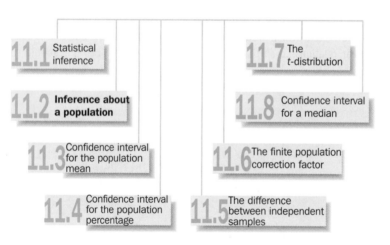

Item	1	2	3	4	5	6	7	8	9	10
Value	10	12	10	14	17	15	14	13	12	13

A quick calculation would tell you that the population parameters are as follows:

$$Mean = 13; \quad Standard\ deviation = 2.160247$$

By clicking on the Answer tab, you can find that, for a sample of 2, the overall mean is 13, with an overall standard deviation of 1.36626. You may wish to compare these answers with those shown, theoretically, later in the chapter. The overall variation for samples of 2 is shown by a histogram in Figure 11.3.

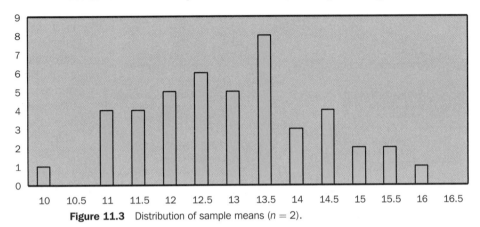

Figure 11.3 Distribution of sample means ($n = 2$).

Look through the spreadsheet for the other answers. Can you find a pattern in the results?

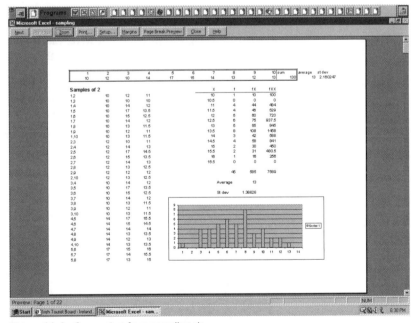

Figure 11.4 Screenshot from sampling.xls.

As we are now dealing with statistics from samples and making inferences to populations we need a notational system to distinguish between the two. Greek letters will be used to refer to population parameters, μ (mu) for the mean and σ (sigma) for the standard deviation, and N for the population size, while ordinary (roman) letters will be used for sample statistics, $\bar{x}$ for the mean, s for the standard deviation, and n for the sample size. In the case of percentages, Π is used for the population and p for the sample.

11.3 Confidence interval for the population mean

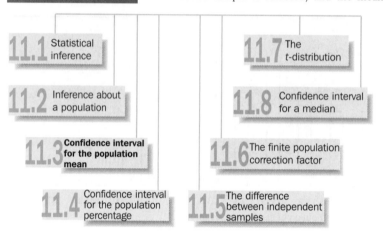

When a sample is selected from a population, the arithmetic mean may be calculated in the usual way, dividing the sum of the values by the size of the sample. If a second sample is selected, and the mean calculated, it is very likely that a different value for the sample mean will be obtained. Further samples will yield more (different) values for the sample mean. Note that the population mean is always the same throughout this process, it is only the different samples which give different answers. This is illustrated in Figure 11.5.

Since we are obtaining different answers from each of the samples, it would not be reasonable to just assume that the population mean was equal to any of the sample means. In fact each sample mean is said to provide a **point estimate** for the population mean, but it has virtually no probability of being exactly right; if it were, this would be purely by chance. We may estimate that the population mean lies within a small interval around the sample mean; this interval represents the **sampling error**. Thus the population mean is estimated to lie in the region:

$$\bar{x} \pm \text{sampling error}$$

Thus, we are attempting to create an **interval estimate** for the population mean.

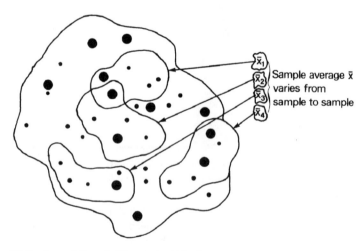

Figure 11.5 A population of different sized 'dots'.

You should recall from Chapter 10 that the area under a distribution curve can be used to represent the probability of a value being within an interval. We are therefore in a position to talk about the population mean being within the interval with a calculated probability.

As we have seen in Section 10.4, the distribution of all sample means will follow a normal distribution, at least for large samples, with a mean equal to the population mean and a standard deviation equal to $\sigma/\sqrt{n}$.

The central limit theorem (for means) states that if a simple random sample of size n ($n > 30$) is taken from a population with mean μ and a standard deviation σ, the sampling distribution of the sample mean is approximately Normal with mean μ and standard deviation $\sigma/\sqrt{n}$.

This standard deviation is usually referred to as the standard error when we are talking about the sampling distribution of the mean. This is a more general result than that shown in Chapter 10, since it does not assume anything about the shape of the population distribution; it could be any shape. Compare this to the result of the sampling.xls spreadsheet. There

$$\frac{\sigma}{\sqrt{2}} = 1.528$$

and the standard deviation obtained from all samples was 1.36626, but remember that here the sample size was only 2. The spreadsheet result is intended only to illustrate that the standard deviation for the distribution of sample means is lower than the population standard deviation.

From our knowledge of the Normal distribution (see Chapter 10 or Appendix C) we know that 95% of the distribution lies within 1.96 standard deviations of the mean. Thus, for the distribution of sample means, 95% of these will lie in the interval

$$\mu \pm 1.96 \frac{\sigma}{\sqrt{n}}$$

as shown in Figure 11.6. This may also be written as a probability statement:

$$P\left(\mu - 1.96 \frac{\sigma}{\sqrt{n}} \leq \bar{x} \leq \mu + 1.96 \frac{\sigma}{\sqrt{n}}\right) = 0.95$$

This is a fairly obvious and uncontentious statement which follows directly from the central limit theorem. As you can see, a larger sample size would narrow the width of the interval (since we are dividing by root n). If we were to increase the percentage of the distribution included, by increasing the 0.95, we would need to increase the 1.96 values, and the interval would get wider.

By rearranging the probability statement we can produce a 95% confidence interval for the population mean:

$$\mu = \bar{x} \pm 1.96 \frac{\sigma}{\sqrt{n}}$$

This is the form of the confidence interval which we will use, but it is worth stating what it says in words:

the true population mean (which we do not know) will lie within 1.96 standard errors of the sample mean with a 95% level of confidence.

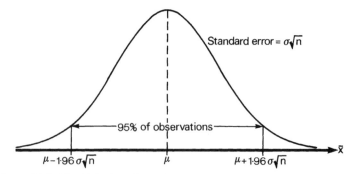

Figure 11.6 The ditribution of sample means.

In practice you would only take a single sample, but this result utilizes the central limit theorem to allow you to make the statement about the population mean. There is also a 5% chance that the true population mean lies outside this confidence interval, for example, the data from sample 3 in Figure 11.7.

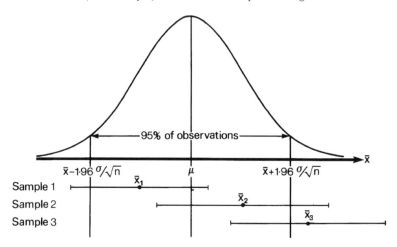

Figure 11.7 Confidence interval from different samples.

Case study 4

In the Arbour Housing Survey (see Case 4) 100 respondents had mortgages, paying on average £253 per month. If it can be assumed that the standard deviation for mortgages in the area of Tonnelle is £70, calculate a 95% confidence interval for the mean.

The sample size is $n = 100$, the sample mean, $\bar{x} = 253$ and the population standard deviation, $\sigma = 70$. By substituting into the formula given above, we have

$$\mu = £253 \pm 1.96 \times \left[£70 \big/ \sqrt{100} \right]$$

$$\mu = £253 \pm £13.72$$

or we could write this as

$$£239.28 \leq \mu \leq £266.72$$

We are fairly sure (95% confident) that the average mortgage for the Tonnelle area is between £239.28 and £266.72. There is a 5% chance that the true population mean lies outside of this interval.

So far our calculations have attempted to estimate the unknown population mean from the known sample mean using a result found directly from the central limit theorem. However, looking again at our formula, we see that it uses the value of the population standard deviation, σ, and if the population mean is unknown *it is highly unlikely* that we would know this value. To overcome this problem we may substitute the sample estimate of the standard deviation, s, but unlike the examples in Chapter 5, here we need to divide by $(n-1)$ rather than n in the formula. This follows from a separate result of sampling theory which states that the sample standard deviation calculated in this way is a better estimator of the population standard deviation than that using a divisor of n. (Note that we do not intend to prove this result which is well documented in a number of mathematical statistics books.)

The structure of the confidence interval is still valid provided that the sample size is fairly large. Thus the 95% confidence interval which we shall use will be:

$$\mu = \bar{x} \pm 1.96s/\sqrt{n}$$

For a 99% confidence interval, the formula would be:

$$\mu = \bar{x} \pm 2.576s/\sqrt{n}$$

Example

1 A sample of 90 people were selected from a large population. If the average amount spent per week on lottery tickets was found to be £5.60 and the sample standard deviation was £2.90, calculate the 95% confidence interval for the mean of the population.

The sample statistics are $n = 90$, $\bar{x} = £5.60$ and $s = £2.90$. By substitution into the formula, the 95% confidence interval is given by:

$$\mu = £5.60 \pm 1.96 \times \frac{£2.90}{\sqrt{90}}$$

$$\mu = £5.60 \pm £0.599$$

or we could write this as: $5.001 \leq \mu \leq £6.199$.

2 A random sample of 300 items from a production line are selected for testing to estimate their average length of life. The sample mean was calculated to be 250 hours and the sample standard deviation found to be 6 hours. Calculate the 95% and 99% confidence intervals for the population mean.

The sample statistics are $n = 300$, $\bar{x} = 250$, and $s = 6$ hours. By substitution into the formula, the 95% confidence interval is:

$$\mu = 250 \pm 1.96 \times \frac{6}{\sqrt{300}}$$

$$\mu = 250 \pm 0.679$$

$$\text{or } 249.321 \leq \mu \leq 250.679$$

Similarly, the 99% confidence interval is given by:

$$\mu = 250 \pm 2.576 \times \frac{6}{\sqrt{300}}$$

$$\mu = 250 \pm 0.892$$

$$\text{or } 249.108 \leq \mu \leq 250.892$$

As this last example illustrates, the more certain we are of the result (i.e. the higher the level of confidence), the wider the interval becomes. That is, the sampling error becomes larger. Sampling error depends on the probability excluded in the extreme tail areas of the Normal distribution, and so, as the confidence level increases, the amount excluded in the tail areas becomes smaller. This example also illustrates a further justification for sampling, since the measurement itself is destructive (length of life), and thus if all items were tested, there would be none left to sell.

11.3.1 Confidence intervals using survey data

It may well be the case that you need to produce a confidence interval on the basis of tabulated data.

Case study

The following example uses the table produced in the Arbour Housing Survey (and reproduced as Table 11.1) showing monthly rent.

Table 11.1 Monthly rent

Rent	Frequency
Under £50	7
£50 but under £100	12
£100 but under £150	15
£150 but under £200	30
£200 but under £250	53
£250 but under £300	38
£300 but under £400	20
£400 or more	5

We can calculate the mean and sample standard deviation using

$$\bar{x} = \frac{\sum fx}{n}$$

$$s = \sqrt{\left[\frac{\sum f(x - \bar{x})^2}{n - 1}\right]} \quad \text{or} \quad s = \sqrt{\left[\frac{\sum fx^2}{n - 1} - \frac{(\sum fx)^2}{n(n - 1)}\right]}$$

The sample standard deviation, s, sometimes denoted by $\hat{\sigma}$, is being used as an estimator of the population standard deviation σ. The sample standard deviation will vary from sample to sample in the same way that the sample mean, $\bar{x}$, varies from sample to sample. The sample mean will sometimes be too high or too low, but on average will equal the population mean μ. You will notice that the distribution of sample means $\bar{x}$, in Figure 11.6 is symmetrical about the population mean μ. In contrast, if we use the divisor n, the sample standard deviation will on average be less than σ. To ensure that the sample standard deviation is large enough to estimate the population standard deviation σ reasonably, we use the divisor $(n - 1)$. The calculations are shown in Tables 11.2 and 11.3.

Confidence intervals are obtained by the substitution of sample statistics, from either Table 11.2 or 11.3 into the expression for the confidence interval.

Table 11.2

Rent	Frequency	x	fx	$f(x-\bar{x})^2$
Under £50	7	25*	175	269 598.437 5
£50 but under £100	12	75	900	256 668.75
£100 but under £150	15	125	1875	138 960.9375
£150 but under £200	30	175	5 250	641 71.875
£200 but under £250	53	225	11 925	745.312 5
£250 but under £300	38	275	10 450	109 784.375
£300 but under £400	20	350	7 000	331 531.25
£400 or more	5	450*	2 250	261 632.812 5
	180		39 825	1 433 093.75

*Assumed mid-point

$$\bar{x} = \frac{\sum fx}{n} = \frac{39\,825}{180} = £221.25$$

$$s = \sqrt{\left[\frac{\sum f(x-\bar{x})^2}{n-1}\right]} = \sqrt{\left[\frac{1\,433\,093.75}{179}\right]} = £89.48$$

Table 11.3

Rent	Frequency	x	fx	fx^2
Under £50	7	25*	175	4 375
£50 but under £100	12	75	900	67 500
£100 but under £150	15	125	1875	234 375
£150 but under £200	30	175	5 250	918 750
£200 but under £250	53	225	11 925	2 683 125
£250 but under £300	38	275	10 450	2 873 750
£300 but under £400	20	350	7 000	2 450 000
£400 or more	5	450*	2 250	1 012 500
	180		39 825	10 244 375

*Assumed mid-point

$$\bar{x} = \frac{\sum fx}{n} = \frac{39\,825}{180} = £221.25$$

$$s = \sqrt{\left[\frac{\sum fx^2}{n-1} - \frac{(\sum fx)^2}{n(n-1)}\right]} = \sqrt{\left[\frac{10\,244\,375}{179} - \frac{(39\,825)^2}{180 \times 179}\right]} = £89.48$$

The 95% confidence interval is:

$$\mu = \bar{x} \pm 1.96\frac{s}{\sqrt{n}}$$

$$\mu = £221.25 \pm 1.96\frac{£89.48}{\sqrt{180}}$$

$$\mu = £221.25 \pm £13.07$$

$$\text{or } £208.18 \leq \mu \leq £234.32$$

11.3.2 Sample size for a mean

As we have seen, the size of the sample selected has a significant bearing on the actual width of the confidence interval that we are able to calculate from the sample results. If this interval is too wide, it may be of little use, for example for a confectionery company to know that the weekly expenditure on a particular type of chocolate was between £0.60 and £3.20 would not help in planning. Users of sample statistics require a level of accuracy in their results. From our calculations above, the confidence interval is given by:

$$\mu = \bar{x} \pm z\frac{s}{\sqrt{n}}$$

where z is the value from the Normal distribution tables (for a 95% interval this is 1.96). We could re-write this as:

$$\mu = \bar{x} \pm e$$

and now

$$e = z \times {}^{s}/{\sqrt{n}}$$

From this we can see that the error, e, is determined by the z value, the standard deviation and the sample size. As the sample size increases, so the error decreases, but to halve the error we would need to quadruple the sample size (since we are dividing by the square root of n).

Rearranging this formula gives:

$$n = \left(\frac{zs}{e}\right)^2$$

and we thus have a method of determining the sample size needed for a specific error level, at a given level of confidence. Note that we would have to estimate the value of the sample standard deviation, either from a previous survey, or from a pilot study.

Example

What sample size would be required to estimate the population mean for a large set of company invoices to within £0.30 with 95% confidence, given that the estimated standard deviation of the value of the invoices is £5.

To determine the sample size for a 95% confidence interval, let $z = 1.96$ and, in this case, $e = £0.30$ and $s = £5$. By substitution, we have:

$$n = \left(\frac{1.96 \times £5}{£0.30}\right)^2 = 1067.11$$

and we would need to select 1068 invoices to be checked, using a random sampling procedure.

11.4 Confidence interval for a population percentage

In the same way that we have used the sample mean ($\bar{x}$) to estimate a confidence interval for the population mean (μ), we can now use the percentage with a certain characteristic in a sample (p) to estimate the percentage with that characteristic in the whole population (Π).

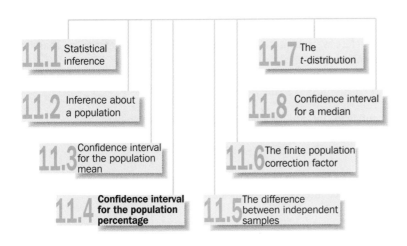

11.1 Statistical inference

11.2 Inference about a population

11.3 Confidence interval for the population mean

11.4 **Confidence interval for the population percentage**

11.7 The t-distribution

11.8 Confidence interval for a median

11.6 The finite population correction factor

11.5 The difference between independent samples

Sample percentages will vary from sample to sample from a given population (in the same way that sample means vary), and for large samples, this will again be in accordance with the central limit theorem. For percentages, this states that if a simple random sample of size n ($n > 30$) is taken from a population with a percentage Π having a particular characteristic, then the sampling distribution of the sample percentage, p, is approximated by a Normal distribution with a mean of Π and a standard error of

$$\sqrt{\frac{\Pi(100 - \Pi)}{n}}$$

The 95% confidence interval for a percentage will be given by:

$$\Pi = p \pm 1.96 \times \sqrt{\frac{\Pi(100 - \Pi)}{n}}$$

as shown in Figure 11.8.

The probability statement would be:

$$P\left(\Pi - 1.96\sqrt{\frac{\Pi(100 - \Pi)}{n}} \leq p \leq \Pi + 1.96\sqrt{\frac{\Pi(100 - \Pi)}{n}}\right) = 0.95$$

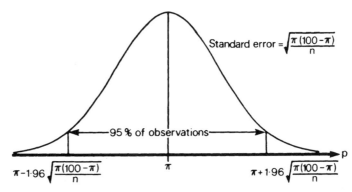

Figure 11.8 The distribution of sample percentages.

but a more usable format is:

$$\Pi = p \pm 1.96\sqrt{\frac{\Pi(100 - \Pi)}{n}}$$

Unfortunately, this contains the value of the population percentage, Π, on the right-hand side of the equation, and this is precisely what we are trying to estimate. Therefore we substitute the value of the sample percentage, p. Therefore the 95% confidence interval that we will use, will be given by:

$$\Pi = p \pm 1.96\sqrt{\frac{p(100 - p)}{n}}$$

A 99% confidence interval for a percentage would be given by:

$$\Pi = p \pm 2.576\sqrt{\frac{p(100 - p)}{n}}$$

Interpretation of these confidence intervals is exactly the same as the interpretation of confidence intervals for the mean.

Example

A random sample of 100 invoices has been selected from a large file of company records. If nine were found to contain errors, calculate a 95% confidence interval for the true percentage of invoices from this company containing errors.

The sample percentage is $p = 9\%$. This sample statistic is used to estimate the population percentage containing errors, Π. By substituting into the formula for a 95% confidence interval, we have:

$$\Pi = 9\% \pm 1.96 \times \sqrt{\frac{9(100 - 9)}{100}} = 9\% \pm 5.609\%$$

or we could write:

$$3.391 \leq \Pi \leq 14.609$$

As you can see, this is rather a wide interval.

Case study

In the Arbour Housing Survey, 246 respondents out of the 300 reported that they had exclusive use of a fixed bath or shower with a hot water supply. Calculate a 95% and a 99% confidence interval.

The sample percentage:

$$p = \frac{246}{300} \times 100 = 82\%$$

The 95% confidence interval:

$$\Pi = 82\% \pm 1.96 \times \sqrt{\frac{82 \times 18}{300}} = 82\% \pm 4.3\%$$

or $77.7\% \leq \Pi \leq 86.3\%$

The 99% confidence interval:

$$\Pi = 82\% \pm 2.576 \times \sqrt{\frac{82 \times 18}{300}} = 82\% \pm 5.7\% \text{ or } 76.3\% \leq \Pi \leq 87.7\%$$

It can be seen that the increased certainty of the stated result (from 95% confident to 99% confident) has also increased the size of the sampling error term.

11.4.1 Sample size for a percentage

As with the confidence interval for the mean, when we are considering percentages, we will often wish to specify the amount of acceptable error in the final result. If we look at the form of the error, we will be able to determine the appropriate sample size. The error is given by:

$$e = z\sqrt{\frac{p(100 - p)}{n}}$$

and rearranging this gives:

$$n = \left(\frac{z}{e}\right)^2 \times p \times (100 - p)$$

The value of p used will either be a reasonable approximation or a value from a previous survey or from a pilot study.

Example

In a pilot survey, 100 invoices are selected randomly from a large file and nine were found to contain errors. What sample size would it be necessary to take if we wish to produce an estimate of the percentage of all invoices with errors to within plus or minus 3% with a 95% level of confidence?

Here we may use the result of the pilot study, $p = 9\%$. The value of z will be 1.96. Substituting into the formula, we have:

$$n = \left(\frac{1.96}{3}\right)^2 \times 9 \times 91 = 349.5856$$

So, to achieve the required level of accuracy, we need a sample of 350 randomly selected invoices.

Where no information is available about the appropriate value of p to use in the calculations, we would use a value of 50%. Looking at the information given in Table 11.4, we can see that at a value of $p = 50\%$ we have the largest possible standard error, and thus the largest sample size requirement. This will be the safest approach where we have no prior knowledge.

Table 11.4 The size of standard error

p	$(100 - p)$	$\sqrt{\left[\frac{p(100 - p)}{n}\right]}$
10	90	$\sqrt{(900/n)}$
20	80	$\sqrt{(1600/n)}$
30	70	$\sqrt{(2100/n)}$
40	60	$\sqrt{(2400/n)}$
50	50	$\sqrt{(2500/n)}$

Example

What sample size would be required to produce an estimate for a population percentage to within plus or minus 3% if no prior information were available?

In this case we would let $p = 50\%$ and assume the 'worst possible case'. By substituting into the formula, we have:

$$n = \left(\frac{1.96}{3}\right)^2 \times 50 \times 50 = 1067.1$$

So, to achieve the required accuracy, we would need a random sample of 1068.

Comparing the last two examples, we see that in both cases the level of confidence specified is 95%, and that the level of acceptable error to be allowed is plus or minus 3%. However, because of the different assumption that we were able to make about the value of p in the formula, we arrive at very different values for the required sample size. This shows the enormous value of having some **prior information**, since, for the cost of a small pilot survey we are able to reduce the main sample size to approximately 35% of the size it would have been without that information. In addition a pilot survey also allows us to test the questionnaire to be used (as discussed in Chapter 3).

An alternative to the usual procedure of a pilot survey followed by the main survey is to use a **sequential sampling procedure**. This involves a relatively small sample being taken first, and then further numbers are added as better and better estimates of the parameters become available. In practice, sequential sampling requires the continuation of interviews until results of sufficient accuracy have been obtained.

11.5 The difference between independent samples

We have so far considered only working with a single sample. In many cases of survey research we also wish to make comparisons between groups in the population, or between seemingly different populations. In other words, we want to make **comparisons between two sets of sample results**. This could, for example, be to test a new machining process in comparison to an existing one by taking a sample of output from each. Similarly we may want to compare consumers in the north with those in the south of a country or region.

In this section we will make these comparisons by calculating the **difference** between the sample statistics derived from each sample. We will also assume that the **two samples are independent** and that we are dealing with **large samples**. (For information on dealing with small samples see Section 11.7.) Although we will not derive the statistical theory behind the results we use, it is important to note that the theory relies on the samples being independent and that the results do not hold if this is not the case. For example, if you took a single sample of people and asked them a series of questions, and then two weeks later asked the same people another series of questions, the samples would not be independent and we could not use the

11.1 Statistical inference

11.2 Inference about a population

11.3 Confidence interval for the population mean

11.4 Confidence interval for the population percentage

11.7 The t-distribution

11.8 Confidence interval for a median

11.6 The finite population correction factor

11.5 The difference between independent samples

confidence intervals shown in this section. (You may recall from Chapter 3 that this methodology is called a **panel** survey.)

One result from statistical sampling theory states that although we are taking the difference between the two sample parameters (the means or percentages), we **add the variances**. This is because the two parameters are themselves variable (see Section 10.3) and thus the measure of variability needs to take into account the variability of both samples.

11.5.1 Confidence interval for the difference of means

The format of a confidence interval remains the same as before:

$$\text{population parameter} = \text{sample statistic} \pm \text{sampling error}$$

but now the population parameter is the difference between the population means $(\mu_1 - \mu_2)$, the sample statistic is the difference between the sample means $(\overline{x_1} - \overline{x_2})$ and the sampling error consists of the z-value from the Normal distribution tables multiplied by the root of the sum of the sample variances divided by their respective sample sizes. This sounds like quite a mouthful (!) but is fairly straightforward to use with a little practice.

The 95% confidence interval for the difference of means is given by the following formula:

$$(\mu_1 - \mu_2) = (\overline{x_1} - \overline{x_2}) \pm 1.96 \sqrt{\left[\frac{s_1^2}{n_1} + \frac{s_2^2}{n_2}\right]}$$

where the subscripts denote sample 1 and sample 2. (Note the relatively obvious point that we must **keep a close check on which sample we are dealing with at any particular time.**)

Case study

It has been decided to compare some of the results from the Arbour Housing Survey with those from the Pelouse Housing Survey. Of particular interest was the level of monthly rent, a summary of which is given below:

The Arbour Housing Survey	The Pelouse Housing Survey
(Survey 1)	(Survey 2)
$n_1 = 180$	$n_2 = 150$
$\overline{x_1} = £221.25$	$\overline{x_2} = £206.38$
$s_1 = £89.48$	$s_2 = £69.88$

By substitution, the 95% confidence interval is:

$$(\mu_1 - \mu_2) = (£221.25 - £206.38) \pm \sqrt{\left[\frac{(89.48)^2}{180} + \frac{(69.88)^2}{150}\right]} = £14.87 \pm £17.20$$

or

$$-£2.30 \le (\mu_1 - \mu_2) \le £32.07$$

As this range **includes zero**, we cannot be 95% confident that there is a difference in rent between the two areas, even though the average rent on the basis of sample information is higher in Tonnelle (the area covered by the Arbour Housing Survey). The observed difference could be explained by inherent variation in sample results.

11.5.2 Confidence interval for the difference of percentages

In this case we only need to know the two sample sizes and the two sample percentages to be able to estimate the difference in the population percentages. The formula for this confidence interval takes the following form:

$$(\Pi_1 - \Pi_2) = (p_1 - p_2) \pm 1.96\sqrt{\left[\frac{p_1(100 - p_1)}{n_1} + \frac{p_2(100 - p_2)}{n_2}\right]}$$

where the subscripts denote sample 1 and sample 2.

Case study

In the Arbour Housing Survey, 234 respondents out of the 300 reported that they had exclusive use of a flush toilet inside the house. In the Pelouse Housing Survey, 135 out of 150 also reported that they had exclusive use of a flush toilet inside the house. Construct a 95% confidence interval for the percentage difference in this housing quality characteristic.

The summary statistics are as follows:

The Arbour Housing Survey (Survey 1)	The Pelouse Housing Survey (Survey 2)
$n_1 = 180$	$n_2 = 150$
$p_1 = \frac{234}{300} \times 100 = 78\%$	$p_2 = \frac{135}{150} \times 100 = 90\%$

By substitution, the 95% confidence interval is:

$$(\Pi_1 - \Pi_2) = (78\% - 90\%) \pm 1.96\sqrt{\left[\frac{78 \times 22}{300} + \frac{90 \times 10}{150}\right]} = -12\% \pm 6.7\%$$

or

$$-18.7\% \leq (\Pi_1 - \Pi_2) \leq -5.3\%$$

This range **does not** include any positive value or zero, suggesting that the percentage from the Arbour Housing Survey is less than that from the Pelouse Housing Survey. In the next chapter we will consider how to test the 'idea' that a real difference exists. For now, we can accept that the sample evidence does suggest such a difference.

The significance of the results will reflect the sample design. The width of the confidence interval (and the chance of including a zero difference) will decrease as:

1 the size of sample or samples is increased
2 the variation is less (a smaller standard deviation)
3 in the case of percentages, the difference from 50% increases (see Table 11.4); and
4 the sample design is improved (e.g. the use of stratification).

The level of confidence still needs to be chosen by the user – 95% being most typical. However, it is not uncommon to see the use of 90%, 99% and 99.9% confidence intervals.

Exercises

1 An operator of fleet vehicles wishes to compare the service costs at two different garages. Records from one garage show that for the 70 vehicles

serviced the mean cost was £55 and the standard deviation £9. Records from the other garage show that for the 50 vehicles serviced the mean cost was £52 and the standard deviation £12. Construct a 95% confidence interval for the difference in servicing costs.

(You should get: 95% confidence interval, $(\mu_1 - \mu_2) = £3 \pm £3.94$.)

2 In a survey of 600 electors, 315 claimed they would vote for party X. A month later, in another survey of 500 electors, 290 claimed they would vote for party X. Construct a 95% confidence interval for the difference in voting.

(You should get: 95% confidence interval, $(\Pi_1 - \Pi_2) = -5.5\% \pm 5.89\%$.)

11.6 The finite population correction factor

In all the previous sections of this chapter we have assumed that we are dealing with samples that are large enough to meet the conditions of the central limit theorem ($n > 30$), but are **small relative to the defined population**. We have seen (Section 11.4.1), for example (making these assumptions), that a sample of just over 1000 is needed to produce a confidence interval with a sampling error of ±3%. However, suppose the population were only a 1000, or 1500 or 2000? Some populations by their nature are small, e.g. specialist retail outlets in a particular region. In some cases we may decide to conduct a census and exclude sampling error. In other cases we may see advantages in sampling.

As the proportion of the population included in the sample increases, the sampling error will decrease. Once we include all the population in the sample – take a census – there will be no

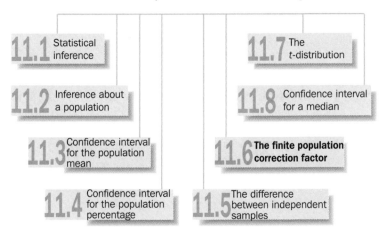

error due to sampling (although errors may still arise due to bias, lying, mistakes, etc.). To take this into account in our calculations we need to correct the estimate of the standard error by multiplying it by the **finite population correction factor**. This is given by the following formula:

$$\sqrt{\left(1 - \frac{n}{N}\right)}$$

where n is the sample size, and N is the population size. As you can see, as the value of n approaches the value of N, the value of the bracket gets closer and closer to zero, thus making the size of the standard error **smaller and smaller**. As the value of n becomes smaller and smaller in relation to N, the value of the bracket gets **nearer and nearer to one**, and the standard error gets closer and closer to the value we used previously.

Where the finite population correction factor is used, the formula for a 95% confidence interval becomes:

$$\mu = \bar{x} \pm 1.96 \times \sqrt{\left(1 - \frac{n}{N}\right)} \times \frac{s}{\sqrt{n}}$$

Example

Suppose a random sample of 30 wholesalers used by a toy producer order, on average 10 000 cartons of crackers each year. The sample showed a standard deviation of 1500 cartons. In total, the manufacturer uses 40 wholesalers. Find a 95% confidence interval for the average size of annual order to this manufacturer.

Here, n = 30, N = 40, $\bar{x}$ = 10 000 and s = 1500. Substituting these values into the formula, we have:

$$\mu = 10,000 \pm 1.96 \sqrt{\left(1 - \frac{30}{40}\right)} \times \frac{1,500}{\sqrt{30}} = 10,000 \pm 268.384$$

which we could write as:

$$9731.616 \leq \mu \leq 10\,268.384$$

If no allowance had been made for the high proportion of the population selected as the sample, the 95% confidence interval would have been:

$$9463.232 \leq \mu \leq 10\,536.768$$

which is considerably wider and would make planning more difficult.

As a 'rule of thumb', we only consider using the finite correction factor if the sample size is 10% or more of the population size. Typically, we don't need to consider the use of this correction factor, because we work with relatively large populations. However, it is worth knowing that 'correction factors' do exist and are seen as a way of reducing sampling error.

Case study

In Case study 4, we could reasonably assume that the number of households (the population for the purpose of these housing surveys) in Tonnelle and Sauterelle was relatively large compared to the sample sizes. In which case, we would not consider making the finite population correction.

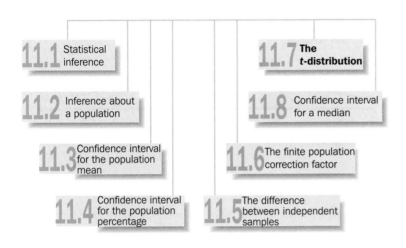

11.7 The t-distribution

We have been assuming that either the population standard deviation (σ) was known (an unlikely event), or that the sample size was sufficiently large so that the sample standard deviation, s, provided a good estimate of the population value (see Section 11.3.1). Where these criteria are not met, we are not able to assume that the sampling distribution is a Normal distribution, and thus the formulae developed so far will not apply. As we have seen in Section 11.3.2, we are able to calculate the standard deviation from sample data, but where we have a small sample, the amount of variability will be higher, and as a result, the confidence interval will need to be wider.

If you consider the case where there is a given amount of variability in any population, when a large sample is taken, it is likely to pick up examples of both high and low values, and thus the variability of the sample will reflect the variability of the population. When a small sample is taken from the same population, the fewer values available make it less likely that all of the variation is reflected in the sample. Thus a given standard deviation in the small sample would imply more variability in the population than the same standard deviation in a large sample.

Even with a small sample, if the population standard deviation is known, then the confidence intervals can be constructed using the Normal distribution as:

$$\mu = \bar{x} \pm z \frac{\sigma}{\sqrt{n}}$$

where z is the critical value taken from the Normal distribution tables.

Where the value of the population standard deviation is not known we will use the t-distribution to calculate a confidence interval:

$$\mu = \bar{x} \pm t \frac{s}{\sqrt{n}}$$

where t is a critical value from the t-distribution. (The derivation of why the t-distribution applies to small samples is beyond the scope of this book, and of most first-year courses, but the shape of this distribution as described below gives an intuitive clue as to its applicability.)

The shape of the t-distribution is shown in Figure 11.9. You can see that it is still a symmetrical distribution about a mean (like the Normal distribution), but that it is wider. In fact it is a misnomer to talk about *the* t-distribution, since the width and height of a particular t-distribution varies with the **number of degrees of freedom**. This new term is related to the size of the sample, being represented by the letter ν (pronounced *new*), and is equal to $n - 1$ (where n is the sample size). As you can see from the diagram, with a small number of degrees of freedom, the t-distribution is wide and flat; but as the number of degrees of freedom increases, the t-distribution becomes taller and narrower. As the

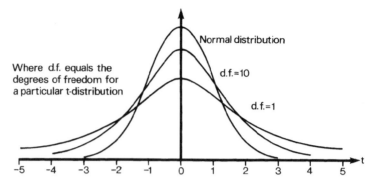

Figure 11.9 The standard Normal distribution (z) and the t-distribution

number of degrees of freedom increases, the t-distribution tends to the Normal distribution.

Values of the t-distribution are tabulated by degrees of freedom and are shown in Appendix D but, to illustrate the point about the relationship to the Normal distribution, consider Table 11.5. We know that to exclude 2.5% of the area of a Normal distribution in the right-hand tail we would use a z-value of 1.96. Table 11.5 shows the comparative values of t for various degrees of freedom.

Before using the t-distribution, let us consider an intuitive explanation of **degrees of freedom**. If a sample were to consist of *one* observation we could estimate the mean (take the average to be that value), but could make no estimate of the variation. If the sample were to consist of *two* observations, we would have only one measure of difference or one degree of freedom. If the sample consisted of *three* values, then we would have two estimates of difference, or two degrees of freedom. Degrees of freedom can be described as the number of independent pieces of information. In estimating variation around a single mean the degrees of freedom will be $n - 1$. If we were estimating the variation around a line on a graph (see Section 15.5) the degrees of freedom would be $n - 2$ since two parameters have been estimated to position the line.

The 95% confidence interval for the mean from sample data when σ is unknown takes the form:

$$\mu = \bar{x} \pm t_{0.025} \frac{s}{\sqrt{n}}$$

where $t_{0.025}$ excludes 2.5% of observations in the extreme right-hand tail area.

Table 11.5 Percentage points of the t-distribution

ν	Percentage excluded in right-hand tail area $2\frac{1}{2}\%$
1	12.706
2	4.303
5	2.571
10	2.228
30	2.042
∞	1.960

Example

A sample of six representatives were selected from a large group to estimate their average daily mileage. The sample mean was 340 miles and the standard deviation 60 miles. Calculate the 95% confidence interval for the population mean.

The summary statistics are: $n = 6$, $\bar{x} = 340$, and $s = 60$.

In this case, the degrees of freedom are $\nu = n - 1 = 5$, and the critical value from the t-distribution is 2.571 (see Appendix D). By substitution, the 95% confidence interval is:

$$\mu = 340 \pm 2.571 \times \frac{60}{\sqrt{6}} = 340 \pm 62.98 \quad \text{or} \quad 277.02 \le \mu \le 402.98$$

If the sampling error is unacceptably large we would need to increase the size of the sample.

Exercises

1 A sample of 15 employees was randomly selected from the workforce to estimate the average travel time to work. If the sample mean was 55 minutes and the standard deviation 13 minutes calculate the 95% and the 99% confidence intervals.

(Answer: 95% confidence interval, $\mu = 55 \pm 7.20$; 99% confidence interval, $\mu = 55 \pm 9.99$.)

2 The sales of a particular product in one week were recorded at five randomly selected shops as 16, 82, 29, 31 and 54. Calculate the sample mean, standard deviation (using a divisor of $n - 1$) and 95% confidence interval.

(Answer: $\bar{x} = 42.4$, $s = 26.02$; 95% confidence interval, $\mu = 42.4 \pm 32.30$.)

We have illustrated the use of the t-distribution for estimating the 95% confidence interval for a population mean from a small sample. Similar reasoning will allow calculation of a 95% confidence interval for a population percentage from a single sample, or variation of the level of confidence by changing the value of t used in the calculation.

Where two small independent samples are involved, and we wish to estimate the difference in either the means or the percentages, we can still use the t-distribution, but now the number of **degrees of freedom** will be related to **both** sample sizes:

$$\nu = n_1 + n_2 - 2$$

and it will also be necessary to allow for the sample sizes in calculating a pooled standard error for the two samples.

In the case of estimating a confidence interval for the difference between two means the pooled standard error is given by:

$$s_p = \sqrt{\frac{(n_1 - 1)s_1^2 + (n_2 - 1)s_2^2}{n_1 + n_2 - 2}}$$

and the confidence interval is:

$$(\mu_1 - \mu_2) = (\overline{x_1} - \overline{x_2}) \pm t \times s_p \times \sqrt{\left[\frac{1}{n_1} + \frac{1}{n_2}\right]}$$

the t value being found from the tables, having

$$\nu = n_1 + n_2 - 2$$

degrees of freedom. A theoretical requirement of this approach is that both samples have variability of the same order of magnitude.

Example

Two processes are being considered by a manufacturer who has been able to obtain the following figures relating to production per hour. Process A produced 110.2 units per hour as the average from a sample of 10 hourly runs. The standard deviation was 4. Process B had 15 hourly runs and gave an average of 105.4 units per hour, with a standard deviation of 3. The summary statistics are as follows:

Process A	Process B
$n_1 = 10$	$n_2 = 15$
$\bar{x}_1 = 110.2$	$\bar{x}_2 = 105.4$
$s_1 = 4$	$s_2 = 3$

Thus the pooled standard error for the two samples is given by

$$s_p = \sqrt{\frac{(10-1)4^2 + (15-1)3^2}{10 + 15 - 2}} = 3.42624$$

There are $\nu = 10 + 15 - 2 = 23$ degrees of freedom, and for a 95% confidence interval, this gives a t-value of 2.069 (see Appendix D). Thus the 95% confidence interval for the difference between the means of the two processes is

$$(\mu_1 - \mu_2) = (110.2 - 105.4) \pm 2.069(3.426\,24)\sqrt{\left[\frac{1}{10} + \frac{1}{15}\right]} = 4.8 \pm 2.894$$

or $1.906 \leq (\mu_1 - \mu_2) \leq 7.694$.

Case study

The sample sizes involved in the Arbour Housing Survey and the Pelouse Housing Survey are well above 30 and we can, for most practical purposes assume the normal distribution. However, if our analysis begins to focus on subsets of the sample, for example those that share a kitchen (Q6(c)) or those paying a monthly rent of under £150 (questions 18–20) then the numbers become small and we need to use the t-distribution. As a guide, we tend to use the confidence intervals based directly on the Normal distribution and assume a large population unless there is good reason to do otherwise.

11.8 Confidence interval for the median – large sample approximation

As we have seen in Chapter 5, the arithmetic mean is not always an appropriate measure of average. Where this is the case, we will often want to use the median. (To remind you, the median is the value of the middle item of a group, when the items are arranged in either ascending or descending order.) Reasons for using a median may be that the data is particularly skewed, for example income or wealth data, or it may lack calibration, for example the ranking of consumer preferences. Having taken a sample, we still need to estimate the errors or variation due to sampling and to express this in terms of a confidence interval, as we did with the arithmetic mean.

Since the median is determined by ranking all of the observations and then counting to locate the middle item, the probability distribution is discrete (the confidence interval for a median can thus be

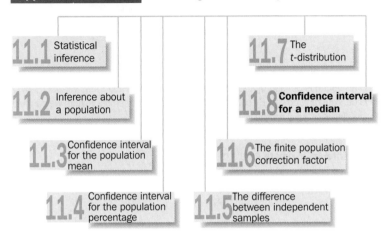

determined directly using the binomial distribution). If the sample is reasonably large ($n > 30$), however, a large sample approximation will give adequate results (see Chapter 10 for the Normal approximation to the binomial distribution).

Consider the ordering of observations by value, as shown below:

$$X_1, X_2, X_3, \ldots, X_n$$

where $X_i \leq X_{i+1}$.

The median is the middle value of this ordered list, corresponding to the $(n+1)/2$ observation. The confidence interval is defined by an upper ordered value (u) and a lower ordered value (l). For a 95% confidence interval, these values are located using:

$$u = \frac{n}{2} + 1.96 \times \frac{\sqrt{n}}{2}$$
$$l = \frac{n}{2} - 1.96 \times \frac{\sqrt{n}}{2} + 1$$

where n is the sample size.

Example

Suppose a random sample of 30 people has been selected to determine the median amount spent on groceries in the last seven days. Results are listed in the table below:

2.50	2.70	3.45	5.72	6.10	6.18
7.58	8.42	8.90	9.14	9.40	10.31
11.40	11.55	11.90	12.14	12.30	12.60
14.37	15.42	17.51	19.20	22.30	30.41
31.43	42.44	54.20	59.37	60.21	65.27

The median will now correspond to the $(30 + 1)/2 = 15\frac{1}{2}$th observation. Its value is found by averaging the 15th and 16th observations:

$$\text{median} = \frac{11.90 + 12.14}{2} = 12.02$$

The sample median is a **point estimate** of the population median. A 95% confidence interval is determined by locating the upper and lower boundaries.

$$u = \frac{30}{2} + 1.96\frac{\sqrt{30}}{2} = 20.368$$

$$l = \frac{30}{2} - 1.96\frac{\sqrt{30}}{2} + 1 = 10.632$$

thus the upper bound is defined by the 21st value (rounding up) and the lower bound by the 10th value (rounding down). By counting through the set of sample results we can find the 95% confidence interval for the median to be:

$$9.14 \leq \text{median} \leq 17.51$$

This is now an **interval estimate** for the median.

Case study

The median will be of particular interest where a skew in the distribution of the data will lead to the mean being pulled upwards (positive skew) or downwards (negative skew). The median would provide a useful contrast to the mean on questions about length of residency (questions 2 and 5) and rent (questions 18–20). Confidence intervals are becoming accepted for the mean but you are less likely to see the use of the median and even less likely to see confidence intervals for medians. The statistics calculated and the types of analysis undertaken will depend on the problem context. You are still likely to find sample data being used in a very descriptive way and may need to make the case for further analysis.

11.9 Conclusions

Sampling is used to improve our knowledge of a defined population. The population may well be defined in terms of people but could also be defined in terms of items produced, retail outlets or in other ways. Essentially the sample statistics are being used to estimate the population parameters. Rather than just use the sample statistic (say the sample mean), as a **point estimate** of the population parameter (say the population mean), we can use an **interval estimate**. In making a point estimate (i.e. using the sample mean as our estimate of the population mean) we will only be *correct by chance*; in fact, there is a very good chance that we will get the answer *wrong*! An interval increases the probability of including the correct answer within our interval estimate; and more than this, we can specify what that probability is by the way in which we construct the confidence interval (i.e. 95% or 99% or any other value we choose).

A confidence interval will normally take the form:

$$\text{population parameter} = \text{sample statistic} \pm \text{a sampling error}$$

where the sampling error is so many (usually 1.96) standard errors.

A summary of the notation and standard errors is given in Table 11.6.

Table 11.6 Notation and standard errors

Sample statistic	Population parameter	Sample estimate of standard error	Degrees of freedom
Large samples:			
$\bar{x}$	μ	$\dfrac{s}{\sqrt{n}}$	
p	π	$\sqrt{\dfrac{p(100-p)}{n}}$	
$\bar{x}_1 - \bar{x}_2$	$\mu_1 - \mu_2$	$\sqrt{\left[\dfrac{s_1^2}{n_1} + \dfrac{s_2^2}{n_2}\right]}$	
$p_1 - p_2$	$\pi_1 - \pi_2$	$\sqrt{\left[\dfrac{p_1(100-p_1)}{n_1} + \dfrac{p_2(100-p_2)}{n_2}\right]}$	
Small samples (using the t-distribution):			
$\bar{x}$	μ	$\dfrac{s}{\sqrt{n}}$	$n-1$
p	π	$\sqrt{\dfrac{p(100-p)}{n}}$	$n-1$
$\bar{x}_1 - \bar{x}_2$	$\mu_1 - \mu_2$	$s_p = \sqrt{\left[\dfrac{(n_1-1)s_1^2 + (n_2-1)s_2^2}{n_1 + n_2 - 2}\right]}$ and $\sigma = s_p \sqrt{\left[\dfrac{1}{n_1} + \dfrac{1}{n_2}\right]}$	$n_1 + n_2 - 2$

11.10 Problems

1 A survey of 50 home buyers with mortgage arrears revealed that the average level of arrears was £1796. Produce a 95% confidence interval for an assumed standard deviation of £500, £1000 and £2000. What would you conclude from your findings?

2 A survey of 180 motorists revealed that they were spending on average £287 per year on car maintenance. Assuming a standard deviation of £94, construct 95% and 99% confidence intervals.

3 A recent investigation had revealed that the average spending on various types of gambling was £6.30 a week with standard deviation of £2.10. Determine a 95% confidence interval assuming sample sizes of 100, 200 and 400. What would you conclude?

4 The mileages recorded for a sample of company vehicles during a given week yielded the following data:

138	164	150	132	144	125	149	157
146	158	140	147	136	148	152	144
168	126	138	176	163	119	154	165
146	173	142	147	135	153	140	135
161	145	135	142	150	156	145	128

(a) Calculate the mean and standard deviation, and construct a 95% confidence interval.

(b) What sample size would be required to estimate the average mileage to within ±3 miles with 95% confidence? State any assumptions made.

5 The number of breakdowns each day on a section of road were recorded for a sample of 250 days as follows:

Number of breakdowns	Number of days
0	100
1	70
2	45
3	20
4	10
5	5
	250

Calculate the 95% and the 99% confidence intervals for the mean. Explain your results.

6 The average weekly overtime earnings from a sample of workers from a particular service industry were recorded as follows:

Average weekly overtime earnings (£)	Number of workers
Under 1	10
1 but under 2	29
2 but under 4	17
5 but under 10	12
10 or more	3
	80

(a) Calculate the mean, standard deviation and the 95% confidence interval for the mean.
(b) What sample size would be required to estimate the average overtime earnings to within $\pm£0.50$ with a 95% confidence interval?

7 A survey of 300 home buyers in a particular area found that 18 had mortgage payments in arrears. Construct 95% and 99% confidence intervals for the arrears percentage.

8 In a survey of 1000 electors, 20% were found to favour party X.
(a) Construct a 95% confidence interval for the percentage in favour of party X.
(b) What sample size would be required to estimate the percentage in favour of party X to within $\pm1\%$ with a 95% confidence interval?

9 What sample size would be required if you wanted to estimate the percentage of homes with gas central heating to within $\pm5\%$ with a 95% confidence interval if:
(a) a previous survey had shown the percentage to be approximately 42%;
(b) no prior information were available?

10 A sample of 75 packets of cereals was randomly selected from the production process and found to have a mean of 500 g and standard deviation of 20 g. A week later a second sample of 50 packets of cereal was selected, using the same procedure, and found to have a mean of 505 g and standard deviation of 16 g. Construct a 95% confidence interval for the change in the average weight of cereal packets.

11 The number of breakdowns each day on two sections of road, section A and section B, were recorded independently for a sample of 250 days as follows:

Number of breakdowns	Number of days Section A	Section B
0	100	80
1	70	65
2	45	57
3	20	31
4	10	11
5	5	6
	250	250

Construct a 95% confidence interval for the difference in the average number of breakdowns on the two sections of road and comment on the structure of the test.

12 A sample of 120 housewives was randomly selected from those reading a particular magazine, and 18 were found to have purchased a new household product. Another sample of 150 housewives was randomly selected from those not reading the particular magazine, and only six were found to have purchased the product. Construct a 95% confidence interval for the difference in the purchasing behaviour.

13 In a survey of 1000 electors, 600 in the North and 400 in the South, 22% were found to favour party X in the North and 18% to favour party X in the South. Construct a 95% confidence interval to show the regional difference.

14 In a sample of 200 cars produced by company A, 42 were found to have faults whilst in a sample of 230 cars produced by company B, 46 were found to have faults. Is there any evidence to suggest a significant difference in the percentage of cars with faults produced by company A and company B?

15 A sample of 35 workers was randomly selected from a workforce of 110 to estimate the average amount spent weekly at the canteen. The sample mean was £5.40 and the sample standard deviation was £2.24.
(a) Calculate a 95% confidence interval for the mean.
(b) What sample size would be required to estimate the mean to within ±£0.50 with a 95% confidence interval?

16 A small club has 50 members and the committee wishes to find their views on the introduction of a life membership fee. They are able to randomly select twenty members and find that only three of these support the idea. Construct a 95% confidence interval for the percentage of members who are in favour of a life membership fee.

17 Two hundred people work for a small engineering company and 49 out of a random sample of 50 workers are in favour of a new bonus scheme. Construct a 95% confidence interval for the percentage of the whole workforce who are in favour of the scheme. Would it surprise you to learn that 8 people were against the scheme?

18 To estimate the average cost of window replacement, 11 quotes were obtained for a typical semi-detached house. The mean was £1259 and the standard deviation was £153. Construct a 95% confidence interval for the mean.

19 The time taken to complete the same task was recorded for seven participants in a training exercise as follows:

Participant	1	2	3	4	5	6	7
Time taken (in minutes)	8	7	8	9	7	7	9

Construct a 95% confidence interval for the average time taken to complete the task.

20 A survey of expenditure on a manufacturer's product has found that the average amount spent in the South is £28 per month with a variance of £5.30. In the North the average was £24 per month with a variance of £3.40. The sample sizes were 10 and 14 respectively. Construct a 95% confidence interval for the difference in the average amount spent in the two regions.

21 A company claims that women take more time off through illness than men. The union wish to test this claim and are able to select two random samples from the workforce: a sample of 17 women and a sample of 11 men. From the personnel records it is found that for the sample of women, the mean number of days illness is 6.3 with a variance of 5.8. For the men the mean is 5.1 days with a variance of 2.6. To help the union in arguing its case, construct a 99% confidence interval for the difference in time lost through illness between men and women.

22 Using the car mileage data from question 4, determine the median and the 95% confidence interval for the median. Compare your answer with that obtained in question 4.

23 Describe how you would design a survey to estimate the percentage of homes in the UK in need of major repairs and what the likely costs would be.

12 Significance testing

Significance testing aims to make **statements** about a **population parameter**, or parameters, on the basis of **sample evidence**. It is sometimes called **point estimation**. The emphasis here is on testing whether a set of sample results support, or are consistent with, some fact or supposition about the population. We are thus concerned to get a 'Yes' or 'No' answer from a **significance test**; either the sample results do support the supposition, or they do not. As you might expect, the two ideas of significance testing and confidence intervals are so closely linked, all of the assumptions that we have made in Chapter 11 about the underlying sampling distribution remain the same for significance tests.

Since we are dealing with samples, and not a census, we can never be 100% sure of our results (since sample results will vary from sample to sample and from the population parameter in a known way – see Figure 11.5). However, by testing an idea (or hypothesis) and showing that it is very, very unlikely to be true we can make useful statements about the population which can form the basis for business decision making.

Significance testing is about putting forward an idea or **testable proposition that captures the properties of the problem** that we are interested in. The idea we are testing may have been obtained from a past census or survey, or it may be an assertion that we put forward, or has been put forward by someone else, for example, an assertion that a majority of people support legislation to enhance animal welfare. Similarly, it could be a critical value in terms of product planning; for example, if a new machine is to be purchased but the process is only viable with a certain level of throughput, we would want to be sure that the company can sell, or use, that quantity before commitment to the purchase of that particular type of machine. Advertising claims could also be tested using this method by taking a sample and testing if the advertised characteristics, for example strength, length of life, or percentage of people preferring this brand, are supported by the evidence of the sample. In developing statistical measures we may want to test if a certain parameter is significantly different from zero; this has useful applications in the development of regression and correlation models (see Chapters 15–17).

Taking the concept a stage further, we may have a set of sample results from two regions giving the percentage of people purchasing a product, and want to test whether there is a **significant difference** between the two percentages. Similarly, we may have a survey which is carried out on an annual basis and wish to test whether there has been a significant shift in the figures since last year. In the same way that we created a confidence interval for the differences between two samples, we can also test for such differences.

Finally we will also look at the situation where only small samples are available and consider the application of the t-distribution to significance testing; another concept to which we will return in Chapters 15 and 16.

After working through this chapter, you will be able to:

- understand and apply the concept of a significance test
- use a hypothesis test on a population mean or percentage
- construct one-sided tests of hypotheses
- understand the different types of error
- test the difference in population means and population percentages
- construct tests given small sample sizes.

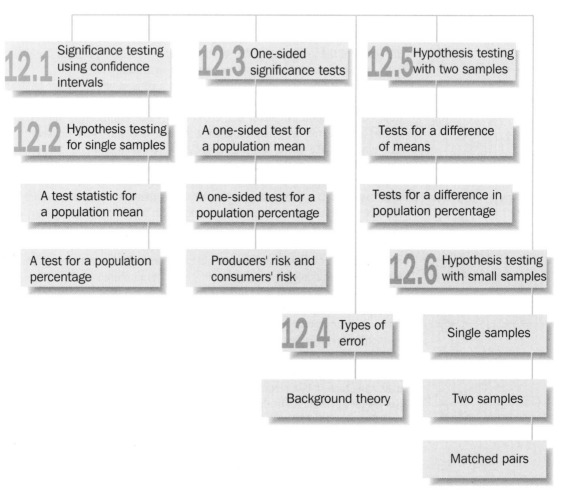

Figure 12.1 Structure of Chapter 12.

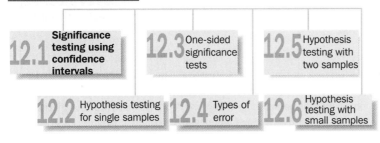

12.1 Significance testing using confidence intervals

Significance testing is concerned with **accepting or rejecting** ideas. These ideas are known as **hypotheses**. If we wish to test one in particular, we refer to it as the **null hypothesis**. The term 'null' can be thought of as meaning no change or no difference from the assertion.

As a procedure, we would first state a null hypothesis; something we wish to judge as true or false on the basis of statistical evidence. We would then check whether or not the null hypothesis was consistent with the confidence interval. If the null hypothesis is contained **within** the confidence interval it will be **accepted**, otherwise, it will be rejected. A confidence interval can be regarded as a **set of acceptable hypotheses**.

To illustrate significance testing consider an example from Chapter 11.

Case study

In the Arbour Housing Survey, it was found that the average monthly rent was £221.25 and that the standard deviation was £89.48, on the basis of 180 responses (see Section 11.3.1). The 95% confidence interval was constructed as follows:

$$\mu = \bar{x} \pm 1.96 \frac{s}{\sqrt{n}}$$

$$= £221.25 \pm 1.96 \times \frac{£89.48}{\sqrt{180}}$$

$$= £221.25 \pm £13.07$$

or

$$£208.18 \leq \mu \leq £234.32$$

Now suppose the purpose of the survey was to test the view that the amount paid in monthly rent was about £200. Given that the confidence interval can be regarded as a set of acceptable hypotheses, the null hypothesis (denoted by H_0) would be written as:

$$H_0 : \mu = £200$$

As the value of £200 is not included within the confidence interval the null hypothesis must be rejected.

The values of the null hypothesis that we can accept or reject are shown on Figure 12.2.

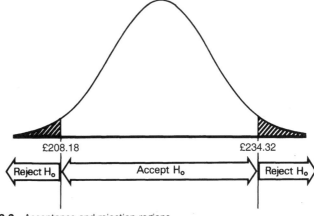

Figure 12.2 Acceptance and rejection regions.

In rejecting the view that the average rent is £200 we must also accept that *there is a chance that our decision could be wrong*. There is a 2.5% chance that the average is less than £208.18 (which would include an average of £200) and a 2.5% chance that the average is greater than £234.32. As we shall see (Section 12.4) there is a probability of making a wrong decision, which we need to balance against the probability of making the correct decision.

12.2 Hypothesis testing for single samples

Hypothesis testing is merely an alternative name for significance tests, the two being used interchangeably. This name does stress that we are testing some supposition about the population, which we can write down as a hypothesis. In fact, we will have two hypotheses whenever we conduct a significance test: one relating to the supposition that we are testing, and one that describes *the alternative situation*.

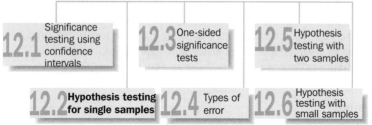

The first hypothesis relates to the claim, supposition or previous situation and is usually called the null hypothesis and labeled as H_0. It implies that there has been no change in the value of the parameter that we are testing from that which previously existed; for example, if the average spending per week on beer last year among 18–25 year olds was £15.53, then the null hypothesis would be that it is still £15.53. If it has been claimed that 75% of consumers prefer a certain flavour of icecream, then the null hypothesis would be that the population percentage preferring that flavour is equal to 75%.

The null hypothesis could be written out as a sentence, but it is more usual to abbreviate it to:

$$H_0 : \mu = \mu_0$$

for a mean where μ_0 is the claimed or previous population mean.

For a percentage:

$$H_0 : \pi = \pi_0$$

where π_0 is the claimed or previous population percentage.

The second hypothesis summarizes what will be the case if the null hypothesis is **not true**. It is usually called the alternative hypothesis (*fairly obviously!*), and is labelled as H_A or as H_1 depending on which text you follow: we will use the H_1 notation here. This alternative hypothesis is usually not specific, in that it does not usually specify the exact alternative value for the population parameter, but rather, it just says that some other value is appropriate on the basis of the sample evidence; for example, the mean amount spent is not equal to £15.53, or the percentage preferring this flavour of icecream is not 75%.

As before, this hypothesis could be written out as a sentence, but a shorter notation is usually preferred:

$$H_1 : \mu \neq \mu_0$$

for a mean where μ_0 is the claimed or previous population mean.

For a percentage:

$$H_1 : \pi \neq \pi_0$$

where π_0 is the claimed or previous population percentage.

Whenever we conduct a hypothesis test, **we assume that the null hypothesis is true while we are doing the test,** and then come to a conclusion on the basis of the figures that we calculate during the test.

Since the sampling distribution of a mean or a percentage (for large samples) is given by the Normal distribution, to conduct a test, we compare the sample evidence (sample statistic) with the null hypothesis (what is assumed to be true). This difference is measured by a **test statistic.** A test statistic of zero or reasonably near to zero would suggest the null hypothesis is correct. The larger the value of the test statistic the larger the difference between the sample evidence and the null hypothesis. If the difference is sufficiently large we conclude that the null hypothesis is incorrect and accept the alternative hypothesis. To make this decision we need to decide on a **critical value** or **critical values.** The critical value or values define the point at which the chance of the null hypothesis being true is at a small, predetermined level, usually 5% or 1% (called the significance level). In this chapter you will see the z-value being used as a test statistic. It is more usual to divide the Normal distribution diagram into sections or areas, and to see whether the z-value falls into a particular section; then we may either accept or reject the null hypothesis.

However, it is worth noting that some **statistical packages** just calculate the probability of the null hypothesis being true when you use them to conduct tests. The reason for doing this is to see how likely or unlikely the result is. If the result is particularly unlikely when the null hypothesis is true, then we might begin to question whether this is, in fact, the case. Obviously we will need to define more exactly the phrase **'particularly unlikely'** in this context before we can proceed with hypothesis tests.

Most tests are conducted at the 5% level of significance, and you should recall that in the normal distribution, the z-values of $+1.96$ and -1.96 cut off a total of 5% of the distribution, 2.5% in each tail. If a calculated z-value is between -1.96 and $+1.96$, then we accept the null hypothesis; if the calculated z-value is below -1.96 or above $+1.96$, we reject the null hypothesis in favour of the alternative hypothesis. This situation is illustrated in Figure 12.3.

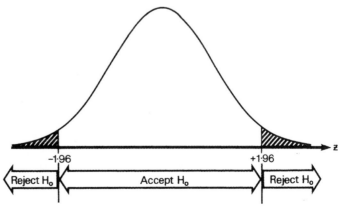

Figure 12.3 Acceptance and rejection regions.

If we were to conduct the test at the 1% level of significance, then the two values used to cut off the tail areas of the distribution would be $+2.576$ and -2.576.

For each test that we wish to conduct, the basic layout and procedure will remain the same, although some of the details will change, depending on what exactly we are testing. A proposed layout is given below, and we suggest that by following this you will present clear and understandable significance tests (and not leave out any steps).

Step	Example
1 State hypotheses.	$H_0: \mu = \mu_0$
	$H_1: \mu \neq \mu_0$
2 State significance level.	5%
3 State critical (cut-off) values.	-1.96
	$+1.96$
4 Calculate the test statistic (z).	Answer varies for each test, but say 2.5 for example.
5 Compare the z-value to the critical values	In this case it is above $+1.96$.
6 Come to a conclusion.	Here we would reject H_0.
7 Put your conclusion into English.	The sample evidence does not support the original claim that the mean was the specified value.

While the significance level may change, which will lead to a change in the critical values, the major difference as we conduct different types of hypothesis test is in the way in which we calculate the z-value at step 4.

12.2.1 A test statistic for a population mean

In the case of testing for a particular value for a population mean, the formula used to calculate z in step 4 is:

$$z = \frac{\bar{x} - \mu}{\sigma / \sqrt{n}}$$

As we saw in Chapter 11, the population standard deviation (σ) is rarely known, and we use the sample standard deviation (s) in its place. We are assuming that the null hypothesis is true, and so $\mu = \mu_0$ The formula will become:

$$z = \frac{\bar{x} - \mu_0}{s / \sqrt{n}}$$

We are now in a position to carry out a test.

Example

A production manager claims that an average of 50 boxes per hour are filled with finished goods at the final stage of a production line. A random sampling of 48 different workers, at different times, working at the end of identical production lines shows an average number of boxes filled as 47.5 with a standard deviation of 0.7 boxes. Does this evidence support the assertion by the production manager at the 5% level of significance?

Step 1:   The null hypothesis is based on the production manager's assertion:

$$H_0 : \mu = 50 \text{ boxes per hour}$$

The alternative hypothesis is any other answer:

$$H_1 : \mu \neq 50 \text{ boxes per hour}$$

Step 2:   As stated in the question, this level is 5%

Step 3:   The critical values are −1.96 and +1.96 (from Appendix C)

Step 4:   Using the formula given above, the z value can be calculated as:

$$z = \frac{47.5 - 50}{0.7 / \sqrt{48}} = \frac{-2.5}{0.101036} = -24.744$$

Step 5:   The calculate value (−24.744) is **below** the lower critical value (−1.96)

Step 6:   We may therefore **reject the null hypothesis** at the 5% level of significance

Step 7:   The sample evidence does not support the production manager's assertion that the average number of boxes filled per hour is 50. The number is **significantly different** from 50.

Although there is only a 2.5 difference between the claim and the sample mean in this example, the test shows that the sample result is significantly different from the claimed value. In all hypothesis testing, it is not the absolute difference between the values which is important, but the number of standard errors that the sample value is away from the claimed value. Statistical significance should not be confused with the notion of business significance or importance.

12.2.2 A test for a population percentage

Here the process will be identical to the one followed above, except that the formula used to calculate the z-value will need to be changed. (You should recall from Chapter 11 that the standard error for a percentage is different from the standard error for a mean.)

The formula will be:

$$z = \frac{p - \Pi_0}{\sqrt{\dfrac{\Pi_0(100 - \Pi_0)}{n}}}$$

where p is the sample percentage and Π_0 is the claimed population percentage (remember that we are assuming that the null hypothesis is true).

Example

An auditor claims that 10% of invoices for a company are incorrect. To test this claim a random sample of 100 invoices are checked, and 12 are found to be incorrect. Test, at the 5% significance level, if the auditor's claim is supported by the sample evidence.

Step 1:   The hypotheses can be stated as:

$$H_0: \pi_0 = 10\%$$
$$H_1: \pi_0 \neq 10\%$$

Step 2:   The significance level is 5%.

Step 3:   The critical values are −1.96 and +1.96.

Step 4:   The sample percentage is $\frac{12}{100} \times 100 = 12\%$.

$$z = \frac{12 - 10}{\sqrt{\dfrac{10(100 - 10)}{100}}} = 0.67$$

Step 5: The calculated value falls between the two critical values.
Step 6: We therefore cannot reject the null hypothesis.
Step 7: The evidence from the sample is consistent with the auditor's claim that 10% of the invoices are incorrect.

In this example, when the calculated value falls between the two critical values. The answer is to **not reject** the claim of 10% rather than to accept the claim. This is because we only have the sample evidence to work from, and we are aware that it is subject to sampling error. The **only way** to firmly accept the claim would be to check every invoice (i.e. carry out a **census**) and work out the actual population percentage which are incorrect.

When we calculated a confidence interval for the population percentage (in Section 11.4) we used a slightly different formulation of the sampling error. There we used

$$\sqrt{\left[\frac{p(100 - p)}{n}\right]}$$

because we **did not know** the value of the population percentage; we were using the sample percentage as our best estimate, and also as the only available value. When we come to significance tests, we have already made the **assumption** that the null hypothesis is true (while we are conducting the test), and so we can use the hypothesized value of the population percentage in the calculations. The formula for the sampling error will thus be

$$\sqrt{\left[\frac{\Pi_0(100 - \Pi_0)}{n}\right]}$$

In many cases there would be very little difference in the answers obtained from the two different formulae, but it is **good practice** (and shows that you know what you are doing) to use the correct formulation.

Case study

It has been claimed on the basis of census results that 87% of households in Tonnelle now have exclusive use of a fixed bath or shower with a hot water supply. In the Arbour Housing Survey of this area, 246 respondents out of the 300 interviewed reported this exclusive usage. Test at the 5% significance level whether this claim is supported by the sample data.

Step 1: The hypothesis can be stated as:

$$H_0 : \pi_0 = 87\%, \qquad H_1 : \pi_0 \neq 87\%$$

We are assuming that the claim being made is correct.
Step 2: The significance level is 5% (but could have been set at a different level if required).
Step 3: The critical values are −1.96 and +1.96.
Step 4: The sample percentage is $(246/300) \times 100 = 82\%$

$$z = \frac{82 - 87}{\sqrt{\left[\frac{87(100 - 87)}{300}\right]}} = -2.58$$

Step 5:   The calculated value falls below the critical value of –1.96.
Step 6:   We therefore reject the null hypothesis at the 5% significance level.
Step 7:   The sample evidence does not **support** the view that 87% of households in the Tonnelle area have exclusive use of a fixed bath or shower with a hot water supply. Given a sample percentage of 82% we could be tempted to conclude that the percentage was lower – the sample does suggest this, but the test was not structured in this way and we must accept the alternative hypothesis that the population percentage is not likely to be 87%. We will consider this issue again when we look at one-sided tests.

12.3 One-sided significance tests

The significance tests shown in the last section may be useful if all we are trying to do is test whether or not a claimed value could be true, but in most cases we want to be able to go one step further. In general, we would want to specify whether the real value is **above** or **below** the claimed value in those cases where we are able to reject the null hypothesis.

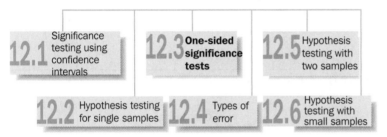

One-sided tests will allow us to do exactly this. The method employed, and the appropriate test statistic which we calculate, will **remain exactly the same**; it is the **alternative hypothesis** and the **interpretation** of the answer which will change. Suppose that we are investigating the purchase of cigarettes, and know that the percentage of the adult population who regularly purchased last year was 34%. If a sample is selected, we do not want to know only whether the percentage purchasing has changed, but rather whether it has decreased (or increased). **Before** carrying out the test it is necessary to decide which of these two propositions you wish to test.

If we wish to test whether or not the percentage has decreased, then our hypotheses would be:

$$\begin{array}{ll} \text{null hypothesis} & H_0 : \pi = \pi_0 \\ \text{alternative hypothesis} & H_1 : \pi < \pi_0 \end{array}$$

where π_0 is the actual percentage in the population last year. This may be an appropriate hypothesis test if you were working for a health lobby.

If we wanted to test if the percentage had increased, then our hypotheses would be:

$$\begin{array}{ll} \text{null hypothesis} & H_0 : \pi = \pi_0 \\ \text{alternative hypothesis} & H_1 : \pi > \pi_0 \end{array}$$

This could be an appropriate test if you were working for a manufacturer in the tobacco industry and were concerned with the effects of improved packaging and presentation.

To carry out the test, we will want to concentrate the chance of rejecting the null hypothesis at **one end** of the Normal distribution. Where the significance level is 5%, then the critical value will be -1.645 (i.e. the cut off value taken from the Normal distribution tables) for the hypotheses $H_0: \pi = \pi_0$, $H_1: \pi < \pi_0$; and $+1.645$ for the hypotheses $H_0: \pi = \pi_0$, $H_1: \pi > \pi_0$. (Check these figures from Appendix C.) Where the significance level is set at 1%, then the critical value becomes either -2.33 or $+2.33$. In terms of answering examination questions, it is important to *read the wording very carefully* to determine which type of test you are required to perform.

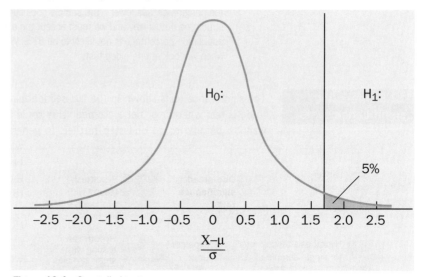

Figure 12.4 One-tailed test.

The interpretation of the calculated z-value is now merely a question of deciding into which of two sections of the Normal distribution it falls.

12.3.1 A one-sided test for a population mean

Here the hypotheses will be in terms of the population mean, or the claimed value. Consider the following example.

Example

A manufacturer of batteries has assumed that the average expected life is 299 hours. As a result of recent changes to the filling of the batteries, the manufacturer now wishes to test if the average life has increased.

A sample of 200 batteries was taken at random from the production line and tested. Their average life was found to be 300 hours with a standard deviation of 8 hours. You have been asked to carry out the appropriate hypothesis test at the 5% significance level.

Step 1: $H_0: \mu = 299$

           $H_1: \mu > 299$

Step 2:   The significance level is 5%.
Step 3:   The critical value will be $+1.645$.

Step 4:

$$z = \frac{300 - 299}{8/\sqrt{200}} = 1.77$$

(Note that we are still assuming the null hypothesis to be true while the test is conducted.)

Step 5:   The calculated value is larger than the critical value.

Step 6:   We may therefore reject the null hypothesis.

      (Note that had we been conducting a two-sided hypothesis test, then we would have been unable to reject the null hypothesis, and so the conclusion would have been that the average life of the batteries had not changed.)

Step 7:   The sample evidence supports the supposition that the average life of the batteries has increased by a significant amount.

Although the significance test has shown that there has been a significant increase in the average length of life of the batteries, this **may not be an important conclusion** for the manufacturer. For instance, it would be quite misleading to use it to back up an advertising campaign which claimed that 'our batteries now last even longer!'. Whenever hypothesis tests are used it is important to **distinguish between statistical significance and importance**. This distinction is often ignored.

Case study

It has been argued that, because of the types of property and the age of the population in Tonnelle, average mortgages are likely to be around £200 a month. However, the Arbour Housing Trust believe the figure is higher because many of the mortgages are relatively recent and many were taken out during the house price boom. Test this proposition, given the result from the Arbour Housing Survey that 100 respondents were paying an average monthly mortgage of £254. The standard deviation calculated from the sample is £72.05 (it was assumed to be £70 in an earlier example). Use a 5% significance level.

1 $H_0: \mu_0 = £200$
   $H_1: \mu_0 > £200$

2 The significance level is 5%.

3 The critical value will be +1.645.

4

$$z = \frac{£254 - £200}{£72.05/\sqrt{100}} = 7.49$$

5 The test statistic value of 7.49 is greater than the critical value of +1.645.

6 On this basis, we can **reject the null hypothesis** that monthly mortgages are around £200 in favour of the alternative that they are likely to be more.

7 A test provides a means to clarify issues and resolve different arguments. In this case the average monthly mortgage payment is higher than had been argued but this **does not necessarily** mean that the reasons put forward by the Arbour Housing Trust for the higher levels are correct. In most cases, further investigation is required. It is always worth checking that the sample structure is the same as the population structure. In an area like Tonnelle we would expect a range of housing and it can be difficult for a small sample to reflect this. The conclusions also refer to the average. The pattern of mortgage payments could vary considerably between different groups of residents. It is known that the Victorian housing is again attracting a more affluent group of residents, and they could have the effect of pulling the mean upwards. Finally, the argument for the null hypothesis could be based on dated information, e.g. the last census.

12.3.2 A one-sided test for a population percentage

Here the methodology is exactly the same as that employed above, and so we just provide two examples of its use.

Example

A small political party expects to gain 20% of the vote in by-elections. A particular candidate from the party is about to stand in a by-election in Derbyshire South East, and has commissioned a survey of 200 randomly selected voters in the constituency. If 44 of those interviewed said that they would vote for this candidate in the forthcoming by-election, test whether this would be significantly above the national party's claim. Use a test at the 5% significance level.

1 $H_0: \pi = 20\%$

   $H_1: \pi > 20\%$

2 The significance level is 5%.

3 The critical value will be $= +1.645$.

4 Sample percentage $= \frac{44}{200} \times 100 = 22\%$

$$z = \frac{22 - 20}{\sqrt{\left[\frac{20(100 - 20)}{200}\right]}} = 0.7071$$

5 $0.7071 < 1.645$.

6 Cannot reject H_0.

7 There is no evidence that the candidate will do better than the national party's claim.

Here there may not be evidence of a statistically significant vote above the national party's claim (a sample size of 1083 would have made this 2% difference statistically significant), but if the candidate does manage to achieve 22% of the vote, and it is a close contest over who wins the seat between two other candidates, then the extra 2% could be very important.

Example

A fizzy drink manufacturer claims that over 40% of teenagers prefer its drink to any other. In order to test this claim, a rival manufacturer commissions a market research company to conduct a survey of 200 teenagers and ascertain their preferences in fizzy drinks. Of the randomly selected sample, 75 preferred the first manufacturers drink. Test at the 5% level if this survey supports the claim.

Sample percentage is:

$$\frac{75}{200} \times 100 = 37.5\%$$

1 $H_0: \pi = 40\%; \qquad H_1: \pi < 40\%$

2 The significance level is 5%

3 The critical value will be -1.645

4 $z = \dfrac{37.5 - 40}{\sqrt{\dfrac{40(100 - 40)}{250}}} = -0.807$

5 $-1.645 < -0.807$

6 Cannot reject H_0

7 There is insufficient evidence to show that less than 40% of teenagers prefer that particular fizzy drink.

Case study

The Arbour Housing Survey was an attempt to describe local housing conditions on the basis of responses to a questionnaire. Many of the characteristics will be summarized in percentage terms, e.g. type of property, access to baths, showers, flush toilets, kitchen facilities. Such enquiries investigate whether these percentages have increased or decreased over time, and whether they are higher or lower than in other areas. If this increase or decrease, or higher or lower value is specified in terms of an alternative hypothesis, the appropriate test will be one-sided.

12.3.3 Producers' risk and consumers' risk

One-sided tests are sometimes referred to as testing producers' risks or consumers' risks. If we are looking at the amount of a product in a given packet or lot size, then the reaction to variations in the amount may be different from the two groups. Say, for instance, that the packet is sold as containing 100 items. If there are more than 100 items per packet, then the producer is effectively giving away some items, since only 100 are being charged for. The producer, therefore, is concerned not to overfill the packets but still meet legal requirements. In this situation, we might presume that the consumer is quite happy, since some items come free of charge. In the opposite situation of less than 100 items per packet, the producer is supplying less than 100 but being paid for 100 (this, of course, can have damaging consequences for the producer in terms of lost future sales), while the consumers receive less than expected, and are therefore likely to be less than happy. Given this scenario, one would expect the producer to conduct a one-sided test using H_1: $\mu > \mu_0$, and a consumer group to conduct a one-sided test using H_1: $\mu < \mu_0$. There may, of course, be legal implications for the producer in selling packets of 100 which actually contain less than 100. (In practice most producers will, in fact, play it safe and aim to meet any minimum requirements.)

12.4 Types of error

We have already seen, throughout this section of the book, that samples can only give us a partial view of a population; there will always be some chance that the true population value really does lie outside of the confidence interval, or that we will come to the wrong decision when conducting a significance test. In fact these probabilities are specified in the names that we have already used – a 95% confidence interval and a test at the 5% level of significance. Both imply a 5% chance of being wrong. You might want to argue that this number is too big, since it gives a 1 in 20 chance of making the wrong decision, but it has been accepted in both business and social research over many years. One reason is that we are dealing with a situation where we rely on people telling us the 'truth' which they may not always do! Using a smaller number, say 1%, would also imply a spurious level of accuracy in our results. (Even in medical research, the figure of 5% may be used.)

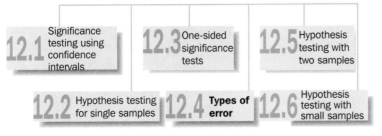

If you consider significance tests a little further, however, you will see that there are, in fact, two different ways of getting the wrong answer. You could throw out the claim when it is, in fact, true; or you could fail to reject it when it is, in fact, false. It becomes important to distinguish between these two types of error. As you can see from Table 12.1, the two different types of error are referred to as Type I and Type II.

Table 12.1 Possible results of a hypothesis test

	Accept H_0	Reject H_0
If H_0 is correct	Correct decision	Type I error
If H_0 is not correct	Type II error	Correct decision

A Type I error is the rejection of a null hypothesis when it is true. The probability of this is known as the **significance level of the test**. This is usually set at either 5% or 1% for most business applications, and is decided on before the test is conducted.

A Type II error is the failure to reject a null hypothesis which is false. The probability of this error **cannot be determined** before the test is conducted.

Consider a more everyday situation. You are standing at the curb, waiting to cross a busy main road. You have four possible outcomes:

1 If you decide not to cross at this moment, and there is something coming, then you have made a **correct** decision.
2 If you decide not to cross, and there is nothing coming, you have wasted a little time, and made a **Type II** error.
3 If you decide to cross and the road is clear, you have made a **correct** decision.
4 If you decide to cross and the road isn't clear, as well as the likelihood of injury, you have made a **Type I** error.

The error which **can be controlled** is the Type I error; and this is the one which is set before we conduct the test.

12.4.1 Background theory

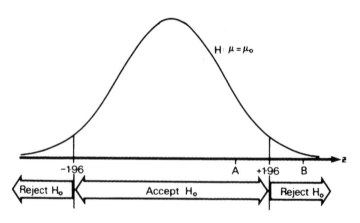

Figure 12.5 The sampling distribution assuming the null hypothesis to be correct.

Consider the sampling distribution of z (which we calculate in step 4), consistent with the null hypothesis, H_0: $\mu = \mu_0$ illustrated as Figure 12.5. The two values for the test-statistic, A and B, shown in Figure 12.5, are both possible, with B being less likely than A. The construction of this test, with a 5% significance level, would mean the acceptance of the null hypothesis when A was obtained and its rejection when B was obtained. Both values could be attributed to an alternative hypothesis, H_1: $\mu = \mu_1$, which we accept in the case of B and reject for A. It is worth noting that if the test statistic follows a sampling distribution we can never be certain about the correctness of our decision. What we can do, having fixed a significance level (Type I error), is construct the rejection region to minimize Type II error. Suppose the alternative hypothesis was that the mean for the population, μ, was not μ_0 but a larger value μ_1. This we would state as:

$$H_1 : \mu = \mu_1 \text{ where } \mu_1 > \mu_0$$

or, in the more general form:

$$H_1 : \mu > \mu_0$$

If we keep the same acceptance and rejection regions as before (Figure 12.5), the Type II error is as shown in Figure 12.6.

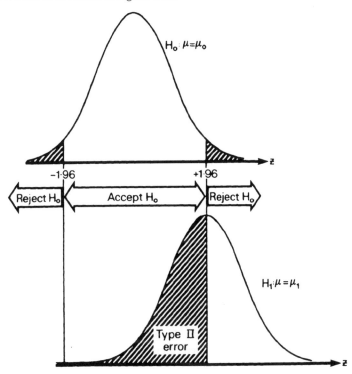

Figure 12.6   The Type II error resulting from a two-tailed test.

As we can see, the probability of accepting H_0 when H_1 is correct can be relatively large. If we test the null hypothesis H_0: $\mu = \mu_0$ against an alternative hypothesis of the form H_1: $\mu < \mu_0$ or H_1: $\mu > \mu_0$, we can reduce the size of the Type II error by careful definition of the rejection region. If the alternative hypothesis is of the form H_1: $\mu < \mu_0$, a critical value of $z = -1.645$ will define a 5% rejection region in the left-hand tail, and if the alternative hypothesis is of the form H_1: $\mu > \mu_0$, a critical value of $z = 1.645$, will define a 5% rejection region in the right-hand tail. The reduction in Type II error is illustrated in Figure 12.7.

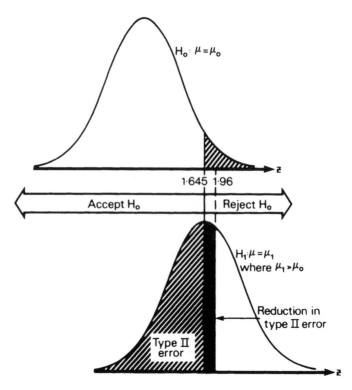

Figure 12.7 Type II error corresponding to a one-sided test.

It can be seen from Figure 12.7 that the test statistic now rejects the null hypothesis in the range 1.645–1.96 as well as values greater than 1.96. If we construct one-sided tests, there is **more chance** that we will reject the null hypothesis in favour of a **more radical** alternative that the parameter has a larger value (or has a smaller value) than specified when that alternative is true. Note that if the alternative hypothesis was of the form H_1: $\mu < \mu_0$, the rejection region would be defined by the critical value $z = -1.645$ and we would reject the null hypothesis if the test statistic took this value or less.

12.5 Hypothesis testing with two samples

All of the tests so far have made comparisons between some known, or claimed, population value and a set of sample results. In many cases we may know little about the actual population value, but will have available the results of another survey. Here we want to find if there is a difference between the two sets of results.

This could be a situation where we have two surveys conducted in different parts of the country, or at different points in time. Our concern now is to determine if the two sets of results are consistent with each other (come from the same parent population) or whether there is a difference between them. Such comparisons may be

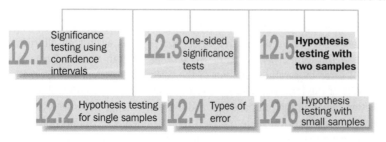

important in a marketing context, for instance, looking at whether or not there are different perceptions of the product in different regions of the country. They could be used in performance appraisal of employees to answer questions on whether the average performance in different regions is, in fact, different. They could be used in assessing a forecasting model, to see if the values have changed significantly from one year to the next.

Again we can use the concepts and methodology developed in previous sections of this chapter. Tests may be two-sided to look for a difference, or one-sided to look for a significant increase (or decrease). Similarly, tests may be carried out at the 5% or the 1% level, and the interpretation of the results will depend upon the calculated value of the test statistic and its comparison to a critical value. It will be necessary to rewrite the hypotheses to take account of the two samples and to find a new formulation for the test statistic (step 4 of the process).

12.5.1 Tests for a difference of means

In order to be clear about which sample we are talking about, we will use the suffixes 1 and 2 to refer to sample 1 and sample 2. Our **assumption** is that the two samples do, in fact, come from the **same population**, and thus the population means associated with each sample result will be the same. This assumption will give us our null hypothesis:

$$H_0 : \mu_1 = \mu_2 \quad \text{or} \quad (\mu_1 - \mu_2) = 0$$

The alternative hypothesis will depend on whether we are conducting a two-sided test or a one-sided test. If it is a two-sided test, the alternative hypothesis will be:

$$H_1 : \mu_1 \neq \mu_2 \quad \text{or} \quad \mu_1 - \mu_2 \neq 0$$

and if it is a one-sided test, it would be either

$$H_1 : \mu_1 > \mu_2 \quad \text{or} \quad \mu_1 - \mu_2 > 0$$

or

$$H_1 : \mu_1 < \mu_2 \quad \text{or} \quad \mu_1 - \mu_2 < 0$$

We could also test whether the difference between the means could be assigned to a specific value, for instance that the difference in take home pay between two groups of workers was £25; here we would use the hypotheses:

$$H_0 : \mu_1 - \mu_2 = £25 \qquad H_1 : \mu_1 - \mu_2 \neq £25$$

The test statistic used is closely related to the formula developed in Section 11.5.1 for the confidence interval for the difference between two means. It will be:

$$z = \frac{(\overline{x_1} - \overline{x_2}) - (\mu_1 - \mu_2)}{\sqrt{\left[\dfrac{s_1^2}{n_1} + \dfrac{s_2^2}{n_2}\right]}}$$

but if we are using a null hypothesis which says that the two population means are equal, then the second bracket on the top line **will be equal to zero**, and the formula for z will be simplified to:

$$z = \frac{(\bar{x_1} - \bar{x_2})}{\sqrt{\left[\frac{s_1^2}{n_1} + \frac{s_2^2}{n_2}\right]}}$$

The example below illustrates the use of this test.

Example

A union claims that the standard of living for its members in the UK is below that of employees of the same company in Spain. A survey of 60 employees in the UK showed an average income of £895 per week with a standard deviation of £120. A survey of 100 workers in Spain, after making adjustments for various differences between the two countries and converting to sterling, gave an average income of £914 with a standard deviation of £ 90. Test at the 1% level if the Spanish workers earn more than their British counterparts.

1 H_0: $(\mu_1 - \mu_2) = 0$ assuming no difference.
   H_1: $(\mu_1 - \mu_2) < 0$ we are testing that one is less than the other.
2 Significance level is 1%.
3 Critical value is −2.33.
4 $z = \dfrac{895 - 914}{\sqrt{\left[\dfrac{120^2}{60} + \dfrac{90^2}{100}\right]}} = -1.06$
5 −2.33 < −1.06.
6 Therefore we cannot reject H_0.
7 There is no evidence that the Spanish earn more than their British counterparts.

12.5.2 Tests for a difference in population percentage

Again, adopting the approach used for testing the difference of means, we will modify the hypotheses, using π instead of μ for the population values, and redefine the test statistic. Both one- and two-sided tests may be carried out.

The test statistic will be:

$$z = \frac{(p_1 - p_2) - (\Pi_1 - \Pi_2)}{\sqrt{\left[\frac{\Pi_1(100 - \Pi_1)}{n_1} + \frac{\Pi_2(100 - \Pi_2)}{n_2}\right]}}$$

but if the null hypothesis is $\pi_1 - \pi_2 = 0$, then this is simplified to:

$$z = \frac{(p_1 - p_2)}{\sqrt{\left[\frac{\Pi_1(100 - \Pi_1)}{n_1} + \frac{\Pi_2(100 - \Pi_2)}{n_2}\right]}}$$

Example

A manufacturer believes that a new promotional campaign will increase favourable perceptions of the product by 10%. Before the campaign, a random sample of 500 consumers showed that 20% had favourable attitudes towards the product. After the campaign, a second random sample of 400 consumers found favourable reactions amongst 28%. Use an appropriate test at the 5% level of significance to find if there has been a 10% improvement in favourable perceptions.

1 Since we are testing for a specific increase, the hypotheses will be:

$$H_0: (\pi_2 - \pi_1) = 10\%$$

$$H_1: (\pi_2 - \pi_1) < 10\%$$

2 The significance level is 5%.
3 The critical value is -1.645.

4 $z = \dfrac{(28 - 20) - (10)}{\sqrt{\left[\dfrac{20(100 - 20)}{500} + \dfrac{28(100 - 28)}{400}\right]}} = -0.697$

5 $-1.645 < -0.697$.
6 We cannot reject H_0.
7 The sample supports a 10% increase in favourable views.

12.6 Hypothesis testing with small samples

As we saw in Chapter 11, the basic assumption that the sampling distribution for the sample parameters is a Normal distribution only holds when the samples are large ($n > 30$). Once we turn to small samples, we need to use a different sampling distribution: the t-distribution (see Section 11.7). Apart from this difference, which implies a change to the formula for the test statistic (step 4 again!) and also to the table of critical values, all of the tests developed so far in this chapter may be applied to small samples.

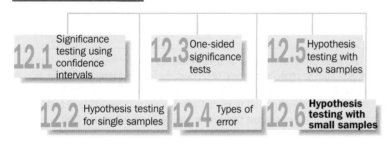

12.6.1 Single samples

For a single sample, the test statistic for a population mean is calculated by using:

$$t = \frac{\bar{x} - \mu}{s/\sqrt{n}}$$

with $(n - 1)$ degrees of freedom.

For a single sample, the test statistic for a population percentage will be calculated using:

$$t = \frac{p - \Pi}{\sqrt{\dfrac{\Pi(100 - \Pi)}{n}}}$$

and, again there will be $(n - 1)$ degrees of freedom. (Note that, as with the large sample test, we use the null hypothesis value of the population percentage in the formula.)

Below are two examples to illustrate the use of the t-distribution in hypothesis testing.

Example

A lorry manufacturer claims that the average annual maintenance cost for its vehicles is £500. The maintenance department of one of their customers believes it to be higher, and to test this randomly selects a sample of six lorries from their large fleet. From this sample, the mean annual maintenance cost was found to be £555, with a standard deviation of £75. Use an appropriate hypothesis test, at the 5% level, to find if the manufacturer's claim is valid.

1 H_0: $\mu = £500$

H_1: $\mu > £500$

2 Significance level is 5%.
3 Degrees of freedom $= 6 - 1 = 5$.
Critical value $= 2.015$.

4 $t = \dfrac{555 - 500}{75/\sqrt{6}} = 1.796$

5 $1.796 < 2.015$.
6 Therefore we cannot reject H_0.
7 The sample evidence does not suggest that maintenance costs are more than £500 per annum.

Example

A company had a 50% market share for a newly developed product last year. It believes that as more entrants start to produce this type of product, its market share will decline, and in order to monitor this situation, decides to select a random sample of 15 customers. Of these, six have bought the company's product. Carry out a test at the 5% level to find if there is sufficient evidence to suggest that their market share is now below 50%.

1 H_0: $\mu = 50\%$

H_1: $\mu < 50\%$

2 Significance level $= 5\%$.
3 Degrees of freedom $= 15 - 1 = 14$.
Critical value $= -1.761$.
4 The sample percentage is $\frac{6}{15} \times 100 = 40\%$

$$t = \dfrac{40 - 50}{\sqrt{\left[\dfrac{50(100 - 50)}{15}\right]}} = -0.775$$

5 $-1.761 < -0.775$.
6 We cannot reject H_0.
7 It appears that the company still has a 50% market share.

12.6.2 Two samples

Where we have two samples, we will again use similar tests to Section 12.5 but with formulae developed using the sampling errors developed in Chapter 11.

In the case of estimating a confidence interval for the difference between two means the pooled standard error (assuming that both samples have similar variability) was given by:

$$s_p = \sqrt{\left[\frac{(n_1 - 1)s_1^2 + (n_2 - 1)s_2^2}{n_1 + n_2 - 2}\right]}$$

Using this, we have the test statistic:

$$t = \frac{(\bar{X_1} - \bar{X_2}) - (\mu_1 - \mu_2)}{s_p\sqrt{\left[\frac{1}{n_1} + \frac{1}{n_2}\right]}}$$

with $(n_1 + n_2 - 2)$ degrees of freedom.

Example

A company has two factories, one in the UK and one in Germany. The head office staff feel that the German factory is more efficient than the British one, and to test this select two random samples. The British sample consists of 20 workers who take an average of 25 minutes to complete a standard task. Their standard deviation is 5 minutes. The German sample has 10 workers who take an average of 20 minutes to complete the same task, and the sample has a standard deviation of 4 minutes. Use an appropriate hypothesis test, at the 1% level, to find if the German workers are more efficient.

1 $H_0: (\mu_1 - \mu_2) = 0$

   $H_1: (\mu_1 - \mu_2) > 0$

2 Significance level $= 1\%$.
3 Degrees of freedom $= 20 + 10 - 2 = 28$.
   Critical value $= 2.467$

$$s_p = \sqrt{\left[\frac{(19 \times 5^2) + (9 \times 4^2)}{20 + 10 - 2}\right]} = 4.70182$$

$$t = \frac{25 - 20}{4.70182\sqrt{\left[\frac{1}{20} + \frac{1}{10}\right]}} = 2.746$$

5 $2.746 > 2.467$.
6 We therefore reject H_0.
7 It appears that the German workers are more efficient at this particular task.

12.6.3 Matched pairs   A special case arises when we are considering tests of the difference of means if the two samples are related in such a way that we may pair the observations. This may arise when two different people assess the same series of situations, for example different interviewers assessing the same group of candidates for a job. The example below illustrates the situation of matched pairs.

Example

Seven applicants for a job were interviewed by two personnel officers who were asked to give marks on a scale of 1 to 10 to one aspect of the candidates' performance. A summary of the marks given is shown below.

		Marks given to candidate:					
	I	II	III	IV	V	VI	VII
Interviewer A	8	7	6	9	7	5	8
Interviewer B	7	4	6	8	5	6	7

Test if there is a significant difference between the standards being applied by the two interviewers at the 5% level.

1 Since we are working with matched pairs, we need only look at the differences between the two scores. The hypotheses will be:

$$H_0: \mu = 0$$

$$H_1: \mu \neq 0$$

2 The significance level is 5%.

3 The number of degrees of freedom is $(7 - 1) = 6$ and the critical value is 2.447.

The calculation of summary statistics for recorded differences of marks

Interviewer A	Interviewer B	Difference (x)	$(x - \bar{x})^2$
8	7	1	0
7	4	3	4
6	6	0	1
9	8	1	0
7	5	2	1
5	6	−1	4
8	7	1	0
		7	10

4 We now need to calculate the summary statistics from the paired samples.

$$\bar{x} = \frac{\sum x}{n} = \frac{7}{7} = 1$$

$$s = \sqrt{\left[\frac{\sum(x - \bar{x})^2}{n - 1}\right]} = \sqrt{\frac{10}{6}} = 1.2910$$

$$t = \frac{1 - 0}{1.2910/\sqrt{7}} = 2.0494$$

5 Now $2.0494 < 2.447$.

6 We cannot reject the null hypothesis.

7 There is no evidence that the two interviewers are using different standards in their assessment of this aspect of the candidates.

The use of such matched pairs is common in market research and psychology. If, for reasons of time and cost, we need to work with small samples, results can be improved using paired samples since it reduces the between-sample variation.

12.7 Conclusions

The ideas developed in this chapter allow us to think about statistical data as evidence. If we consider the practice of law as an analogy, then a set of circumstances require a case to be made, an idea is proposed (that of innocence), an alternative is constructed (that of guilt) and evidence is construed to arrive at a decision. In statistical testing, a null hypothesis is proposed, usually in terms of no change or no difference, an alternative is constructed and data is used to accept or reject the null hypothesis. Strangely, perhaps, these tests are not necessarily acceptable in law.

Given the size of many data sets, we could propose many hypotheses and soon begin to find that we were rejecting the more conservative null hypotheses and accepting the more radical alternatives. This does not, in itself, mean that there are differences or that we have identified change or that we have followed good statistical practice. We need to accept that with a known probability (usually 5%), we can reject the null hypothesis when it is correct – Type I error. We also need to recognize that if we keep looking for relationships we will eventually find some. Testing is not merely about meeting the technical requirements of the test but constructing hypotheses that are meaningful in the problem context. In an ideal world, before the data or evidence was collected, the problem context would be clarified and meaningful hypotheses specified. It is true that we can discover the unexpected from data and identify important and previously unknown relationships. The process of enquiry or research is therefore complex, requiring a balance between the testing of ideas we already have and allowing new ideas to emerge. It should also be noted that a statistically significant result may not have business significance. It could be that statistically significant differences between regions, for example, were acceptable because of regional differences in product promotion and competitor activity.

Since we are inevitably dealing with sample data when we conduct tests of significance, we can never be 100% sure of our answer – remember Type I and Type II error (and other types of error). What significance testing does do is provide a way of quantifying error and informing judgement.

Case study

The Arbour Housing Survey was part of a review of housing conditions in Tonnelle. There are clearly a number of issues such a survey could address, either directly or indirectly for other users. A similar survey conducted by the Pelouse Housing Trust can be used indirectly for comparative purposes. Given the range of matters covered, the data can come in a variety of forms: **categorical, ordinal and interval data**. We can generate hypotheses that require the testing of a single value (we hypothesize that a parameter is equal to a particular value) or those that consider differences (we hypothesize that the difference of parameters is equal to zero or some other value). In this chapter we have only considered those hypotheses that can be specified in terms of parameters; these are often referred to as **parametric tests**. There are tests that are not specified in terms of parameters (see Chapter 13).

We need to ensure that technically the statistical testing is correct and **also** that the research approach is correct. In any survey like this it is important to clarify the purpose and what is to be achieved.

12.8 Problems

1 To what extent is significance testing a science or an art?

2 A survey of 50 home-buyers with mortgage arrears revealed that the average level of arrears was £1796. It had been thought that the average level of arrears would have been about £1500. Produce 95% confidence intervals for assumed standard deviations of £500, £1000 and £2000 and use each of these to test the view that the average is £1500. What do you conclude?

3 A survey of 180 motorists revealed that they were spending on average £287 per year on car maintenance, with a standard deviation of £94. Given that 95% and 99% confidence intervals have already been constructed (see question 2, section 11.10) how would you view the hypothesis that this spending was really about £300?

4 The management of a company claim that the average weekly earnings (including overtime payments) of their employees is £850. The local trade union official disputes this figure, believing that the average earnings are lower. In order to test this claim, a random sample of 150 employees was selected from the payroll. The sample average was calculated to be £828 and the sample standard deviation to be £183.13. Use an appropriate test at the 5% level to determine which side of the argument you agree with.

5 A machine is supposed to be adjusted to produce components to a dimension of 2.000 inches. In a sample of 50 components the mean was found to be 2.001 inches and the standard deviation to be 0.003 inches. Is there evidence to suggest that the machine is set too high?

6 A production manager believes that 100 hours are required to complete a particular task. The accountant, however, disagrees and believes that the task takes less time. To test the view of the accountant, the time taken to complete the task was recorded on 60 occasions by an independent observer. The mean was found to be 96 hours and the standard deviation to be 2.5 hours. Formulate and perform an appropriate test at the 5% significance level.

7 A manufacturer sells a product by weight and states on the packaging that the weight is 375 grams. The quality control inspector suspects that the average weight of packets is, in fact, higher than this and institutes a survey to check these suspicions. In a random sample of 100 packets, the average weight is found to be 379 grams with a standard deviation of 14.2 grams. Test at the 5% level of significance whether this result confirms the suspicions of the quality control inspector.

8 A supermarket sets standards for its checkout operators of serving 50 customers per shift. The evening shift claim that they are not able to meet this standard since customers tend to buy more on evening shopping expeditions. A survey of 40 evening operator shifts finds an average of 47.2 customers served with a standard deviation of 1.3. Test at the 5% level of significance if the operators are significantly below the standard set by the company.

9 A manufacturer claims that only 2% of the items produced are defective. If seven defective items were found in a sample of 200 would you accept or reject the manufacturer's claim?

10 A photocopying machine produces copies, 18% of which are faulty. The supplier of the machine claims that by using a different type of paper the percentage of faulty copies will be reduced. If 45 are found to be faulty from a sample of 300 using the new paper, would you accept the claim of the supplier?

11 Directors of a company claim that 90% of the workforce support a new shift pattern which they have suggested. A random survey of 100 people in the workforce finds 85 in favour of the new scheme. Test at the 5% level if there is a significant difference between the survey results and the claim made by the directors. If there is a statistical difference, does it really matter to the company?

12 When introducing new products to the market place a particular company has a policy that a minimum of 40% of those trying the product at the test market stage should express their approval of it. Testing of a new product has just been completed with a sample of 200 people, of whom 78 expressed their approval of the product. Does this result suggest that significantly less than 40% of people approve of the product? (Conduct your test at the 5% level of significance.)

13 A dispute exists between workers on two production lines. The workers on production line A claim that they are paid less than those on production line B. The company investigates the claim by examining the pay of 70 workers from each production line. The results were as follows:

Sample statistics	Production line	
	A	B
Mean	£393	£394.50
Standard deviation	£6	£7.50

Formulate and perform an appropriate test.

14 A manufacturer of soft drinks currently operates in a segment of the French market and wishes to expand production to enter the Italian market. The manufacturer has been advised that the typical consumer in the target segment of the Italian market is likely to purchase more. To test this view, a survey was conducted in each of the defined markets. In France the sample of 50 gave an average of five purchases a week with a variance of 2.3 and in Italy a sample of 100 gave an average of 5.6 purchases a week with a variance of 1.4. What conclusions could the manufacturer draw from these survey results?

15 Market awareness of a new chocolate bar has been tested by two surveys, one in the Midlands and one in the South East. In the Midlands of 150 people questioned, 23 were aware of the product, whilst in the South East 20 out of 100 people were aware of the chocolate bar. Test at the 5% level of significance if the level of awareness is higher in the South East.

16 Two models of washing machine, the Washit and the Supervat, were tested for a specified range of faults. In a sample of 200 Washit washing machines, 60 were found to have such faults and in a sample of 250 Supervat washing machines, 52 were found to have such faults. Do you consider the two models of washing machine to be equally prone to such faults?

17 A sample of 10 job applicants was asked to complete a mathematics test. The average time taken was 28 minutes and the standard deviation was 7 minutes. If the test had previously taken job applicants on average 30 minutes, is there any evidence to suggest that the job applicants are now able to complete the test more quickly?

18 Records were kept for seven cars to test whether they could achieve a fuel consumption of 56 miles to a gallon of petrol. The results obtained were 52, 49, 57, 53, 54, 55 and 53 miles to the gallon. Formulate and perform an appropriate test at the 5% significance level.

19 A claim has been made that 20% of people are capable of benefiting from a new language learning method developed by a certain company. A consumer group selected a random sample of 14 people who try the new method and, of these, only four are seen to benefit. Test the company's claim on the basis of these survey results.

20 A slot machine manufacturer claims that there is a pay-out on half of the occasions that the machine is used. After incurring substantial losses it is decided to check the machine installed in the 'Man of War' public house. The machine is played 18 times and a pay-out given on eleven of these. Test if the machine is set to give more than the claimed percentage of pay-outs.

21 The times taken to complete a task of a particular type were recorded for a sample of eight employees before and after a period of training as follows:

Employee	Time to complete task (minutes)	
	Before training	After training
1	15	13
2	14	15
3	18	15
4	14	13
5	15	13
6	17	16
7	13	14
8	12	12

Test whether the training is effective.

22 The hypothesis $H_0: \mu = 20$ is to be tested against the hypothesis $H_1: \mu = 21$. The test is based on a sample of 100 at the 5% significance level and the sample standard deviation is 4. What is the Type II error if the rejection region is (a) both tails of the Normal distribution, and (b) only the right tail of the Normal distribution?

13 Non-parametric tests

When looking at hypothesis testing in the previous chapter we were concerned with specific statistics (parameters), which represent statements about the population, and these are then tested by using further statistics derived from the sample. Such **parametric tests** are extremely important in the development of statistical theory and for testing of many sample results, but they do not cover all types of data, particularly when parameters cannot be calculated in a meaningful way. We therefore need to develop other tests which will be able to deal with such situations; a small range of these **non-parametric tests** are covered in this chapter.

Parametric tests require the following conditions to be satisfied:

1 A null hypothesis can be stated in terms of parameters.
2 A level of measurement has been achieved that gives validity to differences.
3 The test statistic follows a known distribution.

It is not always possible to define a meaningful parameter to represent the aspect of the population in which we are interested. For instance, what is an average eye-colour? Equally it is not always possible to give meaning to differences in values, for instance if brands of soft drink are ranked in terms of value for money or taste.

Where these conditions cannot be met, non-parametric tests may be appropriate, but note that for some circumstances, there **may be no suitable test**. As with all tests of hypothesis, it must be remembered that even when a test result is significant in statistical terms, there may be situations where it has no importance in practice. A non-parametric test is **still a hypothesis test**, but rather than considering just a single parameter of the sample data, it looks at the **overall distribution** and compares this to some known or expected value, usually based upon the null hypothesis.

Objectives

After working through this chapter, you should be able to:

- understand when it is more appropriate to use a non-parametric test
- understand and apply chi-square tests
- understand and apply the Mann–Whitney U test
- understand and apply the Wilcoxon test
- understand and apply the Runs test.

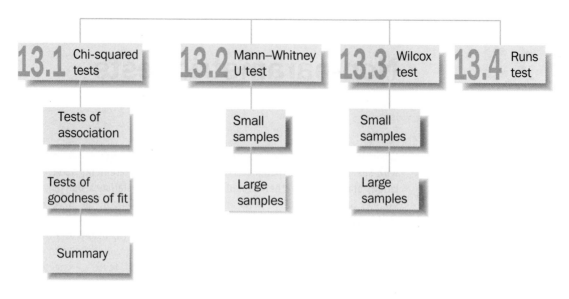

Figure 13.1 Structure of Chapter 13.

13.1 Chi-squared tests

The format of this test is similar to the parametric tests already considered (see Chapter 12). As before, we will define hypotheses, calculate a test statistic, and compare this to a value from tables in order to decide whether or not to reject the null hypothesis. As the name may suggest, the statistic calculated involves squaring values, and thus the result can only be positive. This is by far the most widely used non-parametric hypothesis test and is almost invariably used in the early stages of the analysis of questionnaire results. As statistical programs become easier and easier to use, more people are able to conduct this test, which, in the past, took a long time to construct and calculate.

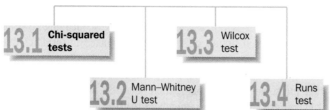

The shape of the chi-squared distribution is determined by the number of degrees of freedom (cf. the t-distribution used in the previous chapter). In general, for relatively low degrees of freedom, the distribution is skewed to the left, as shown in Figure 13.2. As the number of degrees of freedom approaches infinity, then shape of the distribution approaches a Normal distribution.

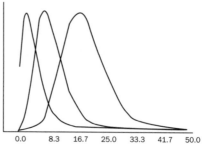

Figure 13.2 χ^2 distribution.

We shall look at two particular applications of the chi-squared (χ^2) test. The first considers survey data, usually from questionnaires, and tries to find if there is an **association** between the answers given to a pair of questions. Secondly, we will use a chi-squared test to check whether a particular set of data follows a known statistical distribution.

13.1.1 Tests of association

When analysing results, typically from a questionnaire, the first step is usually to find out how many responses there were to each alternative in any given question or classification. Such collation of the data allows the calculation of percentages and other descriptive statistics.

Case study

In the Arbour Housing Survey we might be interested in how the responses to question 4 on type of property and question 10 on use of the local post office relate. Looking directly at each question would give us the following:

Question 4   How would your property be best described?

House	150
Flat	100
Bedsit	45
Other	5
Total	300

Question 10   How often do you use the local post office?

Once a month	40
Once a week	200
Twice a week	50
More often	10
Total	300

In addition to looking at one variable at a time, we can construct tables to show how the answers to one question relate to the answers to another; these are commonly referred to as **cross-tabulations**. The single tabulations tell us that 150 respondents (or 50%) live in a house and that 40 respondents (or 13.3%) use their local post office 'once a month'. They do not tell us how often people who live in a house use the local post office and whether their pattern of usage is different from those that live in a flat. To begin to answer these questions we need to complete the table (see Table 13.1).

Table 13.1

	Type of property				
How often	House	Flat	Bedsit	Other	Total
Once a month					40
Once a week					200
Twice a week					50
More often					10
Total	150	100	45	5	300

It would be an extremely boring and time-consuming job to manually fill a table like Table 13.1, even with a small sample such as this. In fact, in some cases we might want to cross-tabulate three or more questions. However, most statistical packages will produce this type of table very quickly. For relatively small sets of data you could use Excel, but for larger scale surveys it would be advantageous to use a more specialist package such as SPSS (the Statistical Package for the Social Sciences). With this type of program it can take rather longer to prepare and enter the data, but the range of analysis and the flexibility offered make this well worthwhile.

The cross-tabulation of questions 4 and 10 will produce the type of table shown as Table 13.2.

Table 13.2

| How often | Type of property | | | | |
	House	Flat	Bedsit	Other	Total
Once a month	30	5	4	1	40
Once a week	110	80	8	2	200
Twice a week	5	10	33	2	50
More often	5	5	0	0	10
Total	150	100	45	5	300

We are now in a better position to relate the two answers, but, because different numbers of people live in each of the types of accommodation, it is not immediately obvious if different behaviours are associated with their type of residence. A chi-squared test will allow us to find if there is a **statistical association** between the two sets of answers; and this, together with other information, may allow the development of a **proposition** that there is a causal link between the two.

To carry out the test we will follow the seven steps used in Chapter 12.

Step 1 State the hypotheses.
H_0: There is no association between the two sets of answers.
H_1: There is an association between the two sets of answers.

Step 2 State the significance level.
As with a parametric test, the significance level can be set at various values, but for most business data it is usually 5%.

Step 3 State the critical value.
The chi-squared distribution varies in shape with the number of **degrees of freedom** (in a similar way to the *t*-distribution), and thus we need to find this value before we can look up the appropriate critical value.

Degrees of freedom

Consider Table 13.1. There are four rows and four columns, giving a total of 16 cells. Each of the row and column totals is **fixed** (i.e. these are the actual numbers given by the frequency count for each question), and thus the individual cell values **must add up** to the appropriate totals. In the first row, we have freedom to put any numbers into three of the cells, but the fourth is then fixed because all four must add to the (fixed) total (i.e. three degrees of freedom). The same will apply to the second row (i.e. three more degrees of freedom). And again to the third row (three more degrees of freedom). Now all of the values on the fourth row are fixed because of the totals (zero degrees of freedom). Totaling these, we have $3 + 3 + 3 + 0 = 9$ degrees of freedom for this table. This is illustrated in Table 13.3. As you can see, you can choose any three cells on the first row, not necessarily the first three.

Table 13.3   Degrees of freedom for a 4×4 table

Free	Free	Free	Fixed	40
Free	Free	Free	Fixed	200
Free	Free	Free	Fixed	50
Fixed	Fixed	Fixed	Fixed	10
150	100	45	5	300

There is a short cut! If you take the number of rows minus one and multiply by the number of columns minus one you get the number of degrees of freedom

$$\nu = (r - 1) \times (c - 1)$$

Using the tables in Appendix E, we can now state the critical value as 16.9.

Step 4 Calculate the test statistic.
The chi-squared statistic is given by the following formula:

$$\chi^2 = \sum \left[\frac{(O - E)^2}{E} \right]$$

where O is the observed cell frequencies (the actual answers) and E is the expected cell frequencies (if the null hypothesis is true).

Finding the expected cell frequencies takes us back to some simple probability rules, since the null hypothesis makes the assumption that the two sets of answers are independent of each other. If this is true, then the cell frequencies will depend only on the totals of each row and column.

Calculating the expected values

Consider the first cell of the table (i.e. the first row and the first column). The number of people living in houses is 150 out of a total of 300, and thus the probability of someone living in a house is:

$$\frac{150}{300} = 0.5$$

The probability of 'once a month' is:

$$\frac{40}{300} = 0.13333$$

Thus the probability of living in a house and 'once a month' is:

$$0.5 \times 0.13333 = 0.066665$$

Since there are 300 people in the sample, one would expect there to be

$$0.066665 \times 300 = 19.9995$$

people who fit the category of the first cell. (Note that the observed value was 30.)

Again there is a short cut! Look at the way in which we found the expected value.

$$19.9995 = \frac{40}{300} \times \frac{150}{300} \times 300$$

$$Expected \text{ cell frequency} = \frac{(\text{row total})}{(\text{grand total})} \times \frac{(\text{column total})}{(\text{grand total})} \times (\text{grand total})$$

$$Expected \text{ cell frequency} = \frac{(\text{row total}) \times (\text{column total})}{(\text{grand total})}$$

We need to complete this process for the other cells in the table, but remember that, because of the degrees of freedom, you only need to calculate nine of them, the rest being found by subtraction.

Statistical packages will, of course, find these expected cell frequencies very quickly.

The expected cell frequencies are shown in Table 13.4.

Table 13.4

How often	House	Type of property Flat	Bedsit	Other	Total
Once a month	20	13.3	6	0.7	40
Once a week	100	66.7	30	3.3	200
Twice a week	25	16.7	7.5	0.8	50
More often	5	3.3	1.5	0.2	10
Total	150	100	45	5	300

(Note that all numbers have been rounded for ease of presentation.)

If we were to continue with the chi-squared test by hand calculation we would need to produce the type of table shown as Table 13.5.

Step 5 Compare the calculated value and the critical value.
The calculated χ^2 value of $140.875 > 16.9$.

Step 6 Come to a conclusion.
We already know that chi-squared cannot be below zero. If all of the expected cell frequencies were exactly equal to the observed cell frequencies, then the value of chi-squared would be zero. Any differences between the observed and expected cell frequencies may be due to either sampling error or to an association between the answers; the larger the differences, the more likely it is that there is an association. Thus, if the calculated value is below the critical value, we will be unable to reject the null hypothesis, but if it is above the critical value, we reject the null hypothesis.

In this example, the calculated value is above the critical value, and thus we reject the null hypothesis.

Step 7 Put the conclusion into English.
There appears to be an association between the type of property people are living in and the frequency of using the local post office. We need now to examine whether such an association is meaningful within the problem context and

Table 13.5

O	E	$(O - E)$	$(O - E)^2$	$\dfrac{(O - E)^2}{E}$
30	20	10.0	100.00	5.00
5	13.3	−8.3	68.89	5.18
4	6	−2.0	4.00	0.67
1	0.7	0.3	0.09	0.13
110	100	10.0	100.00	1.00
80	66.7	13.3	176.89	2.65
8	30	−22.0	484.00	16.13
2	3.3	−1.3	1.69	0.51
5	25	−20.0	400.00	16.00
10	16.7	−6.7	44.89	2.69
33	7.5	25.5	650.25	86.70
2	0.8	1.2	1.44	1.80
5	5	0.0	0.00	0.00
5	3.3	1.7	2.89	0.88
0	1.5	−1.5	2.25	1.50
0	0.2	−0.2	0.04	0.20

chi-squared = 141.04

(Note that the rounding has caused a slight error here. If the original table is analysed using SPSS, the chi-squared value is 140.875.)

the extent to which the association can be explained by other factors. The chi-squared test is only telling you that the association (a word we use in this context in preference to relationship) is likely to exist but not what it is.

An adjustment

In fact, although the basic methodology of the test is correct, there is a problem. One of the basic conditions for the chi-squared test is that all of the expected frequencies must be above five. This is not true for our example! In order to make this condition true, we need to combine adjacent categories until their expected frequencies are equal to five or more.

To do this, we will combine the two categories 'Bedsit' and 'Other' to represent all non-house or flat dwellers; it will also be necessary to combine 'Twice a week' with 'More often' to represent anything above once a week. The new three-by-three cross-tabulation is shown as Table 13.6. The number of degrees of freedom now becomes $(3 - 1) \times (3 - 1) = 4$, and the critical value (from tables) is 9.49.

Re-computing the value of chi-squared from Step 4, we have a value of approximately 107.6, which is still substantially above the critical value of chi-squared. However, in other examples, the amalgamation of categories may affect the decision. In practice, one of the problems of meeting this condition is deciding which categories to combine, and deciding what, if anything, the new category represents. (One of the most difficult examples is the need to combine ethnic groups in a sample.)

This has been a particularly long example since we have been explaining each step as we have gone along. Performing the tests is much quicker in practice, even if a computer package is not used.

Table 13.6

| | | Type of property | |
How often	House	Flat	Other
Observed frequencies:			
Once a month	30	5	5
Once a week	110	80	10
More often	10	15	35
Expected frequencies:			
Once a month	20	13.3	6.7
Once a week	100	66.7	33.3
More often	30	20	10

Example

Purchases of different strengths of lager are thought to be associated with the gender of the drinker and a brewery has commissioned a survey to find if this is true. Summary results are shown below.

| | | Strength | |
	High	Medium	Low
Male	20	50	30
Female	10	55	35

1 H_0: No association between gender and strength bought.
   H_1: An association between the two.
2 Significance level is 5%.
3 Degrees of freedom $= (2 - 1) \times (3 - 1) = 2$.
   Critical value $= 5.99$.
4 Find totals:

| | | Strength | | |
	High	Medium	Low	Total
Male	20	50	30	100
Female	10	55	35	100
Total	30	105	65	200

Expected frequency for Male and High Strength is

$$\frac{100 \times 30}{200} = 15$$

Expected frequency for Male and Medium Strength is

$$\frac{100 \times 105}{200} = 52.5$$

Continuing in this way the expected frequency table can be completed as follows:

	High	Strength	Low	Total
		Medium		
Male	15	52.5	32.5	100
Female	15	52.5	32.5	100
Total	30	105	65	200

Calculating chi-squared:

O	E	$(O-E)$	$(O-E)^2$	$\dfrac{(O-E)^2}{E}$
20	15	5	25	1.667
10	15	−5	25	1.667
50	52.5	−2.5	6.25	0.119
55	52.5	2.5	6.25	0.119
30	32.5	−2.5	6.25	0.192
35	32.5	2.5	6.25	0.192

Chi-squared $= 3.956$

5 $3.956 < 5.99$.

6 Therefore we *cannot reject* the null hypothesis.

7 There appears to be no association between the gender of the drinker and the strength of lager purchased at the 5% level of significance.

13.1.2 Tests of goodness-of-fit

If the data has been collected and seems to follow some pattern, it would be useful to identify that pattern and to determine whether it follows some (already) known statistical distribution. If this is the case, then many more conclusions can be drawn about the data. (We have seen a selection of statistical distributions in Chapters 9 and 10.) The chi-squared test provides a suitable method for deciding if the data follows a particular distribution, since we have the observed values and the expected values can be calculated from tables (or by simple arithmetic). For example, do the sales of whisky follow a Poisson distribution? If the answer is 'yes', then sales forecasting might become a much easier process.

Again, we will work our way through examples to clarify the various steps taken in conducting goodness-of-fit tests. The statistic used will remain as:

$$\chi^2 = \sum \left[\frac{(O-E)^2}{E} \right]$$

where O is the observed frequencies and E is the expected (theoretical) frequencies.

Test for a uniform distribution

You may recall the uniform distribution from Chapter 9; it implies that each item or value occurs the same number of times. Such a test would be useful where we want to find if several individuals are all working at the same rate, or if sales of various 'reps' are the same. Suppose we are examining the number of tasks completed in a set time by five machine operators and have available the data shown in Table 13.7.

Table 13.7

Machine operator	Number of tasks completed
Alf	27
Bernard	31
Chris	29
Dawn	27
Eric	26
Total	140

We can again follow the seven steps:

1 State the hypotheses:
H_0: All operators complete the same number of tasks.
H_1: All operators do not complete the same number of tasks.
(Note that the null hypothesis is just another way of saying that the data follows a uniform distribution.)

2 The significance level will be taken as 5%.

3 The degrees of freedom will be the number of cells minus the number of parameters required to calculate the expected frequencies minus one. Here $\nu = 5 - 0 - 1 = 4$. Therefore (from tables) the **critical value** is 9.49.

4 Since the null hypothesis proposes a uniform distribution, we would expect all of the operators to complete the same number of tasks in the allotted time. This number is:

$$\frac{\text{Total tasks completed}}{\text{Number of operators}} = \frac{140}{5} = 28$$

We can then complete the table:

O	E	$(O - E)$	$(O - E)^2$	$\dfrac{(O-E)^2}{E}$
27	28	−1	1	0.0357
31	28	3	9	0.3214
29	28	1	1	0.0357
27	28	−1	1	0.0357
26	28	−2	4	0.1429
			chi-squared =	**0.5714**

5 $0.5714 < 9.49$.

6 Therefore we do not reject H_0.

7 There is no evidence that the operators work at different rates in the completion of these tasks.

Tests for binomial and Poisson distributions

A similar procedure may be used in this case, but in order to find the theoretical values we will need to know one parameter of the distribution. In the case of a Binomial distribution this will be the probability of success, p (see Chapter 9). For a Poisson distribution it will be the mean (λ).

Example

The components produced by a certain manufacturing process have been monitored to find the number of defective items produced over a period of 96 days. A summary of the results is contained in the following table:

Number of defective items:	0	1	2	3	4	5
Number of days:	15	20	20	18	13	10

You have been asked to establish whether or not the rate of production of defective items follows a Binomial distribution, testing at the 1% level of significance.

1 H_0: the distribution of defectives is Binomial.
   H_1: the distribution is not Binomial.
2 The significance level is 1%.
3 The number of degrees of freedom is:

$$\text{No. of cells} - \text{No. of parameters} - 1 = 6 - 1 - 1 = 4$$

From tables (Appendix E) the critical value is 13.3 (but see below for modification of this value).

4 In order to find the expected frequencies, we need to know the probability of a defective item. This may be found by first calculating the average of the sample results.

$$\bar{x} = \frac{(0 \times 15) + (1 \times 20) + (2 \times 20) + (3 \times 18) + (4 \times 13) + (5 \times 10)}{96}$$

$$= \frac{216}{96} = 2.25$$

From Chapter 9, we know that the mean of a Binomial distribution is equal to np, where n is maximum number of defectives, and so

$$p = \frac{\bar{x}}{n} = \frac{2.25}{5} = 0.45$$

Using this value and the Binomial formula we can now work out the theoretical values, and hence the expected frequencies. The Binomial formula states

$$P(r) = \binom{n}{r} p^r (1-p)^{n-r}$$

r	$P(r)$	Expected frequency = $96 \times P(r)$	
0	0.0503	4.83 ⎫	24.6
1	0.2059	19.77 ⎭	
2	0.3369	32.34	
3	0.2757	26.47	
4	0.1127	10.82 ⎫	12.59
5	0.0185	1.77 ⎭	
	1.0000	96.00	

Note that because two of the expected frequencies are less than five, it has been necessary to combine adjacent groups. This means that we must modify the number of degrees of freedom, and hence the critical value.

$$\text{Degrees of freedom} = 4 - 1 - 1 = 2$$

$$\text{Critical value} = 9.21$$

We are now in a position to calculate chi-squared.

r	0	E	$(0 - E)$	$(0 - E)^2$	$\dfrac{(0 - E)^2}{E}$
0, 1	35	24.60	10.40	108.16	4.3967
2	20	32.34	−12.34	152.28	4.7086
3	18	26.47	−8.47	71.74	2.7103
4, 5	23	12.59	10.41	108.37	8.6075

chi-squared = 20.4231

5 $20.4231 > 9.21$.
6 We therefore reject the null hypothesis.
7 There is no evidence to suggest that the production of defective items follows a Binomial distribution at the 1% level of significance.

Example

Items are taken from the stockroom of a jewellery business for use in producing goods for sale. The owner wishes to model the number of items taken per day, and has recorded the actual numbers for a hundred day period. The results are shown below.

Number of items	Number of days
0	7
1	17
2	26
3	22
4	17
5	9
6	2

If you were to build a model for the owner, would it be reasonable to assume that withdrawals from stock follow a Poisson distribution? (Use a significance level of 5%.)

1 H_0: the distribution of withdrawals is Poisson.
   H_1: the distribution is not Poisson.
2 Significance level is 5%.
3 Degrees of freedom $= 7 - 1 - 1 = 5$. Critical value $= 11.1$ (but see below).
4 The parameter required for a Poisson distribution is the mean, and this can be found from the sample data.

$$\bar{x} = \frac{(0 \times 7) + (1 \times 17) + (2 \times 26) + (3 \times 22) + (4 \times 17) + (5 \times 9) + (6 \times 2)}{100}$$

$$= \frac{260}{100} = 2.6 = \lambda \text{(lambda)}$$

We can now use the Poisson formula to find the various probabilities and hence the expected frequencies:

$$P(x) = \frac{\lambda^x e^{-\lambda}}{x!}$$

x	$P(x)$	Expected frequency $P(x) \times 100$
0	0.0743	7.43
1	0.1931	19.31
2	0.2510	25.10
3	0.2176	21.76
4	0.1414	14.14
5	0.0736	7.36 ⎫ 12.26
*6 or more	0.0490	4.90 ⎭
	1.0000	100.00

(*Note that, since the Poisson distribution goes off to infinity, in theory, we need to account for all of the possibilities. This probability is found by summing the probabilities from 0 to 5, and subtracting the result from one.)

Since one expected frequency is below 5, it has been combined with the adjacent one. The degrees of freedom therefore are now $6 - 1 - 1 = 4$, and the critical value is 9.49. Calculating chi-squared, we have:

(x)	O	E	$(O - E)$	$(O - E)^2$	$\dfrac{(O - E)^2}{E}$
0	7	7.43	−0.43	0.1849	0.0249
1	17	19.31	−2.31	5.3361	0.2763
2	26	25.10	0.90	0.8100	0.0323
3	22	21.76	0.24	0.0576	0.0027
4	17	14.14	2.86	8.1796	0.5785
5+	11	12.26	−1.26	1.5876	0.1295

chi-squared $= 1.0442$

5 $1.0442 < 9.49$.
6 We therefore cannot reject H_0.
7 The evidence from the monitoring suggests that withdrawals from stock follow a Poisson distribution.

Test for the Normal distribution

This test can involve more data manipulations since it will require grouped (tabulated) data and the calculation of two parameters (the mean and the standard deviation) before expected frequencies can be determined. (In some cases, the data may already be arranged into groups.)

Example

The results of a survey of 150 people contain details of their income levels which are summarized here. Test at the 5% significance level if this data follows a Normal distribution.

Weekly income	Number of people
Under £100	30
£100 but under £200	40
£200 but under £300	45
£300 but under £500	20
£500 but under £900	10
Over £900	5
	150

1 H_0: The distribution is Normal.
   H_1: The distribution is not Normal.
2 Significance level is 5%.
3 Degrees of freedom $= 6 - 2 - 1 = 3$.
   Critical value (from tables) $= 7.81$.
4 The mean of the sample $= £266$ and the standard deviation $= £239.02$ calculated from original figures. (N.B. see Section 11.2.2 for formulae.)
   To find the expected values we need to:

   (a) convert the original group limits into z-scores (by subtracting the mean and then dividing by the standard deviation)
   (b) find the probabilities by using the Normal distribution tables (Appendix C); and
   (c) find our expected frequencies.

   This is shown in the table below.

Weekly income	z	Probability	Expected frequency $= prob \times 150$
Under £100	−0.70	0.2420	36.3
£100 but under £200	−0.28	0.1471	22.065
£200 but under £300	0.14	0.1666	24.99
£300 but under £500	0.98	0.2708	40.62
£500 but under £900	2.65	0.15948	23.922 ⎫ 24.525
Over £900		0.00402	0.603 ⎭
		1.00000	150

(Note that the last two groups are combined since the expected frequency of the last group is less than five. This changes the degrees of freedom to $5 - 2 - 1 = 2$, and the critical value to 5.99.)

Thus we can now calculate the chi-squared value:

O	E	$(O - E)$	$(O - E)^2$	$\dfrac{(O - E)^2}{E}$
30	36.3	−6.300	39.6900	1.0934
40	22.065	17.935	321.6642	14.5780
45	24.99	20.01	400.4001	16.0224
20	40.62	−20.62	425.1844	10.4674
15	24.525	−9.525	90.7256	3.6993

$$\text{chi-squared} = 45.8605$$

5 45.8605 > 5.99.
6 We therefore reject the null hypothesis.
7 There is no evidence that the income distribution is Normal.

13.1.3 Summary note   It is worth noting the following characteristics of chi-squared:

- χ^2 is only a **symbol**; the square root of χ^2 has no meaning.
- χ^2 can never be less than zero. The squared term in the formula ensures **positive values.**
- χ^2 is concerned with comparisons of observed and expected frequencies (or counts). We therefore only need a **classification** of data to allow such counts and not the more stringent requirements of measurement.
- If there is a **close correspondence** between the observed and expected frequencies, χ^2 will tend to a low value attributable to sampling error, suggesting the correctness of the null hypothesis.
- If the observed and expected frequencies are **very different** we would expect a large positive value (not explicable by sampling errors alone), which would suggest that we reject the null hypothesis in favour of the alternative hypothesis.

13.2 Mann–Whitney U test

This non-parametric test deals with two samples that are independent and may be of different sizes. It is the equivalent of the *t*-test that we considered in Chapter 12. Where the samples are small (<30) we need to use tables of critical values (Appendix G) to find whether or not to reject the null hypothesis; but where the sample size is large, we can use a test based on the Normal distribution.

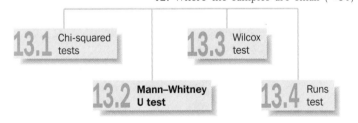

The basic premise of the test is that once all of the values in the two samples are put into a single ordered list, if they come from the same parent population, then the rank at which values from Sample 1 and Sample 2 appear will be by chance (at random). If the two samples come from different populations, then the rank at which sample values appear will not be random and there will be a tendency for values from one of the samples to have lower ranks than values from the other sample. We are therefore testing whether the positioning of values from the two samples in an ordered list follows the same pattern or is different.

While we will show how to conduct this test by hand calculation, packages such as SPSS and MINITAB will perform the test by using the appropriate commands.

13.2.1 Small sample test

Consider the situation where samples have been taken from two branches of a chain of stores. The samples relate to the daily takings and both branches are situated in city centres. We wish to find if there is any difference in turnover between the two branches.

Branch 1: £235, £255, £355, £195, £244, £240, £236, £259, £260
Branch 2: £240, £198, £220, £215, £245

1   H_0: the two samples come from the same population.
    H_1: the two samples come from different populations.
2   We will use a significance level of 5%.
3   To find the critical level for the test statistic, we look in the tables (Appendix G) and locate the value from the two sample sizes. Here the sizes are 9 and 5, and so the critical value of U is 8.
4   To calculate the value of the test statistic, we need to rank all of the sample values, keeping a note of which sample each value came from.

Rank	Value	Sample
1	195	1
2	198	2
3	215	2
4	220	2
5	235	1
6	236	1
7.5	240	1
7.5	240	2
9	244	1
10	245	2
11	255	1
12	259	1
13	260	1
14	355	1

(Note that in the event of a tie in ranks, an average is used.)

We now sum the ranks for each sample:

Sum of ranks for Sample 1 = 78.5

Sum of ranks for Sample 2 = 26.5

We select the **smallest** of these two, i.e. 26.5 and put it into the following formula:

$$T = S - \frac{n_1(n_1 + 1)}{2}$$

where S is the smallest sum of ranks, and n_1 is the number in the sample whose ranks we have summed.

$$T = 26.5 - \frac{5(5+1)}{2} = 11.5$$

5 $11.5 > 8$.
6 Therefore we reject H_0.
7 Thus we conclude that the two samples come from different populations.

11.2.2 Large sample test

Consider the situation where the awareness of a company's product has been measured amongst groups of people in two different countries.

Measurement is on a scale of 0–100, and each group has given a score in this range; the scores are shown below.

Country A: 21, 34, 56, 45, 45, 58, 80, 32, 46, 50, 21, 11, 18, 89, 46, 39, 29, 67, 75, 31, 48
Country B: 68, 77, 51, 51, 64, 43, 41, 20, 44, 57, 60

Test to discover whether the level of awareness is the same in both countries.

1 H_0: the levels of awareness are the same.
  H_1: the levels of awareness are different.
2 We will take a significance level of 5%.
3 Since we are using an approximation based on the Normal distribution, the critical values will be ± 1.96 for this two-sided test.
4 Ranking the values, we have:

Rank	Value	Sample	Rank	Value	Sample
1	11	A	16.5	46	A
2	18	A	18	48	A
3	20	B	19	50	A
4.5	21	A	20.5	51	B
4.5	21	A	20.5	51	B
6	29	A	22	56	A
7	31	A	23	57	B
8	32	A	24	58	A
9	34	A	25	60	B
10	39	A	26	64	B
11	41	B	27	67	A
12	43	B	28	68	B
13	44	B	29	75	A
14.5	45	A	30	77	B
14.5	45	A	31	80	A
16.5	46	A	32	89	A

Sum of ranks of A = 316.

Sum of ranks of B = 212 (minimum)

Therefore

$$T = 212 - \frac{11(11+1)}{2}$$

$$= 146$$

The approximation based on the Normal distribution requires the calculation of the mean and standard deviation as follows:

$$\text{Mean} = \frac{n_1 n_2}{2}$$

$$= \frac{21 \times 11}{2}$$

$$= 115.5$$

$$\text{Standard error} = \sqrt{\frac{n_1 \times n_2 (n_1 + n_2 + 1)}{12}}$$

$$= \sqrt{\frac{21 \times 11 \times (21 + 11 + 1)}{12}}$$

$$= 25.204$$

We can now calculate the z value which gives the number of standard errors from the mean and can be compared to the critical value from the Normal distribution.

$$z = \frac{146 - 115.5}{25.204}$$

$$= 1.21$$

5 $1.21 < 1.96$.
6   Therefore we cannot reject the null hypothesis.
7   There appears to be no difference in the awareness of the product between the two countries.

13.3 Wilcoxon test

This test is the non-parametric equivalent of the t-test for matched pairs and is used to identify if there has been a change in behaviour. This could be used when analysing a set of panel results, where information is collected both before and after some event (for example, an advertising campaign) from the same people.

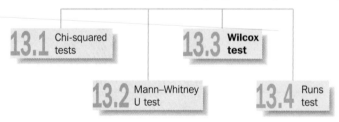

Here the basic premise is that while there will be changes in behaviour, or opinions, the ranking of these changes will be random if there has been no overall change (since the positive and negative changes will cancel each other out). Where there has been an overall change, then the ranking of those who have moved in a positive direction will be different from the ranking of those who have moved in a negative direction.

As with the Mann–Whitney test, where the sample size is small we shall need to consult tables to find the critical value (Appendix H); but where the sample size is large we can use a test based on the Normal distribution.

While we will show how to conduct this test by hand calculation, again packages such as SPSS and MINITAB will perform the test by using the appropriate commands.

13.3.1 Small sample test

Consider the situation where a small panel of eight members have been asked about their perception of a product before and after they have had an opportunity to try it. Their perceptions have been measured on a scale, and the results are given below.

Panel member	Before	After
A	8	9
B	3	4
C	6	4
D	4	1
E	5	6
F	7	7
G	6	9
H	7	2

You have been asked to test if the favourable perception (shown by a high score) has changed after trying the product.

1 H_0: There is no difference in the perceptions.
  H_1: There is a difference in the perceptions.
2   We will take a significance level of 5%, although others could be used.
3   The critical value is found from tables (Appendix H); here it will be 2. We use the number of pairs (8) minus the number of draws (1). Here the critical value is for $n = 8 - 1 = 7$.
4   To calculate the test statistic we find the differences between the two scores and rank them by absolute size (i.e. ignoring the sign). Any ties are ignored.

Before	After	Difference	Rank
8	9	+1	2
3	4	+1	2
6	4	-2	4
4	1	-3	5.5
5	6	+1	2
7	7	0	ignore
6	9	+3	5.5
7	2	-5	7

Sum of ranks of positive differences = 11.5.
Sum of ranks of negative differences = 16.5.

We select the **minimum** of these (i.e. 11.5) as our test statistic.
5   11.5 > 2.
6   Therefore we cannot reject the null hypothesis. (This may seem an unusual result, but you need to look carefully at the structure of the test.)
7   There has been a change in perception after trying the product.

13.3.2 Large sample test

Consider the following example. A group of workers has been packing items into boxes for some time and their productivity has been noted. A training scheme is initiated and the workers' productivity is noted again one month after the completion of the training. The results are shown below.

Person	Before	After	Person	Before	After
A	10	21	N	40	41
B	20	19	O	21	25
C	30	30	P	11	16
D	25	26	Q	19	17
E	27	21	R	27	25
F	19	22	S	32	33
G	8	20	T	41	40
H	17	16	U	33	39
I	14	25	V	18	22
J	18	16	W	25	24
K	21	24	X	24	30
L	23	24	Y	16	12
M	32	31	Z	25	24

1   H_0: there has been no change in productivity.
    H_1: there has been a change in productivity.
2   We will use a significance level of 5%.
3   The critical value will be ± 1.96, since the large sample test is based on a Normal approximation.
4   To find the test statistic, we must rank the differences, as shown below:

Person	Before	After	Difference	Rank
A	10	21	+11	23.5
B	20	19	−1	5.5
C	30	30	0	ignore
D	25	26	+1	5.5
E	27	21	−6	21
F	19	22	+3	14.5
G	8	20	+12	25
H	17	16	−1	5.5
I	14	25	+11	23.5
J	18	16	−2	12
K	21	24	+3	14.5
L	23	24	+1	5.5
M	32	31	−1	5.5
N	40	41	+1	5.5
O	21	25	+4	17
P	11	16	+5	19
Q	19	17	−2	12
R	27	25	−2	12
S	32	33	+1	5.5
T	41	40	−1	5.5
U	33	39	+6	21
V	18	22	+4	17
W	25	24	−1	5.5
X	24	30	+6	21
Y	16	12	−4	17
Z	25	24	−1	5.5

(Note the treatment of ties in absolute values when ranking. Also note that n is now equal to 25.)

Sum of positive ranks $= 218$

Sum of negative ranks $= 107$ (minimum)

The mean is given by:

$$\frac{n(n + 1)}{4} = \frac{25(25 + 1)}{4} = 162.5$$

The standard error is given by:

$$\sqrt{\frac{n(n + 1)(2n + 1)}{24}} = \sqrt{\left[\frac{25(25 + 1)(50 + 1)}{24}\right]} = 37.165$$

Therefore the value of z is given by

$$z = \frac{107 - 162.5}{37.165} = -1.493$$

5 $-1.96 < -1.493$.
6 Therefore we cannot reject the null hypothesis.
7 A month after the training there has been no change in the productivity of the workers.

13.4 Runs test

This is a test for randomness in a dichotomized variable, for example gender is either male or female. The basic assumption is that if gender is unimportant then the sequence of occurrence will be random and there will be no long runs of either male or female in the data. Care needs to be taken over the order of the data when using this test, since if this is changed it will affect the result. The sequence could be chronological, for example the date on which someone was appointed if you were checking on a claimed equal opportunity policy. The runs test is also used in the development of statistical theory, for example looking at residuals in time series analysis.

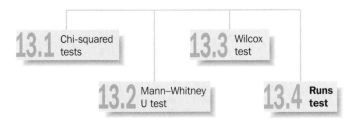

While we will show how to conduct this test by hand calculation, packages such as SPSS and MINITAB will perform the test by using the appropriate commands.

Example

With equal numbers of men and women employed in a department there have been claims that there is discrimination in choosing who should attend conferences. During April, May and June of last year the gender of those going to conferences was noted and is shown below.

Date	Person attending	
April 10	Male	
April 12	Female	
April 16	Female	
April 25	Male	
May 10	Male	*(continues overleaf)*
May 14	Male	
May 16	Male	
June 2	Female	
June 10	Female	
June 14	Male	
June 28	Female	

1 H_0: there is no pattern in the data (i.e. random order).
   H_1: there is a pattern in the data.
2 We will use a significance level of 5%.
3 Critical values are found from tables (Appendix I). Here, since there are six men and five women and we are conducting a two-tailed test, we can find the values of 3 and 10.
4 We now find the number of runs in the data:

Date	Person attending	Run no.
April 10	Male	1
April 12	Female	
April 16	Female	2
April 25	Male	
May 10	Male	
May 14	Male	
May 16	Male	3
June 2	Female	
June 10	Female	4
June 14	Male	5
June 28	Female	6

Note that a run may constitute just a single occurrence (as in run 5) or may be a series of values (as in run 3). The total number of runs is 6.
5 $3 < 6 < 10$.
6 Therefore we cannot reject the null hypothesis.
7 There is no evidence that there is discrimination in the order in which people are chosen to attend conferences.

13.5 Conclusions

Many of the tests which we have considered use the ranking of values as a basis for deciding whether or not to reject the null hypothesis, and this means that we can use non-parametric tests where only **ordinal data** is available. Such tests do not suggest that the parameters (e.g. the mean and variance), are unimportant, but that we do not need to know the underlying distribution in order to carry out the test. They are also called **distribution free tests**. In general, non-parametric tests are less efficient than parametric tests (since they require larger samples to achieve the same level of Type II error). However, we can apply non-parametric tests to a wide range of applications and in many cases they may be the only type of test available. This chapter has only considered a few of the many non-parametric tests but should have given you an understanding of how they differ from the more common, parametric tests and how they can be constructed.

Case study

Typically, one would see parametric tests being used with the type of survey conducted by the Arbour Housing Trust. If the questions are concerned with time or monetary amounts then the responses are easily summarized in terms of a parameter or parameters and hypotheses tested using the Normal distribution. However, when the data lacks the qualities of interval measurement or is unlikely to follow a known distribution, like the Normal distribution, then non-parametric tests become important. The chi-squared test is particularly important in market research where questions are often cross-tabulated and concerned with characteristics or opinions.

13.6 Problems

1   The respondents of a survey were classified by magazine read and income as follows:

Magazine read	Annual income (£)		
	under 10 000	10 000 and under 15 000	15 000 and over
A	28	60	57
B	12	40	53

Test the hypothesis that magazine choice is independent of the level of income using a 5% level of significance.

2   In a survey concerned with changes in working procedures the following table was produced:

	Opinion on changes in working procedures		
	in favour	opposed	undecided
Skilled workers	21	36	30
Unskilled workers	48	26	19

Test the hypothesis that the opinion on working procedures is independent of whether workers are classified as skilled or unskilled.

3   The table below gives the number of claims made in the last year by the 9650 motorists insured with a particular insurance company:

	Insurance groups			
Number of claims	I	II	III	IV
0	900	2800	2100	800
1	200	950	750	450
2 or more	50	300	200	150

Is there an association between the number of claims and the insurance group?

4 A production process uses four machines in its three-shift operation. A random sample of breakdowns was classified according to machine and the shift in which the breakdown occurred:

		Machine		
Shift	A	B	C	D
1	10	11	8	9
2	16	9	13	11
3	12	9	14	9

Is there reason to doubt the independence of shift and machine breakdown?

5 A random sample of 500 units is taken from each day's production and inspected for defective units. The number of defectives recorded in the last working week were as follows:

Day	Number of defectives
Monday	15
Tuesday	8
Wednesday	5
Thursday	5
Friday	12

Test the hypothesis that the difference between the days is due to chance.

6 The number of breakdowns each day on a section of road were recorded for a sample of 250 days as follows:

Number of breakdowns	Number of days
0	100
1	70
2	45
3	20
4	10
5	5
	250

Test whether a Poisson distribution describes the data.

7 The number of car repairs completed each month in the last year were recorded as follows:

Month	Number of repairs	Month	Number of repairs
January	95	July	108
February	98	August	95
March	92	September	94
April	90	October	92
May	83	November	97
June	102	December	94

Is there reason to believe that the number of car repairs does not follow a uniform distribution?

8 The demand for hire cars from a specialist company has been tabulated for the last 100 working weeks.

Demand for hire cars	Number of weeks
0	39
1	32
2	19
3	10

Does demand follow a known distribution?

9 The average weekly overtime earnings from a sample of workers from a particular service industry were recorded as follows:

Average weekly overtime earnings (£)	Number of workers
Under 1	19
1 but under 2	29
2 but under 5	17
5 but under 10	12
10 or more	3
	80

Do average weekly overtime earnings follow a Normal distribution?

10 Eggs are packed into cartons of six. A sample of 90 cartons is randomly selected and the number of damaged eggs in each carton counted.

Number of damaged eggs	Number of cartons
0	52
1	15
2	8
3	5
4	4
5	3
6	3

Does the number of damaged eggs in a carton follow a Binomial distribution?

11 Applications for posts in a major company are received from both men and women. The company has an equal opportunities policy but suspects that women are inhibited from applying in some cases. It has therefore monitored the gender of applicants for a series of posts and the results are shown below.

	Number of applicants
Female	5, 10, 11, 21, 15, 17, 21, 10, 14, 0, 0, 0
Male	30, 8, 8, 12, 20, 22, 32, 25, 8, 6, 12, 6

(Note that some posts had no women applicants.)

Use a Mann–Whitney test at the 5% level of significance to find if there is a significant difference in the number of applications from men and women.

12 Appointments depend on qualifications, experience and personal qualities. In an effort to distinguish the role of experience, a placement agency has noted the number of years of management experience held by people sent to interviews for senior posts. It has tabulated this against whether or not they obtained the job. The results are shown in the following table.

	Number of years of management experience
Appointed	14, 16, 19, 40, 21, 10, 6, 11, 30, 35
Not appointed	17, 23, 3, 2, 1, 7, 5, 24, 15, 7, 5, 20, 1, 3, 9, 12, 8, 1, 4, 2, 13

Use a Mann–Whitney test to find if the experience of those appointed is greater than that of those not appointed.

13 Various positions on a committee became vacant and elections were held. Votes could be split into 'Right' and 'Left', and those given to the 9 successful candidates are shown below.

'Right'	10	30	27	42	39	16	14	15	28
'Left'	17	12	31	29	50	11	18	33	9

Use a Mann–Whitney test to find if the new committee is mainly supported by the 'Right'.

14 Attitudes to 'Green' issues can be characterized into conservative and radical. During the building of a new airport, the percentages who held radical views on the possible harmful effects on the environment were canvassed in local towns and villages, and are recorded in the following table.

	Percentages holding radical views
Villages	31, 31, 50, 10, 12, 17, 19, 17, 22, 22, 23, 27, 42, 17, 5, 6, 8, 24, 31, 15
Towns	30, 18, 25, 41, 37, 30, 29, 43, 51

Test at the 1% level of significance if there is a difference in the level of support for a radical view between the villages and the towns.

15 A national charity has recently held a recruitment drive and to test if this has been successful has monitored the membership of ten area groups before the campaign and three months after the campaign.

Local area group:	A	B	C	D	E	F	G	H	I	J
Before	25	34	78	49	39	17	102	87	65	48
After	30	38	100	48	39	16	120	90	60	45

Use a Wilcoxon test, at the 5% level of significance, to find if there has been an increase in local group membership.

16 Local public support for a company's environmental policy was tested in the

areas close to its various factories before and after a national advertising campaign aimed at increasing the company's environmentally friendly image. The results from the twenty factories are shown in the accompanying table.

Factory	Before	After	Factory	Before	After
1	50	53	11	63	66
2	48	53	12	62	63
3	30	28	13	70	70
4	27	25	14	61	60
5	49	50	15	57	60
6	52	56	16	51	50
7	48	47	17	44	40
8	54	59	18	42	41
9	58	59	19	47	50
10	60	62	20	30	24

Use a Wilcoxon test at the 5% significance level to find if there has been a change in local support for the company's policy. (N.B. Use the Normal approximation.)

17 Employees from a company are tested before and after a training course. The results for ten employees are shown below.

Employee	Before	After
A	10	14
B	12	13
C	13	14
D	15	14
E	17	18
F	17	19
G	18	16
H	9	15
I	5	4
J	3	1

Test at the 2.5% level of significance if there has been a general increase in the employees' abilities following the training course.

18 A panel survey asked participants to rate the appearance of a company's product before a change in packaging and again after the change had been introduced. Each participant rated the product on a scale of 0–100.

Participant	Before	After	Participant	Before	After
A	80	85	K	37	40
B	75	82	L	55	68
C	90	91	M	80	88
D	65	68	N	85	95
E	40	34	O	17	5
F	72	79	P	12	5
G	41	30	Q	15	14
H	10	0	R	23	25
I	16	12	S	34	45
J	22	16	T	61	80

Carry out a Wilcoxon test at the 5% level of significance to find if the new packaging has a more favourable reaction amongst consumers.

19 To test the claimed equal opportunities policy of a company the ethnic origin of successful candidates has been monitored for one year. (Candidates are classified merely as 'English' and 'Non-English'.)

Date of appointment	Ethnic origin
1st February	E
7th February	Non-E
4th March	Non-E
5th March	Non-E
30th March	E
5th May	Non-E
10th June	E
18th June	E
22nd August	Non-E
29th August	E
1st September	Non-E
14th November	Non-E
18th December	E

Use a runs test at the 5% level of significance to test if there is any bias in selection of candidates. What criticisms would you have about the way in which the monitoring was undertaken?

20 Items produced by a certain process can be classified as acceptable (A) or defective (D). Twenty-three such items from a particular machine are checked and the sequence below was obtained.

D, D, D, D, A, A, A, A, A, A, A, D, D, D, D, A, A, A, A, D, D, D, D

Test if the sequence is random.

Part 4   Conclusions

This part of the book has considered the ideas of generalizing the results of a survey to the whole population, and the testing of assertions about that population on the basis of sample data. Both of these concepts are fundamental to the use of statistics in making business decisions. They are essential if you want to take a set of sample results, say from market testing a new product, and make predictions about it's likely sales success, or otherwise. They need to be used in quality control systems which monitor items of production (or service) and compare them to set levels (derived from confidence intervals). We saw in Part 1 that sampling was necessary in order to report meaningful results in a timely fashion. Now we have a tool which can take these results and generalize them to the parent population.

As we already knew, some data sets cannot be adequately described by statistical parameters such as the mean and standard deviation. In these cases we are still able to test assertions about the population from the sample results by using non-parametric tests. In much survey research, the chi-squared test is the most widely used test since the ideas and attitudes being measured do not use a cardinal scale.

There will always be some dispute about the meaning of testing results and a formal mechanism such as the hypothesis test is not going to completely dispel this. What it can do, however, give some agreement on the method used for testing whilst allowing much more discussion of the hypotheses, the means of deriving the sample, and the ways of interpreting the result. This might even lead to more considered decisions being taken!

Part 5 Relating variables and predicting outcomes

Introduction

Most business problems and situation are not the simple, isolated sets of events that you find in textbooks. While we may be able to describe the data on one aspect of a market – for example, the age profile of current and potential purchasers of our product – this is unlikely to provide us with a complete explanation of their behaviour. To do this, we may need to describe many other aspects of the market and the consumers – for example, the trends in the overall size of the market, the segment in the market, consumers' lifestyles, consumers' wealth, factors which influence consumer attitudes, and the behaviour of competing producers. Even when we have been able to measure and describe factors such as these, we know that each will have a different impact on our sales.

To be useful, our analysis needs to be able to do two things. It needs to describe the relationships that currently exist between the variables, and it needs to predict the outcomes, if one or more of the factors change. In practice, we are unlikely to be able to measure or model all of the factors which might just possibly have an effect on what we are trying to predict. We therefore need some method of choosing between those variables which are available. Inevitably, we will miss some data which is relevant, and thus our models are unlikely to ever be perfect. ('There is one thing that is certain about predictions. They will be wrong!')

It may be that we can just relate the behaviour of a variable to the passage of time, noting that it is always higher at certain points in the year, and lower at others. Such behaviour is referred to as seasonality. Identifying such a pattern, and then quantifying it, will be necessary for production control, the hiring of casual labour, and the stock control policies which will be appropriate for the company. In the service sector, such seasonality will identify where the peaks of demand will occur and allow policies to be developed to cope with these situations.

Part 5 looks at ways of modelling data to allow us to make **predictions** about other situations, or about the future. It also looks at ways of assessing such predictions before the actual events occur. Such predictions are essential in business to allow planning for production, stock control or the development of marketing strategies.

Case 5: Jasbeer, Rajan & Co.

The company has been established for about 10 years and has both its production facilities and its management offices at a site on an industrial estate to the south of Leicester, and close to Junction 21 of the M1. (This is also the junction with the M69.) Jasbeer and Rajan originally set up the company as a very small operation in a 'low rent' unit within a converted factory in Leicester city centre, but business expanded, and they moved to the current premises 5 years ago. It was also at this time that a new injection of capital into the business was made by taking on three new partners. The company then changed its status from a partnership to a limited company.

This company manufactures decorations and party items and is faced by a very variable demand for its products through the year. It is in a competitive market where there are several other suppliers, as well as imports, and wishes to understand its current market position and to be able to make predictions about the future.

Data on the company's sales together with other relevant data from the past $7\frac{1}{2}$ years is available to you to help Jasbeer, Rajan & Co deal with their problems. (This data is available on the website.)

Quick start	**Relating variables and predicting outcomes**

Prediction is about making some assertion about the future behaviour of a variable or variables. It may rely on patterns of past behaviour of that single variable, or it might try to bring together the effects of a combination of variables.

Time series
The basic models are:

■ additive
$$A = T + C + S + R$$

■ multiplicative
$$A = T * C * S * R$$

T = trend; C = cyclical; S = seasonal; R = residual.

Trend can be found by:

■ moving average
■ linear regression
■ exponentially weighted moving average.

Correlation and regression
Correlation looks for association between data sets.
Measured by:

$$r = \frac{n \sum XY - \sum X \sum Y}{\sqrt{\left[n \sum X^2 - (\sum X)^2\right]\left[n \sum Y^2 - (\sum Y)^2\right]}}$$

r varies from -1 to $+1$.

Regression produces an equation of a line, and this then allows you to make predictions.

The line is $y = a + bx$, where

$$b = \frac{n \sum XY - \sum X \sum Y}{n \sum X^2 - (\sum X)^2}$$

$$a = \frac{\sum Y}{n} - b \times \frac{\sum X}{n}$$

In multiple regression you need to check MALTHUS.

14 Time series

Very little in the business world is static. The argument is often put forward that the world is becoming more dynamic as the speed of communication increases and transport services become more flexible. Information has become a major resource for most companies. The behaviour of variables over time has been noted and recorded over centuries, but the volume of such data recording, and the detail that is kept, has increased markedly in the past 20 years, partly as a result of the easy accessibility of information technology. This technology also means that much of the data collected by other people or organizations can be accessed via telecommunications links such as the Internet.

This variability may take the form of a gradual movement continually in the same direction, or, more usually, as a series of apparently haphazard oscillations. A few of the haphazard movements seem to move and change direction purely in relation to things that are happening at the present time, for example the *Financial Times* 100 share index. Others seem to be proceeding in some general direction when viewed over fairly lengthy time periods, even though short-term fluctuations are frequent. Still other variables tend to behave in a particular way at certain points in time, perhaps every spring, or once every nine years. These observations, while interesting in themselves for a particular variable that affects a company, are purely descriptive of what has happened in the past.

The challenge is to use this data and such observations to draw more general conclusions about a particular set of data. There are three possible ways to use the historical data:

1 record the events in the form of a table of data or as a chart;
2 explain the behaviour of the series over time;
3 predict what is likely to happen in the future.

Records of events are descriptive and may be held for administrative convenience, or public record. Such records of data collected by the government are easily available from both libraries and through computer databases. (See Part 1 for details of secondary data sources.)

An explanation of the behaviour of a variable may highlight particular events which caused the variable to take on an unrepresentative value – for example, during a postal strike, the amount of business done by mail-order catalogues is likely to be particularly low, or birth rates may suddenly increase, nine months after a series of power cuts. Such explanations would also look at the various elements which combine to bring about the actual behaviour of the data. This may be particularly useful for devising policies on stock control.

Predictions about the future behaviour of the variable, or variables, will be essential when a company or organization begins to look at the future. This could be a short-term situation in looking at production control and planning, or it could be longer-term in looking at future strategy for the organization as a whole.

Such predictions will involve identifying the patterns which have been present in the past, and then projecting these into the future. This involves a fundamental assumption that these patterns will still be relevant in the future. For the near future, this assumption is likely to be true, but for the distant future there are so many things that could, and will, change to affect the particular variable in which we are interested that the assumption can only be viewed as a rough guide to likely events.

An example of this problem may be drawn from population studies: we need to project population figures into the future in order to plan the provision of housing, schools, hospitals, roads and other public utilities. How far ahead do we need to project the figures? For schools we need to plan 5–6 years ahead, to allow time to design the buildings, acquire the land and train teachers in the case of expansion of provision. For contraction, the planning horizon is somewhat shorter. In the case of other social service needs, for example the increasing number of elderly in the UK over the next 30–40 years, advance warning will allow consideration of how these needs are to be met, and who is to finance the provision.

If we are willing to accept the assumption of a continuing pattern, plus the restriction of only limited prediction, then we may build models of the behaviour of a variable over time and use them to project into the future.

Objectives

After working through this chapter, you will be able to:

- state the factors which make up a time series
- state the two basic models
- calculate the trend
- identify the appropriate model to use
- calculate the seasonal factors
- calculate the residual factors
- calculate an EWMA model.

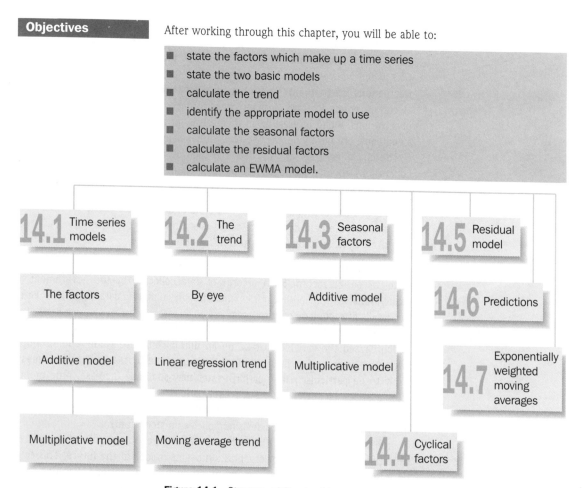

Figure 14.1 Structure of Chapter 14.

Case 5

Despite the apparently random nature of a graph of a variable over time, there is often an underlying pattern. Looking at a simple plot of the sales volume data for Jasbeer, Rajan & Co., as shown in Figure 14.2, we can just see a collection of points. By joining them together, with straight lines, we immediately begin to notice a pattern in the data – see Figure 14.3.

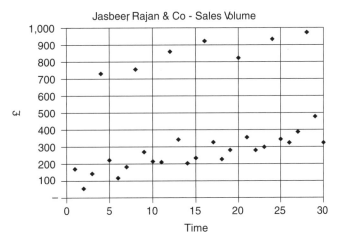

Figure 14.2

This pattern can be shown to consist of various elements, or factors, which are combined together in some way to reproduce the original time series. In this section we will discuss the four factors which make up a time series and the two methods of combining them into a model. There is no single model which will be perfect in every situation.

14.1.1 The factors

For a time series, it is usually possible to identify at least three of the following factors:

1 Trend (T) – this is a broad, underlying movement in the data which represents the general direction in which the figures are moving. It can be identified in a number of different ways, as shown in Section 14.2 below.

2 Seasonal factors (S) – these are regular fluctuations which take place within one complete period. If the data are quarterly, then they are fluctuations specifically associated with each quarter; if the data are daily, then fluc-

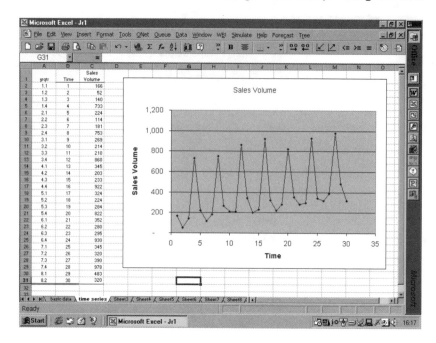

Figure 14.3

tuations are associated with each day. An example would be the demand for electronic games. If we have quarterly data for sales, then there will usually be the highest level of sales in the fourth quarter each year (because of Christmas), with perhaps the lowest levels in the third quarter each year (because families are on holiday). These are discussed in Section 14.3.

3 **Cyclical factors** (C) – this is a longer-term regular fluctuation which may take several years to complete. To identify this factor we would need to have annual data. The most famous example of a cycle is the trade cycle in economic activity in the UK that was observed in the late nineteenth century. While this cycle, which lasted approximately 9 years, does not exist at the start of the twenty-first century, there are other cycles which do affect businesses and the economy in general. These are discussed in Section 14.4.

4 **Random factors** (R)– many other factors affect a time series, and their overall effect is usually small. However, from time to time they do have a significant, but unpredictable, effect on the data. For example, if we are interested in new house starts, then occasionally there will be a particularly low figure due to a more than usually severe winter. Despite advances in weather forecasting, these are not yet predictable. The effects of these nonpredictable factors will be gathered together in this random, or residual, factor. There is further discussion of these in Section 14.5.

These factors may be combined together in several different ways, and two models are discussed below.

14.1.2 The additive model

In the additive model all of the elements are added together to give the original or actual data (A).

$$A = T + S + C + R$$

For many models there will not be sufficient data to identify the cyclical element, and thus the model will be reduced to

$$A = T + S + R$$

Since the random element is unpredictable, we shall make a *working assumption* that its overall value, or average value, is 0.

The additive model will be most appropriate where the variations about the trend are of similar magnitude in the same period of each year or week, as in Figure 14.4.

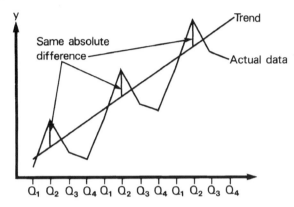

Figure 14.4

14.1.3 The multiplicative model

In the multiplicative model the main, or predictable, elements of the model are multiplied together; the random element may be either multiplied together with this product:

$$A = T \times S \times C \times R$$

(here A and T are actual quantities, while S, C and R are ratios) or may be added to the product:

$$A = T \times S \times C + R$$

(here A, T and R are actual quantities while S and C are ratios).

In the second case, the random element is still assumed to have an average value of 0, but in the former case the assumption is that this average value is 1 Again, the lack of data will often mean that the cyclical element cannot be identified, and thus the models will become:

$$A = T \times S \times R \quad \text{and} \quad A = T \times S + R$$

The multiplicative model will be most appropriate for situations where the variations about the trend are the same proportionate size (or percentage) of the trend in the same period of each year or week, as in Figure 14.5.

Note that the illustrations to this section have used linear trends for clarity, but the arguments apply equally to non-linear trends. In the special case where the trend is a horizontal line, then the same absolute deviation from the trend in a particular quarter will be identical to looking at the percentage change from the trend in that quarter; i.e. with a constant (or stationary) trend, both models will give the same result, as illustrated in Figure 14.6.

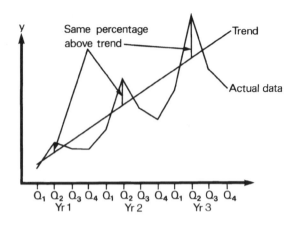

Figure 14.5

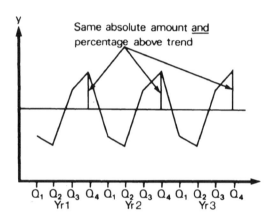

Figure 14.6

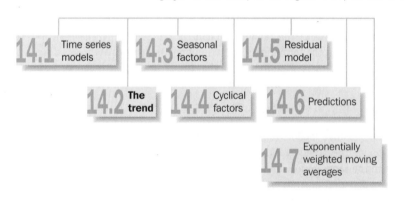

14.2 The trend

Most series follow some sort of long-term movements, upwards, downwards or at some constant level. The first step in analysing a time series is to construct a graph of the data, as in Figure 14.7, to see if there is any obvious underlying movement. In the diagram, we see that the data values are, generally, increasing with time. For some purposes, where all we need is an over-all impression of how the data is behaving, such a graph may be all that is necessary. Where we need to go further and make predictions of the series into the future, then we will need to identify the trend as a series of values.

There are several methods of identifying a trend within a time series, and we will consider three of them here. The graph will give some guidance on which method to use to identify the trend. If the broad underlying movement appears to be linear, then a regression trend would be appropriate; however, if there appears to be a curvilinear trend, then the moving average might provide better answers.

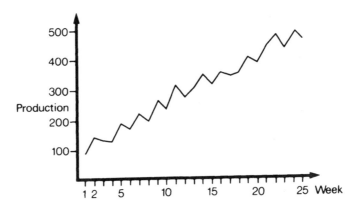

Figure 14.7

14.2.1 Trend by eye

If all that is required is a general idea of where the trend is going, then using your judgement to draw a trend line onto the graph may be sufficient. If at some stage you required values for the trend, then these could be read off the graph. The problem with this method is that several people all drawing such a trend line will tend to come to (slightly) different lines, and thus a discussion may well arise over who has got the best line, rather than the implications of the direction which the trend is taking. In addition, estimation by eye does not provide an approach that would work for more complex analysis (e.g. see Chapter 17 on multiple regression).

Use the data on the website together with either MICROSTATS or a spread-sheet such as EXCEL to obtain the graph of sales volume for Jasbeer, Rajan & Co. and then draw on a trend line by eye.

14.2.2 A linear regression trend

A regression line is a way of drawing an 'average' line through a set of data; fuller details are given in Chapter 16. For the purposes of this section we will just quote the relevant formulae for finding a straight line through our time series data. In terms of notation, we will treat the time series data as the Y variable, since this is the one which we will want to predict, and will treat the time variable as the X values.

The X values will be a dummy variable where 1 will represent the first time period, 2 the second time period, and so on until the final time period. This is illustrated in Table 14.1 below.

Having established the values for the X and Y variables, we can now use the following formulae to identify the trend line through the data:

$$\hat{y} = a + bx$$

where

$$b = \frac{n \sum xy - \sum x \sum y}{n \sum x^2 - \left(\sum x\right)^2}$$

and

$$a = \bar{y} - b\bar{x}$$

Table 14.1   Sales volume for Jasbeer, Rajan & Co.

Yr. qtr.	Time (X)	Volume (Y)	X^2	XY
1.1	1	166	1	166
1.2	2	52	4	104
1.3	3	140	9	420
1.4	4	733	16	2 932
2.1	5	224	25	1 120
2.2	6	114	36	684
2.3	7	181	49	1 267
2.4	8	753	64	6 024
3.1	9	269	81	2 421
3.2	10	214	100	2 140
3.3	11	210	121	2 310
3.4	12	860	144	10 320
4.1	13	345	169	4 485
4.2	14	203	196	2 842
4.3	15	233	225	3 495
4.4	16	922	256	14 752
5.1	17	324	289	5 508
5.2	18	224	324	4 032
5.3	19	284	361	5 396
5.4	20	822	400	16 440
6.1	21	352	441	7 392
6.2	22	280	484	6 160
6.3	23	295	529	6 785
6.4	24	930	576	22 320
7.1	25	345	625	8 625
7.2	26	320	676	8 320
7.3	27	390	729	10 530
7.4	28	978	784	27 384
8.1	29	483	841	14 007
8.2	30	320	900	9 600
Totals	465	11 966	9 455	207 981

Case 5

Using the data from Jasbeer, Rajan & Co. we can illustrate the method of finding the linear trend line. (Note that there is rather too much data here to normally attempt the calculations by hand (!), but since we have the totals already worked out, we can quickly find the required results.)

First of all we will note the various totals that will be needed in the calculations:

$$n = 30 \qquad \sum X = 465$$
$$\sum Y = 11\,966 \qquad \sum X^2 = 9455$$
$$\sum XY = 207\,981$$

Putting these numbers into the formulae will give us the following results:

$$b = \frac{30 \times 207\,981 - 465 \times 11\,966}{30 \times 9455 - (465)^2}$$

$$= \frac{6\,239\,430 - 5\,564\,190}{283\,650 - 216\,225}$$

$$= \frac{675\,240}{67\,425}$$

$$= 10.014\,682\,981\,09 \quad \text{say } 10.015$$

$$a = \frac{11\,966}{30} - 10.014\,682\,981\,09 \times \frac{465}{30}$$

$$= 398.866\,666\,667 - 10.014\,682\,981\,09 \times 15.5$$

$$= 398.866\,666\,667 - 155.227\,586\,206\,9$$

$$= 243.690\,804\,598 \quad \text{say } 243.69$$

Thus the trend line through this data is:

$$\text{Trend} = 243.69 + 10.015X$$

To find the actual values for the trend at various points in time, we now substitute the appropriate X values into this equation.

For $X = 1$ we have:

$$\text{Trend} = 243.69 + 10.015(1) = 253.705$$

For $X = 2$ we have:

$$\text{Trend} = 243.69 + 10.015(2) = 263.72$$

And so on. Obviously, these calculations are very much easier to perform on a spreadsheet such as Excel.

The results of these calculations are shown in Table 14.2 but differ slightly (253.65, 263.67) because the spreadsheet retains the accuracy of the original calculations. The results can now be placed onto the graph of the data, and this is shown in Figure 14.8.

Use the data on the website together with a spreadsheet such as Excel to obtain the graph of sales volume for Jasbeer, Rajan & Co. and calculate the trend line and place it onto your graph.

14.2.3 A moving average trend

This type of trend tries to smooth out the fluctuations in the original series by looking at relatively small sections, finding an average, and then moving on to another section. The size of the small section will often be related to the type of data that we are looking at; if it is quarterly data, we would use subsets of 4; if monthly data, subsets of 12; if daily data, subsets of 5 or 7.

In Table 14.3 we have daily data on the sickness and absence records of staff at Jasbeer, Rajan & Co. and the appropriate subset size is 5, representing one complete cycle, which here is one week. Taking the first 5 days' absences, we find that the total is 48, and dividing by 5 (the number of days involved) gives an average of 9.6. Since both of these figures relate to the first five days, it will be most appropriate to record them opposite the middle day (i.e. Wednesday) of the data subset. If we now move the subset forward in line by one day, we will have

Table 14.2 Linear trend values

Yr. qtr.	Time (X)	Sales volume (Y)	Linear trend
1.1	1	166	253.65
1.2	2	52	263.67
1.3	3	140	273.68
1.4	4	733	283.70
2.1	5	224	293.71
2.2	6	114	303.73
2.3	7	181	313.74
2.4	8	753	323.76
3.1	9	269	333.77
3.2	10	214	343.79
3.3	11	210	353.80
3.4	12	860	363.82
4.1	13	345	373.83
4.2	14	203	383.84
4.3	15	233	393.86
4.4	16	922	403.87
5.1	17	324	413.89
5.2	18	224	423.90
5.3	19	284	433.92
5.4	20	822	443.93
6.1	21	352	453.95
6.2	22	280	463.96
6.3	23	295	473.98
6.4	24	930	483.99
7.1	25	345	494.01
7.2	26	320	504.02
7.3	27	390	514.04
7.4	28	978	524.05
8.1	29	483	534.06
8.2	30	320	544.08

another subset of 5 days (from Week 1: Tuesday, Wednesday, Thursday, Friday, and from Week 2, Monday). For this group, the total number of absences is 47, which gives an average of 9.4 days. These two results will be recorded opposite the middle of this subset (i.e. Thursday of Week 1). This process will continue until we reach the last subset of 5 days.

Use the data on the website (file JR2.XLS) with a spreadsheet such as Excel to obtain the graph of absence and sickness records for Jasbeer, Rajan & Co. Calculate the moving average trend line and place it onto your graph. (You should get the remaining trend figures are: 12.6, 12.4, 12.6, 12.4, 11.6, 10.8, 10.4, 9.6, 9.0, 8.6, 8.2.)

This data is illustrated in Figure 14.9.

Once each of the averages is recorded opposite the middle day of the subset to which it relates, we have found the moving-average trend. Two points should be noted about this trend, firstly that there are no trend figures for the *first* two data points, nor for the *last* two data points. Secondly, that the extension of this trend into the future to make predictions will be more difficult than from the linear trend calculated above. However, the trend is not limited to a small group of functional shapes and will thus be able to follow data where the trend does change direction.

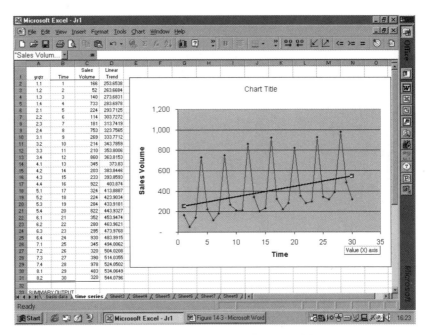

Figure 14.8 Screenshot of trend and data from an Excel spreadsheet.

Table 14.3 Absence and sickness records of Jasbeer, Rajan & Co.

Week	Day	Absence and sickness	$\sum 5s$	Average
1	Monday	4		
	Tuesday	7		
	Wednesday	8	48	9.6
	Thursday	11	47	9.4
	Friday	18	48	9.6
2	Monday	3	50	10.0
	Tuesday	8	52	10.4
	Wednesday	10	55	11.0
	Thursday	13	58	11.6
	Friday	21	59	11.8
3	Monday	6	62	12.4
	Tuesday	9	66	13.2
	Wednesday	13	73	14.6
	Thursday	17	71	14.2
	Friday	28	70	14.0
4	Monday	4	68	13.6
	Tuesday	8	67	13.4
	Wednesday	11		
	Thursday	16		
	Friday	24		
5	Monday	3		
	Tuesday	9		
	Wednesday	10		
	Thursday	12		
	Friday	20		
6	Monday	1		
	Tuesday	5		
	Wednesday	7		
	Thursday	10		
	Friday	18		

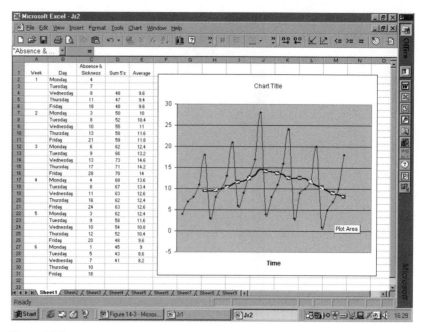

Figure 14.9

A feature of using a subset with an odd number of data points is that there will be a middle item opposite which to record the answers; if we use subsets with an even number of data points, there will be **no middle item**. Does this matter? If the only thing that we want to do is to identify the trend then the answer is no! However, we usually want to go on and look at other aspects of the time series, and in this case we will want each trend value to be associated with a particular data point. To do this we need to use an extra step in the calculations to centre the average we are finding.

The data in Table 14.4 are quarterly sales, and thus the appropriate subset will be 4 (or one year). Summing sales for the first four quarters gives a total of 1091 sales which is again recorded in the middle of the subset (here, between quarters 2 and 3 of year 1). Moving the subset forward by one quarter and summing gives a total of 1149 sales, recorded between quarters 3 and 4 of year 1. As above, this process continues up to and including the final subset of four quarters. None of these sums of four numbers is directly opposite any of the original data points. To bring a total number of sales (and hence the average too) opposite a data point, we may now add the first column of totals in *pairs*, putting the new total between them, and hence opposite a particular time period. (In the first case this will be opposite quarter 3 of year 1.) Each of these new totals is the sum of two sums of four numbers, i.e. a sum of eight numbers, so we need to divide by 8 to obtain the average.

This set of figures is a centred four point moving-average trend. To graph this trend, we plot each of the trend values at the time period where it appears in our calculations, as in Figure 14.10. As with the previous example, there are no trend values for the first two data points, nor the last two. If the size of the subset had been larger, say 12s for monthly data, we would again need to add the $\sum 12$s in pairs to give a sum of 24 and then divide by 24 to get the centred moving average trend.

Table 14.4 Sales volume for Jasbeer, Rajan & Co.

Yr. qtr.	Sales volume	∑ 4s	∑ 8s	MA TREND
1.1	166			
1.2	52	1091		
1.3	140	1149	2240	280.000
1.4	733	1211	2360	295.000
2.1	224	1252	2463	307.875
2.2	114	1272	2524	315.500
2.3	181	1317	2589	323.625
2.4	753	1417	2734	341.750
3.1	269	1446	2863	357.875
3.2	214	1553	2999	374.875
3.3	210	1629	3182	397.750
3.4	860	1618	3247	405.875
4.1	345	1641	3259	407.375
4.2	203	1703	3344	418.000
4.3	233	1682	3385	423.125
4.4	922	1703	3385	423.125
5.1	324	1754	3457	432.125
5.2	224	1654	3408	426.000
5.3	284	1682	3336	417.000
5.4	822	1738	3420	427.500
6.1	352	1749	3487	
6.2	280	1857	3606	
6.3	295	1850	3707	
6.4	930	1890	3740	
7.1	345	1985	3875	
7.2	320	2033	4018	
7.3	390	2171	4204	
7.4	978	2171	4342	
8.1	483			
8.2	320			

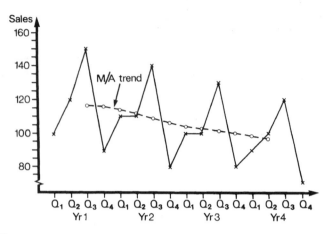

Figure 14.10

Use the data on the website (file JR.XLS) together with a spreadsheet such as Excel to obtain the graph of Sales Volume for Jasbeer, Rajan & Co., calculate the centred moving average trend line and place it onto your graph. (The remaining trend figures are: 435.875, 450.750, 463.375, 467.500, 484.375, 502.250, 525.500, 542.750.)

Exercise

> If we had monthly data where we used a subset of 12, how many data points would not have a trend value associated with them?

14.3 The seasonal factors

Seasonal effects on sales, absenteeism, production, etc. are familiar to most businesses as they are often higher than the average in one part of the year and lower in another period. Many government statistics are quoted as '**seasonally adjusted**', with statements such as 'the figure for unemployment increased last month, but, taking into account seasonal factors, the underlying trend is downwards'. Our aim when analysing a time series is to identify when the actual figures differ from the average (or trend) and also to find the magnitude of the variations. Knowing such information on sales, for example, would be of considerable benefit in both production planning and stock control policies for the company.

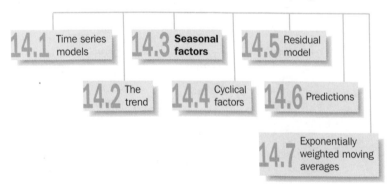

As we saw in Section 14.1, there are two basic models (the additive and the multiplicative) which are used when analysing time series data and the method of calculation of the seasonal factors will depend on which of these models is chosen.

14.3.1 Using an additive model

The additive model for relatively small amounts of data was given as:

$$A = T + S + R$$

The actual data (A) is known, and the trend (T) can be isolated by one of the methods outlined above. As a working hypothesis, we can assume that $R = 0$, since the average value of R must be zero for the model to fully work, and the values random. This leaves us with:

$$A = T + S \quad \text{or} \quad S = A - T$$

This subtraction may be applied to every data point for which we have both an actual data value **and** a trend value. (You see now why we were looking at the number of data points that didn't have a trend value when we used the moving average trend.)

In Table 14.5 we continue to analyse the situation faced by Jasbeer, Rajan & Co. We have taken the linear regression trend through this data (as identified in Section 14.2.2) and performed the appropriate subtraction. This sort of calculation is ideally done on a spreadsheet. Exactly the same process would be used if we were to use the moving average trend identified in Section 14.2.3.

Table 14.5 Sales volume, trend and seasonals–additive model

Yr. qtr.	Sales volume	Linear trend	$A - T$
1.1	166	253.654	−87.654
1.2	52	263.668	−211.668
1.3	140	273.683	−133.683
1.4	733	283.698	449.302
2.1	224	293.712	−69.712
2.2	114	303.727	−189.727
2.3	181	313.742	−132.742
2.4	753	323.757	429.243
3.1	269	333.771	−64.771
3.2	214	343.786	−129.786
3.3	210	353.801	−143.801
3.4	860	363.815	496.185
4.1	345	373.830	−28.830
4.2	203	383.845	−180.845
4.3	233	393.859	−160.859
4.4	922	403.874	518.126
5.1	324	413.889	−89.889
5.2	224	423.903	−199.903
5.3	284	433.918	−149.918
5.4	822	443.933	378.067
6.1	352	453.947	−101.947
6.2	280	463.962	−183.962
6.3	295	473.977	−178.977
6.4	930	483.991	446.009
7.1	345	494.006	−149.006
7.2	320	504.021	−184.021
7.3	390	514.036	−124.036
7.4	978	524.050	453.950
8.1	483	534.065	−51.065
8.2	320	544.080	−224.080

We are now faced by a very large number of 'seasonal factors', but we know that there are only four seasons to the year! However, if we look at the final column of Table 14.5, we can see a pattern within the numbers. There are eight different values associated with quarter 1, but they are all negative (i.e. below the trend) and are within a range of −28 to −150. Similar observations may also be made about the other quarters.

Looking at each quarter in turn, we may treat each appropriate value from the final column of Table 14.5 as an estimate of the actual seasonal factor, and from these find an average which would be a single estimate of that seasonal factor. Perhaps the easiest way of doing this process is to construct an additional table which simply rewrites the final column of Table 14.5 to bring together the various estimates for each quarter. This is done in Table 14.6 below.

Table 14.6 Seasonal estimates using an additive model

Year	Q1	Q2	Q3	Q4
1	−87.654	−211.668	−133.683	449.302
2	−69.712	−189.727	−132.742	429.243
3	−64.771	−129.786	−143.801	496.185
4	−28.830	−180.845	−160.859	518.126
5	−89.889	−199.903	−149.918	378.067
6	−101.947	−183.962	−178.977	446.009
7	−149.006	−184.021	−124.036	453.950
8	−51.065	−224.080		
Total	−642.875	−1503.992	−1024.015	3170.882
Average	−80.359	−187.999	−146.288	452.983

These final four figures represent the additive seasonal factors for this set of sales volume data. (Note that in some cases we have divided by eight, and in others we divided by seven to find the average.) Since the seasonal factors were defined as movements within a year, one would **expect that they would cancel each other out over a year** – in other words, they should add up to zero. Here they do not. If the total of the seasonal factors were small, relative to the data we are using, then we could, in practice, ignore this, but here the total is 38.337, almost as large as some of the data points. The failure of these figures to add up to zero is partly because of the limited number of seasonal estimates and partly because there are not equal numbers of estimates for each quarter. To do this we will divide the total of the seasonal estimates by four (since there are four quarters) and distribute this to each quarter. (And since the total is **positive**, we need to subtract the total divided by four from each seasonal factor.)

This will then give us the following seasonal factors:

Q1   −89.944

Q2   −197.583

Q3   −155.872

Q4   443.399

which do sum to zero.

Use the data on the website (file JR2.XLS), together with a spreadsheet such as Excel, to find the seasonal factors for the absence and sickness data for Jasbeer, Rajan & Co. (You should get: Monday -8.08; Tuesday -3.64; Wednesday -1.3; Thursday 2.2; Friday 10.68. Sum of seasonals $= -0.14$, which we can probably ignore, since it is so small.)

14.3.2 Using a multiplicative model

The multiplicative model for relatively small amounts of data was given as:

$$A = T \times S \times R$$

The actual data (A) is known, and the trend (T) can be isolated by one of the methods outlined above. As a working hypothesis we can assume that $R = 1$, since the average value of R must be 1 for the model to fully work, and the values random. This leaves us with:

$$A = T \times S \qquad \text{or} \qquad S = \frac{A}{T}$$

This division may be applied to every data point for which we have both an actual data value and a trend value, as we did with the additive model.

Again we will use the sales volume data from Jasbeer, Rajan & Co. with a linear trend to illustrate the use of this methodology. This is shown in Table 14.7.

The numbers which we have obtained this time are ratios, rather than actual values, and show how far above or below the trend the actual figures are. (Again, this set of calculations is ideally suited to work on a spreadsheet.) We can approach the calculation in exactly the same way that we did for the additive seasonal factors, by treating each value as an estimate, and then constructing a working table to get an average estimate for each season. Since there are four quarters, then the sum of these seasonal factors should be 4. This is illustrated in Table 14.8.

The sum of these seasonal factors is 4.085, which is near enough to 4 that we may decide not to try to correct so that their sum is exactly 4. Seasonal factors are often quoted as percentages. Here the values would be:

Q1	78.8%
Q2	51.0%
Q3	61.6%
Q4	217.1%

Use the data on the website (file JR2.XLS) together with a spreadsheet such as Excel to find the seasonal factors for the absence and sickness data for Jasbeer, Rajan & Co. (You should get: Monday 0.286; Tuesday 0.681; Wednesday 0.881; Thursday 1.186; Friday 1.929 Sum of seasonals $= 4.963$, which we can probably ignore, the difference from 5, since it is so small.)

14.3.3 Seasonal adjustment of time series

Many published series are quoted as being 'seasonally adjusted'. The aim is to show how the trends in the figures are moving, without the hindrance of seasonal variation, which may tend to obscure such trends. This is more, however, than just quoting a trend value, since we still have the effects of the cyclical and random variations retained in the quoted figures.

Table 14.7 Sales volume, trend and seasonals–multiplicative model

Yr. qtr.	Sales volume	Linear trend	A/T
1.1	166	253.654	0.654
1.2	52	263.668	0.197
1.3	140	273.683	0.512
1.4	733	283.698	2.584
2.1	224	293.712	0.763
2.2	114	303.727	0.375
2.3	181	313.742	0.577
2.4	753	323.757	2.326
3.1	269	333.771	0.806
3.2	214	343.786	0.622
3.3	210	353.801	0.594
3.4	860	363.815	2.364
4.1	345	373.830	0.923
4.2	203	383.845	0.529
4.3	233	393.859	0.592
4.4	922	403.874	2.283
5.1	324	413.889	0.783
5.2	224	423.903	0.528
5.3	284	433.918	0.655
5.4	822	443.933	1.852
6.1	352	453.947	0.775
6.2	280	463.962	0.603
6.3	295	473.977	0.622
6.4	930	483.991	1.922
7.1	345	494.006	0.698
7.2	320	504.021	0.635
7.3	390	514.036	0.759
7.4	978	524.050	1.866
8.1	483	534.065	0.904
8.2	320	544.080	0.588

Table 14.8 Seasonal estimates using a multiplicative model

Year	Q1	Q2	Q3	Q4
1	0.654	0.197	0.512	2.584
2	0.763	0.375	0.577	2.326
3	0.806	0.622	0.594	2.364
4	0.923	0.529	0.592	2.283
5	0.783	0.528	0.655	1.852
6	0.775	0.603	0.622	1.922
7	0.698	0.635	0.759	1.866
8	0.904	0.588		
Total	6.307	4.079	4.309	15.196
Average	0.788	0.510	0.616	2.171

Taking the additive model, we have:

$$A = T + C + S + R \qquad \text{and} \qquad A - S = T + C + R$$

Constructing such an adjusted series relies heavily upon having correctly identified the seasonal factors from the historic data; that is, having used the appropriate model over a sufficient time period. There is also a heroic assumption that seasonal factors identified from past data will still apply to current and future data. This may be a workable solution in the short term, but the seasonal factors should be re-calculated as new data becomes available.

By creating a new column in your spreadsheet which has the four seasonal factors repeated through time, and then subtracting this from the column containing the original data, we will create the seasonally adjusted series.

14.4 The cyclical factors

Although we can talk about there being a cyclical factor in time series data and can try to identify it by using *annual* data, these cycles are rarely of consistent lengths. A further problem is that we would need six or seven full cycles of data to be sure that the cycle was there, and for some proposed cycles this would mean obtaining 140 years of data!

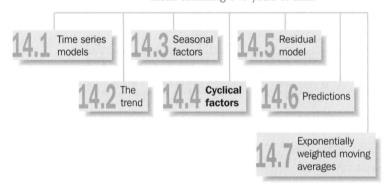

Several cycles have been proposed and a brief outline of a few of these is given below.

1 Kondratieff Long Wave: 1920s, a 40–60-year cycle; there seems to be very little evidence to support the existence of this cycle.
2 Kuznets Long Wave: a 20-year cycle; there seems to be some evidence to support this from studies of GNP and migration.
3 Building cycle: a 15–20-year cycle; some agreement that it exists in various countries.
4 Major and minor cycles: Hansen 6–11-year major cycles, 2–4-year minor cycles; *cf*. Schumpeter inventory cycles. Schumpeter: change in rate of innovations leads to changes in the system.
5 Business cycles: recurrent but not periodic, 1–12 years, *Cf*. minor cycles, trade cycle.

At this stage we can construct graphs of the annual data and look for **patterns** which match one of these cycles. Since one cycle may be superimposed upon another, this identification is likely to prove difficult. Removing the trend from the data may help, so that we consider graphs of $(A - T) = C + R$ rather than graphs of the original time series. (Note that there is no seasonal factor since we are dealing with annual data.)

To illustrate this procedure, see Figure 14.11 on New Car Registrations in the UK for a 40-year period. Although it is often claimed that this data is cyclical, the graph of the original data does not highlight any obvious cycle. Looking at the graph of $A - T$ (Figure 14.12), it is possible to identify that the data exhibits some cyclical variation, but the period of the cycle is rather more difficult to identify.

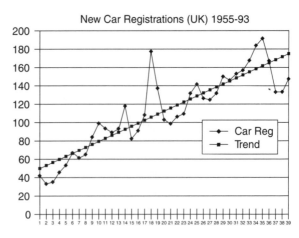

Figure 14.11

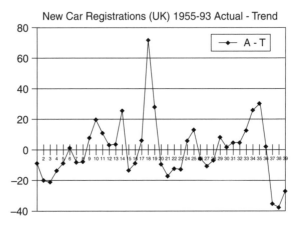

Figure 14.12

14.5 The residual or random factor

In building the time series models that we have developed in this chapter, we have had to make simplifying assumptions about this factor. With the additive model, we assumed it to be zero; with the multiplicative model we assumed it to be one. Now that we have identified the other elements of the time series, we can come back to the residual to find if our assumptions were justified. Checking the values of the residual factor will suggest whether or not the model that we have used is a good fit to the actual data.

The calculation of the individual residual elements in the additive model will be by using the formula:

$$R = A - T - S$$

where S is the average seasonal factor for each period.

In the case of the multiplicative model, the calculation will be:

$$R = \frac{A}{(T \times S)}$$

where S is the average seasonal index for each period.

Taking the sales volume data for Jasbeer, Rajan & Co., with the linear trend we can obtain the following figures for the residual element in both the additive and the multiplicative models (shown in Table 14.9).

On the basis of the data in Table 14.9, we would conclude that the linear trend and the additive model is a fairly poor fit to the data, whilst the linear trend with the multiplicative model is a better fit, since the average residual factor is equal to one. This is shown in Figures 14.13 and 14.14.

Table 14.9   Residual factors for Jasbeer, Rajan & Co.

Yr. Qtr.	Additive model, R	Multiplicative model, R
1.1	2.290	0.830
1.2	−14.085	0.387
1.3	22.189	0.831
1.4	5.903	1.190
2.1	20.231	0.967
2.2	7.856	0.736
2.3	23.130	0.937
2.4	−14.155	1.071
3.1	25.172	1.022
3.2	67.797	1.221
3.3	12.072	0.964
3.4	52.786	1.089
4.1	61.114	1.171
4.2	16.739	1.037
4.3	−4.987	0.961
4.4	74.727	1.052
5.1	0.055	0.993
5.2	−2.320	1.036
5.3	5.954	1.063
5.4	−65.332	0.853
6.1	−12.004	0.984
6.2	13.621	1.184
6.3	−23.105	1.011
6.4	2.610	0.885
7.1	−59.063	0.886
7.2	13.562	1.245
7.3	31.837	1.232
7.4	10.551	0.860
8.1	38.879	1.147
8.2	−26.496	1.154
Total	287.527	30.000
Average	9.584	1.000

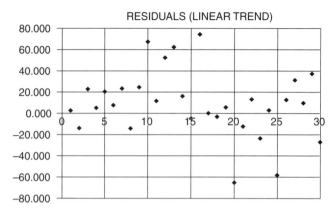

Figure 14.13

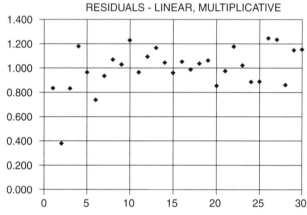

Figure 14.14

14.6 Predictions

Predicting from these models is a matter of **recombining the various elements** which we have been able to identify. This will involve extending the trend into the future, and then applying the appropriate seasonal factor. (We do not include the residual in this process.) For a linear trend we have seen (Section 14.2.1) that extension of the trend is achieved by substituting appropriate values for x; however, the problem is somewhat more difficult from a moving-average trend. Predictions on the trend are done by extending its graph in an appropriate direction, consistent with its past behaviour. There are two problems here: there is a considerable amount of judgement (or assumption) used in the process and, since the first few predictions are the best (being closest to the data), these are being used to find a trend for past data. (This is because the moving-average trend ends before the end of the data – Section 14.2.3.)

Case 5

To illustrate this process we have extended the linear trend for the sales volume data for Jasbeer, Rajan & Co. in Table 14.10 and applied the appropriate seasonal factors.

Table 14.10   Prediction of sales volume for Jasbeer, Rajan & Co.

(a) The additive model

Yr. Qtr.	Trend	Seasonal	Prediction
8.3	554.094	−155.872	398.222
8.4	564.109	443.399	1007.508
9.1	574.124	−89.944	484.18
9.2	584.138	−197.583	386.555
9.3	594.153	−155.872	438.281
9.4	604.168	443.399	1047.567

(b) Multiplicative model

Yr. Qtr.	Trend	Seasonal	Prediction
8.3	554.094	0.616	341.322
8.4	564.109	2.171	1224.681
9.1	574.124	0.788	452.410
9.2	584.138	0.510	297.910
9.3	594.153	0.616	365.998
9.4	604.168	2.171	1311.649

Time series models are often judged on the accuracy of the predictions.

Seasonal or other factors are often removed from past or present time series data in order to highlight salient features. For example:

$$\text{deseasonalized data} = A - S \text{ (for appropriate period)}$$
$$\text{detrended data} \quad = A - T$$

14.7 Exponentially weighted moving averages

Where we require short-term forecasts, for example in stock control situations, we may use a smoothing method known as **exponentially-weighted moving averages** (EWMA). This takes into account the movements in the data that have occurred in the past, but weights the more recent figures more heavily than those in the distant past. The prediction of the next value of y is shown in the equation given below:

$$\hat{y}_{t+1} = y_t + (1-\alpha)y_{t-1}$$
$$+ (1-\alpha)^2 y_{t-2}$$
$$+ (1-\alpha)^3 y_{t-3} + \cdots$$

where α is a **smoothing con-**stant. Further analysis of this equation shows that the right-

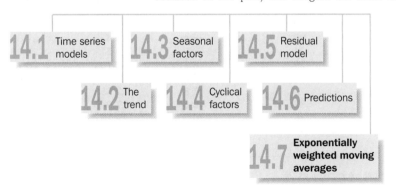

hand side can be reduced to the sum of only two terms:

$$\hat{y}_{t+1} = \alpha y_t + (1 - \alpha)\hat{y}_t$$

provided that we have an initial prediction for y_t. (In practice we often use the first actual value of y as the first predicted value $(\hat{y}_1 = y_1)$ since the effects of any error in predicting the first actual value will very quickly be minimized.)

The choice of alpha will depend upon experience with the data and how quickly we wish the predictions to react to changes in the data. If we choose a *high* value for alpha, the predictions will be *very sensitive* to changes in the data. Where the data is subject to random shocks, then these will be passed on to the predictions. Where a *low* value of alpha is used, the predictions will be *slow to react* to changes in the data, but will be less affected by random shocks. Thus if we are dealing with data which rarely experiences random shocks, we would want to use a high value for alpha so that genuine changes of direction in the data are reflected in the predictions as soon as possible. In practice, values of alpha are usually chosen from within the range 0.1–0.6.

Case 5

In the case of Jasbeer, Rajan & Co. they wish to be able to predict the number of new customers that they are likely to attract in the next month, and have been able to provide the data shown in Table 14.11. Two predictive models have been constructed, the first using a value of 0.3 for alpha, and the second using a value of 0.65 for alpha.

Working with the data from January, we have an actual figure of 25, and have taken 25 as our initial prediction. (Note that it is usual to use the first data point as the initial prediction.) Alpha times the actual figure gives 7.5. Taking one minus alpha (0.7) times the initial prediction gives 17.5. Adding these two figures together gives 25, which is the prediction for February. Taking the data for February, we have an actual value of 30, so 0.3 times this gives 9, and 0.7 times the prediction for February (25) gives 17.5. Adding these two numbers gives a prediction for March of 26.5. This process continues to the end of the table.

Using the same process, but with an alpha value of 0.65, gives Table 14.12.

Table 14.11   EWMA model of new customers with Alpha = 0.3

Month	New C's (Y_t)	0.3 Y_t	0.7 $\hat{Y}_t$	$\hat{Y}_{t+1}$
J	25	7.5	17.5	
F	30	9	17.5	25.000
M	25	7.5	18.55	26.500
A	32	9.6	18.235	26.050
M	28	8.4	19.485	27.835
J	24	7.2	19.519	27.885
J	29	8.7	18.703	26.719
A	21	6.3	19.182	27.403
S	26	7.8	17.838	25.482
O	32	9.6	17.946	25.638
N	35	10.5	19.282	27.546
D	38	11.4	20.848	29.782
J				32.248

NB: Initial value taken as 25.

Table 14.12 EWMA model of new customers with Alpha = 0.65

Month	New C's (Y_t)	0.65 Y_t	0.35 $\hat{Y}_t$	$\hat{Y}_{t+1}$
J	25	16.25	8.75	
F	30	19.5	8.75	25.000
M	25	16.25	9.888	28.250
A	32	20.8	9.148	26.138
M	28	18.2	10.482	29.948
J	24	15.6	10.039	28.682
J	29	18.85	8.974	25.639
A	21	13.65	9.738	27.824
S	26	16.9	8.186	23.388
O	32	20.8	8.780	25.086
N	35	22.75	10.353	29.580
D	38	24.7	11.586	33.103
J				36.286

NB: Initial value taken as 25.

This second model appears to be better than the first, since the predictions are closer to the actual data. One way of deciding on the optimum value for alpha is to draw a comparison between the actual and predicted values. To do this, we take the difference between the two figures for each data point and then square the result (to get rid of any negative signs). We then find the average of these squared deviations (called mean squared error). The model with the lowest average would be the best on this criterion. Table 14.13 was produced using an alpha value of 0.3.

Using the same process gives a value of 261.488 for the sum of the squared deviations when alpha is equal to 0.65.

As you can see from the layout of Tables 14.11 and 14.12, the building of EWMA models is a natural candidate for the use of spreadsheets. We suggest that you use individual columns for each calculation (as in Table 14.11) when you first build such models, but that the whole formula can be put into a single cell once you see the way in which the model works. Building such spreadsheet

Table 14.13 Sum of squared differences using $\alpha = 0.30$

Month	New C's	αY_t	$1 - \alpha \hat{Y}$	$\hat{Y}_{t+1}$	Difference	Squared difference
J	25	7.5	17.5			
F	30	9	17.5	25.000	5.000	25
M	25	7.5	18.55	26.500	−1.500	2.25
A	32	9.6	18.235	26.050	5.950	35.402 5
M	28	8.4	19.484 5	27.835	0.165	0.027 225
J	24	7.2	19.519 15	27.885	−3.885	15.089 34
J	29	8.7	18.703 41	26.719	2.281	5.202 277
A	21	6.3	19.182 38	27.403	−6.403	41.003 6
S	26	7.8	17.837 67	25.482	0.518	0.267 927
O	32	9.6	17.946 37	25.638	6.362	40.479 26
N	35	10.5	19.282 46	27.546	7.454	55.556 63
D	38	11.4	20.847 72	29.782	8.218	67.528
				32.248		
Total						287.806 8

models will allow you to experiment with different values of alpha to find the 'best' value for a particular set of data.

There are many extensions to this basic EWMA model, and for those who are interested in extending their knowledge, we recommend the book by Lewis (1981), *Scientific Inventory Control*, Butterworths.

14.8 Summary and conclusions

In this section we will draw together the predictions that have been made for Jasbeer, Rajan & Co. and then make some more general conclusions on time series analysis.

Case 5

By taking a time series approach to the sales volume data for Jasbeer, Rajan & Co. we have been able to identify two sets of predictions: those with a linear trend and those with a moving average trend. In each case we have also been able to use either an additive or a multiplicative model. This gives four sets of predictions, shown in Table 14.14. For the linear trend models, we identified the multiplicative model as being the better in terms of the residual values. It is left to you to carry out a similar analysis for the moving average models. (Note that the moving average models may differ slightly from your predictions, since they are very dependent on how you have extended the trend line.)

Table 14.14   Summary predictions

Yr. qtr.	Linear trend additive model	Linear trend multiplicative model	MA trend additive model	MA trend multiplicative model
8.3	398.222	341.322	436	356.06
8.4	1007.508	1224.681	1052	1282.58
9.1	484.18	452.410	533	480.83
9.2	386.555	297.910	455	344.079
9.3	438.281	365.998	504	662
9.4	1047.567	1311.649	1120	1425.38

Which predictions are best will only be obvious with hindsight, when we are able to compare the actual sales volume with our predictions.

Any analysis of time series data must inevitably make the heroic assumption that the behaviour of the data in the past will be a good guide to its behaviour in the future. For many series, this assumption will be valid, but for others, no amount of analysis will predict the future behaviour of the data. Where the data reacts quickly to external information (or random shocks) then time series analysis, on its own, will not be able to predict what the new figure will be.

Even for more 'well-behaved' data, shocks to the system will lead to variations in the data which cannot be predicted, and thus the predictions should be treated as a guide to the future and not as a statement of exactly what will happen. For many businesses, the process of attempting to analyse the past behaviour of data may be more valuable than the actual predictions obtained from the analysis, since it will highlight both trends and seasonal effects which can then be taken into account in planning for the future.

14.9 Problems

1 The level of economic activity in a region has been recorded over a period of four years, and the data is presented below.

Yr	Qtr	Activity level	Yr	Qtr	Activity level
1	1	105	3	1	118
	2	99		2	109
	3	90		3	96
	4	110		4	127
2	1	111	4	1	126
	2	104		2	115
	3	93		3	100
	4	119		4	135

(a) Construct a graph for this data.
(b) Find a centred four-point moving-average trend and place it on your graph.
(c) Calculate the corresponding additive seasonal components.
(d) Use your results to predict the level of economic activity for the next 2 years.
(e) Would you have any reservations about using this model?

2 The number of customers in a shop has been recorded over the past few weeks. The results are shown below.

Wk	Day	Cust	Wk	Day	Cust	Wk	Day	Cust
1	M	120	3	M	105	5	M	85
	T	140		T	130		T	100
	W	160		W	160		W	150
	TH	204		TH	220		TH	180
	F	230		F	275		F	280
	S	340		S	400		S	460
	SU	210		SU	320		SU	340
2	M	130	4	M	100			
	T	145		T	120			
	W	170		W	150			
	TH	200		TH	210			
	F	250		F	260			
	S	380		S	450			
	SU	300		SU	330			

(a) Graph the data.
(b) Find a seven point moving average trend and place it on your graph.
(c) Calculate the additive 'seasonal' factors.
(d) Calculate the residual factor and construct a graph.
(e) Explain why this model is of little use and how you would make it better.

3 Sales from company JCR's motor division have been monitored over the past four years and are presented below:

Year	Quarter 1	Quarter 2	Quarter 3	Quarter 4
1	20	30	39	60
2	40	51	62	81
3	50	64	74	95
4	55	68	77	96

(a)   Construct a graph of this data.
(b)   Find a centred four-point moving average trend and place it on your graph.
(c)   Calculate the four seasonal components using an additive model.
(d)   Use your model to predict sales for years 5 and 6.
(e)   Prepare a brief report to the company on future sales.

4   Use the data in question 3 to calculate a linear regression trend through the data, calculate the seasonal components, again using an additive model, and predict sales for years 5 and 6. Does this model represent an improvement upon the previous model?

5   The number of calls per day to a local authority department have been logged for a four-year period and are presented below:

Year	Quarter 1	Quarter 2	Quarter 3	Quarter 4
1	20	10	4	11
2	33	17	9	18
3	45	23	11	25
4	60	30	13	29

(a)   Construct a graph of this data.
(b)   Find a linear regression trend through the data.
(c)   Calculate the four seasonal components using a multiplicative model.
(d)   Predict the number of calls for the next 2 years.

6   Use the data and the trend from question 5 to find the average residual component for the time series using:
(a)   an additive model;
(b)   a multiplicative model.

7   A company's advertising expenditure has been monitored for 3 years, giving the following information:

| | Advertising expenditure | | | |
Year	Q1	Q2	Q3	Q4
1	10	15	18	20
2	14	16	19	23
3	16	18	20	25

(a)   Calculate a linear regression trend for this data.
(b)   Graph the data and the trend.
(c)   Find the additive seasonal components for each quarter.
(d)   Predict the level of advertising expenditure for each quarter of year 4.

8   The level of economic activity in a county of England has been recorded for 15 years as follows:

Year	Activity rate	Year	Activity rate
1	52.7	8	52.6
2	54.4	9	50.7
3	54.7	10	49.8
4	55.4	11	48.3
5	53.8	12	43.8
6	53.5	13	40.3
7	53.4	14	37.8
		15	35.1

(a) Graph the data.
(b) Find a linear regression trend through this data and place it on your graph.
(c) Predict the activity rate for year 25.
(d) How confident are you of your prediction?

9 The sales of a product were monitored over a 20-month period; the results are shown below.

Month	Sales	Month	Sales
1	10	11	136
2	22	12	132
3	84	13	124
4	113	14	118
5	132	15	116
6	137	16	112
7	139	17	107
8	140	18	95
9	140	19	80
10	140	20	68

(a) Graph the data.
(b) Find the linear trend through the data and place this on the graph.
(c) Find the three-point moving average trend through the data and place this on the graph.
(d) Comment on your results.

10 Sales over the past 24 months have been recorded for a certain company, and are shown in the following table. Find the exponentially weighted moving averages for this data using the following values of alpha.
(a) 0.2
(b) 0.4
(c) 0.6

Month:	1	2	3	4	5	6	7	8
Sales:	225	230	226	240	245	260	280	310

Month:	9	10	11	12	13	14	15	16
Sales:	320	280	240	220	230	238	239	251

Month:	17	18	19	20	21	22	23	24
Sales:	255	265	300	325	356	300	264	231

Plot the series and the EWMAs and decide which of these values is most appropriate in this case.

11 A package holiday company sends out brochures to prospective clients. At certain times of the year it employs extra staff to cope with the demand. Monthly enquiries over the past 3 years have been recorded and are shown below.

Month:	1	2	3	4	5
Enquiries:	2000	2500	2000	1800	1400
Month:	6	7	8	9	10
Enquiries:	800	200	100	900	1400
Month:	11	12	13	14	15
Enquiries:	1600	1800	2300	2800	2400
Month:	16	17	18	19	20
Enquiries:	2000	1400	600	200	50
Month:	21	22	23	24	25
Enquiries:	1000	1600	1900	2000	2600
Month:	26	27	28	29	30
Enquiries:	3000	2800	1900	1100	400
Month:	31	32	33	34	35
Enquiries:	100	50	1100	1800	2000
Month:	36	37	38	39	40
Enquiries:	2300	2800	3100	2900	1800

Use this information to find the EWMA and hence predict the number of enquiries in the following month. (Use an alpha value of 0.3.) Attempt to find a better value of alpha to use with this data.

12 A help desk was set up when a new computer system was introduced into a company. The number of requests for help was logged over the first 15 days and these numbers are shown below.

Day	Requests
1	15
2	20
3	18
4	19
5	21
6	25
7	23
8	28
9	28
10	30
11	30
12	28
13	25
14	26
15	22

Use EWMA with alpha values of
(a) 0.2
(b) 0.4
(c) 0.6
(d) 0.8
to make predictions of the number of requests for help on day 16.

15 Correlation

The term **correlation** has become a common word in everyday news reports and commentaries; so much so, that you will already have some idea of its meaning. In a general sense, it is about two or more sets of things being related. Often, however, it is a word which is used in a rather vague sense, with an implication that the (two) things move together in the same direction. In a statistical context, we will need to define correlation rather more exactly than this, and then establish some way of measuring it. Once we can do this we can try to decide what the result actually means, assessing its significance in both statistical terms and in the context of the problem at which we are looking.

Considerable data is collected by both the government and by companies. The government data is published in a series of journals, as we saw in Part 1 and is available for businesses to use in a variety of ways, including looking for correlations with their own data. It seems obvious that there will be relationships between some sets of data, and it is our aim in this chapter to explore if a relationship does exist between two sets of data, and, if so, how strong it is. We will also be interested in whether this relationship is in some sense 'better' than some other relationship.

This chapter will look only at the situation where we are trying to relate two variables together; the situation of many variables is dealt with in Chapter 17. Again we will use the data from Jasbeer, Rajan & Co. to illustrate correlation analysis.

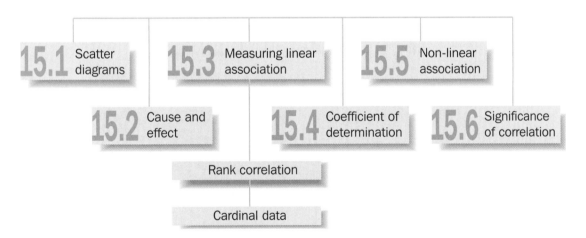

Figure 15.1 Structure of Chapter 15.

Objectives

After working through this chapter, you should be able to:

- describe what is meant by correlation
- argue whether or not this is causal
- calculate a correlation measure for ordinal data
- calculate a correlation measure for cardinal data
- select appropriate transformations to linearity
- test the significance of the correlation.

15.1 Scatter diagrams

As we saw in Chapter 4, visual representation can give an immediate impression of a set of data. Where we are dealing with two variables, the appropriate method of illustration is the scatter diagram. A scatter diagram allows us to show two variables together, one on each axis, each pair being represented by a cross on a graph. Where there is only a very limited number of pairs of data, little can be inferred from such a diagram, but in most business situations

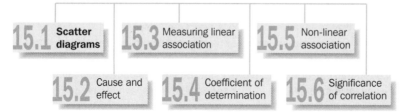

15.1 Scatter diagrams

15.3 Measuring linear association

15.5 Non-linear association

15.2 Cause and effect

15.4 Coefficient of determination

15.6 Significance of correlation

there will be a large number of observations and we may distinguish some pattern in the picture obtained. The lack of a pattern, however, may be just as significant.

Case 5

Taking data from the case study example of Jasbeer, Rajan & Co., we can create a scatter diagram for sales volume against promotional spending. (This is relatively simple to do from a spreadsheet.) This is shown in Figure 15.2.

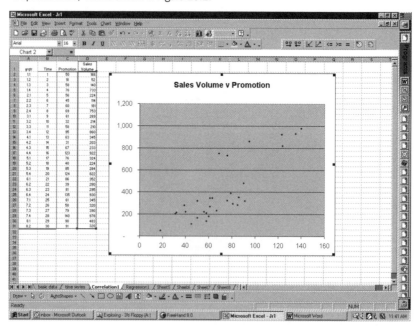

Figure 15.2

From this diagram we can infer that, in general, the more that is spent on promotion, the higher the level of sales volume. This is by **no means a deterministic relationship**, however, and **does not imply a cause and effect** between the two variables; it merely means that times of high promotional spending were also times of high sales volumes.

Figure 15.3 shows the relationship which is likely to exist between the number employed with a fixed level of capital investment, and the total output from such a plant. Here we see that there is somewhat more of a relationship than in the previous example, since the points lie within a much narrower band. This type of result is expected if the law of diminishing returns is true. The law says that as more and more of one factor of production is used, with fixed amounts of the other factors, there will come a point when total output will fall.

Having established that some relationships can be seen from scatter diagrams, we now need to find some 'extreme' relationships so that we may compare future diagrams with given standards.

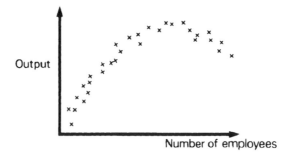

Figure 15.3

Looking at the two diagrams in Figure 15.4, we see situations where all of the data points fall on a straight line. Such a relationship is known as a *perfect* or *deterministic* linear relationship. Part (a) of the diagram shows a positive linear relationship, since as the value of x increases, so does the value of y; in fact, for a one unit rise in x, there is always a given increase in y. A negative linear relationship is illustrated in part (b) of Figure 15.4, where an increase in the value of x is matched by a decrease in the value of y.

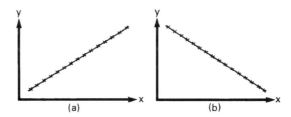

Figure 15.4

Although the signs of the relationships shown in Figure 15.4 are different, they both represent a perfect relationship, and thus can be thought of as one extreme. The opposite of a perfect relationship between x and y would be one where there is no relationship between the two variables, and this is illustrated in Figure 15.5.

These three cases form a basis for comparison with actual data which we may collect, or which may be generated by a business. It is unlikely that any

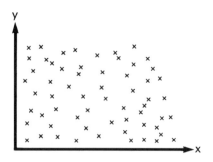

Figure 15.5

real situation will give an exact match to any one of these scatter diagrams, but they will allow us to begin our interpretation of the relationship before going into detailed calculations. (They may also help to avoid silly mistakes such as having a scatter diagram which shows a positive relationship and, through a small error, producing a numerical calculation which gives a negative answer.)

Scatter diagrams are often dismissed in favour of numerical calculations, but they do have a significant role to play in the analysis of bivariate relationships. Where the scatter diagram is similar to Figure 15.5 there may be little point in going on to work out a numerical relationship since no amount of information on likely values of x would allow you to make predictions of the behaviour of y. Similarly, if a particular scatter diagram is similar to Figure 15.3, there may be no point in calculating a linear relationship since the diagram shows a non-linear situation. Thus scatter diagrams can save considerably the time needed to analyse a situation.

Scatter diagrams can fairly easily be obtained from a spreadsheet such as Lotus or Excel.

Exercise

> Obtain a scatter diagram of the data from Jasbeer, Rajan & Co. for sales and promotional spend, and also for sales and advertising. What can you conclude from these scatter diagrams?

15.2 Cause and effect relationships

It is very tempting, having plotted a scatter diagram such as Figure 15.6 which suggests a strong linear relationship, then to go on to say that wage rises cause rises in prices, i.e. inflation. Others, however, would argue from this diagram that it is price rises that cause wage rises.

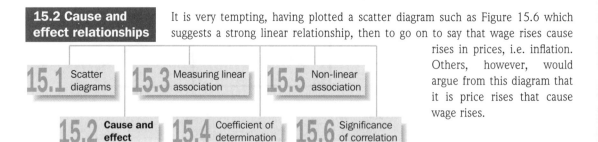

Exercise

> Is either assertion correct? Are they both correct?

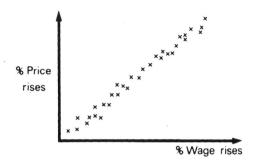

Figure 15.6

While it is unlikely that there will ever be an answer to these questions that everyone can agree with, some consideration of cause and effect will help in understanding the relationship.

Only one of the variables can be a cause at a particular point in time, but they could both be effects of a common cause! If you increase the standard of maternity care in a particular region, you will, in general, increase the survival rate of babies: it is easy to identify the direction of the relationship as it is extremely unlikely that an increasing survival rate would encourage health authorities to increase spending and improve standards! But would you agree on the direction of causation in Figure 15.7?

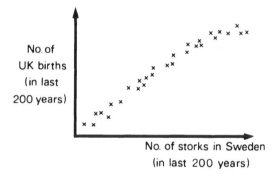

Figure 15.7

Here there appears to be a correlation, but it is spurious as *both* variables are changing over *time*. Time itself cannot be a cause, except perhaps in the ageing process, but where a situation of this type occurs, or where we have two effects of a **common cause**, there will be no way of controlling, or predicting, the behaviour of one variable by taking action on the other variable. An example of two effects of a common cause would be ice-cream sales and the grain harvest, both being affected by the weather.

While it cannot be the intention of this chapter to look deeply into the philosophy of cause and effect it will help to pick out a few conclusions from that subject. David Hume, in the eighteenth century, suggested that 'all reasonings concerning matters of fact seem to be founded on the relation of cause and effect'. He then went on to prove that this is not rationally justified. Which matters are justified in using cause and effect has interested many philosophers.

Conditions, or variables, can be divided into two types, **stable** and **differential**, and it is the second type which is most likely to be identified as a cause.

New roadworks on the M1 at Watford Gap will be cited as a cause of traffic jams tailing back to Coventry, since the roadworks are seen as a changed situation from the normally open three lane road. Drug abuse will be cited as a cause of death and illness, since only taking the recommended dosage is seen as normal (in some cases this will be a nil intake, e.g. heroin), and thus the alteration of normal behaviour is seen as causing the illness or death. There is also the temporal ordering of events; an 'effect should not precede a cause' and usually the events should be close together in time. A long time interval leads us to think of chains of causality. It is important to note the timing of events, since their recording must take place at the same time; for example, the studies by Doll and Hill of smoking and deaths from lung cancer conducted in the UK in the 1950s asked how many cigarettes per day people had smoked in the past and not just how many they were currently smoking. Some events may at first sight appear to be the 'wrong' way around. If most newspapers and commentators are predicting a sharp rise in VAT at a forthcoming budget in, say, six weeks' time, then many people will go out to buy fridges, washing machines and video recorders before budget day. If asked, they will say that the reason for their purchase is the rise in VAT in the budget in 6 weeks' time; however, if we look a little more closely, we see that it is their expectation of a VAT rise, and not the rise itself, which has caused them to go out and buy consumer durables. This, however, leads us to a further problem: it is extremely difficult to measure expectations!

A final but fundamental question that must be answered is '**does the relationship make sense?**'

Returning now to the problem in Figure 15.6, we could perhaps get nearer to cause and effect if we were to lag one of the variables, so that we relate wage rises in one month to price rises in the previous month: this would give some evidence of prices as a cause. The exercise could be repeated with wage rises lagged one month behind price increases – giving some evidence of wages as a cause. In a British context of the 1970s and 1980s both lagged relationships will give evidence of a strong association. This is because the wage–price–wage spiral is quite mature and likely to continue, partly through cause and effect, partly through expectations and partly through institutionalized factors.

Exercises

> 1 Give two other examples of variables between which there is a spurious correlation apart from those mentioned in the text.
> 2 What are the stable and differential conditions in relation to a forest fire?

Attempts to identify causes and their effects will also be concerned with which factors are controllable by the business, and which are not. For example, even if we could establish a causal relationship between Jasbeer, Rajan & Co.'s sales value and the weather, this would be of little value, since it seems unlikely that they will be able to control the weather. Such a relationship becomes more of a historical curiosity, than a useful business analysis model.

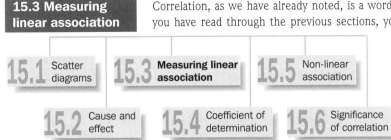

Correlation, as we have already noted, is a word which is now in common use. If you have read through the previous sections, you will now be able to produce a pictorial representation of the association between two sets of data, and this will considerably enhance your ability to talk about any correlation between two variables. In many circumstances this may be sufficient, but for a more detailed analysis we need to take the next step and find a way of measuring how strongly two variables are associated (the strength of the relationship). Such measurement will enable us to make comparisons between different models or proposed explanations of the behaviour of the variables. The measure that we will calculate is called **the coefficient of correlation**.

As we have already seen in Figure 15.4, an extreme type of relationship is one where all of the points of the scatter diagram lie on a straight line. Data which gave the scatter diagram shown in Figure 15.4(a) would have a correlation coefficient of $+1$, since y increases as x increases, while the correlation coefficient from Figure 15.4(b) would be -1. Where there is no relationship between the variables, the coefficient of correlation would be equal to zero (for example in Figure 15.5). Each of these situations is very unlikely to exist in practice. Even if there were to be a perfect, linear relationship between two variables, there would probably be some measurement errors and the data which we collected would give an answer between -1 and $+1$. Data generated by businesses and the government is often affected by the same underlying movements and disturbances and this will mean that even where there is no real relationship between two variables (i.e. no cause and effect), there may still be a correlation coefficient that differs from zero.

Finally, before we begin the process of calculation, we should make a distinction between cardinal and ordinal data. Although the method of calculation is basically the same, we must remember that the **type of data will affect the way in which we interpret the answers calculated**. Note that when we are using a statistical package we may be able to use exactly the same commands whichever type of data we have; if, however, you have to perform the calculation by hand, there is a simplified formula which applies only to the case of ordinal data. (Note also that certain corrections may be applied to the calculation of the correlation coefficient from ordinal data by some statistical packages.)

15.3.1 Rank correlation

Ordinal data consists of values defined by the **position of the data in an ordered list** (a rank) and may be applied to situations where no numerical measure can be made, but where best and worst or most favoured and least favoured can be identified. They may also be used for international comparisons where the use of exchange rates to bring all of the data into a common currency would not be justified. Rankings are often applied where people or companies are asked to express preferences or to put a series of items into an order. In these situations is is no longer possible to say how much better first is than second; it may be just fractionally better, or it may be outstandingly better. We therefore need to exercise caution when interpreting a rank correlation coefficient (or Spearman's correlation coefficient) since similar ranking, and hence fairly high correlation, may represent different views of the situation.

Case 5

In Table 15.1 we show the popularity ranking of ten of the various products produced by Jasbeer, Rajan & Co. for last year and this year.

Table 15.1 Ranking of Jasbeer, Rajan & Co.'s products

Product	Last year	This year
Crackers	1	3
Joke String	2	4
Hats	3	1
Masks	4	2
Joke Food	5	6
Balloons	6	10
Whistles	7	9
Streamers	8	7
Flags	9	8
Mini Joke Book	10	5

We can go ahead and calculate a rank correlation coefficient between these two sets of data; in doing so we are trying to answer the question 'Are there similarities in the rankings of the products from one year to the next?', or 'Is there an association between the rankings?' To do this we need to use a formula known as **Spearman's coefficient of rank correlation**:

$$r = 1 - \frac{6 \sum d^2}{n(n^2 - 1)}$$

The first step is to find the difference in the rank given to each product (d = last year − this year), which gives us a series of positive and negative numbers which add up to zero. The way of overcoming this is to square each difference, and then sum these squared values to use in the formula. This calculation has been done in Table 15.2. Such calculations are easily performed by hand, but could be done on a spreadsheet. (Note that, if you are using a spreadsheet, you can just use the standard regression command to obtain the correlation coefficient.)

Table 15.2 Calculation of rank correlation

Product	Last year	This year	d	d^2
Crackers	1	3	−2	4
Joke String	2	4	−2	4
Hats	3	1	2	4
Masks	4	2	2	4
Joke Food	5	6	−1	1
Balloons	6	10	−4	16
Whistles	7	9	−2	4
Streamers	8	7	1	1
Flags	9	8	1	1
Mini Joke Books	10	5	5	25
Total				64

Using the formula given above, we now have:

$$r_s = 1 - \frac{6 \times 64}{10(100-1)}$$

$$= 1 - \frac{384}{10 \times 99} = 1 - \frac{384}{990}$$

$$= 1 - 0.387878$$

$$= 0.61212$$

This answer suggests that there is some degree of association between the rankings from one year to the next, but probably not enough to be useful to the business. The limitations mentioned above about rank correlation must be borne in mind when interpreting this answer. Rank correlation would not, normally, be used with cardinal data since information would be lost by moving from actual measured values to simple ranks. However, if one variable was measured on the cardinal scale and the other was measured on the ordinal scale, ranking both and then calculating the rank correlation coefficient may prove the only practical way of dealing with these mixed measurements. Rank correlation also finds applications where the data is cardinal but the relationship is non-linear.

15.3.2 Correlation for cardinal data

Cardinal data does not rely on subjective judgement, but on some form of objective measurement to obtain the data. Since this measurement produces a scale where it is possible to say by how much two items of data differ (which it wasn't possible to do for ranked data), then we may have rather more confidence in our results. Results close to either $+1$ or -1 will indicate a high degree of association between two sets of data, but as mentioned earlier, this does not necessarily imply a cause and effect relationship. The statistic which we will calculate is known as **Pearson's correlation coefficient** and is given by the following formula:

$$r = \frac{n\sum xy - \sum x \sum y}{\sqrt{\{(n\sum x^2 - [\sum x]^2)(n\sum y^2 - [\sum y]^2)\}}}$$

Although this formula looks very different from the Spearman's formula, they can be shown to be the same (see Section 15.10). For those who are more technically minded, the formula is the covariance of x and y (how much they vary together), divided by the root of the product of the variance of x and the variance of y (how much they each, individually, vary). The derivation is shown in Section 15.9.

Case 5

Continuing our example from Jasbeer, Rajan & Co., we have already seen that there appears to be an association between Sales Volume and Promotional Spend (see Figure 15.1). We will now work through the process of finding the numerical value of the correlation coefficient, doing this by hand. In most cases you would just have the data in a spreadsheet and issue the command to find the correlation coefficient. The data are shown in Table 15.3.

Table 15.3   Sales volume and promotion for Jasbeer, Rajan & Co.

Yr. qtr	Sales volume ('000s)	Promotion (£'000s)
1.1	166	50
1.2	52	18
1.3	140	58
1.4	733	76
2.1	224	56
2.2	114	45
2.3	181	60
2.4	753	69
3.1	269	61
3.2	214	32
3.3	210	58
3.4	860	95
4.1	345	63
4.2	203	31
4.3	233	67
4.4	922	123
5.1	324	76
5.2	224	40
5.3	284	85
5.4	822	124
6.1	352	86
6.2	280	39
6.3	295	81
6.4	930	135
7.1	345	61
7.2	320	50
7.3	390	79
7.4	978	140
8.1	483	90
8.2	320	91

Our first task is to assign labels to the two variables. As a general rule, we usually give the label y to the variable which we are trying to predict, or the one which we cannot control. Here, this would be the sales volume. The other variable is then labelled as x. Once we have done this, we can look at the formula which was given above, and notice that we need to calculate a series of summations. These are:

$$\sum x \quad \sum y \quad \sum x^2 \quad \sum y^2 \quad \sum xy$$

We also need to know the number of pairs of data (n).

The appropriate calculations are shown in Table 15.4.

Table 15.4 Correlation calculations

Yr. qtr	Sales volume y	Promotion x	x^2	y^2	xy
1.1	166	50	2500	27556	8300
1.2	52	18	324	2704	936
1.3	140	58	3364	19600	8120
1.4	733	76	5776	537289	55708
2.1	224	56	3136	50176	12544
2.2	114	45	2025	12996	5130
2.3	181	60	3600	32761	10860
2.4	753	69	4761	567009	51957
3.1	269	61	3721	72361	16409
3.2	214	32	1024	45796	6848
3.3	210	58	3364	44100	12180
3.4	860	95	9025	739600	81700
4.1	345	63	3969	119025	21735
4.2	203	31	961	41209	6293
4.3	233	67	4489	54289	15611
4.4	922	123	15129	850084	113406
5.1	324	76	5776	104976	24624
5.2	224	40	1600	50176	8960
5.3	284	85	7225	80656	24140
5.4	822	124	15376	675684	101928
6.1	352	86	7396	123904	30272
6.2	280	39	1521	78400	10920
6.3	295	81	6561	87025	23895
6.4	930	135	18225	864900	125550
7.1	345	61	3721	119025	21045
7.2	320	50	2500	102400	16000
7.3	390	79	6241	152100	30810
7.4	978	140	19600	956484	136920
8.1	483	90	8100	233289	43470
8.2	320	91	8281	102400	29120
Totals	11966	2139	179291	6947974	1055391

We already know that $n = 30$. If we now substitute the various totals into the formula, we get:

$$r = \frac{30 \times 1\,055\,391 - (2139)(11\,966)}{\sqrt{\{[30 \times 179\,291 - (2139)^2][30 \times 6\,947\,974 - (11\,966)^2]\}}}$$

$$= \frac{31\,661\,730 - 25\,595\,274}{\sqrt{\{[5\,378\,730 - 4\,575\,321][208\,439\,220 - 143\,185\,156]\}}}$$

$$= \frac{6\,066\,456}{\sqrt{[803\,409 \times 65\,254\,064]}}$$

$$= \frac{6\,066\,456}{7\,240\,559.5}$$

$$= 0.837\,843\,536$$

This confirms that there is a fairly high level of association between sales volume and promotional spend for Jasbeer, Rajan & Co. While we know nothing of cause and effect here, since promotional spend is a controllable variable, the company may wish to use it in developing a strategic policy for the marketing of its products.

Exercise

Using a spreadsheet obtain the correlation coefficient for this data.

15.4 The coefficient of determination

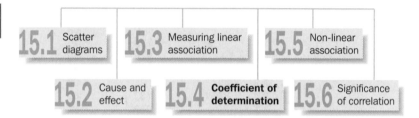

While the correlation coefficient is used as a measure of the association between two sets of data, probably a more common statistic is the **coefficient of determination**. This statistic will allow us to take an extra step in analysing the relationship. It is defined to be

$$\frac{\text{variation in the variable explained by the association}}{\text{total variation in the variable}}$$

Fortunately this is not difficult to find since it can be shown to be equal to the value of the coefficient of correlation squared. (Note that this only makes sense for cardinal data.) The value obtained is usually multiplied by 100 so that the answer may be quoted as a percentage. Many statistical packages will automatically give the value of r^2 when told to find correlation and regression for a set of data.

The coefficient of determination will **always be positive**, and can only take values between 0 and 1 (or 0 and 100 if quoted as a percentage), and will have a lower numerical value than the coefficient of correlation. For example, if the correlation were $r = 0.93$, then the coefficient of determination would be $r^2 = 0.8649$. This would be interpreted as 86.49% of the variation in one of the variables being explained by the association between them, while the remaining 13.51% is explained by other factors. (Merely stating this is useful when answering examination questions where you are asked to comment on your answer.)

The value of the coefficient of determination is often used in deciding whether or not to continue the analysis of a particular set of data. Where it has been shown that one variable only explains 10% or 20% of the behaviour of another variable, there seems little point in trying to predict the future behaviour of the second from the behaviour of the first. This point can easily be missed with the simplicity of making predictions using computer packages. However, if x explains 95% of the behaviour of y, then predictions are likely to be fairly accurate.

Interpretation of the value of r^2 should be made with two provisos in mind. Firstly, the value has been obtained from a specific set of data over a given range of values. This means that it does not necessarily apply to other ranges of values. In general, it may well apply to data close to the range of values used, but is much less likely to apply to data well away from this range. Inspection of the scatter diagram may help in determining where it does apply. Secondly, the value obtained for r^2 does not give evidence of cause and effect, despite the fact that we talk about *explained* variations. Explained here refers to *explained by the analysis* and is purely an arithmetic answer. In the case of Jasbeer, Rajan & Co. we can simply square the correlation coefficient which we obtained in the last section to get the answer 0.7019, or approximately 70% of the variation in sales volume is *explained* by variations in promotional spend.

Exercises

> 1 Why is it important to know the number of observations when interpreting a correlation coefficient?
> 2 If a firm has studied the relationship between its sales and the price which it charges for its product over a number of years and finds that $r^2 = 0.75$, how would you interpret this result? Is there enough evidence to suggest a cause and effect relationship?

15.5 Measuring non-linear association

In looking at the idea of correlation, so far we have limited ourselves to considering only the case of linear correlation, i.e. only scatter diagrams which give an approximate straight line will give relatively high values when we calculate the correlation coefficient. Obviously this is **not the only possible case** where a correlation may exist. Many situations exist where, as one variable increases, the other increases at an increasing rate, e.g. growth against time – or increases at a decreasing rate, e.g. sales growth reaching maturity. Other types of relationship are also likely to occur in business situations. In order to measure association, we will still use Pearson's correlation coefficient, but will adjust it to allow for the non-linearity of the data. We must return to the scatter diagram in order to decide upon the type of non-linearity present. (It is beyond the scope of this book to look at every type of non-linearity, but a few of the most commonly met types are given below.)

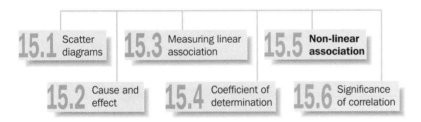

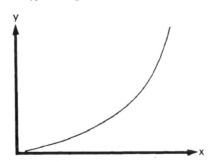

Figure 15.8

If the scatter diagram shows the relationship in Figure 15.8, then we will find that by taking the log of *y*, we get a linear function as in Figure 15.9. When performing the calculations, we will take the log of *y* and then use this value exclusively, as in Table 15.5. (Note that we could use either natural logs or logs to base 10. Here logs to base 10 are used.)

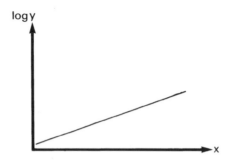

Figure 15.9

Table 15.5 Non-linear correlation

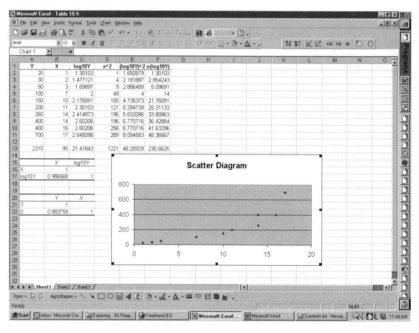

$$r = \frac{(10)(230.79) - (95)(21.43)}{\sqrt{\{[(10)(1221) - 95^2][(10)(48.3117) - 21.43^2]\}}}$$

$$= \frac{272.05}{\sqrt{\{3185 \times 23.8721\}}}$$

$$= 0.9866 \qquad \text{(with } x \text{ and } y, \, r = 0.884)$$

We have carried out a **transformation to linearity**.

Deciding on the appropriate transformation to use may be fairly straightforward where there is a high degree of correlation, since the scatter diagram will be approximately the shape of some 'known' function. To help you make a selection, Figure 15.10 illustrates some of the more widely used transformations and their appropriate scatter diagrams below.

When the correlation is lower, it may be a case of trial and error in finding an appropriate transformation. This is rather easier than it sounds, since we presume that you would be working on a computer. In a spreadsheet you transform

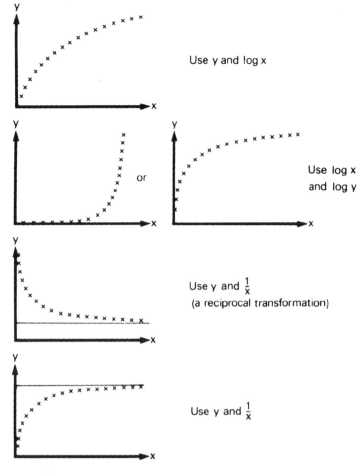

Figure 15.10

the column that interests you and then just use the transformed columns in the commands you give for correlation to be carried out.

It is worth a note here about the use of computers in calculating correlation. At one time, using transformations was a time-consuming and fairly difficult process, and was only undertaken after considerable thought and analysis. Now, however, it is so simple that there is considerable danger that transformations will be done just for the sake of it. With most business-type data there is likely to be some relationship between variables, even if it does not make sense, and thus the ease of performing transformations may lead to spurious correlations being mistaken for useful ones, just because they exist.

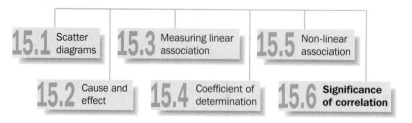

15.6 Testing the significance of the correlation

We have talked about 'high' and 'low' correlations, but the statistical significance of a particular numerical value will vary with the number of pairs of observations available to us. The smaller the number of observations, the higher must be the value of the correlation coefficient to establish association between the two sets of data, and thus, more generally, between the two variables under consideration. Showing that there is a significant correlation will still not be conclusive evidence for cause and effect but it will be a necessary condition before we propound a theory of this type.

As we have seen in Chapter 12, a series of values can be thought of as a sample from a wider population, and thus we can use a test of hypothesis to evaluate the results we have obtained. In the same way that we were able to test a sample mean, we may also test a correlation coefficient, here the test being to find if the value obtained is significantly greater than zero. The correlation coefficient is treated as the sample value, r, and the population (or true) value of correlation is represented by the letter ρ (rho). Statistical theory shows that while the distribution of the test statistic for r (see step 4 below) is symmetrical, it is narrower than a Normal distribution, and in fact follows a t-distribution.

Using the methodology proposed in Chapter 12, we may conduct the test. Suppose that we have obtained a linear correlation coefficient of 0.65 from ten pairs of data.

1. State hypotheses: $H_0: \rho = 0$, $H_1: \rho > 0$.
   (Note that we are using a one-tailed test.)
2. State the significance level: we will use 5%.
3. State the critical value: 1.86.
   (Note that there are $(n - 2) = 8$ degrees of freedom for this test, and that the critical value is obtained from Appendix D.)
4. Calculate the test statistic:

$$t = \frac{r\sqrt{(n-2)}}{\sqrt{(1-r^2)}}$$

$$= \frac{0.65\sqrt{(10-2)}}{\sqrt{(1-0.65^2)}}$$

$$= \frac{1.8385}{0.7599}$$

$$= 2.419$$

5. Compare the values: $2.419 > 1.86$.
6. Conclusion: we may reject H_0.
7. Put the conclusion into English. Since the value of the calculated statistic is above the critical value we may conclude that there is a correlation between the two sets of data which is significantly above zero.

It would be possible to use a two-tailed test to try to show that there was a correlation between the two sets of data, but in most circumstances it is preferable to use a one-tailed test.

With the increasing use of computer packages it has become very easy to construct models of data and to obtain correlation coefficients between pairs of data. A situation frequently faced is where we are trying to find a variable associated with some variable we are trying to predict. Data is obtained and the correlation calculated but the question arises as to whether a model with a slightly higher correlation is significantly better than one with a slightly lower correlation.

Similarly, we may have data from different regions or countries and want to know if there is more correlation between two specific variables in one of these.

Suppose that we have obtained the following information:

Sample 1: From a sample of ten companies in Italy, the correlation between turn-over and profit is 0.76.

Sample 2: From a sample of 15 companies in France, the correlation between turnover and profit is 0.81.

Is there more association between these two variables in French companies than in Italian companies?

Using our established testing procedure:

1 Hypotheses: H_0: $\rho_1 = \rho_2$, H_1: $\rho_1 < \rho_2$.
2 Significance level is 5%.
3 Critical value is -1.645. (Note that here we revert to using values from the Normal distribution.)
4 Calculate the test statistic (*this is fairly lengthy*).
   (a) Calculate the values of z_r

$$z_{r1} = 0.5 \ln \left[\frac{(1 + r_1)}{(1 - r_1)} \right]$$

$$z_{r2} = 0.5 \ln \left[\frac{(1 + r_2)}{(1 - r_2)} \right]$$

where ln is the natural log.
For our data this gives:

$$z_{r1} = 0.5 \ln \left[\frac{(1 + 0.76)}{(1 - 0.76)} \right] = 0.9962$$

$$z_{r2} = 0.5 \ln \left[\frac{(1 + 0.81)}{(1 - 0.81)} \right] = 1.1270$$

   (b) Calculate the values of the standard deviations:

$$\sigma_{z1} = \sqrt{\frac{1}{(n_1 - 3)}}$$

$$\sigma_{z2} = \sqrt{\frac{1}{(n_2 - 3)}}$$

For our data this gives:

$$\sigma_{z1} = \sqrt{\frac{1}{(10 - 3)}} = 0.377\,96$$

$$\sigma_{z2} = \sqrt{\frac{1}{(15 - 3)}} = 0.288\,68$$

(c) Use the formula for testing:

$$z = \frac{z_{r1} - z_{r2}}{\sqrt{(\sigma_{z1}^2 + \sigma_{z2}^2)}}$$

Here

$$z = \frac{0.9962 - 1.1270}{\sqrt{(0.37796^2 + 0.28868^2)}}$$

$$= \frac{-0.1308}{0.47559}$$

$$= -0.275$$

5 Compare the values: $-0.275 > -1.645$.
6 Conclusion: we cannot reject H_0.
7 Put conclusion into English. Since the calculated value is above the critical value we are unable to reject the null hypothesis, and must therefore conclude that the difference between the two correlation coefficients is due to sampling error.

Some statistical packages may work out the significance test for you, or just give a probability value, and if this is below 0.05, then the value of r is significantly above zero (single value test).

15.7 Conclusions

Finding whether or not there is a correlation between two variables can be very useful to a business, and correlation analysis is a useful first step when linking two variables together. Where it can be shown that there is an association between two sets of data it will be reasonable to take further steps in the analysis, particularly for cardinal data. While a high correlation cannot be said to provide evidence of a causal relationship between the two variables, some would argue that it is a necessary, rather than a sufficient, condition for proposing such a relationship to exist. In other words, having a high correlation does not mean that you have a cause and effect relationship, but if you think that there should be such a relationship (maybe from some suggested theory), then you should be able to find a high correlation between the variables. If, further, the business can 'control' one of the variables, then it might want to take steps to do this in the light of the existence of such a correlation. Where no correlation exists, it might reduce its efforts to control that variable — for example, if it were shown that new book signing sessions by authors were not related to sales of the books, publishers might abandon them.

Exercise

Using a spreadsheet obtain the correlation coefficients for each of the 'controllable' variables with sales volume and then with sales value for Jasbeer, Rajan & Co. from the data on the website. Which are the key variables for the company?

15.8 Problems

1 Calculate Spearman's coefficient of correlation for the following data:

x	1	2	3	4	5
y	3	1	4	5	2

2 Find a correlation coefficient using the following ordinal data:

x	1	2	3	4	5	6	7	8
y	3	8	7	5	6	1	4	2

3 Calculate the rank correlation coefficient for the following data:

x	1	2	3	4	5
y	2	1	4	5	3

4 Eight brands of washing powder have been ranked by groups of people in the North and the South. Their rankings are as follows:

Brand	Rank in North	Rank in South
A	1	2
B	4	6
C	8	8
D	3	1
E	6	5
F	2	4
G	5	7
H	7	3

Is there an association between the rankings?

5 A group of people have been ranked in terms of both their mechanical/technical ability and their tact in handling people and their rankings are given below. Find if there is a correlation between the two rankings.

Person	Mechanical/technical	Tact
A	1	7
B	4	8
C	7	1
D	8	5
E	5	6
F	2	4
G	3	2
H	6	3

6 A group of athletes were ranked before a recent race and their positions in the race noted. Details are given below. Find the correlations between the rankings and the positions.

Ranking	1	2	3	4	5	6	7	8	9	10
Finishing position	3	5	2	1	10	4	9	7	8	6

7 Ranking of individuals or events can be a very personal activity. Suggest two characteristics, or events, which individuals would rank (a) similarly, (b) differently.

8 Using the cardinal data given below, find Pearson's coefficient of correlation.

x	1	2	3	4	5
y	3	6	10	12	14

9 Take the data given below and construct a scatter diagram. Find the correlation coefficient and the coefficient of determination for this data.

x	10	12	14	16	18	20	22	24	26	28
y	25	24	22	20	19	17	13	12	11	10

10 Use the following data to calculate the coefficient of correlation and the coefficient of determination. Can you draw any conclusions from your results?

x	4	2	6	7	8	5	2	4
y	10	5	15	16	19	14	8	11

11 A farmer has recorded the number of fertilizer applications to each of the fields in one section of the farm and, at harvest time, records the weight of crop per acre. The results are given in the accompanying table.

x	1	2	4	5	6	8	10
y	2	3	4	7	12	10	7

Find the correlation between fertilizer applications and weight of crop per acre. Draw a scatter diagram from the data. Using your two results, what advice, if any, could you give to the farmer?

12 Determine the correlation coefficient for the following data by inspection.

x	11	12	13	14	15	16	17	18	19	20
y	4	4	4	4	4	4	4	4	4	4

If you are unable to see the answer, use a spreadsheet to find the correlation coefficient.

13 The personnel department of a company has conducted a pre-interview test of aptitude on candidates for some time. It is now in possession of annual appraisal interview reports from line managers on the successful candidates. You have been asked to assess the validity of the results from the pre-interview tests on the basis of the following evidence:

Person	Pre-test score	Report score
A	50	67
B	62	70
C	85	80
D	91	79
E	74	68
F	53	67
G	74	81
H	59	67
I	84	90
J	67	75
K	41	40
L	85	80
M	68	71
N	79	82
O	83	76
P	67	78
Q	81	86
R	75	78
S	82	64
T	72	67

14 Workers on the shop floor are paid on a piecework basis for producing items. The union claims that this seriously discriminates against newer workers, since there is a fairly steep learning curve which workers follow, with the result that more experienced workers can perform the task in about half of the time taken by a new employee. You have been asked to find out if there is any basis for this claim. To do this, you have observed ten workers on the shop floor, timing how long it takes them to produce the item. It was then possible for you to match these times with the length of the workers' experience. The results obtained are shown below.

Person	Months' experience	Time taken
A	2	27
B	5	26
C	3	30
D	8	20
E	5	22
F	9	20
G	12	16
H	16	15
I	1	30
J	6	19

(a) Construct a scatter diagram for this data.

(b) Find the coefficient of determination and interpret the result obtained.

15 During the making of certain electrical components each item goes through a series of heat processes. The length of time spent in this heat treatment is related to the useful life of the component. To find the nature of this relationship a sample of twenty components are selected from the process and tested to destruction and the results are presented below.

Time in process (minutes)	Length of life (hours)
25	2005
27	2157
25	2347
26	2239
31	2889
30	2942
32	3048
29	3002
30	2943
44	3844
41	3759
42	3810
41	3814
44	3927
31	3110
30	2999
55	4005
52	3992
49	4107
50	3987

(a) Construct a scatter diagram from this data.

(b) Use a spreadsheet to find the correlation between the time spent in the heat processes and the length of useful life. Hence find the coefficient of determination.

(c) Compare your scatter diagram with those in Figure 15.10 and thus decide upon a suitable transformation for the data. Calculate the new correlation coefficient and the new coefficient of determination.

(d) Comment on the results that you have found.

16 Construct a scatter diagram for the following data:

x	1	2	3	4	5	6	7	8	9	10
y	10	10	11	12	12	13	15	18	21	25
x	11	12	13	14	15	16	17	18	19	20
y	26	29	33	39	46	60	79	88	100	130

Find the coefficient of correlation and the coefficient of determination. Now find the log of y and re-calculate the two statistics. How would you interpret your results?

17 Costs of production have been monitored for some time within a company and the following data found:

Production level ('000)	Average total cost (£'000)
1	70
2	65
3	50
4	40
5	30
6	25
7	20
8	21
9	20
10	19
11	17
12	18
13	18
14	19
15	20

(a) Construct a scatter diagram for the data.

(b) Calculate the coefficient of determination and explain its significance for the company.

(c) Is there a better model than the simple linear relationship which would increase the value of the coefficient of determination? If your answer is 'yes', calculate the new coefficient of determination.

(d) What factors would affect the average total cost other than the production level?

18 Find the correlation coefficient between the following sets of data and test its significance.

x	25	30	60	45	40	80	70	90	15	20
y	100	95	50	60	65	30	35	10	120	110

19 Find the correlation coefficient and the coefficient of determination for the following data.

x	y
5	5
10	6
15	7
20	8
25	10
30	12
35	14
40	18
45	22
50	28
55	34
60	40
65	48
70	60
75	75

Find ways to increase the correlation coefficient.

15.9 Derivation of the correlation coefficient

Pearson's correlation coefficient is

$$\frac{\text{covariance of } x \text{ and } y}{\sqrt{[(\text{variance of } x)(\text{variance of } y)]}}$$

$$= \frac{\frac{1}{n}\sum(x-\bar{x})(y-\bar{y})}{\sqrt{\left[\frac{1}{n}\sum(x-\bar{x})^2\frac{1}{n}\sum(y-\bar{y})^2\right]}}$$

$$= \frac{\frac{1}{n}[\sum(xy-x\bar{y}-\bar{x}y+\bar{x}\bar{y})]}{\frac{1}{n}\sqrt{[\sum(x^2-2x\bar{x}+\bar{x}^2)\sum(y^2-2y\bar{y}+\bar{y}^2)]}}$$

$$= \frac{\sum xy - \frac{1}{n}\sum x\sum y - \frac{1}{n}\sum x\sum y + \frac{n}{n^2}\sum x\sum y}{\sqrt{\left\{\left[\sum x^2 - \frac{2}{n}(\sum x)^2 + \frac{n}{n^2}(\sum x)^2\right]\times\left[\sum y^2 - \frac{2}{n}(\sum y)^2 + \frac{n}{n^2}(\sum y)^2\right]\right\}}}$$

$$= \frac{\sum xy - \frac{1}{n}(\sum x)(\sum y)}{\sqrt{\left\{\left[\sum x^2 - \frac{1}{n}(\sum x)^2\right]\left[\sum y^2 - \frac{1}{n}(\sum y)^2\right]\right\}}}$$

or if we multiply top and bottom by n, we have:

$$r = \frac{n\sum xy - \sum x\sum y}{\sqrt{\{[n\sum x^2 - (\sum x)^2][n\sum y^2 - (\sum y)^2]\}}}$$

15.10 Algebraic link between Spearman's and Pearson's coefficients

Pearson's coefficient is:

$$r = \frac{n\sum xy - \sum x\sum y}{\sqrt{\{[n\sum x^2 - (\sum x)^2][n\sum y^2 - (\sum y)^2]\}}}$$

Now ranked data (both x and y) will only take on integer values from 1 to n and there are formulae for the sum of the first n natural numbers and the sum of their squares. These are:

$$\sum x = \sum y = \frac{n(n+1)}{n}$$

and

$$\sum x^2 = \sum y^2 = \frac{n(n+1)(2n+1)}{6}$$

and therefore

$$r = \frac{n\sum xy - \left(\sum x\right)^2}{n\sum x^2 - \left(\sum x\right)^2}$$

Also

$$\sum d = \sum (x - y)$$

so

$$\sum d^2 = \sum (x - y)^2 = \sum x^2 + \sum y^2 - 2\sum xy$$
$$= 2\sum x^2 - 2\sum xy$$

Rearranging gives

$$\sum xy = \sum x^2 - 0.5\sum d^2$$
$$= \frac{n(n+1)(2n+1)}{6} - \frac{\sum d^2}{2}$$

Putting all of these together gives:

$$r = \frac{n\dfrac{n(n+1)(2n+1)}{6} - \dfrac{\sum d^2}{2} - \dfrac{n^2(n+1)^2}{4}}{\dfrac{n(n(n+1)(2n+1))}{6} - \dfrac{n^2(n+1)^2}{4}}$$

Multiplying the top and bottom by 12 gives:

$$r = \frac{2n^2(n+1)(2n+1) - 6n\sum d^2 - 3n^2(n+1)^2}{2n^2(n+1)(2n+1) - 3n^2(n+1)^2}$$

Collecting terms gives:

$$r = \frac{2n^2(n+1)(2n+1) - 3n^2(n+1)^2 - 6n\sum d^2}{2n^2(n+1)(2n+1) - 3n^2(n+1)^2}$$
$$= 1 - \frac{6n\sum d^2}{2n^2(n+1)(2n+1) - 3n^2(n+1)^2}$$

The bottom line of the fraction can be simplified to $n^2(n^2 - 1)$ so

$$r = 1 - \frac{6n\sum d^2}{n^2(n^2 - 1)}$$
$$= 1 - \frac{6\sum d^2}{n(n^2 - 1)}$$

16 Regression

From our discussions in the previous chapter, we were able to find a way of determining whether or not a relationship existed between two variables. You should also have discovered that by using appropriate software, the calculations of the correlation coefficient can be done very quickly. Where we are able to identify a correlation which we are prepared to describe as 'high' in our context, then the next, obvious, question will be: 'If a relationship exists, what is it?' If we are able to specify the relationship in the form of an equation, then a business will be able to make **predictions** about the behaviour of an uncontrollable variable from actions which it takes to fix the value of a controllable variable. Of course, such an equation is unlikely to be deterministic (except in the very unlikely event of the correlation coefficient being equal to plus or minus one), and so we may be asked to specify a range of values for the prediction and attach a confidence level to this interval. The values of the coefficients of the equation may also help a business to understand how a relationship works, and the effects of changing the value of a variable. For example, if a reduction in price is known to be related to an increase in sales, the business could decide by how much to reduce its price. In some markets a 5% reduction may lead to vastly increased sales, whilst in other markets where there is strong brand loyalty, a 5% price reduction may hardly affect sales at all.

Spreadsheet software makes it very easy to find the equation of a relationship between two variables, but it must be remembered that such an equation will be useless unless there is a relatively strong correlation between the variables.

It is worth noting that **regression only applies to cardinal data.**

Objectives

After working through this chapter, you should be able to:

- describe the link between regression and correlation
- calculate the coefficients of a regression line
- make predictions from a regression line
- identify the problems of extrapolation
- apply regression to non-linear situations.

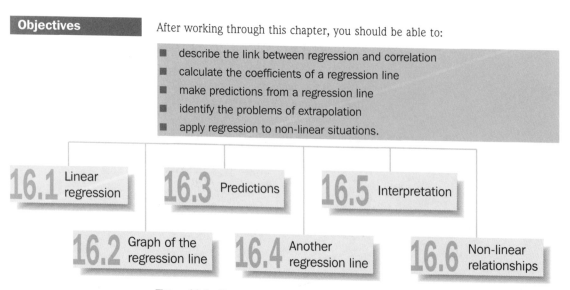

16.1 Linear regression

16.2 Graph of the regression line

16.3 Predictions

16.4 Another regression line

16.5 Interpretation

16.6 Non-linear relationships

Figure 16.1 The structure of Chapter 16.

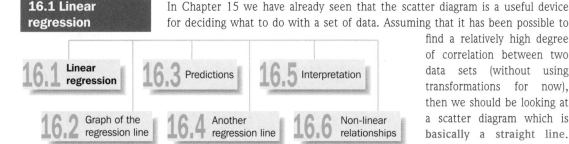

16.1 Linear regression

In Chapter 15 we have already seen that the scatter diagram is a useful device for deciding what to do with a set of data. Assuming that it has been possible to find a relatively high degree of correlation between two data sets (without using transformations for now), then we should be looking at a scatter diagram which is basically a straight line. Such straight lines are remarkably powerful tools of analysis in business situations and have a wide range of application. (Even where a relationship is not linear, it may be that certain parts of it can be modelled as if they were linear, i.e. between certain values.)

Given a particular scatter diagram, it will be possible to draw many different straight lines through the data (unless the correlation is equal to plus or minus 1), and we therefore need to establish a way of finding the best line through the data. Of course, to do this we need to establish what we mean by 'best'!

To define best we may reconsider the reason for trying to estimate the regression line. Our objective is to be able to predict the behaviour of one of the variables, say y, from the behaviour of the other, say x. So the best line will be the one which gives us the best prediction of y. Best will now refer to how close the predictions of y are to the actual values of y. Since we only have the data pairs which have given us the scatter diagram we will have to use those values of x to predict the corresponding values of y, and then make an assumption that if the relationship holds for the current data, that it will hold for other values of x.

If you consider Figure 16.2, which shows part of a scatter diagram, you will see that a straight line has been superimposed on to the scatter of points and the **vertical** distance, d, has been marked between one of the points and the line. This shows the difference, at one particular value of x, between what actually happened to the other variable, y, and what would be predicted to happen if we use the linear function, $\hat{y}$ (pronounced y hat). This is the error that would be made in using the line for prediction.

To get the best line for predicting y we want to make all of these errors as small as possible, and it can be shown that to minimize something we can use calculus (see Section 16.9 for complete derivation of formulae). Working through this process will give values for a and b and thus we will have identified a specific

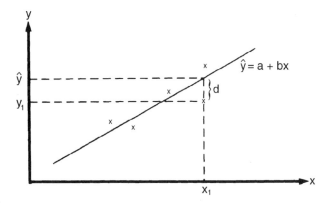

Figure 16.2

linear function through the data. The line we have identified may be called by a number of names, the line of best fit, the least squares line (since it minimizes the sum of squared differences of observed values from the regression line – see Section 16.9), or the regression line of y **on** x (since it is the line which was derived from a desire to predict y values from x values).

The formulae for estimating a and b are:

$$b = \frac{n\sum xy - \sum x \sum y}{n\sum x^2 - (\sum x)^2}$$

$$a = \bar{y} - b\bar{x}$$

giving the regression line: $\hat{y} = a + bx$.

If we compare the formula for b with the correlation formula for cardinal data given in Chapter 15:

$$r = \frac{n\sum xy - \sum x \sum y}{\sqrt{\{[n\sum x^2 - (\sum x)^2][n\sum y^2 - (\sum y)^2]\}}}$$

then we see that their numerators are identical, and the denominator of b is equal to the first bracket of the denominator for r. Thus the calculation of b, after having calculated the correlation coefficient, will be a simple matter.

Returning to our ongoing example from Jasbeer, Rajan & Co., we calculated the correlation coefficient between sales volume and promotional spend to be 0.8378 and had the following totals from our hand calculations:

$$\sum x = 2139 \quad \sum y = 11\,966 \quad \sum x^2 = 179\,291$$
$$\sum y^2 = 6\,947\,974 \quad \sum xy = 1\,055\,391$$

Case 5

Using this data, we may now find the equation of sales volume on promotional spend for Jasbeer, Rajan & Co., substituting into the formulae given above.

$$b = \frac{30 \times 1\,055\,391 - (2139)(11\,966)}{30 \times 179\,291 - (2139)^2}$$
$$= \frac{31\,661\,730 - 25\,595\,274}{5\,378\,730 - 4\,575\,321}$$
$$= \frac{6\,066\,456}{803\,409}$$
$$= 7.550\,894$$

$$a = \frac{11\,966}{30} - 7.550\,894 \times \frac{2139}{30}$$
$$= 398.866\,667 - 7.550\,894 \times 71.3$$
$$= 398.866\,667 - 538.378\,74$$
$$= -139.512\,08$$

Thus the equation of the regression line is:

$$\hat{y} = -139.512\,08 + 7.550\,894x$$

or

$$\text{sales volume} = -139.51 + 7.55 \times \text{promotional spend}.$$

Exercise

Using a spreadsheet obtain the regression line for this data.

16.2 The graph of the regression line

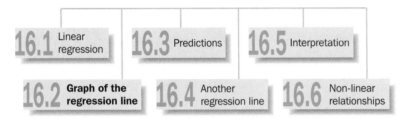

Having identified the specific regression line that applies to a particular set of points, it is usual to place this on the scatter diagram. Since we are dealing with a linear function, we only need two points on the line, and these may then be joined using a ruler. The two points could be found by substituting values of x into the regression equation and working out the values of $\hat{y}$, but there is an easier way. For a straight line, $y = a + bx$, the value of a is the intercept on the y-axis, so this value may be plotted. From the formula for calculating a (or from the proof in Section 16.9) we see that the line goes through the point $(\bar{x}, \bar{y})$, which has already been calculated, so that this may be plotted. The two points are then joined together (Figure 16.3).

Exercise

Using a spreadsheet obtain the scatter diagram of sales volume (as y) and promotional spend (as x) for Jaspeer, Rajan & Co. and place the regression line onto it.

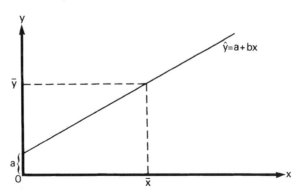

Figure 16.3

16.3 Predictions from the regression line

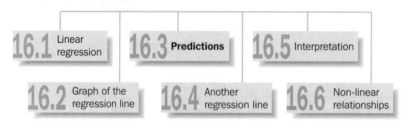

One of the major aims of regression analysis is to make predictions. We are trying to predict what will happen to one variable when the other one changes. Usually we will be interested in changing a controllable variable (such as price or production level) and seeing the effect on some uncontrollable variable, such as sales volume or level of profit. Such prediction may be in the future, or may relate to values of the x variable for which we do not, currently, have any y variable values. In deriving the regression line given in Section 16.1 we emphasized that it was for predicting y values, so to make the prediction we can simply substitute x values into this equation.

Taking the regression line which we have found for Jasbeer, Rajan & Co. in an earlier section, we can now make a prediction for some values of promotional spend which do not occur in the original data. Let's assume that the company decides to spend £150 000 on promotion. Then:

$$Sales\ volume = -139.51 + 7.55 \times Promotion$$
$$= -139.51 + 7.55 \times 150$$
$$= -139.51 + 1132.5$$
$$= 992.99$$

So the predicted sales volume will be 992 units.

How good this prediction will be depends upon two factors, the value of the correlation coefficient and the value of x used. If the correlation coefficient is close to 1 (or −1), then all of the points on the scatter diagram are close to the regression line and thus we would expect the prediction to be fairly accurate. However, with a low correlation coefficient, the points will be widely scattered away from the regression line, and thus we will have less faith in the prediction. For this reason it may not be worth calculating a regression line if the correlation coefficient is small.

The value of x used to make the prediction will also affect the accuracy of that prediction. If the value is close to the average value of x, then the prediction is likely to be more accurate than if the value used is remote from $\bar{x}$. In Chapter 11 we looked at confidence intervals for certain statistics, and the same idea can be applied to the regression line when we consider prediction. Note that we will have to calculate the confidence interval for *each* value of x that we use in prediction since the width of the interval depends on a standard deviation for the prediction which itself varies. At first sight these formulae look extremely daunting, but they use numbers that have already been calculated!

The 95% confidence interval will be:

$$\hat{y}_0 \pm t_{\nu,2\frac{1}{2}\%}\hat{\sigma}_p$$

where $\hat{y}_0$ is the value of y obtained by putting x_0 into the regression equation, $\hat{\sigma}_p$ is the standard deviation of the prediction (see below) and $t_{\nu,2\frac{1}{2}\%}$ is a value from the t-distribution, obtained from tables and using both tails of the distribution with $\nu = n - 2$, the number of degrees of freedom.

To find the standard deviation of the prediction, $\hat{\sigma}_p$, we will use the following formulae:

$$\hat{\sigma}^2 = \frac{1}{n(n-2)}$$

$$\times \frac{[n\sum x^2 - \sum x)^2][n\sum y^2 - (\sum y)^2] - (n\sum xy - \sum x\sum y)^2}{n\sum x^2 - (\sum x)^2}$$

and

$$\hat{\sigma}_p = \hat{\sigma} \times \sqrt{\left[\frac{1}{n} + \frac{n(x_0 - \bar{x})^2}{n\sum x^2 - (\sum x)^2}\right]}$$

These formulae can be simplified if we look back to the formula for correlation:

$$r = \frac{n\sum xy - \sum x\sum y}{\sqrt{\{[n\sum x^2 - (\sum x)^2][n\sum y^2 - (\sum y)^2]\}}}$$

let A equal the numerator, B equal the first bracket of the denominator and C equal the second bracket of the denominator. Then

$$r = \frac{A}{\sqrt{(B \times C)}}$$

Using the same notation, we have:

$$\hat{\sigma}^2 = \frac{1}{n(n-2)}\left(\frac{B \times C - A^2}{B}\right)$$

and

$$\hat{\sigma}_p = \hat{\sigma} \times \sqrt{\left[\frac{1}{n} + \frac{n(x_0 - \bar{x})^2}{B}\right]}$$

Before going on to look at an example which uses these formulae, consider the bracket $(x_0 - \bar{x})$ which is used in the second formula. x_0 is the value of x that we are using the make the prediction, and thus the value of this bracket gets larger the farther we are away from the mean value of x; this in turn will change the value of $\hat{\sigma}_p$ which explains why the confidence intervals will vary in width, with the value of x used.

Case 5

Coming back to our prediction for sales volume for Jasbeer, Rajan & Co., we can now see that our prediction of a single (point) value is not of tremendous use to the company. Normally you would only work out the confidence intervals for the actual prediction that you were making, but we will use this regression line to illustrate the nature of the confidence intervals, and work out the values for a very wide range of values of x.

Looking at the prediction of sales volume when promotional spend is £150 000, we have:

$$r = \frac{6\,066\,456}{\sqrt{[803\,409 \times 65\,254\,064]}} = 0.837\,843\,536$$

so that:

$$A = 6\,066\,456; \quad B = 803\,409; \quad C = 65\,254\,064$$

$$n = 30; \quad \sum x = 2139 \text{ (from previous calculations)}.$$

Using the formula which we stated above, we can now work out the variance of the regression as follows:

$$\hat{\sigma}^2 = \frac{1}{30(30-2)} \times \left[\frac{(803\,409)(65\,254\,064) - 6\,066\,456^2}{803\,409}\right]$$

$$= \frac{1}{840} \times \frac{15\,623\,813\,904\,240}{803\,409}$$

$$= \frac{1}{840} \times 19\,446\,899.28$$

$$= 23\,151.070\,572\,31$$

$$\hat{\sigma} = 152.154\,76$$

This is the standard deviation of the regression, and we can now use this result to find the standard deviation of the prediction.

$$\hat{\sigma}_p = 152.154\,76 \times \sqrt{\left[\frac{1}{30} + \frac{30 \times (150 - 71.3)^2}{803\,409}\right]}$$

$$= 152.154\,76 \times \sqrt{\left[\frac{1}{30} + \frac{30 \times 6193.69}{803\,409}\right]}$$

$$= 152.154\,76 \times \sqrt{\left[\frac{1}{30} + \frac{185\,810.7}{803\,409}\right]}$$

$$= 152.154\,76 \times \sqrt{[0.033\,333\,3 + 0.231\,277\,8]}$$

$$= 152.154\,76 \times \sqrt{0.264\,611\,18}$$

$$= 152.154\,76 \times 0.514\,403\,7$$

$$= 78.268\,97$$

This is the standard deviation of the prediction where $x = 150$. To find the confidence interval, we need to know the t-value from tables (Appendix D) for $n - 2$ degrees of freedom ($30 - 2 = 28$). For a 95% confidence interval, we need to look up $t_{28,2.5\%}$ and this is equal to 2.048, so that the confidence interval is:

$$992.99 \pm 2.048 \times 78.268\,97$$

$$992.99 \pm 160.294\,85$$

$$832.695 \text{ to } 1153.285$$

Case 5

Since these figures are in thousands, this translates into 'We are 95% confident that the sales volume level for Jasbeer, Rajan & Co., if they choose to spend £150 000 on promotion, will be between 832 thousand and 1.15 million units.'

Obviously, doing these sorts of sums by hand is something to be avoided, and you would normally use either a dedicated statistical package or a spreadsheet. Table 16.1 presents the results of a spreadsheet which has calculated the upper and lower confidence limits for this equation. Figure 16.4 shows the results graphically.

You should note from the table and the diagram how the upper and lower limits diverge from the point prediction (i.e. the regression line) as we move further and further from the average of x. A stylized representation is shown in Figure 16.5 to emphasize the point.

Prediction is often broken down into two sections:

1 predictions from x values that are within the original range of x values are called **interpolation**
2 predictions from outside this range are called **extrapolation**.

Interpolation is seen as being relatively accurate (depending upon the value of the correlation coefficient), whilst extrapolation is seen as less accurate. Even where the width of the confidence interval is deemed acceptable, we must remember that in extrapolation we are assuming that the linear regression which we have calculated will apply outside of the original range of x values. This may not be the case!

Table 16.1 Prediction for Jasbeer, Rajan & Co.

Promo	Sales	t	sp	Lower	Upper
10	−64.01	2.048	63.404 7	−193.863	65.842 82
20	11.49	2.048	55.197 37	−101.554	124.534 2
30	86.99	2.048	47.394 51	−10.074	184.054
40	162.49	2.048	40.232 15	80.094 55	244.885 4
50	237.99	2.048	34.116 11	168.120 2	307.859 8
60	313.49	2.048	29.699 97	252.664 5	374.315 5
70	388.99	2.048	27.805 81	332.043 7	445.936 3
80	464.49	2.048	28.933 28	405.234 6	523.745 4
90	539.99	2.048	32.771 98	472.873	607.107
100	615.49	2.048	38.519 7	536.601 7	694.378 3
110	690.99	2.048	45.457 98	597.892 1	784.087 9
120	766.49	2.048	53.122 35	657.695 4	875.284 6
130	841.99	2.048	61.240 81	716.568 8	967.411 2
140	917.49	2.048	69.654 77	774.837	1060.143
150	992.99	2.048	78.268 97	832.695 1	1153.285
160	1068.49	2.048	87.023 98	890.264 9	1246.715
170	1143.99	2.048	95.881 24	947.625 2	1340.355
180	1219.49	2.048	104.814 8	1004.829	1434.151
190	1294.99	2.048	113.806 8	1061.914	1528.066
200	1370.49	2.048	122.844 2	1118.905	1622.075

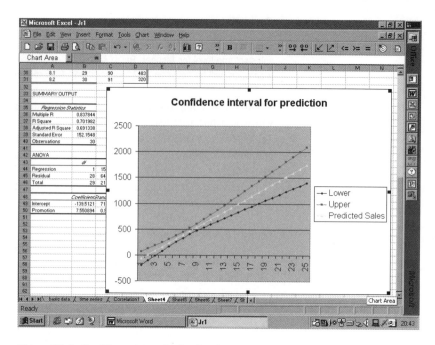

Figure 16.4   Confidence interval using Excel.

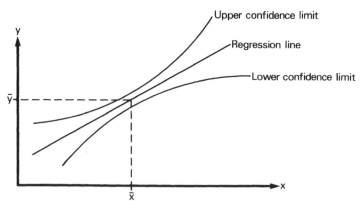

Figure 16.5

16.4 Another regression line

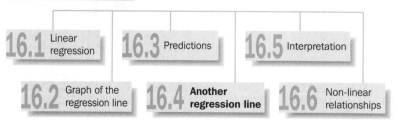

When we were trying to decide how to define the best line through the scatter diagram we used the vertical distance between the points and the linear function. What would happen if we had used horizontal distances? Doing this would mean that we were trying to predict x values from a series of y values, but again we would want to minimize the errors in prediction. Following through a similar line of

logic using horizontal distances gives:
if

$$\hat{x} = c + my$$

then

$$m = \frac{n \sum xy - \sum x \sum y}{n \sum y^2 - (\sum y)^2}$$

and

$$c = \bar{x} - m\bar{y}$$

The result of this calculation is called the regression line of x on y. Note that this also goes through the point $(\bar{x}, \bar{y})$, but in most circumstances, the line will be different from the regression line of y on x. The only exception is when the correlation coefficient is equal to $+1$ or -1, since there we have a completely deterministic relationship and all of the points on the scatter diagram are on a single straight line.

For many sets of data it is only necessary to calculate one of the regression lines, since it will only make sense to predict one of the two variables. If we are using data on advertising expenditure and sales of a product, then we wish to be able to predict sales for a particular level of advertising expenditure.

As an example of the two regression lines, consider the example shown in Table 16.2.

The correlation from this data is 0.8444, so we would expect the two regression lines to be quite close together. Working out the regression line of y on x gives:

$$y = 29.667 + 0.507\,88x$$

Using the formula given above, we have, for x on y:

Table 16.2 An example of the two regression lines

	x	y	x^2	y^2	xy
	10	30	100	900	300
	20	25	400	625	500
	30	61	900	3721	1830
	40	57	1600	3249	2280
	50	60	2500	3600	3000
	60	64	3600	4096	3840
	70	55	4900	3025	3850
	80	64	6400	4096	5120
	90	85	8100	7225	7650
	100	75	10000	5625	7500
Totals	550	576	38500	36162	35870

$$m = \frac{10 \times 35\,870 - 550 \times 576}{10 \times 36\,162 - 576^2}$$
$$= \frac{358\,700 - 316\,800}{361\,620 - 331\,776}$$
$$= \frac{41\,900}{29\,844}$$
$$= 1.403\,967\,3$$

and

$$c = 55 - 1.403\,967\,3 \times 57.6$$
$$= 55 - 80.868\,52$$
$$= -25.868\,52$$

Therefore the equation of the regression line of x on y is

$$x = 1.403\,97 - 25.869y$$

The original data and the two regression lines are shown in Figure 16.6.

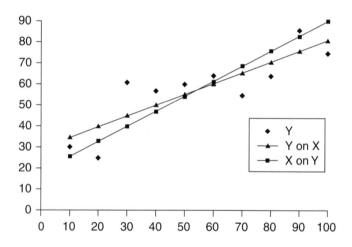

Figure 16.6

16.5 Interpretation

As stated above, the major reason for wishing to identify a regression relationship is to make predictions about what will happen to one of the variables (the endogenous variable) at some value of the other variable (often called the exogenous variable). However, it is also possible to interpret the parts of the regression equation. According to the simple Keynesian model of the economy, consumption expenditure depends on the level of disposable

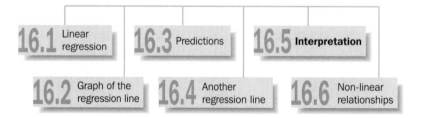

16.1 Linear regression

16.2 Graph of the regression line

16.3 Predictions

16.4 Another regression line

16.5 **Interpretation**

16.6 Non-linear relationships

income. If data are collected on each of these variables over a period of time, then a regression relationship can be calculated. For example:

$$\text{consumption expenditure} = 2000 + 0.85(\text{disposable income})$$

From this we see that even if there is no disposable income, then consumption expenditure is 2000 to meet basic needs, this being financed from savings or from government transfer payments. Also, we can see that consumption expenditure increases by 0.85 for every increase in disposable income of 1. If the units used were pounds, then 85 pence of each pound of disposable income is spent on consumption and 15 pence is saved. Thus 0.85 is the average propensity to consume (APC). This type of relationship will work well for fairly narrow ranges of disposable income, but the correlation coefficient is likely to be small if all households within an economy are considered together, since those on low disposable incomes will tend to spend a higher proportion of that income on consumption than those on very high disposable incomes.

16.6 Non-linear relationships

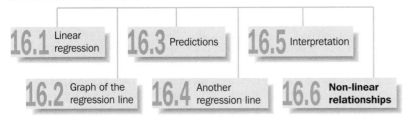

As we saw in Section 15.5, some relationships are **non-linear**, and thus we will need to be able to calculate a regression relationship that will allow prediction from a non-linear function. If the correlation coefficient has already been calculated, then little additional calculation is needed to obtain the regression equation.

Extending the example in Section 15.5, we continue to use the log of y instead of y:

$$r = \frac{272.05}{\sqrt{(3185 \times 23.8721)}}$$

$$= 0.9866$$

For $\log y$ on x

$$b = \frac{272.05}{3185}$$

$$= 0.085\,42$$

and

$$a = \frac{\sum(\log_{10} y)}{n} - b\bar{x}$$

$$= \frac{21.43}{10} - 0.085\,42(9.5)$$

$$= 2.143 - 0.811\,49$$

$$= 1.331\,51$$

So the log-linear regression relationship is

$$\widehat{\log_{10} y} = 1.331\,51 + 0.085\,42x$$

NB. Although this is a linear relationship, if you wish to plot it on the original scatter diagram, you will have to work out several individual values of $\hat{y}$ and join these together to form a curve.

To predict from this relationship, if $x_0 = 9$ then

$$\widehat{\log_{10} y} = 1.331\,51 + 0.085\,42 \times 9$$
$$= 1.331\,51 + 0.768\,78$$
$$= 2.100\,29$$

and by taking the antilog

$$\hat{y} = 125.98$$

Finding the regression line for other non-linear transformations will follow a similar pattern.

16.7 Conclusions

The methods shown in this chapter and the previous one are probably the most widely used models in business. They are relatively easy to use, and provided that there is a fairly high correlation in the factors involved, give reasonably good predictions, provided we don't go too far from the original data. (Extrapolation from x values that are a very long distance from the mean will always give poor, if not meaningless, results.) Computer packages, especially spreadsheets, have made these techniques widely available, and even easier to use, but this also makes them easier to misuse.

Exercise

Using a spreadsheet obtain the regression equations for each of the 'controllable' variables with sales volume and then with sales value for Jasbeer, Rajan & Co. from the data on the website. Which are the key variables for the company?

16.8 Problems

1 Use the data given below to create a scatter diagram.

x	1	2	3	4	5
y	3	6	10	12	14

Find the regression line of y on x and place this on to the scatter diagram.

2 Find the regression line of y on x for the data given below.

x	10	12	14	16	18	20	22	24	26	28
y	25	24	22	20	19	17	13	12	11	10

Construct a scatter diagram and place your regression line on to the graph.

3 Find the regression line of y on x for the following data.

x	4	2	6	7	8	5	2	4
y	10	5	15	16	19	14	8	11

4 A farmer wishes to predict the number of tons per acre of crop which will result from a given number of applications of fertilizer. Data has been collected and is shown below.

Fertilizer applications	1	2	4	5	6	8	10
Tons per acre	2	3	4	7	12	10	7

Find a suitable regression relationship to help the farmer in making the required prediction, and from your result predict the number of tons per acre from seven fertilizer applications.

5 Determine the regression line of y on x from the following data without calculation.

x	11	12	13	14	15	16	17	18	19	20
y	4	4	4	4	4	4	4	4	4	4

6 Use the data given below to find the regression line of y on x and the regression line of log y on x.

x	1	2	3	4	5	6	7	8	9	10
y	10	10	11	12	12	13	15	18	21	25
x	11	12	13	14	15	16	17	18	19	20
y	26	29	33	39	46	60	79	88	100	130

Construct a scatter diagram and place both of your regression lines on to it.

7 During the making of certain electrical components each item goes through a series of heat processes. The length of time spent in this heat treatment is related to the useful life of the component. To find the nature of this relationship a sample of 20 components are selected from the process and tested to destruction and the results are presented below.

Time in process (minutes)	Length of life (hours)	Time in process (minutes)	Length of life (hours)
25	2005	41	3759
27	2157	42	3810
25	2347	41	3814
26	2239	44	3927
31	2889	31	3110
30	2942	30	2999
32	3048	55	4005
29	3002	52	3992
30	2943	49	4107
44	3844	50	3987

(a) Find the regression line of useful life on time spent in process.
(b) Predict the useful life of a component which spends 33 minutes in process.

(c) Predict the useful life of a component which spends 60 minutes in process.

(d) Using a suitable transformation, find a regression relationship which has a higher coefficient of determination. (Hint: look back to question 15 at the end of Chapter 15.)

(e) From this new relationship, predict the useful life of a component which spends 33 minutes in process.

(f) From this new relationship, predict the useful life of a component which spends 60 minutes in process.

8 Costs of production have been monitored for some time within a company and the following data found:

Production level (000s)	Average total cost (£000s)
1	70
2	65
3	50
4	40
5	30
6	25
7	20
8	21
9	20
10	19
11	17
12	18
13	18
14	19
15	20

(a) Find the regression line of average total cost on production level.

(b) Place this on to a scatter diagram.

(c) Predict the average total cost if the production level were:
(i)   8 500 units;
(ii) 16 000 units;
(iii) 20 000 units.

(d) Find an alternative model to relate average total cost to production level and make the same predictions again using the new model.

9 Find the regression line of y on x for the following data and test the significance of the coefficients.

x	25	30	60	45	40	80	70	90	15	20
y	100	95	50	60	65	30	35	10	120	110

10 Find the regression line through the following data.

x	y
5	5
10	6
15	7
20	8
25	10
30	12
35	14
40	18
45	22
50	28
55	34
60	40
65	48
70	60
75	75

Assess if this is a good fit to the data; if you judge that it is not, find a regression line with a better fit.

11 The personnel department of a company has conducted a pre-interview test of aptitude on candidates for some time. It is now in possession of annual appraisal interview reports from line managers on the successful candidates. You have been asked build a model to predict the appraisal score of candidates on the basis of the following evidence:

Person	Pre-test score	Report score
A	50	67
B	62	70
C	85	80
D	91	79
E	74	68
F	53	67
G	74	81
H	59	67
I	84	90
J	67	75
K	41	40
L	85	80
M	68	71
N	79	82
O	83	76
P	67	78
Q	81	86
R	75	78
S	82	64
T	72	67

12 Workers on the shop floor are paid on a piecework basis for producing items. The union claims that this seriously discriminates against newer workers, since there is a fairly steep learning curve which workers follow with the result that more experienced workers can perform the task in about half of the time taken by a new employee. You have been asked to find out if there is any basis for this claim. To do this, you have observed 10 workers on the shop floor, timing how long it takes them to produce an item. It was then possible for you to match these times with the length of the workers' experience. The results obtained are shown below.

Person	Months' experience	Time taken
A	2	27
B	5	26
C	3	30
D	8	20
E	5	22
F	9	20
G	12	16
H	16	15
I	1	30
J	6	19

(a) Construct a scatter diagram for this data.
(b) Find the regression line of time taken on months' experience.
(c) Place the regression line on the scatter diagram.
(d) Predict the time taken for a worker with:
    (i) 4 months' experience
    (ii) 5 years' experience
(e) Comment on the union's claim.

16.9 Appendix

Equation of the straight line to predict y values is:

$$\hat{y} = a + bx$$

Thus for a value x_1, we have

$$\hat{y}_1 = a + bx_1$$

and

$$d_1 = y_1 - \hat{y}_1$$
$$= y_1 - (a + bx_1)$$
$$= y_1 - a - bx_1$$

where d is the difference between an observed y and its value predicted by the regression equation.

This vertical distance may be defined, and calculated as above, for each point on the scatter diagram. To minimize the prediction error, we may sum these vertical distances; however, if the observed y values deviate from the regression line in a random manner, then:

$$\sum d = 0$$

To overcome this problem, we may square each value of d, and then minimize the sum of these squares:

$$d_1 = y_1 - a - bx_1$$

$$d_1^2 = (y_1 - a - bx_1)^2$$

$$S = \sum_{i=1}^{n} d_i^2$$

$$= \sum_{i=1}^{n} (y_i - a - bx_i)^2$$

To minimize S we need to differentiate the function with respect to a and b (since the values of x and y are fixed by the problem we are trying to solve). Thus:

$$\frac{\partial S}{\partial a} = -2 \sum_{i=1}^{n} (y - a - bx_i) = 0$$

$$\frac{\partial S}{\partial b} = -2 \sum_{i=1}^{n} x_i(y_i - a - bx_i) = 0$$

These are the **first-order conditions**, and if they are rearranged, we have the normal equations:

$$\sum y = na + b \sum x \qquad\qquad 16.1$$

$$\sum xy = a \sum x + b \sum x^2 \qquad\qquad 16.2$$

These equations could be used as a pair of simultaneous equations each time that we wanted to identify the values of a and b, but it is usually more convenient to find the general solution for a and b to give the following formulae.

Multiply equations 16.1 by $(\sum x)/n$:

$$\frac{\sum x \sum y}{n} = a \sum x + \frac{b}{n} (\sum x)^2 \qquad\qquad 16.3$$

Subtracting equation 16.3 from equation 16.2:

$$\sum xy - \frac{\sum x \sum y}{n} = b \sum x^2 - \frac{b}{n}(\sum x)^2$$

$$= b \left[\sum x^2 - \frac{(\sum x)^2}{n} \right]$$

So

$$n \sum xy - \sum x \sum y = b \left[n \sum x^2 - (\sum x)^2 \right]$$

or

$$b = \frac{n \sum xy - \sum x \sum y}{n \sum x^2 - (\sum x)^2}$$

Rearranging equation 16.1 gives:

$$a = \frac{\sum y - b \sum x}{n}$$

or

$$a = \bar{y} - b\bar{x}$$

17 Multiple regression and correlation

In the last two chapters we have developed the ideas of regression and correlation. Correlation allowed us to measure the relationship between two variables, and to put a value on the strength of that relationship. Regression determines the parameters of an equation which relates the variables, and in that way provides a method to predict the behaviour of one variable from the behaviour of the other.

In this chapter we consider the effects of adding one or more variables to the equation, so that the dependent variable (usually *y*) is now predicted by the behaviour of two or more variables. In most cases this will increase the amount of correlation in the model and will tend to give 'better' predictions. An obvious question to ask now is that if two independent variables give us better results than one, then why not try three, or four, or five, or even more? For most relationships in business and economics our answer would be 'Yes!'. Since we are dealing with complex relationships which are not fully explained by theory, using extra, relevant variables may well give us better results.

Once we move to several variables on the right-hand side of the regression equation, difficulties can arise which may mean that the answer we obtain does not necessarily give us the best results. We will consider some of these issues within this chapter.

Even where there is some theoretical background which suggests that several variables should be used to explain the behaviour of the dependent variable, it may not be that easy to get the necessary data. Take, for instance, the sales of a product. Economic theory suggests that the price of that product will explain some of the variations in sales; and this should be easy to find. However, the simple economic model is based on *ceteris paribus*, and once we allow for variations in other variables we will be looking for data on the prices of other products, the tastes and preferences of consumers, the income levels of consumers, advertising and promotion of the product, and advertising and promotion of competitive products. There may also be a role for the level of activity in the economy including the growth rate, levels of unemployment, rate of inflation, and expectations of future levels and rates of these variables. Some of this data, such as that on tastes and preferences, is likely to be very difficult to obtain.

It is not feasible to develop multiple regression relationships by hand for more than very simple data, and so we will assume that you have at least some access to a computer system running a suitable package. Print-outs are obtainable from spreadsheets such as Excel, MINITAB and SPSS. While, in the past, this topic was excluded from introductory courses, the wide use of computer packages now makes it easily accessible to most students.

In this chapter we will begin by looking again at the two-variable regression and correlation results, but particularly looking at some of the theory behind the basic equation, and at the sort of results produced by statistical packages. We will then move on to building more complex models, and the problems that may arise when using business-type data.

After working through this chapter, you should be able to:

- describe why multiple regression may give better results when making a forecast
- interpret a print-out of a multiple regression relationship
- state the MALTHUS problems
- establish if a model exhibits the MALTHUS problems
- create a multiple regression model.

Figure 17.1   Structure of Chapter 17.

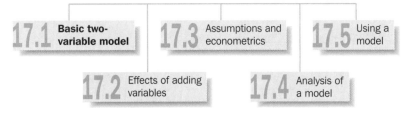

17.1 The basic two-variable model

As a first step we will re-work a two-variable regression model, adding some extra comments on the type of print-out obtained from typical statistical packages. Taking the data given in Table 17.1 and running it through the regression analysis part of Excel will give the print-out shown in Figure 17.2 The equivalent print-out from SPSS for Windows is shown in Figure 17.3.

As you can see, there are differences between the two print-outs, but, more importantly, they both contain rather more information than just the correlation coefficient and the equation of the regression line. The SPSS print-out (Figure 17.3) gives the mean and standard deviation for the data, which can be useful when you are selecting variables from among a large number in your data set. It helps to make sure that you have instructed the machine to use the correct variables. It also shows the correlation matrix for all of the variables – not very useful at this stage, but, as you will see, a particularly useful table later in our analysis. Both print-outs then include the same information.

Table 17.1 Simple data

x	60	85	110	95	140	160	80	40	55	90	115	120	180	95
y	25	20	35	40	60	55	45	15	20	30	40	50	70	45

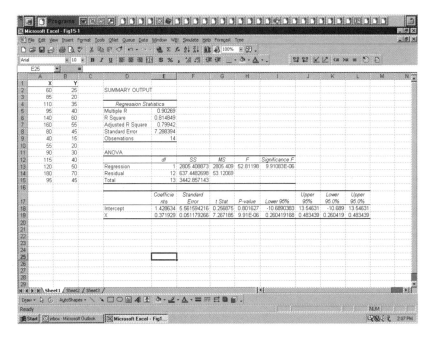

Figure 17.2 Excel output from regression.

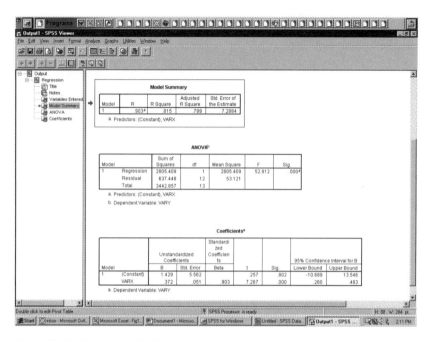

Figure 17.3 Output from SPSS for regression.

Multiple R:	0.90269	the correlation coefficient.
R Square:	0.81485	the coefficient of determination.
Adjusted R Square	0.79942	this is an unbiased estimate of the coefficient of determination – the theory says that R^2 is biased upwards.
Standard Error	7.28839	the overall variation in the model.

Most of this will be recognizable from the ideas put forward in the last chapter. It is worth noting that the coefficient of determination is, in fact, biased upwards (so giving a 'too good' answer), and that the packages work out an unbiased value. When we begin to compare models, or decide if a model is useful, it is this **unbiased** figure which should be used.

We will ignore the ANOVA (analysis of variance) section for now.

From the final part of either print-out, we can read off the equation of the regression line – labelled 'Coefficients' in the Excel print-out, and 'B' in the SPSS print-out. This gives:

$$y = 1.428634 + 0.371929x$$

So far, there is nothing really new, but if we look a little further through the print-outs, we see that each goes part of the way to conducting tests of hypotheses on the coefficients. At the end of the Excel print-out we can see that for both the intercept (a) and the slope (b) the table shows the coefficient and the standard deviation. Dividing the coefficient by the standard deviation gives the t-statistic, and the associated probability is given by the P-value column. We are interested in whether or not the coefficients are zero (i.e. does the relationship we have found make statistical sense!), and so we carry out a one-tailed test. For the coefficient of x, this will be:

$$H_0: \beta_1 = 0$$
$$H_1: \beta_1 > 0$$

At the 5% significance level, we need the figure in the column headed P-value to be below 0.05, and here it is very tiny at 0.00000991 (9.91E−06), and so the coefficient passes the t-test (can reject H_0), and we can say that the variable (x) adds significantly to the predictive power of the model. (Hardly surprising since this is the only variable!) In the case of the intercept, we cannot reject H_0, and thus we cannot be sure that the intercept is non-zero. In general, less importance tends to be attached to intercepts failing the t-test, than to variables failing. There may, however, be some special cases, where the need for the intercept to be non-zero is crucial. For example, if you were looking at personal spending and income, then there must be some subsistence level which people adhere to (maybe above or below the appropriate benefit level), and thus we would expect the intercept to be non-zero.

The other columns give the lower and upper 95% confidence limits on the coefficients. Where these are of opposite signs, this confirms the decision of the hypothesis test to not be able to reject H_0, or to say that the value may be zero.

The SPSS print-out gives the coefficients and, at the far right, the t-statistic. You would need to work out the number of degrees of freedom (here $n - 2$), and then look up the critical value in the t-tables (Appendix D).

17.2 The effects of adding variables

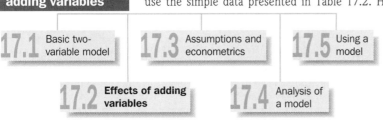

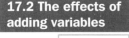

To demonstrate the effects of building up a model by adding variables we will use the simple data presented in Table 17.2. Here the premise is that the sales of a product are dependent on its price, the marketing spend of the company, the level of economic activity in the economy and the unit cost.

Building such models assumes that you have access to software such as Excel or SPSS (we will use the Excel print-out to illustrate this section). These packages make a series of assumptions (which are detailed in Section 17.3), and we shall only look at the results of calculations in these sections. As a manager you would be unlikely to be asked to actually do the calculations! Using the data given in Table 17.2 we will initially build a series of two-variable models, in order to assess which variable has the highest correlation with sales. A summary of the results is shown in Table 17.3.

Using the simple criteria of **highest coefficient of determination**, we would select the index of economic activity as the variable which, on its own, explains most of the variability in sales (i.e. 69.99% of the variability).

The next step would be to add one of the other variables in order to increase the explanatory power of the model. There are six ways of selecting two from four, so there are six different models which can be built. Looking at the set of results shown in Table 17.4, we can see the differences in the extra explanatory power of the various models.

Table 17.2 Company data

Time	Sales (Y)	Price (X2)	Marketing spend (X3)	Index of economic activity (X4)	Index of unit cost (X5)
1	986	1.8	0.4	100	100.0
2	1025	1.9	0.4	103	101.1
3	1057	2.1	0.5	104	101.1
4	1248	2.2	0.7	106	106.2
5	1142	2.2	0.6	102	106.3
6	1150	2.3	0.7	103	108.4
7	1247	2.4	0.9	107	108.7
8	1684	2.2	1.1	110	114.2
9	1472	2.3	0.9	108	113.8
10	1385	2.5	1.1	107	114.1
11	1421	2.5	1.2	104	115.3
12	1210	2.6	1.4	99	120.4
13	987	2.7	1.1	97	121.7
14	940	2.8	0.8	98	119.8
15	1001	2.9	0.7	101	118.7
16	1025	2.4	0.9	104	121.4
17	1042	2.4	0.7	102	121.3
18	1210	2.5	0.9	104	120.7
19	1472	2.7	1.2	107	122.1
20	1643	2.8	1.3	111	124.7

Table 17.3 Two-variable models

Equation	r^2	Adjusted r^2
Sales = 965.4 + 104.54 Price	0.01866	−0.03586
Sales = 786.3 + 492.66 Marketing spend	0.40426	0.371164
Sales = −3988.9 + 50.13 Index of economic activity	0.71570	0.699907
Sales = 448.1 + 6.75 Index of unit costs	0.05623	0.003795

Note that −0.03 in the first line is merely the result of using a formula and is not intended to be a serious result since r^2 cannot be negative.

Table 17.4 Three-variable models

Equation	R^2	Adjusted R^2
$S = 1476.69 - 377.59P + 743.61M$	0.542789	0.488999
$S = -4447.31 + 152.83P + 50.999E$	0.755366	0.726586
$S = 218.51 - 188.85P + 12.75U$	0.072568	−0.03654
$S = -3526.92 + 332.94M + 42.88E$	0.885348	0.87186
$S = 2187.94 + 784.96M - 14.54U$	0.522971	0.46685
$S = -4694.57 + 49.93E + 6.38U$	0.765911	0.738371

Where P = price, M = marketing spend, E = index of economic activity, U = index of unit cost.

Table 17.5 Four-variable models

Equation	R^2	Adjusted R^2
$S = -2950.78 - 142.78P + 439.5M + 39.74E$	0.902589	0.884324
$S = 1901.56 - 262.0P + 799.47M - 6.6U$	0.554265	0.470689
$S = -2641.57 + 469.67M + 40.1E - 6.29U$	0.905524	0.88781
$S = -4684.27 + 19.62P + 50.06E + 5.75U$	0.766082	0.722222

From Table 17.4 we can see that the 'best' model (i.e. the one with the highest adjusted R^2) is that where sales depend on marketing spend and economic activity (Adj. $R^2 = 0.87$).

The third step might be to add a third explanatory variable to our models, and there are four ways of selecting three explanatory variables. The results of doing this are shown in Table 17.5.

From Table 17.5 we can see that the 'best' model (i.e. the one with the highest adjusted R^2) is that where sales depend on marketing spend, economic activity and unit costs (Adj $R^2 = 0.887$).

Finally we can put all of the variables into the model to give the results shown in Figure 17.4.

Fairly obviously we would not normally go through this rather long process of building every possible model from the data which we have available. One would

Sales & all of them									
SUMMARY OUTPUT									
Regression Statistics									
Multiple R	0.952656								
R Square	0.907553								
Adjusted R Square	0.882901								
Standard Error	77.36381								
Observations	20								
ANOVA									
df	*SS*	*MS*	*F*	*F*	*Significance*				
Regression		4	881347.2	220336.8	36.81386	1.37E-07			
Residual		15	89777.38	5985.159					
Total		19	971124.6						
Coefficients	*Standard Error 95%*	*value t Stat Upper 95%*		*Lo P-v*	*Upper 95.000%*		*r 95.000% e*	*w*	*Lo*
Intercept	-2638.85	693.4829	-3.80521	0.001725	-4116.97	-1160.72	-4116.97	-1160.72
X Variable 1	-68.3059	119.0526	-0.57375	0.574643	-322.061	185.4489	-322.061	185.4489
X Variable 2	478.5033	99.87372	4.791084	0.000238	265.6274	691.3793	265.6274	691.3793
X Variable 3	39.45952	5.211796	7.571194	1.69E-06	28.35083	50.56821	28.35083	50.56821
X Variable 4	-4.34868	4.845328	-0.8975	0.383632	-14.6763	5.978903	-14.6763	5.978903

Figure 17.4

Table 17.6   Summary of the models

No. of explanatory variables	No. of models	Best r^2	Best adjusted r^2	No. of variables passing t-test at 5% level
1	4	0.715701	0.699907	1
2	6	0.885348	0.87186	2
3	4	0.905524	0.88781	2
4	1	0.907553	0.882901	2

normally attempt to build a model which made both logical and statistical sense, and this would often start by using all of the variables available and then reducing the size of the model (or its shape) in order to meet the various assumptions made by multiple regression. This process is the subject of the next section.

The large number of models built in this section allow us to illustrate several features of multiple regression, and these are summarized in Table 17.6.

As you can see, by adding extra variables into the model we have, in every case, increased the value of the correlation coefficient, but the explanatory power of the model, as shown by the adjusted r^2 figure, **does not necessarily increase** as more variables are added. Even though the models have higher correlations, not necessarily all of their explanatory variables pass the t-test, i.e. the model says that the variables are not relevant, but the R^2 figure has increased! This is the sort of anomaly which means that building multiple regression models is not just a case of putting the data into a spreadsheet and pressing a couple of buttons.

Exercise

Using a spreadsheet obtain the various multiple regression equations for some of the models shown in this section to ensure that you can replicate our results.

17.3 Assumptions and econometric problems

There are five basic assumptions that are normally made when calculating a multiple regression relationship:

1 that we are dealing with a **linear function** of the independent variables plus a disturbance term, $\hat{y} = \beta_1 + \beta_2 x_2 + \beta_3 x_3 + \cdots + \beta_n x_n + u$

2 that this disturbance term has a mean of zero

3 that there is a constant variance in the model

4 that the independent variables are fixed, i.e. non-stochastic

5 that there is no significant linear relationship between the independent variables.

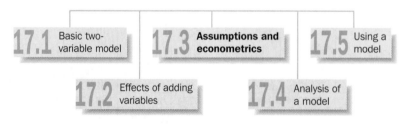

Along with the increasing complexity of multiple regression models over simple regression models, we find that meeting all of these assumptions is particularly difficult, especially when using business-type data. This is partly because we find that whilst there may be a strong relationship with the variable that we are trying to explain, there is also, often, a strong relationship between at least some of the 'explanatory variables'. These problems can be summarized under the mnemonic 'MALTHUS', as shown in Figure 17.5.

Within this section we will look briefly at each of these problems and indicate ways in which you may be able to identify whether or not your particular model is suffering from one or more of them. At this level we cannot go into sufficient depth to show all of the possible solution methods that may be needed, but you should know enough by the end of this section to decide whether or not you can feel confident about the model you have calculated. (Should you need to take this subject further, then we would recommend that you consult a book on econometrics.)

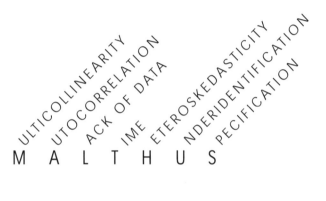

Figure 17.5

17.3.1
Multicollinearity

This term refers to the interrelatedness of the variables on the right-hand side of the equation, the independent variables. In an ideal world, there would be no interrelationship between the independent variables, but since we are not in an ideal world, but are dealing with business and economic problems, some interrelationships will exist. Consider the situation shown in Figure 17.6.

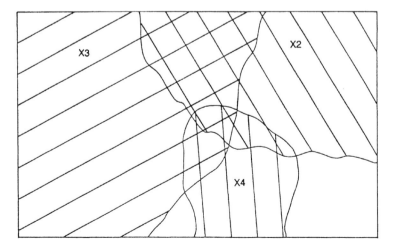

Figure 17.6

The outer box represents the *total* variation in the dependent variable and the shaded areas represent the variation 'explained' by the individual independent variables X2, X3 and X4. As we can see, there is considerable overlap between X2 and X3, and this overlap represents multicollinearity. There is little overlap between X4 and X2, showing that there is little multicollinearity. Spreadsheet programs do not automatically produce such a diagram, but packages such as SPSS do produce a correlation matrix which relates all of the variables together. Multicollinearity is where there is a high correlation between two or more independent variables. Table 17.7 shows such a matrix.

Table 17.7   Correlation matrix

	Y	X2	X3	X4
Y	1.00	0.75	0.65	0.45
X2	0.75	1.00	0.86	0.23
X3	0.65	0.86	1.00	0.33
X4	0.45	0.23	0.33	1.00

From this we see that we have perfect correlation on the diagonal from top left to bottom right (1.00) since this is a variable correlated with itself. We need high correlations in the first column since this represents the correlation of each variable, individually, with Y. The other values should, ideally, be low, but the 0.86 shows that there is multicollinearity between variables X2 and X3. There is no specific value at which we would say multicollinearity exists; it is a matter of judgement.

If multicollinearity exists in a model then the coefficients of the independent variables may be unstable, especially if we try to use the equation to forecast after there has been a policy change. Such a change is likely to change the multicollinearity between the independent variables, and thus make the model invalid.

Where multicollinearity exists we could delete some of the independent variables from the model in order to remove the effects of the correlations, or we could try adding more data in an attempt to find the underlying relationships. Providing that there is no policy change, multicollinearity will not seriously affect the predictions from the model in the short run.

17.3.2
Autocorrelation

When building a multiple regression model we assume that the disturbance terms are all independent of each other. If this is not the case, then the model is said to suffer from autocorrelation. We will use the error terms generated by the computer program to look for autocorrelation with regression models.

Autocorrelation may arise when we use quarterly or monthly data, since there will be a seasonal effect which is similar in successive years, and this will mean that there is some correlation between the error terms. The basic test for autocorrelation is the **Durbin–Watson test**, which is automatically calculated by most computer programs. This statistic can only take values between 0 and 4, with an ideal value being 2 indicating the absence of autocorrelation. Since we are unlikely to get a value exactly equal to 2, we need to consult tables (see Appendix J) as we analyse a multiple regression model.

Suppose that we have an equation calculated from 25 observations which has five independent variables. Using the steps set out in Chapter 12 for conducting hypothesis tests we have:

1 State hypotheses: H_0: $\rho = 0$; no autocorrelation.

$\qquad\qquad\qquad$ H_1: $\rho > 0$; positive autocorrelation.

Note that the test is *always* a one-tailed test and that we therefore can test for either positive autocorrelation or negative autocorrelation, but not both at once.

2 State the significance level: 5%, say.

3 State the critical value(s):

These will vary with the number of observations ($n = 25$) and the number of independent variables (usually labelled k, here equal to 5). From the tables we find two values: $d_l = 0.95$ and $d_u = 1.89$.

4 Calculate the test statistic:

This is done automatically by the computer program.

5 Compare the test statistic to the table values: It is easiest here to use a line to represent the distribution, as in the diagram below:

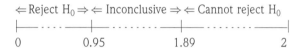

$\Leftarrow$ Reject $H_0 \Rightarrow \Leftarrow$ Inconclusive $\Rightarrow \Leftarrow$ Cannot reject H_0

| 0 | 0.95 | 1.89 | 2 |

6 Come to a conclusion:

If the calculated value is below the lower limit (0.95) then we reject the null hypothesis and if it is above the upper limit we cannot reject the null hypothesis. Any value between the limits leads to an inconclusive result, and we are unable to say if autocorrelation exists.

7 Put the conclusion into English:

Where we reject the null hypothesis, then we say that autocorrelation exists in the model and that this may lead to errors in prediction.

It is possible to take a series of extra steps which will help to remove autocorrelation from the model, but we refer you to more advanced texts for details of these methods.

17.3.3 Lack of data

Many data sets are **incomplete or subject to review and alteration**. Lack of data may represent a situation where data has only been collected for a short period, for example, a company which has only been trading for a year. When new statistics are calculated by government these series will not be usable in modelling for some years.

A related problem is that of unsuitable data. Data may be available on the variable which you are trying to model, but the definition may be different from the one which you wish to use. An example of this situation would be the official statistics on income and expenditure of the personal sector of the economy where the definition includes unincorporated businesses, charities and trade unions.

17.3.4 Time and cost constraints

In an educational context, there may be little cost pressure to complete the modelling process, at least in money terms. In a business context there is considerably more **pressure**, since predictions are needed to plan production and distribution decisions give a time constraint whilst the overall budget and potential savings give a cost constraint.

17.3.5 Heteroskedasticity

We have assumed that the variation in the data remains of the same order throughout the model, **i.e. a constant variance**. If this is not the case, then the model is said to suffer from heteroskedasticity. It is most prevalent in time series models which deal with long periods of data and usually has an increasing variance over time. This is illustrated by the scatter diagram in Figure 17.7.

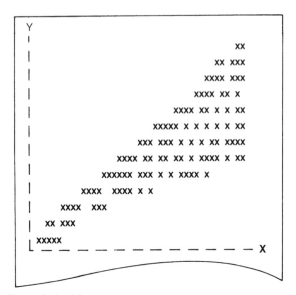

Figure 17.7 Heteroskedasticity.

Heteroskedasticity in a model will lead to problems of prediction since the equation predicts the mean value, and as we can see from the diagram, the variation about the mean is increasing.

17.3.6 Under-identification

This is a problem which affects **multiple equation models**, and as such is outside the scope of this book. It represents a situation where it is not possible to identify which equation has been estimated by the regression analysis, for example in market analysis both the demand function and the supply function have the quantity as a function of price, tastes and preferences, etc.

17.3.7 Specification

This problem relates to the **choice of variables** in the equation and to the 'shape' of the equation. When constructing business models, we rarely know for certain which variables are relevant. We may be sure of some, and unsure about others. If we build a model and leave out relevant variables, then the equation will give biased results. If we build a model which includes variables which are not relevant in this case, then the variance will be higher than it would otherwise be. Neither case is ideal.

Similar problems arise over the 'shape' of the equation; should we use a linear function, or should we try a non-linear transformation?

In practice, we never know whether we have the exact specification of the model! It is usually prudent to build a model with too many variables, rather than too few, since the problem of increased variance may be easier to deal with than the problem of biased predictions.

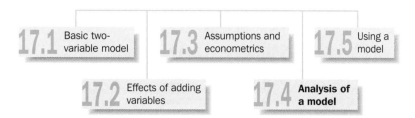

17.4 Analysis of a multiple regression model

17.1 Basic two-variable model

17.3 Assumptions and econometrics

17.5 Using a model

17.2 Effects of adding variables

17.4 Analysis of a model

Case 5

As an example of the analysis of a multiple regression model we will extend the case study of Jasbeer, Rajan & Co. by incorporating advertising spend, price, growth in the economy and unemployment level with promotional spend as our explanatory variables. The resulting model is shown as an SPSS for Windows printout in Figure 17.8.

Looking at the print-out in Figure 17.8 we can see that the Adjusted R Square value is fairly high, at 0.69297. The independent variables are 'explaining' nearly 70% of the variation in sales volume. However, when we look at the *t*-statistics (and their significance) we can see that there is only one variable which passes the *t*-test, i.e. promotional spend, with price passing the *t*-test at the 10% level. This suggests that sales volume seems to be 'controllable' by variables which the company can manipulate, rather than by things which are beyond their control. (Which is probably good news for the company!)

Going through MALTHUS, we can now look through the print-out to see which, if any, of these econometric problems may exist in this model.

Multicollinearity

We need to look at the correlation matrix, where we can see that the correlation between promotion and sales is high (at 0.838) but that the other correlations with sales are fairly low. The indications of multicollinearity are the high correlations between unemployment and growth, unemployment and price and, to a lesser extent, growth and advertising spend. Each of these will have an effect on the stability of the coefficients in the model.

Autocorrelation

To test for this problem we need to find the Durbin–Watson statistic, at the end of the print-out. Here it is 3.33816. Our hypotheses would be:

H_0: $\rho = 0$; no autocorrelation

H_1: $\rho < 0$; negative autocorrelation (since DW > 2)

The critical values from the tables (Appendix J) are 1.14 and 1.74, but we need to subtract from 4 to get values for a test of negative autocorrelation; this gives us 2.86 and 2.26. The value on the print-out is above the upper critical value, and we must therefore conclude that we reject the null hypothesis. There appears to be negative autocorrelation in the model. This will affect the predictive ability of the model (likely to be bad news for the company). You might also want to look at the plot of the residuals from the model – this is shown

19 Jul 94 SPSS for MS WINDOWS Release 6.0   Page 1

This software is functional through October 31, 1995.

         * * * *   M U L T I P L E   R E G R E S S I O N   * * * *

Listwise Deletion of Missing Data

	Mean	Std Dev	Label
SV	398.867	273.870	Sales Volume
ADS	99.800	11.463	Advertising Spend
G	1.523	1.107	Growth
P	1.254	.056	Price
PRO	71.300	30.388	Promotional Spend
UE	9.350	1.906	Unemployment

N of Cases =   30

Correlation, 1-tailed Sig:

	SV	ADS	G	P	PRO	UE
SV	1.000	.053	.138	.260	.838	-.288
	.	.390	.234	.083	.000	.062
ADS	.053	1.000	-.507	.389	.094	.055
	.390	.	.002	.017	.310	.387
G	.138	-.507	1.000	.292	.124	-.739
	.234	.002	.	.058	.257	.000
P	.260	.389	.292	1.000	.456	-.773
	.083	.017	.058	.	.006	.000
PRO	.838	.094	.124	.456	1.000	-.375
	.000	.310	.257	.006	.	.020
UE	-.288	.055	-.739	-.773	-.375	1.000
	.062	.387	.000	.000	.020	.

         * * * *   M U L T I P L E   R E G R E S S I O N   * * * *

Equation Number 1   Dependent Variable..   SV   Sales Volume

  Descriptive Statistics are printed on Page   2

Block Number 1. Method: Enter
   ADS   G   P   PRO   UE

Variable(s) Entered on Step Number
   1..   UE   Unemployment
   2..   ADS   Advertising Spend
   3..   PRO   Promotional Spend
   4..   G   Growth
   5..   P   Price

Multiple R   .86366
R Square   .74590
Adjusted R Square   .69297
Standard Error   151.75300

Analysis of Variance
	DF	Sum of Squares	Mean Square
Regression	5	1622440.10754	324488.02151
Residual	24	552695.35913	23028.97330

F =   14.09043   Signif F =   .0000

Figure 17.8   (Continues opposite)

```
--------------------- Variables in the Equation ---------------------

Variable              B           SE B      95% Confdnce Intrvl B          Beta

ADS            4.349150      3.804322     -3.502586    12.200885        .182041
G             17.069751     56.303873    -99.135733   133.275235        .068973
P          -2097.750006  1168.295411  -4508.993237   313.493224       -.431745
PRO            8.298066      1.058637      6.113147    10.482985        .920749
UE           -33.703653     43.703607   -123.903466    56.496160       -.234561
(Constant)  2293.584660  1735.876883  -1289.089162  5876.258482


---------- Variables in the Equation ----------

Variable    Tolerance        VIF         T  Sig T

ADS          .417546        2.395     1.143  .2642
G            .204551        4.889      .303  .7644
P            .183120        5.461    -1.796  .0852
PRO          .767300        1.303     7.838  .0000
UE           .114445        8.738     -.771  .4481
(Constant)                  1.321      .1989


Rediduals Statistics:

               Min       Max      Mean   Std Dev    N

*PRED       5.6472  888.1078  398.8667  236.5295   30
*RESID   -212.1160  238.6306     .0000  138.0524   30
*ZPRED     -1.6625    2.0684     .0000    1.0000   30
*ZRESID    -1.3978    1.5725     .0000     .9097   30

Total Cases =        30


Durbin-Watson Test =    3.22816
```

in Figure 17.9. You can see the effects of negative autocorrelation, where one residual is almost always of the opposite sign to the previous one.

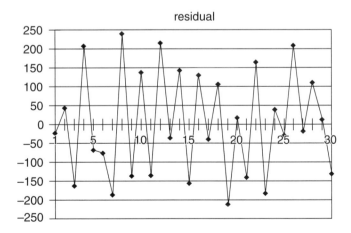

Figure 17.9

Lack of data

Our data sets are complete, but there may still be problems since the growth and unemployment figures are percentages rounded up to one decimal place. Small changes will thus not be reflected in the data which we have used.

Time and cost   The company will have the internal data available (at least in some form) but will
                                   need to gather the externally held data. It would probably not have the expertise
                                   internally to construct such a model.

Under-identification   Not a problem since this is a single equation model.

Specification   With only one variable seeming to affect the sales volume, this would appear to
                                   be the basic problem with the model. We could look to other variables internally
                                   which might help to explain our sales, or we could search for other macroeco-
                                   nomic variables which may model the effects of the economy as a whole on the
                                   company's performance. We could look at non-linear models and see if these give
                                   'better' results. There is one other area, however, in this model which may be
                                   the key to the 'poor performance' of the explanatory variables. If you have read
                                   Chapter 14, you may recognize this problem. The dependent variable is heavily
                                   seasonal, and while the promotional spend is also seasonal, the other explanatory
                                   variables are not. Maybe we need to either build a model using annual data, or
                                   we need to take away the seasonal effects, where these exist, and look for the
                                   underlying relationships and model. Whichever route we decide to take, we are
                                   changing the specification of the model.

17.5 Using multiple regression models

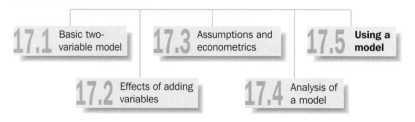

Having established a multiple regression relationship from historical data, we
usually wish to use this to make predictions about the future. As in any model of
this type, if the value of R^2 is low, or the specification of the model is very poor, then the predictions will be worthless.

Such predictions will depend on having suitable data available on the explanatory variables in order to predict the future of the dependent variable. This would not necessarily be the case in the model which we have just analysed above, since, whilst we can set the levels of price, promotional spend and advertising internally, we will not know the values of growth or unemployment for next year, in advance. (In practical terms, the data on the current period is often not published for three to six months!) Modellers overcome this problem by building time lags into their models, so that they are using data from six months or a year before to predict current and future behaviour of the dependent variable. (Thus ensuring that the data is available!) Building in such time lags, however, leads to other problems, especially with autocorrelation, and testing for it.

Having built time lags into the model, we merely substitute values of the explanatory variables into the equation to get predictions. This is illustrated in Table 17.8.

In Table 17.8 we are at time period 24 with all data available up until that point. The data values marked with a star (*) are those which are used to make the prediction of Y for period 24. If we now substitute the next value of X2 (5.5), X3 (102.2) and X4 (2.8) into the equation, then we find a prediction for

Table 17.8 $Y_t = 20 + 6.3X2_{t-2} + 4.7X3_{t-3} - 2.7X4_{t-4}$

Time	Y	Prediction	X2	X3	X4
20	520		4.7	103.7	2.7*
21	527		4.9	102.6*	2.8
22	531		5.3*	102.2	2.9
23	535		5.5	101.4	2.9
24	530	528.32	5.7	103.1	2.6
25		527.43			
26		524.66			

period 25 (527.43). A similar process will give the prediction for period 26. No further predictions can be made from this model until more data becomes available (i.e. in period 25).

Exercise

> Working with the data for Jasbeer, Rajan & Co. build further multiple regression models and assess their validity.

17.6 Conclusions

Multiple regression analysis is a very powerful statistical technique which will enable us to make better predictions about the behaviour of the dependent variable than simple two-variable models. The increased complexity, however, raises problems of interpretation of the results since in business situations, not all of the underlying assumptions are met. Multiple regression should not be used blindly; it is necessary to perform a series of tests on the model before it can be used to make predictions. As with any technique, it is necessary to ask questions about the reliability and accuracy of the data before accepting the results of the model. The increasing use of spreadsheets means that more and more people will have the technical ability to create such models, but they are unlikely to have the level of understanding necessary to ask the right questions about the model. Having read through this chapter, you should be in a position to help them.

17.7 Problems

1

Y	10	12	15	17	19	22	24	27	29	30
X_2	1	1	2	2	3	4	4	5	5	6
X_3	10	9	8	7	6	5	4	3	2	1

(a) Use the data given above to find the regression relationship between Y and X_2.
(b) Use the data given above to find the regression relationship between Y and X_3.
(c) Use the data given above to find the regression relationship between Y and X_2 and X_3.
(d) From your answer to part (c) test each of the coefficients to find if they are non-zero.
(e) Find the coefficient of multiple determination.

2 Use the data given below to find the multiple regression relationship which allows you to predict Y. Assess this model.

Y	X_1	X_2	X_3	X_4
100	25	87	14	1003
120	45	60	16	2510
125	50	61	15	3610
130	60	55	12	4875
140	50	52	9	6951
136	85	45	5	5241
145	90	40	21	7842
86.4	30	120	17	845
131	68	39	16	6543
116	33.6	59	14	2215
125	36	63	24	3945
123.2	90	65	11	3512
103.3	60	83	18	1562
141	95	37	15	8747
109	47.4	60	14	2546

3 The following data represent a particular problem when using multiple regression analysis. Identify the problem, and construct a model which overcomes this issue.

Y	X_1	X_2	X_3
150	44	22	1
190	56	28	2
240	64	32	3
290	68	34	4
350	76	38	5
400	80	40	6
450	86	43	7
500	90	45	8
650	100	50	9
800	120	60	10

4 A brewer is interested in the amount spent per week on alcohol and the factors which influence this. Data has been collected on the amount spent, the income of the head of household (in thousands per year) and the household size for 20 families and the results are presented below.

Amount spent per week	Income of head of household	Household size
20	6	1
17	5	2
5	10	1
0	14	4
3	25	2
8	10	5
14	21	1
19	17	1
32	29	2
17	14	3
9	7	1
8	9	3
4	14	2
20	19	1
10	13	1
9	10	2
7	9	3
14	11	3
59	34	6
7	10	2

(a) Find the regression relationship of amount spent on income and family size.

(b) Carry out a statistical analysis of this result.

(c) What would be the problems for the brewer in interpreting this answer?

5 A personnel and recruitment company wishes to build a model of likely income level and identifies three factors which it feels are important. Data collected on 20 clients gives the following results.

Income level	Years of post-16 education	Years in post	No. of previous jobs
15	2	5	0
20	5	3	1
17	5	7	2
9	2	2	0
18	5	8	2
24	7	4	3
37	10	11	2
24	5	7	1
19	6	4	0
21	2	8	4
39	7	12	2
24	8	8	1
22	5	6	2
27	6	9	1
19	4	4	1
20	4	5	2
24	5	2	3
23	5	6	1
17	4	3	4
21	7	4	1

(a) Find the regression relationship for predicting the income level from the other variables.

(b) Assess the statistical quality of this model.

(c) Would you expect the company to find the model useful?

(d) What other factors would you wish to build into such a model?

6 A manufacturer of diesel engines for small boats wished to predict the overall demand for the product. This demand was a derived demand from the sales of new boats and sales of replacement engines. Data was collected on a series of macroeconomic variables and built into an econometric model, the results of which are presented in Figure 17.10 opposite.

UKBOAT	=	UK boat sales;
TIME	=	dummy variable;
SWF4	=	exchange rate to Swiss franc with lag of one year;
MLR4	=	minimum lending rate with lag of one year;
GDP4	=	percentage change in gross domestic product of UK with lag of one year.

Assess this model in both statistical terms and in terms of its likely usefulness to the company.

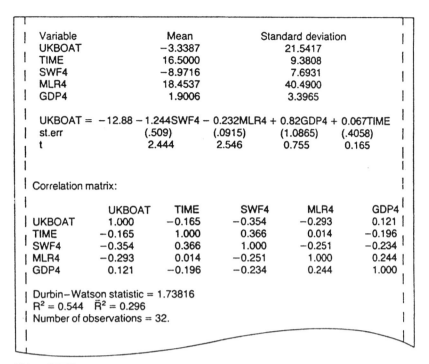

```
| Variable                Mean              Standard deviation |
| UKBOAT                 -3.3387                 21.5417       |
| TIME                   16.5000                  9.3808       |
| SWF4                   -8.9716                  7.6931       |
| MLR4                   18.4537                 40.4900       |
| GDP4                    1.9006                  3.3965       |
|                                                             |
| UKBOAT = -12.88 - 1.244SWF4 - 0.232MLR4 + 0.82GDP4 + 0.067TIME |
| st.err              (.509)       (.0915)     (1.0865)   (.4058) |
| t                    2.444        2.546        0.755      0.165 |
|                                                             |
|                                                             |
| Correlation matrix:                                         |
|                                                             |
|            UKBOAT     TIME       SWF4       MLR4       GDP4  |
| UKBOAT      1.000    -0.165     -0.354     -0.293      0.121 |
| TIME       -0.165     1.000      0.366      0.014     -0.196 |
| SWF4       -0.354     0.366      1.000     -0.251     -0.234 |
| MLR4       -0.293     0.014     -0.251      1.000      0.244 |
| GDP4        0.121    -0.196     -0.234      0.244      1.000 |
|                                                             |
| Durbin-Watson statistic = 1.73816                           |
| R² = 0.544   R̄² = 0.296                                      |
| Number of observations = 32.                                |
```

Figure 17.10

7 A company has developed a model of its sales and asks you to build a spreadsheet model to allow predictions to be made. The specification of the model is as follows:

Sales value = Sales volume × Price

Sales volume =	10 000	
	−20 000	(Price − average price in the industry)
	+30	× Advertising expenditure
	+20	× Promotional expenditure
	+	a seasonal factor
	+25 000	× Growth rate in the economy
	−250 000	× No. of companies in the industry
	−100 000	× Unemployment rate in the economy
	−2	× Advertising expenditure of other companies
	−1.5	× Promotional spending of other companies

The following data is available for the first 10 time periods:

Time	Price	Average price	Advertising expenditure	Promotional advertising
1	1.35	1.25	145 000.00	80 000.00
2	1.35	1.25	146 450.00	81 600.00
3	1.45	1.25	147 914.50	83 232.00
4	1.45	1.30	149 393.65	84 896.64
5	1.45	1.30	150 887.58	86 594.57
6	1.45	1.30	152 396.46	88 326.46
7	1.45	1.30	153 920.42	90 092.99
8	1.45	1.30	155 459.63	91 894.85
9	1.45	1.35	157 014.22	93 732.75
10	1.45	1.35	158 584.36	95 607.41

The seasonal factors are

$-100\,000$

$-200\,000$

$-150\,000$

$+450\,000$

repeated through time. The number of companies is:

 8 in periods 1 to 9

 9 in periods 10 to 15

10 in periods 16 to 25

Data on external factors for the first 10 periods:

Time	Growth	U/E	Others' advertising	Others' promotion
1	2.90	14	600 000.00	250 000.00
2	2.80	14.10	612 000.00	260 000.00
3	2.70	14.10	624 240.00	270 400.00
4	2.40	14.30	636 724.80	281 216.00
5	2.10	14.30	649 459.30	292 464.64
6	1.40	14.70	662 448.48	304 163.23
7	0.30	15.10	675 697.45	316 329.75
8	0.30	14	689 211.40	328 982.94
9	−0.10	14.40	702 995.63	342 142.26
10	−0.70	14.60	717 055.54	355 827.95

(a) Use this information to build a spreadsheet model and obtain a graph of sales value for the 10 periods.

(b) Use the information provided below to extend the spreadsheet to predict sales for 25 time periods, obtaining a graph of sales value for the whole period.

(c) What are the implications of your predictions?
Write a brief report (400 words maximum).

(d) Vary price, advertising spend and promotional spend (within reasonable limits) to smooth out the sales value figures. Write a brief report on your suggested policy for the company (400 words maximum).

Extra information:

Time	Price	Average price	Growth
11	1.50	1.35	−1.30
12	1.50	1.35	−1.40
13	1.50	1.45	−0.09
14	1.50	1.45	−0.50
15	1.60	1.45	−0.10
16	1.60	1.45	0.10
17	1.60	1.55	0.30
18	1.60	1.55	0.70
19	1.65	1.55	1.30
20	1.65	1.55	1.50
21	1.70	1.55	1.70
22	1.70	1.65	2.20
23	1.70	1.65	2.30
24	1.70	1.65	2.20
25	1.70	1.65	2.00

Note also the following:
(a) Advertising expenditure continues to grow at 1% per time period.
(b) promotional spending continues to grow at 2% per time period.
(c) Unemployment $= 14 + (\text{Growth}_t - \text{Growth}_{t-1})$
(d) Others' advertising spend increases by 2% per time period.
(e) Others' promotional spend increases by 4% per time period.

Part 5   Conclusions

If you have been through the whole of this part you will have gained considerable knowledge of techniques for relating variables and predicting outcomes. We hope that you will also have acquired a certain cynicism about the blind use of these techniques just because a particular bit of computer software is able to perform the calculations.

Overall, there is no, one, 'correct' method of approaching the problem of prediction, which is so vital to businesses. In practice some combination of the ideas and techniques shown in this part are likely to be useful to businesses. Many will prefer some kind of prediction, even when they know it to be very vague, to no prediction at all. Using the ideas presented here, you should be able to assess the validity of the predictions which you make, and also make rational choices about which techniques to try first. Sometimes you may be wrong! Perhaps more importantly, you may now be less likely to be conned by other people when they make forecasts about the future.

Case 5

Bring together all of the work you have done in this part on the case study of Jasbeer, Rajan & Co. and try to come to some overall conclusions about the future of the business, in terms of sales volume.

Using the other data which is available on the website on costs, behaviour of competitors, etc. build further models to predict sales volume for up to five years ahead, and then work out the profit position of the business. (Assume that there is no stockholding from year to year.)

A list of the variables available on the website is given in the table below.

Column	Name	Description
1	yrqtr	Year and quarter
2	Time	Dummy variable, yr 1, qtr $1 = 1$
3	Sales Volume	Number of units sold
4	Ads	Amount spent on advertising
5	Promotion	Amount spent on promotion
6	Price	Selling price per unit
7	Growth	Growth in the economy
8	U/E	Unemployment rate in the economy
9	Av. Price	Average price in the industry sector
10	Others Ads	Total advertising spend by industry
11	Others Promotion	Total promotional spend by industry
12	No. Co's	Number of companies in the industry

Manufacturing cost data is as follows:

Year	1	2	3	4	5	6	7	8
Cost per item	0.26	0.28	0.29	0.30	0.32	0.34	0.36	0.39

and thereafter increases at 8% per annum.

Number of items produced times the manufacturing cost gives cost of goods sold (COGS).

Future predictions of growth are as follows:

Period	8.3	8.4	9.1	9.2	9.3	9.4	10.1	10.2	10.3
Growth	−1.3	−1.4	−0.9	−0.5	−0.1	0.1	0.3	0.7	1.3
Period	10.4	11.1	11.2	11.3	11.4	12.1	12.2	12.3	12.4
Growth	1.5	1.7	2.2	2.3	2.2	2.0	1.9	1.8	1.8

The number of companies in future is expected to be:

From–To	8.2–9.3	9.4–12.1	12.2 onwards
Number	9	10	11

Note that U/E is defined as a function of growth.

$$\text{Expenses} = \text{advertising cost} + \text{promotion} + 20\% \text{ of COGS.}$$

The following accounting relationships maybe helpful.

- Sales revenue = number of items sold × price.
- Gross profit = sales revenue − cost of goods sold (COGS).
- Net profit = gross profit − expenses.

Profits are normally reported on an annual basis.

Part 6   Modelling

Models can provide a description of situations or phenomena that are difficult to examine in any other way. We are all familiar with how scaled-down versions of a car or aeroplane can be used to demonstrate the characteristics of the real thing. We are also familiar with how a flight simulator can model situations for experienced and trainee pilots so that they are prepared for the types of real-life events they may encounter. This type of modelling is referred to as **iconic** and is only included here to illustrate the kind of insight we look for from a model. A second kind of modelling is where one factor is used to describe another; this is called **analogue** modelling. Speed being represented on a dial (speedometer) or workflow by a liquid are both examples of analogue modelling. There is a third kind of modelling, **symbolic**, which is of particular interest because it describes how mathematics and other notation can be used to represent problem situations. In the following chapters we will try to capture characteristics of all three kinds of model by the use of scaling, representation by measurement and representation by equations.

Most business situations of significance will be complex. Our need to understand this complexity will depend on our role within the business and how we choose to solve problems. As a manager we may need to model operational matters like stock control; as a senior executive we may need to model the strategic position of the company in the market place. In order to understand a problem situation, it is likely that we will need to make a number of simplifying assumptions (rationality is usually one!). To understand the implications of any proposed change, it is necessary to explore how changes to one factor impact on other factors. Mathematical models attempt to describe real problems in abstract terms, yet terms which capture the essential characteristics. A mathematical model will usually involve the use of equations with some values kept fixed as parameters and other allowed to vary as input (independent) variables or output (dependent) variables.

Models are unlikely to give perfect answers. What models allow the user to do is work through various assumptions about the business situation and see the possible consequences. We can try out all sorts of ideas on mathematical models, at little or no cost. Models can even be used to plan for disasters. For example, what will happen to cash flow next month if we lose a major order? We can always test and improve our models by simulating the problem and then returning to the original problem situation.

The models described in the following chapters have found a wide range of business applications.

Case 6: The Ressembler Group

The Ressembler Group has a wide range of interests in a number of small businesses including the production of made to measure doors and door frames, imported kitchen units and kitchen refurbishment. Recently they have diversified into holiday home development in mainland Europe and video hire. The group has the general characteristics of a holding company, where the individual business units are regarded as autonomous, and little effort is made to co-ordinate their activities.

The group has been very successful over the past 10 years in financial terms, and has been seen as investing and divesting at opportune times. The individual businesses have been set tough financial targets and then left alone to achieve them. Managers can expect considerable rewards if targets are met or exceeded, but have reason to fear for their jobs and their business if targets are not met, even in the short term. The business units are not represented at board level and the level of central corporate staff and services has been kept to a minimum.

Recently, trading conditions have been seen as more difficult and are expected to get worse. It is known that a number of the managers in the business units would like to see a more consistent approach to strategy and improved central services. The board are reluctant to increase the level of the central services and regard low central overheads as one source of competitive advantage. The group structure has provided a business discipline and cheaper source of finance for the individual business units, and provided the scope for the mangers to act autonomously, subject to the financial controls.

18 The time value of money

Money facilitates the exchange of goods and services. It can take the physical form of coins or notes, or the more representational form of a promise to pay. Money provides a base of reference for a wide range of transactions, which transcend time and international boundaries.

Money can be used now for immediate consumption or can be held for the purpose of future consumption. As individuals we continue to make decisions on whether to purchase now, perhaps with the help of a loan, or to save. Organizations do exactly the same things and need to decide how to reward stakeholders now and in the future, whether to invest or whether to hold cash surpluses in other forms. This chapter is concerned with how we value money now and how we value both income and expenditure in the future.

Case study

The Ressembler Group has many of the characteristics of an 'investment' company. At a corporate level, it is particularly concerned with the financial returns of individual businesses and is likely to judge group performance mostly in these terms. The group has been able to offer secure and competitive investment finance to its business units but has always insisted on a rigorous analysis of potential benefits. All managers within the business units are expected to be able to make interest rate calculations and understand how the group values future earnings.

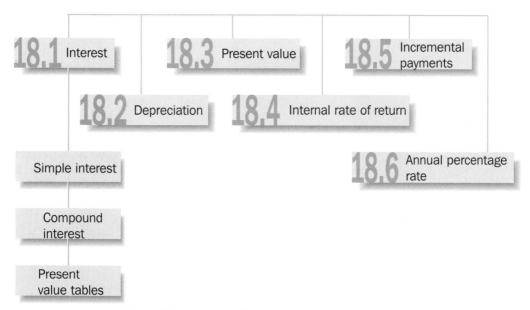

Figure 18.1 Structure of Chapter 18.

Objectives

After working through this chapter, you should be able to:

- calculate simple and compound interest
- understand and apply the concept of depreciation
- understand and apply the concept of present value
- understand and apply the 'internal rate of return'
- apply the 'incremental payments' formula to solve problems involving sinking funds, annuities and mortgages.

18.1 Interest: simple and compound

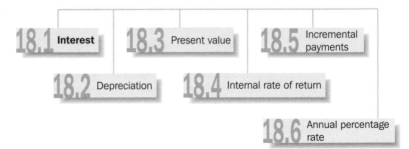

18.1 Interest

18.3 Present value

18.5 Incremental payments

18.2 Depreciation

18.4 Internal rate of return

18.6 Annual percentage rate

An interest rate, usually quoted as a percentage, gives the gain we can expect from each £1 saved. If, for example, we were offered 10% per annum we would expect a gain of 10 pence for each £1 saved. The interest received each year will depend on whether the interest is calculated on the basis of simple interest or compound interest.

18.1.1 Simple interest

If money is invested in a saving scheme where the interest earned each year is paid out and not reinvested, then the saving scheme offers simple interest, The simple interest, I, offered at the end of each year is:

$$I = A_0 \times \frac{r}{100}$$

where A_0 is the initial sum invested and r is the rate of interest as a percentage.

Example

What interest could be expected from an investment of £250 for 1 year at 10% per annum simple interest?

$$I = £250 \times 10/100 = £250 \times 0.1 = £25$$

As simple interest remains the same over the life of the investment, the interest gained over t years is given by $I \times t$ and the value of the investment (including the initial sum A_0) by $A_0 + It$. Few real-world agreements use simple interest.

Example

A sum of £250 is invested at a rate of 12% per annum for 5 years. How much interest would this yield?

$$I = £250 \times 0.12 \times 5 = £150$$

18.1.2 Compound interest

If the interest gained each year is **added to the sum saved**, we are looking at compound interest. The sum carried forward grows larger and larger and the interest gained each year grows larger and larger; the interest is compounded on the initial sum plus interest to date. Given an interest rate of 10%, after 1 year our £1 would be worth £1.10 and after 2 years our £1 would be worth £1.21. Most actual contracts use compound interest.

The sum at the end of the year can be calculated using the following formula:

$$A_t = A_0 \left(1 + \frac{r}{100}\right)^t$$

where A_0 is the initial sum invested, A_t is the sum after t years and r is the rate of interest as a percentage.

Example

What sum would be accumulated if £250 were invested at 9% compound interest for a period of 5 years?

$$A_5 = £250 \left(1 + \frac{9}{100}\right)^5 = £250 \times 1.5386239549 = 384.66$$

After 5 years we would have £384.66.

A proof of this formula will demonstrate the principle involved. Suppose we invest an initial sum of A_0 at an annual interest rate of $r\%$. After 1 year, this sum would be worth

$$A_1 = A_0 + \frac{r}{100} A_0$$

$$= A_0 \left(1 + \frac{r}{100}\right)^1$$

The term in brackets gives the **annual gain** of each £1. An interest rate of 10% would give a multiplicative factor of 1.1 in the first year. If the sum at the end of the first year is then carried forward for one more year, the investment of A_0 at the end of the second year would be worth

$$A_2 = A_1 + \frac{r}{100} A_1$$

$$= A_1 \left(1 + \frac{r}{100}\right)^1$$

We can substitute the value we already have for A_1 to obtain

$$A_2 = A_0 \left(1 + \frac{r}{100}\right)^1 \left(1 + \frac{r}{100}\right)^1$$

$$= A_0 \left(1 + \frac{r}{100}\right)^2$$

If the interest rate is 10%, then the multiplicative factor of 1.1 is used twice to obtain the value of the investment after 2 years.

Continuing in this way, we can obtain the general formula for any number of years t:

$$A_t = A_0 \left(1 + \frac{r}{100} \right)^t$$

The example has assumed that the interest is paid at the end of the year, so that the £1 invested at 10% gains 10p each year. Some investments are of this type, but many others give or charge interest every 6 months (e.g. building societies) or more frequently (e.g. interbank loan interest is charged per day). The concept of compound interest can deal with these situations provided that we can identify how many times per year interest is paid. If, for example, 5% interest is paid every 6 months, during the second 6 months we will be earning interest on more than the initial amount invested. Given an initial sum of £100, after 6 months we have £100 + £5(interest) = £105 and after 1 year we would have £105 + £5.25(interest) = £110.25.

Calculation takes place as follows:

$$A_1 = £100 \left(1 + \frac{0.1}{2} \right)^{2 \times 1} = £110.25$$

i.e. divide the annual rate of interest (10% in this case) by the number of payments each year, and multiply the power of the bracket by the same number of periods.

More generally

$$A_t = A_0 \left(1 + \frac{r}{100 \times m} \right)^{mt}$$

where m is the number of payments per year.

18.1.3 The use of present value tables

The growth of an investment is determined by the **interest rate** and the time scale of the investment as we can see from the multiplicative factor:

$$\left(1 + \frac{r}{100} \right)^t$$

Any mathematical term of this kind can be tabulated to save repeated calculations. In this particular case, the tabulation takes the form of present value factors (see Appendix F), which can be used indirectly to find a compound interest factor. These factors are the reciprocal of what is required:

$$\text{present factor value} = \frac{1}{\left(1 + \frac{r}{100} \right)^t}$$

hence

$$\left(1 + \frac{r}{100} \right)^t = \frac{1}{\text{present value factor}}$$

If we were interested in the growth over 8 years of an investment made at 10% per annum we could first find the present value factor of 0.4665 from tables and then use the reciprocal value of 2.1436 to calculate a corresponding accumulated sum. It should be noted that the use of tables with four significant digits sometimes results in rounding errors. While such tables are very useful, we would often perform these calculations on a spreadsheet, and make it work out the sums for us.

Example

What amount would an initial sum of £150 accumulate to in 8 years if it were invested at a compound interest rate of 10% per annum?

Using tables

$$A_8 = 150 \times \frac{1}{0.4665} = 150 \times 2.1436 = £321.54$$

18.2 Depreciation

In the same way that an investment can increase by a constant percentage each year as given by the interest rate, the book value of an asset can decline by a constant percentage. If we use r as the depreciation rate we can adapt the formula for compound interest to give

$$A_t = A_0 \left(1 - \frac{r}{100}\right)^t$$

where A_t becomes the book value after t years.

Example

Use the 'declining balance method' of depreciation to find the value after 3 years of an asset initially worth £20 000 and whose value declines at 15% per annum.

By substitution we obtain

$$A_3 = £20\,000 \left(1 - \frac{15}{100}\right)^3 = £20\,000(0.85)^3$$

$$= £12\,282.50$$

A manipulation of this formula gives the following expression for the rate of depreciation

$$r = \left(1 - \sqrt[t]{\frac{A_t}{A_0}}\right) \times 100$$

where A_0 is the original cost and A_t is the salvage (or scrap) value after t years (see Figure 18.2).

An alternative to the percentage method of depreciation is to use linear depreciation, where a constant level of value is lost each year. To calculate linear depreciation each year, we divide the total loss of value by the number of years.

Example

The cost of a particular asset is £20 000 and its salvage value is £8000 after 5 years. Determine annual depreciation using the linear method.

The loss of value over 5 years is £20 000 − £8000 = £12 000.

Annual depreciation (using the linear method) = £12 000/5 = £2400

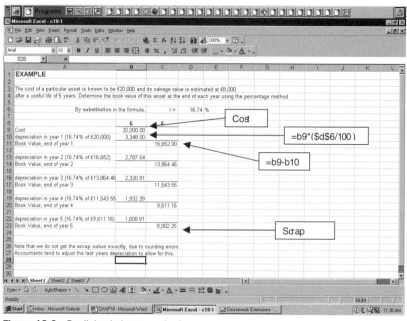

Figure 18.2 Declining balance methods.

A comparison of the two methods shows the extent to which the loss of value is spread across the time period. If a percentage method is used, the loss of value will be mostly accounted for in the early years. It is important to note that there is not a unique correct annual value, but that the calculated value does depend on the model being used.

18.3 Present value

The formula for compound interest can be rearranged to allow the calculation of the amount of money required now to achieve a specific sum at some future point in time given a rate of interest:

$$A_0 = A_t \frac{1}{\left(1 + \dfrac{r}{100}\right)^t}$$

Example

What amount would need to be invested now to provide a sum of £242 in 2 years' time given that the market rate of interest is 10%?

By substitution we obtain

$$A_0 = £242 \times \frac{1}{\left(1 + \dfrac{10}{100}\right)^2}$$

$$= £242 \times 0.826\,446 = £200$$

The amount required now to produce a future sum can be taken as a measure of the worth or the **present value** of the future sum. A choice between £200 now and £200 in 2 years' time would be an easy one for most people. Most would prefer the money in their pockets now. Even if the £200 were intended for a holiday in 2 years' time, it presents the owner with opportunities, one of which is to invest the sum for 2 years and gain £42. The choice between £200 now and £242 in 2 years' time, however, would be rather more difficult. If the interest rate were 10%, the present value of £242 in 2 years' time would be £200. Indeed, if one were concerned *only* with interest rates there would be an **indifference** between the two choices.

Present value provides a method of comparing monies available at different points in time. If different projects or opportunities need to be compared, then looking at the **total** present value for each, provides a rigorous basis for making choices between them. The calculation of the present value of future sums of money is referred to as **discounting**.

Example

You need to decide between two business opportunities. The first opportunity will pay £700 in 4 years' time and the second opportunity will pay £850 in 6 years' time. You have been advised to discount these future sums by using the interest rate of 8%.

By substitution we are able to calculate a present value for each of the future sums. First opportunity:

$$A_0 = £700 \times \frac{1}{\left(1 + \frac{8}{100}\right)^4}$$

$$= £700 \times 0.7350$$

$$= £514.50$$

Second opportunity

$$A_0 = £850 \times \frac{1}{\left(1 + \frac{8}{100}\right)^6}$$

$$= £850 \times 0.6302$$

$$= £535.67$$

On the basis of present value (or discounting) we would choose the second opportunity as the better business proposition. In practice, we would need to consider a range of other factors.

The present value factors of 0.7350 and 0.6302 calculated in this example could have been obtained **directly from tables** (see Section 18.1.3); as you can see from the present value factors given in Appendix F. We can explain the meaning of these factors in two ways. First, if we invest $73\frac{1}{2}$ pence at 8% per annum it will grow to £1 in 4 years and 63 pence (0.6302) will grow to £1 in 6 years. Secondly, if the rate of interest is 8% per annum, £1 in 4 years' time is worth $73\frac{1}{2}$ pence now and £1 in 6 years' time is worth 63 pence now. All this sounds rather cumbersome, but the concept of the present value of a future sum of money is fundamental to investment appraisal decisions made by companies. This is illustrated in the following example.

Example

A business needs to choose between two investment options. It has been decided to discount future returns at 12%. The expected revenues and initial costs are given as follows:

Year	Estimated end of year revenue option 1	option 2
1	300	350
2	350	350
3	410	350
Cost in year 0	300	250

In calculating the present value from different options we generally refer to a discount rate or the rate of return on capital rather than the interest rate. These rates tend to be higher than the market interest rate and reflect the cost of capital to a particular business. Net present value (NPV) for each option is the sum of the constituent present values.

The present value factors can be obtained directly from tables or calculated using $(1 + 0.12)^{-t}$ for the years $t = 1$ to 3. In this example, the costs are immediate and therefore are not discounted.

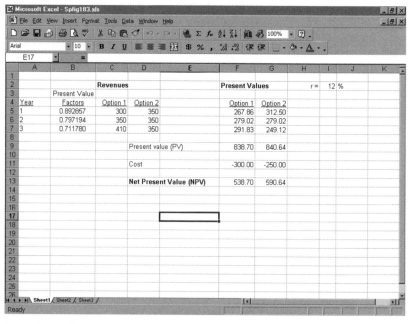

Figure 18.3   Present values.

Although the total revenue over three years is slightly higher with option 1, the value now to a business is higher with option 2. A more immediate revenue presents a business with more immediate opportunities. It can be seen in this example that option 2 offers the business an extra £50 in the first year which can itself be used for additional gain, and is especially useful in maintaining cash flow (see Figure 18.3).

The comparison of these two options depends crucially on the 'time value of money', that is the discount rate, the estimates given for revenue and the completeness of information.

Exercise

> Construct the spreadsheet model for the above example. Use your model to evaluate the two options using a discount rate of (a) 8% and (b) 14%.

18.3.1 Practical problems

This type of exercise looks straightforward in a textbook, but presents a series of problems when it is to be used in business. Initial costs of each project to be considered will be known, but there may be **extra costs** involved in the future which cannot even be estimated at the start, e.g. a change in tariffs in a country to which the company exports. All further cash flow information must be **estimated**, since it is to come in the future, and is thus open to some doubt: if we are dealing with a new product these cash flows are likely to be based on market research (see Chapter 3 on survey methods).

A further practical difficulty is to decide upon which **discount rate** to use in the calculations. This could be:

- the market rate of interest
- the rate of return gained by the company on other projects
- the average rate of return in the industry
- a target rate of return set by the board of directors; or
- one of these plus a factor to allow for the 'riskiness' of the project – high-risk projects being discounted at a higher rate of interest.

High-risk projects are likely to be discriminated against in two ways: by the discount rate used, which is likely to be high, and in the estimated cash flows that are often conservatively estimated.

All attempts to use this type of present value calculation to decide between projects make an **implicit assumption** that the project adopted will succeed.

Net present value is only **an aid** to management in deciding between projects, as it only considers the monetary factors, and those only as far as they are known, or can be estimated: there are many personal, social and political factors which may be highly relevant to the decision. If a company is considering moving its factory to a new site, several sites may be considered, and the net present value of each assessed for the next 5 years. However, if one site is in an area that the managing director (or spouse) dislikes intensely, then it is not likely to be chosen. The workforce may be unwilling to move far, so a site 500 miles away could present difficulties. There may be further **environmental problems**, which are not costed by the company, but are a cost to the community, e.g. smoke, river pollution, extra traffic on country roads.

18.4 The internal rate of return

The method of calculating the internal rate of return is included in this chapter because it provides a useful **alternative** to the net present value method of investment appraisal. The internal rate of return (IRR), sometimes referred to as the **yield**, is the discount rate that produces a net present value of 0. This is the value for r, which makes the net present value of cost **equal** to the net present value of benefits.

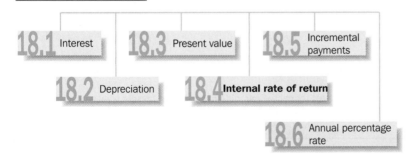

Example

Suppose £1000 is invested now and gives a return of £1360 in one year. The internal rate of return can be calculated as follows:

$$\text{net present value of cost} = \text{net present value of benefits}$$

$$£1000 = \frac{1360}{(1+r)}$$

since cost is incurred now and benefits are received in one year. So,

$$1000(1+r) = 1360$$

$$1000 + 1000r = 1360$$

$$r = 360/1000$$

$$= 0.36 \text{ or } 36\%$$

Example

Suppose £1000 is invested now and gives a return of £800 in one year and a further £560 after two years. We can proceed in the same way:

$$1000 = \frac{800}{(1+r)} + \frac{560}{(1+r)^2}$$

Multiplying by $(1+r)^2$ gives

$$1000(1+r)^2 = 800(1+r) + 560$$

$$1000(1+r)^2 - 800r - 800 - 560 = 0$$

$$1000 + 2000r + 1000r^2 - 800r - 1360 = 0$$

$$1000r^2 + 1200r - 360 = 0$$

The solution to this quadratic equation (see section 21.4.3) is given by $r = -1.4485$ or 0.2485. Only the latter, positive value has a business interpretation, and the internal rate of return in this case is 0.2485 or more simply expressed as 25%.

The internal rate of return cannot be calculated so easily for longer periods of time and is estimated either using a graphical method or by linear interpolation.

18.4.1 The internal rate of return by graphical method

To estimate the internal rate of return graphically we need to determine the net present value of an investment corresponding to two values of the discount rate r. The accuracy of this method is improved if one value of r gives a small positive net present value and the other value of r a small negative value.

Example

Suppose an investment of £460 gives a return of £180, £150, £130 and £100 at the end of years 1, 2, 3 and 4.

Using a spreadsheet format we can determine the net present values of this cash flow for two values of r. In this example, $r = 8\%$ and $r = 10\%$ provide reasonable answers.

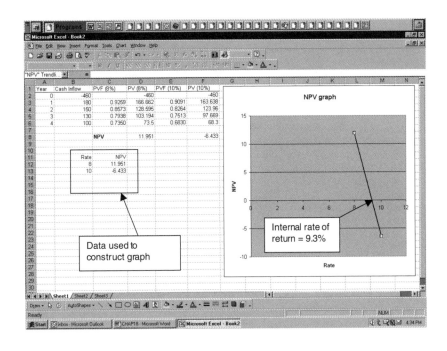

Figure 18.4 Here PVF is the present value factor and PV is the present value.

Exercise

> Construct the preceding spreadsheet in such a way that it can accommodate many different values of r.

We are now able to plot net present value against r, as shown in the screenshot above. By joining the plotted points we can obtain a line showing how net present value decreases as the discount rate increases. When net present value is 0, the corresponding r value is the estimated internal rate of return, 9.3% in this case. You may wish to draw this graph on **graph paper** rather than rely on reading a computer screen (see Figure 18.4).

18.4.2 The internal rate of return by linear interpolation

Again we need to determine the net present value of an investment corresponding to two values of r. Instead of joining points on a graph with a straight line, we calculate a value of r between these two values on a proportionate basis. If

$$r_1 = \text{the lower discount rate,}$$
$$r_2 = \text{the upper discount rate,}$$
$$NVP_1 = \text{the net present value corresponding to } r_1 \text{ and}$$
$$NVP_2 = \text{the net present value corresponding to } r_2$$

then

$$IRR = r_1 + \frac{NPV_1}{(NPV_1 - NPV_2)} \times (r_2 - r_1)$$

Example

Given $r_1 = 8\%$, $r_2 = 10\%$, $NPV_1 = £11.95$ and $NPV_2 = -£6.43$,

$$IRR = 8 + \frac{11.95}{11.95 + 6.43} \times (10 - 8)$$

$$= 8 + 0.65 \times 2$$

$$= 9.3\%$$

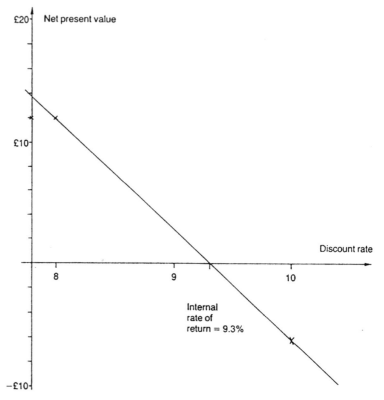

Figure 18.5   Internal rate of return.

Linearity is assumed in these calculations (and the graphical method) but the relationship between net present value and discount rate is non-linear. Provided the two selected values of r are fairly close to the internal rate of return, the estimation procedure produces a result of sufficient accuracy.

The decision of whether to invest or not will depend on how the internal rate of return compares with the discount rate being used by a company (a measure of the cost of capital to the company) or other organization. If the calculations are specified on a spreadsheet, then the effects of changes in parameters are very quickly seen.

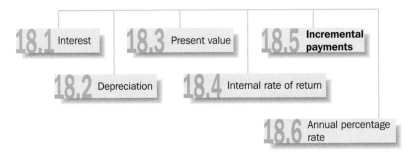

18.5 Incremental payments

In Section 18.1.2 we considered the growth of an initial investment when subject to compound interest. Many saving schemes will involve the same sort of initial investment but will then **add or subtract given amounts at regular intervals**. If x is an amount added at the end of each year, then the sum receivable, S, at the end of t years is given by

$$S = A_0\left(1 + \frac{r}{100}\right)^t$$

$$+ \frac{x\left(1 + \frac{r}{100}\right)^t - x}{r/100}$$

An outline proof of this particular formula is given in Section 18.9.

Examples

1 A savings scheme involves an initial investment of £100 and an additional £50 at the end of each year for the next three years. Calculate the receivable sum at the end of 3 years assuming that the annual rate of interest paid is 10%.

By substitution we obtain

$$S = £100\left(1 + \frac{10}{100}\right)^3 + \frac{£50\left(1 + \frac{10}{100}\right)^3 - £50}{10/100}$$

$$= £133.10 + £165.50$$

$$= £298.60$$

The sum is in two parts, the first being the value of the initial investment (£133.10) and the second being the value of the end of year increments (£165.50).

An alternative is to calculate the growing sum year by year.

	Value of investment
Initial sum	£100
Value at the end of year 1	£110
+ increment of £50	£160
Value at the end of year 2	£176
+ increment of £50	£226
Value at the end of year 3	£248.60
+ increment of £50	£298.60

If the rate of interest or the amount added at the end of the year changed from year to year it would no longer be valid to substitute into the formula given.

In the case of regular withdrawals, we use a negative increment.

2 It has been decided to withdraw £600 at the end of each year for five years from an investment of £30 000 made at 8% per annum compound.

In this example, we have a negative increment of £600. By substitution we obtain:

$$S = £30\,000(1 + 0.08)^5 + \frac{(-£600)(1 + 0.08)^5 - (-£600)}{0.08}$$

$$= £44\,079.842 - £3519.9606$$

$$= £40\,559.881$$

18.5.1 Sinking funds

A business may wish to set aside a fixed sum of money at regular intervals to achieve a specific sum at some future point in time. This sum, known as a sinking fund, may be in anticipation of some future investment need such as the replacement of vehicles or machines.

Example

How much would we need to set aside at the end of each of the following five years to accumulate £20 000, given an interest rate of 12% per annum compound?

We can substitute the following values:

$$S = £20\,000$$

$$A_0 = 0 \text{ (no saving is being made immediately)}$$

$$r = 12\%$$

$$t = 5 \text{ years}$$

to obtain

$$£20\,000 = 0 + \frac{x(1+0.12)^5 - x}{0.12}$$

$$£20\,000 \times 0.12 = 0.7623x$$

$$x = £3148.37$$

where x is the amount we would need to set aside at the end of each year.

18.5.2 Annuities

An annuity is an arrangement whereby a fixed sum is paid in exchange for regular amounts to be received at fixed intervals for a specified time. Such schemes are usually offered by insurance companies and may be particularly attractive to people preparing for retirement.

Example

How much is it worth paying for an annuity of £1000 receivable for the next five years and payable at the end of each year, given interest rates of 11% per annum?

We can substitute the following values:

$$S = 0 \text{ (final value of investment)}$$

$$x = -£1000 \text{ (a negative increment)}$$

$$r = 11\%$$

$$t = 5$$

to obtain

$$0 = A_0(1+0.11)^5 + \frac{(-£1000)(1+0.11)^5 - (-£1000)}{0.11}$$

$$A_0 = £1000 \times \frac{(1+0.11)^5 - 1}{0.11 \times (1+0.11)^5}$$

$$= £1000 \times \frac{1 - \dfrac{1}{(1+0.11)^5}}{0.11} = £1000 \times 3.695\,89 = £3695.89$$

where A_0 is the value of the annuity. (Note that $1/(1+0.11)^5 = 0.5935$ from Appendix F.)

The present value of the annuity is £3695. We could have calculated the value of the annuity by discounting each of £1000 receivable for the next five years by present value factors.

18.5.3 Mortgages

The most common form of mortgage is an agreement to make regular repayments in return for the initial sum borrowed, mostly in buying a house. At the end of the repayment period the outstanding debt is zero.

Example

What annual repayment at the end of each year will be required to repay a mortgage of £25 000 over 25 years if the annual rate of interest is 14%?

We can substitute the following values:

$$S = 0 \text{ (final value of mortgage)}$$
$$A_0 = -£25\,000 \text{ (a negative saving)}$$
$$r = 14\%$$
$$t = 25$$

to obtain

$$0 = -£25\,000(1+0.14)^{25} + \frac{x(1+0.14)^{25} - x}{0.14}$$

$$x = £25\,000\frac{(1+0.14)^{25} \times 0.14}{(1+0.14)^{25} - 1}$$

$$= £25\,000 \times \frac{0.14}{1 - \dfrac{1}{(1+0.14)^{25}}} = 25\,000 \times 0.1455 = £3637.50$$

where x is the annual repayment.

The multiplicative factor of 0.1455 is referred to as the **capital recovery factor** and can be found from tabulations.

18.6 Annual percentage rate (APR)

For an interest rate to be meaningful, it must refer to a period of time, e.g. per annum or per month. Legally, the annual percentage rate (APR), also known as the actual percentage rate, **must be quoted** in many financial transactions. The APR represents the **true cost** of borrowing over the period of a year when the compounding of interest, typically per month, is taken into account.

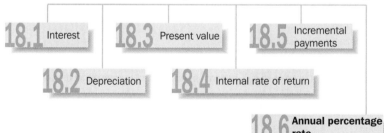

18.1 Interest

18.2 Depreciation

18.3 Present value

18.4 Internal rate of return

18.5 Incremental payments

18.6 **Annual percentage rate**

Example

Suppose a credit card system charges 3% per month compound on the balance outstanding. What is the amount outstanding at the end of one year if £1000 has been borrowed and no payments have been made?

To just multiply the monthly rate of 3% by 12 would give the **nominal rate of interest** of 36% which would underestimate the true cost of borrowing. Using the method outlined in Section 18.1.2:

$$\text{balance outstanding} = £1000(1 + 0.03)^{12} = £1425.76$$

$$\text{interest paid} = £1425.76 - £1000 = £425.76$$

From this we can calculate the APR:

$$\text{APR} = \frac{\text{total interest paid in one year}}{\text{initial balance}} \times 100$$

$$= \frac{£425.76}{£1000} \times 100 = 42.58\%$$

We can develop a formula for the calculation of the APR. The balance outstanding at the end of one year is given by

$$A_0 \left(1 + \frac{r}{100} \right)^m$$

where m is the number of payment periods, usually 12 months, and r has become the monthly rate of interest. The total interest paid is given by

$$A_0 \left(1 + \frac{r}{100} \right)^m - A_0$$

To calculate the annual increase we compare the two:

$$\text{APR} = \frac{A_0 \left(1 + \frac{r}{100} \right)^m - A_0}{A_0} \times 100\% = \left[\left(1 + \frac{r}{100} \right)^m - 1 \right] \times 100\%$$

Examples

If the monthly rate of interest is 1.5%, what is (a) the nominal rate per year and (b) what is the APR?

(a) The nominal rate $= 1.5\% \times 12 = 18\%$
(b) APR $= [(1 + 1.5/100)^{12} - 1] \times 100\% = 19.56\%$

If the nominal rate of interest is 30% per annum but interest is compounded monthly, what is the APR?

The monthly nominal rate $= 30\%/12 = 2.5\%$

APR $= [(1 + 2.5/100)^{12} - 1] \times 100\% = 34.49\%$
What monthly repayment will be required to repay a mortgage of £30 000 over 25 years given an APR of 13%?

To obtain a monthly rate, $r\%$, let

$$[(1+r/100)^{12} - 1] \times 100\% = 13\%$$

$$(1+r/100)^{12} - 1 = 0.13$$

$$r/100 = (1+0.13)^{1/12} - 1$$

$$r = 1.024$$

To calculate the monthly repayment, x (see section 17.5.3), let $A_0 = -£30\,000$, $r = 1.024$ and $t = 25 \times 12 = 300$. Then

$$x = £30\,000 \times \frac{\dfrac{0.010\,24}{1}}{1 - \dfrac{1}{(1.010\,24)^{300}}} = £322.37$$

18.7 Conclusions

Throughout this chapter we have made assumptions about payments and interest, but how realistic have these been? An early assumption was that interest was paid or money received at the end of a year – this is clearly not always the case, but the assumption was made to simplify the calculations (and the algebra!). For compound interest we have shown how to incorporate more frequent payments, and the same principle could be applied to all of the other calculations in this chapter.

The calculations shown have also assumed that the interest rate remains constant for the period of time given, usually several years. As we have seen in the UK and elsewhere, interest rates do fluctuate and, at times, change rapidly. Businesses know and expect interest rates to fluctuate within certain limits and allow for this. When evaluating a project, managers may consider a number of scenarios perhaps using a spreadsheet model. A few contracts do involve fixed interest rates, e.g. hire-purchase agreements, but the vast majority of business contracts have variable interest rates. If we try to incorporate these variable rates into our calculations of, say, net present value, then we will need to estimate or predict future interest rates. These predictions will increase the uncertainty in the figures we calculate. As we have already noted, the higher the interest or discount rate, the less likely we are to invest in projects with a long-term payoff. However, since the interest rates charged to borrowers and lenders tend to change together over time, the opportunity cost of using or borrowing money should not be much affected.

The calculations can be seen as providing a basis for making business decisions but they do assume that projects can be evaluated completely in these financial terms. Management may need to take account of increased concerns about the environment and future legislation. Decisions may be seen as part of a long-term strategy and not taken in isolation. The 'big unknown' remains the future. It is not possible to predict all the changes in markets, competition and technology. Indeed the major challenge is responding to the changes and being flexible to future requirements.

Case study

The way projects are identified and evaluated will vary between businesses and between parts of the same business. In the case of the Ressembler Group, it is likely that the methods shown are used at a corporate level and for major new developments. However, it is not clear what criteria individual managers would use and whether they would be more concerned with meeting short-term targets. In markets where the level of uncertainty is high and potential profits are also high, such as holiday-home development in mainland Europe or video hire, managers would need to explore their attitude to risk and level of return. Some managers may consider only the level of revenue, with little attention to costs; others might use the popular payback method. If the payback method is used, the managers would only look to see how long it takes the cash proceeds to equal the initial outlay. In cases like this, a business would need to consider its strategic requirements and the way managers understand them.

18.8 Problems

1 How would you explain the terms nominal and real to a manager who was unfamiliar with them?

2 What factors do you think should be considered when investment decisions are being made?

3 Should a company try to include environmental (or social) cost in its calculation of net present value. How could these be incorporated into the calculations?

4 At certain times in the past, money lending for interest (usury) has been regarded as morally wrong. How would you justify the practice today for individuals and the business world?

5 A sum of £248 has been invested at an interest rate of 12% per annum for 4 years, What is the value of this investment, if interest is paid (a) as simple interest and (b) compounded each year?

6 An investment of £10 000 has been made on your behalf for the next 5 years. How much will this investment be worth if:
(a) the rate of interest is 10% per annum
(b) interest is paid at 7% per annum for the first £1000, 9% per annum for the next £5000 and 12% per annum for the remainder
(c) the rate of interest is 9% per annum but paid on a 6-monthly basis?

7 How much would an investment of £500 accumulate to in three years if interest were paid at 6% per annum for the first year, 8% per annum for the second year and 12% per annum for the third year?

8 A car is bought for £5680. It loses 15% of its value immediately and 10% per annum thereafter. How much is this car worth after 3 years?

9 A company buys a machine for £7000. If depreciation is allowed for at a rate of 16% per annum, what will be the value of the machine in 4 years' time?

10 You are offered £400 now or £520 at the end of 5 years. You know that 8% per annum is the highest rate of interest you can get. Which offer should you accept?

11 A firm is trying to decide between two projects which have the following cash flows:

| Project | Year | | | |
	1	2	3	4
I	£10 000	£5000	£6000	£4000
II	£12 000	£4000	£4000	£4000

If project I is discounted at 15% and project II at 20%, which project should be chosen?

12 A company has to replace a current production process. The current process is rapidly becoming unreliable whereas demand for the product is growing. The company must choose between alternatives to replace the process. It can buy

(a) either a large capacity process now at a cost of £4 million, or

(b) a medium capacity process at a cost of £2.2 million and an additional medium capacity process, also at a cost of £2.2 million, to be installed after three years.

The contribution to profit per year from operating the two alternatives are:

	Contribution (£m) at year end					
	1	2	3	4	5	6
Large process	2.0	2.3	2.8	2.8	2.8	2.8
2 medium processes	2.0	2.0	2.0	2.4	2.8	2.8

Assume a discount rate of 20%. Present a discounted cash flow analysis of this problem and decide between the alternatives.

Comment on other factors, not taken into account in your discounted cash flow analysis, which you think may be relevant to management's decision.

13 You have decided to save £200 at the end of each year for the next 5 years. How much will you have at the end of the 5 years if you are paid interest of 10% per annum?

14 You have decided to save £200 at the beginning of each year for the next 5 years. How much will you have at the end of the 5 years if you are paid interest of 10% per annum?

15 A sum of £5000 was invested 4 years ago. At the end of each year a further £1000 was added. If the rate of interest paid was 12% per annum, how much is the investment worth now?

16 You require £4000 in 5 years' time. How much will you have to invest at the end of each year if interest charged is 15% per annum?

17 A company has borrowed £5000 to be repaid in equal end of year payments over 10 years. What will the annual repayment be if the interest charged is 15% per annum?

18 A customer credit scheme charges interest at 2% per month compounded. Calculate the true annual interest rate of interest, i.e. the rate which would produce an equivalent result if interest were compounded annually.

19 Calculate the annual percentage rate (APR) of (a) 1.75% per month compound, (b) 5% per quarter compound and (c) 8% per half year compound.

20 Determine the monthly rate of interest compound given an APR of 26%.

21 Given that an investment of £5000 now produces a return of £2000 at the end of years 1, 2 and 3, determine the Internal Rate of Return (IRR) using the discount rates of 9% and 10% as the basis for any calculations.

Proof that

$$S = A_0 \left(1 + \frac{r}{100}\right)^t + \frac{x\left(1 + \frac{r}{100}\right)^t - x}{\frac{r}{100}}$$

where S is the sum at the end of t years, A_0 is the initial investment and x is the amount added at the end of each year.

The value of the investment at the end of the first year is

$$S_1 = A_0 \left(1 + \frac{r}{100}\right)^1 + x$$

The value at the end of the second year is

$$S_2 = \left[A_0 \left(1 + \frac{r}{100}\right)^1 + x\right]\left(1 + \frac{r}{100}\right)^1 + x$$

$$= A_0 \left(1 + \frac{r}{100}\right)^2 + x\left(1 + \frac{r}{100}\right)^1 + x$$

If we continue in this way, it can be shown that the value after t years is

$$S = A_0 \left(1 + \frac{r}{100}\right)^t + x\left(1 + \frac{r}{100}\right)^{t-1}$$

$$+ x\left(1 + \frac{r}{100}\right)^{t-2} + \cdots + x$$

This can be simplified using the summation formula for a geometric progression to give

$$S = A_0 \left(1 + \frac{r}{100}\right)^t + x\left[\frac{1 - \left(1 + \frac{r}{100}\right)^t}{1 - \left(1 + \frac{r}{100}\right)}\right]$$

$$= A_0 \left(1 + \frac{r}{100}\right)^t + x\left[\frac{1 - \left(1 + \frac{r}{100}\right)^t}{\frac{-r}{100}}\right]$$

$$= A_0 \left(1 + \frac{r}{100}\right)^t + \frac{x\left(1 + \frac{r}{100}\right)^t - x}{\frac{r}{100}}$$

19   Linear programming

Linear programming describes graphical and mathematical procedures that seek the optimum allocation of scarce or limited resources to competing products or activities. It is one of the most powerful techniques available to the decision-maker and has found a range of applications in business, government and industry. The determination of an optimum production mix, media selection and portfolio selection are just a few possible examples. They all require definition and, for a numerical solution, mathematical formulation. Typically, the objective is either to maximize the benefits while using limited resources or to minimize costs while meeting certain requirements.

This chapter looks particularly at the **formulation** of linear programming problems and shows the graphical solution to two variable problems. For more complex problems, involving three or more variables, it is more usual to employ computer packages, and we will look at the solution of linear programming problems using Excel.

Objectives   After working through this chapter, you should be able to:

- identify when linear programming provides an appropriate means of analysis
- formulate a problem in linear programming terms
- solve two-variable linear programming problems using the graphical method
- understand computer-based solutions and interpret the print-out from such packages.

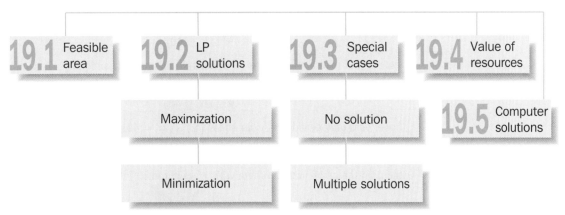

Figure 19.1   Structure of Chapter19.

Case study

The linear programming method could find a number of applications within the Ressembler Group. At a strategic level, the group would be interested in how a portfolio of activities could maximize returns subject to a number of constraints including acceptable levels of overall risk. At an operational level, managers might be interested in how to use available labour time, finance, materials and machine time.

19.1 Definition of a feasible area

If a company needs to decide what to produce, as a matter of good management practice it would want to know all the possible options. In mathematical terms, it would want a definition of a **feasible space** or **feasible area**.

Case study

Suppose a small group of managers from one of the businesses within the Ressembler Group are meeting to discuss how linear programming works and are considering a simple hypothetical problem. The problem involves the production of two products, X and Y. Each unit of X requires 1 hour of labour and each unit of Y requires 2 hours of labour.

Labour hours, in this case, are resource requirements, X and Y the competing products. If all the labour required were available at no cost, there would be no scarcity and no production problem. However, if only 40 hours of labour were available each week then there would be an allocation problem. A decision would have to be taken as to whether only X, or only Y, or some **combination** of the two should be produced. In mathematical terms this **constraint** is written as an inequality:

$$x + 2y \leq 40$$

where $\leq$ is read as '**less than or equal to**'. However, it is more usual to use a capital notation when linear programming:

$$X + 2Y \leq 40$$

If we were only dealing with the equation $X + 2Y = 40$, then all of the points on the line representing this equation on a graph would provide possible solutions. To plot this line, we would typically find **two points** that provide solutions, mark them on the graph and join them with a straight line. Two possible solutions are:

$$X = 0, \quad Y = 20$$
$$X = 40, \quad Y = 0$$

An interpretation of these two solutions would be that if **only Y** is to be produced then 20 units can be made and if **only X** is to be produced then 40 units can be made. Another possible solution would be to produce 10 units of X and 15 units of Y. All three solutions are shown in Figure 19.2.

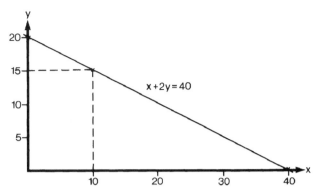

Figure 19.2 The definition of a line.

In the same way that an equation can be represented by a line, an inequality can be represented by an **area**. If we consider the example $X + 2Y \leq 40$, the three points

$$X = 0, \quad Y = 20$$
$$X = 40, \quad Y = 0$$
$$X = 10, \quad Y = 15$$

still provide solutions. In addition to these points, others that give an answer of less than 40 are also acceptable. The point $X = 20$, $Y = 5$ (answer 30) is acceptable whereas the point $X = 20$, $Y = 15$ (answer 50) is not. The 'less than or equal to' inequality defines an area that lies to the **left** of the line as shown in Figure 19.3.

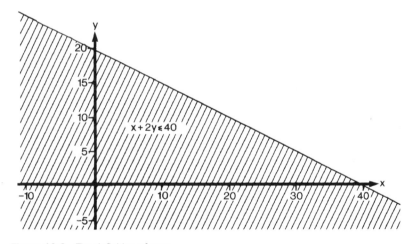

Figure 19.3 The definition of area.

Inequalities can also take the form:

$$X \leq 20, \quad Y \leq 15$$

The area jointly defined by these two inequalities is shown in Figure 19.4.

The definition of what is possible or feasible can be represented mathematically by a number of inequalities and **together** they can define a feasible area.

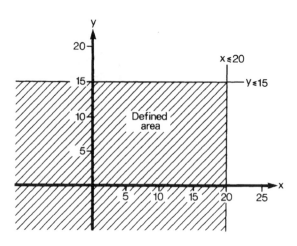

Figure 19.4 Area defined by two inequalities.

19.2 The solution of a linear programming problem

All linear programming problems have three common characteristics:

- a linear objective function
- a set of linear structural constraints
- a set of non-negativity constraints.

The **objective function** is a mathematical statement of what management wishes to achieve. It could be the maximization of profit, minimization of cost or some other measurable objective. Linearity implies that the parameters of the objective function are **fixed**, for example, a constant cost per unit or constant contribution to profit per unit. The structural constraints are the **physical limitations** on the objective function. They could be constraints in terms of budgets, labour, raw materials, markets, social acceptability, legal requirements or contracts. Linearity means that all these constraints have **fixed coefficients** and can be represented by straight lines on a graph. Finally, the **non-negativity** constraints limit the solution to positive (and meaningful) answers only.

19.2.1 The maximization problem

Typically a maximization problem will be concerned with the maximization of profit or contribution to profit but they can also be concerned with production volumes and market share.

Case study

Managers within one of the Ressembler Group of companies want to know how to maximize the profits from two types of frame, X and Y. Each frame X requires 1 hour of labour and 6 litres of moulding material, whereas each unit of Y requires 2 hours of labour and 5 litres of moulding material. The total number of labour hours available each week is 40 and the total amount of moulding material available each week is 150 litres. The profit contribution from frame X is £2 and from frame Y is £3.

In this case, the problem is to maximize the contribution to profit subject to two production constraints. This linear programming problem can be formulated mathematically as:

maximize

$$z = 2X + 3Y \quad \text{the linear objective function}$$

subject to

$$X + 2Y \leq 5$$
$$6X + 5Y \leq 150 \quad \text{the linear structural constraints}$$

and

$$X \geq 0, \quad Y \geq 0 \quad \text{the non-negativity constraints}$$

The **objective function** is the sum of the profit contributions from each product. If we were to decide to produce five units of X and seven units of Y then the total contribution to profit would be $z = 2 \times 5 + 3 \times 7 = £31$. Linear programming provides a method to find what combination of X and Y will maximize the value of z subject to the given constraints.

In this example there are two **structural constraints**, one in terms of available labour and the other in terms of available raw materials. The usefulness of the solutions will depend on how realistically the constraints model the decision problem. If, for example, marketing considerations were ignored the optimum solution may suggest production levels that are incompatible with sales opportunities. Mathematically we need a structural constraint to correspond to each limitation on the objective function. If we consider the labour constraint as an example, the coefficients represent the labour requirements of one hour for product X and 2 hours for product Y. A product mix of five units of X and seven units of Y will require only 19 hours of labour $(1 \times 5 + 2 \times 7)$, does not exceed the available labour time of 40 hours and satisfies the first constraint. A production mix of 18 units of X and 13 units of Y exceeds the available labour time, does not satisfy the constraint and therefore could not provide a possible solution. The product mix of five units of X and seven units of Y also satisfies the remaining structural constraint, $6 \times 5 + 5 \times 7 \leq 150$, and is one of the possible solutions and lies within the feasible area. In this example, the feasible area is defined by the labour and raw material constraints. No account is taken of the many other factors that could affect the optimum production mix.

The feasible area is found graphically by treating each constraint as an equation, plotting the corresponding straight lines and defining an area bounded by the straight lines which satisfies all the inequalities. We proceed as follows.

The **labour constraint**: if we were to use all the labour time available then:

$$X + 2Y = 40$$

$$(X = 0, Y = 20) \quad (X = 40, Y = 0)$$

The two points shown in brackets are the **one product solutions**; we can use the 40 hours of labour to make 20 units of Y each week or 40 units of X. A line joining the two points will show combinations of X and Y that will require 40 hours of labour.

The **raw materials constraint**: if we were to use all the raw material available then:

$$6X + 5Y = 150$$

$$(X = 0, Y = 30) \quad (X = 25, Y = 0)$$

A line joining two possible solutions shown in brackets will show the combinations of X and Y that will require 150 litres of raw materials.

The **non-negativity constraints**: the constraints $x \geq 0$ and $y \geq 0$ exclude any possibility of negative production levels which have no physical counterpart. Together they include the x-axis and the y-axis as possible boundaries of the feasible area.

The feasible area defined by the two structural constraints and the two non-negativity constraints is shown in Figure 19.5. The feasible area is contained within the boundaries of OABC. It is now a matter of deciding which of the points in this area provides an optimum solution. The choice is determined by the objective function. In this example, the choice of whether to produce just X, or just Y or some combination of the two will depend on the **relative profitability** of the two products.

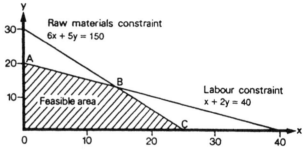

Figure 19.5 The definition of a feasible area.

Profit has been expressed as a mathematical function $z = 2x + 3y$, where z is the profit level. If we were to fix the level of profits, z, the necessary combination of X and Y could be shown graphically as a straight line. This is referred to as a trial profit line. If we let $z = £30$ then:

$$30 = 2X + 3Y$$

$$(X = 0, Y = 10) \quad (X = 15, Y = 0)$$

A profit of £30 can be made by producing 15 units of X, or 10 units of Y, or some combination of the two. This trial profit is shown in Figure 19.6. All the points on the **trial profit line** will produce a profit of £30. The gradient gives the **trade-off** between the two products. In this case, to maintain a profit level you would need to trade two units of Y against three units of X (a loss of £6 against a gain of £6). The trial profit line shown violates none of the constraints so profit can be increased from £30.

Consider a second trial profit line where the profit level is fixed at £60.

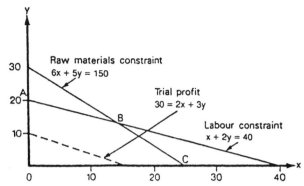

Figure 19.6   A trial profit ($z = £30$).

If we let $z = £60$ then:

$$60 = 2X + 3Y$$

$$(X = 0, Y = 20) \quad (X = 30, Y = 0)$$

This trial profit line is shown in Figure 19.7.

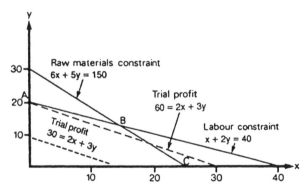

Figure 19.7   Trial profit fixed at £60.

This second trial profit line is higher than and parallel to the first. It can be seen that a profit of £60 can be achieved at point A, and that some points on the trial profit line violate the raw materials constraint. To operate at point A would exhaust all the available labour hours but leave surplus raw materials. This solution can be improved upon by **trading-off** the more labour intensive product Y against the more raw material intensive product X. In terms of the graphical approach, we can note that any line that is higher and parallel to the existing trial profit line represents an improvement. By inspection we can see from Figure 19.7 that higher trial profit lines will eventually lead to the **optimum** point B. This point B $(X = 14\frac{2}{7}, Y = 12\frac{6}{7})$ can be determined directly from the graph, or by **simultaneously solving** the two equations that define lines crossing at point B. The resultant profit is found by substitution into the objective function:

$$z = 2 \times 14\frac{2}{7} + 3 \times 12\frac{6}{7} = £67.14 \text{ per week}$$

19.2.2 The minimization problem

At a departmental level in a profit-making organization or in the non-profit sector the concern may be to deliver a product or service at minimum cost. The approach is illustrated in the following example.

Example

A company operates two types of aircraft, the RS101 and the JC111. The RS101 is capable of carrying 40 passengers and 30 tons of cargo, whereas the JC111 is capable of carrying 60 passengers and 15 tons of cargo. The company is contracted to carry at least 480 passengers and 180 tons of cargo each day. If the cost per journey is £500 for a RS101 and £600 for a JC111, what choice of aircraft will minimize cost?

This linear programming problem can be formulated mathematically as:
Minimize

$$z = 500X + 600Y \quad \text{the linear objective function}$$

where X is the number of RS101s and Y is the number of JC111s subject to the constraints

$$\left.\begin{array}{c} 40X + 60Y \geq 480 \\ 30X + 15Y \geq 180 \end{array}\right\} \quad \text{the linear structural constraints}$$

and

$$X \geq 0, Y \geq 0 \quad \text{the non-negativity constraints}$$

In this case we are attempting to minimize the cost of a service subject to the operational constraints. These structural constraints, the requirement to carry so many passengers and so many tons of cargo, are expressed as 'greater than or equal to'. The inequalities are again used to define the feasible area.

The **passenger constraint**: if we were to carry the minimum number of passengers then

$$40X + 60Y = 480$$

$$(X = 0, Y = 8)(X = 12, Y = 0)$$

We could use 8 JC111s to carry 480 passengers or 12 RS101s or some combination of the two as given by the above equation.

The **cargo constraint**: if we were to carry the minimum amount of cargo then

$$30X + 15Y = 180$$

$$(X = 0, Y = 12)(X = 6, Y = 0)$$

We could use 12 JC111s to carry 180 tons of cargo or 6 RS101s or some combination of the two as given by the above equation.

The **non-negativity constraints**: to ensure that the solution excludes negative numbers of aircraft, $X \geq 0$ and $Y \geq 0$ are included as possible boundaries of the feasible area.

The resultant feasible area is shown in Figure 19.8. The objective is to locate the point of minimum cost within the feasible area.

We proceed as before, by giving a convenient value to z which defines a trial cost line, but in this case attempting to make the line as near to the origin as possible, while retaining at least one point within the feasible area so as to minimize cost. If we let $z = £6000$ then

$$6000 = 500X + 600Y$$

$$(X = 0, Y = 10) \quad (X = 12, Y = 0)$$

A cost of £6000 will be incurred by operating 12 RS101s, or 10 JC111s or a combination of the two as defined by the above equation, including combinations which are non-feasible. This trial cost line is shown in Figure 19.9.

All lines *lower* than and *parallel* to the trial cost line show aircraft combinations that produce lower costs. By inspection we can see that point B is the point of lowest cost ($X = 3, Y = 6$). The level of cost corresponding to the use of 3 RS101s and 6 JC111s can be determined from the objective function:

$$z = 500 \times 3 + 600 \times 6 = £5100 \text{ per day}$$

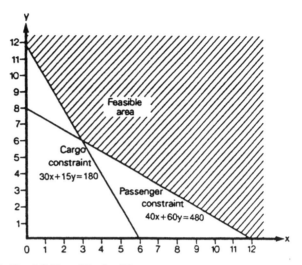

Figure 19.8 The definition of the feasible area.

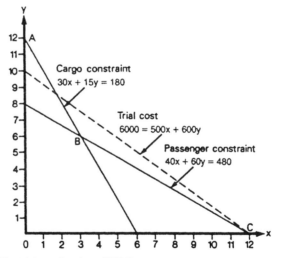

Figure 19.9 The trial cost line ($z = £6000$).

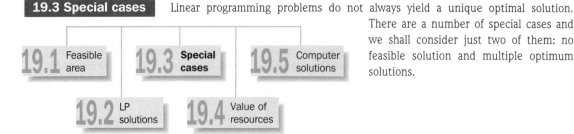

19.3 Special cases Linear programming problems do not always yield a unique optimal solution. There are a number of special cases and we shall consider just two of them: no feasible solution and multiple optimum solutions.

19.3.1 No feasible solution If the constraints are mutually exclusive, no feasible area can be defined and no optimum solution can exist. Consider again the maximization problem.

Maximize

$$z = 2X + 3Y \text{ (the linear objective function)}$$

subject to

$$\left.\begin{array}{r} X + 2Y \leq 40 \\ 6X + 5Y \leq 150 \end{array}\right\} \quad \text{the } \textbf{linear structural constraints}$$

and

$$X \geq 0, Y \geq 0 \quad \text{the non-negativity constraints}$$

The feasible area has been defined by the constraints as shown earlier in Figure 19.5. Suppose that **in addition** to the existing constraints the company is contracted to produce at least 30 units each week. This additional constraint can be written as:

$$X + Y \geq 30$$

As a boundary solution the constraint would be:

$$X + Y = 30$$

$$(X = 0, Y = 30) \quad (X = 30, Y = 0)$$

The three structural constraints are shown in Figure 19.10.

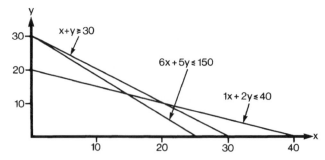

Figure 19.10 No feasible solution.

This case presents the manager with demands that cannot simultaneously be satisfied.

19.3.2 Multiple optimum solutions

A multiple optimum solution results when the objective function is parallel to one of the boundary constraints. Consider the following problem:
Minimize

$$z = 600X + 900Y \text{(the linear objective function)}$$

subject to

$$\left.\begin{array}{c} 40X + 60Y \geq 480 \\ 30X + 15Y \geq 180 \end{array}\right\} \text{ the linear structural constraints}$$

and

$$X \geq 0, Y \geq 0 \quad \text{the non-negativity constraints}$$

This is the aircraft scheduling problem from Section 19.2.2 with different cost parameters in the objective function. If we let $z = \pounds 8100$ then

$$8100 = 600X + 900Y$$

$$(X = 0, Y = 9) \quad (X = 13.5, Y = 0)$$

The resultant trial cost line is shown in Figure 19.11.

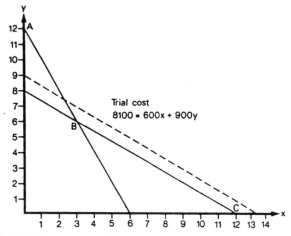

Figure 19.11   Multiple optimum solutions.

This line is **parallel** to the boundary line BC. The lowest acceptable cost solution will be coincidental with the line BC making point B, point C and any other points on the line BC optimal. Multiple optimum solutions present the manager with **choice** and hence some flexibility.

19.4 The value of resources

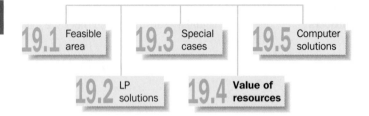

19.1 Feasible area   19.3 Special cases   19.5 Computer solutions

19.2 LP solutions   19.4 **Value of resources**

Linear programming provides a method for evaluating the marginal value of resources. Consider yet again the maximization problem.
Maximize

$$z = 2X + 3Y$$

subject to

$$X + 2Y \leq 40$$

$$6X + 5Y \leq 150$$

and

$$X \geq 0, Y \geq 0$$

In this case, the solution is **limited by** the 40 hours of labour and the 150 litres of moulding material. To assess the value of additional resources we can consider what difference it would make if we could provide an **extra** hour of labour or an **extra** unit of moulding material. The amount added to profit in this case (or more generally, z in the objective function) as a result of the additional unit of resource is seen as the marginal value of the resource and is referred to as the **opportunity cost or the shadow price.**

To determine the shadow price of labour we would increase the hours available from 40 to 41. The linear programming formulation now becomes:
Maximize

$$z = 2X + 3Y$$

subject to

$$X + 2Y \leq 41$$

$$6X + 5Y \leq 150$$

and

$$X \geq 0, Y \geq 0$$

This type of marginal analysis is difficult to show graphically because of the small movements involved. Effectively, the labour constraint has moved **outwards** and can be plotted using the points ($X = 0$, $Y = 20.5$) and ($X = 41$, $Y = 0$). The new solution is $X = 13\frac{4}{7}$ and $Y = 13\frac{5}{7}$. The new level of profit can be found by substitution into the objective function:

$$z = \pounds2 \times 13\tfrac{4}{7} + \pounds3 \times 13\tfrac{5}{7} = \pounds68.29$$

The increase in profit resulting from the additional hour of labour, or shadow price of labour, is the difference between the new profit and the old profit $= \pounds68.29 - \pounds67.14 = \pounds1.15$.

To determine the shadow price of materials we would increase the number of litres available from 150 to 151. The linear programming formulation now becomes:
Maximize

$$z = 2X + 3Y$$

subject to

$$X + 2Y \leq 40$$

$$6X + 5Y \leq 151$$

and

$$X \geq 0, Y \geq 0$$

In this case it is the material constraint that would move outwards while the labour constraint remained unchanged at 40 hours. To plot the new material constraint the points $(X = 0, Y = 30.2)$ and $(X = 25.17, Y = 0)$ may be used. The new solution is $X = 14\frac{4}{7}$ and $Y = 12\frac{5}{7}$. As before, the new level of profit can be found by substitution into the objective function:

$$z = \pounds2 \times 14\tfrac{4}{7} + \pounds3 \times 12\tfrac{5}{7} = \pounds67.29$$

The increase in profit resulting from the additional litre of moulding material, or shadow price of material, is the difference between the new profit and the old profit $= \pounds67.29 - \pounds67.14 = \pounds0.15$.

If the manager were to pay below £1.15 for the additional hour of labour (unlikely to be available at these rates!), then profits could be increased, and if the manager were to pay above this figure then profits would decrease. Similarly, if the manager can pay below £0.15 for an additional unit of moulding material, then profits can be increased but if the manager were to pay above this level then profits would decrease.

It is useful to see the effect of increasing both labour and materials by one unit. The linear programming formulation now becomes:
Maximize

$$z = 2X + 3Y$$

subject to

$$X + 2Y \leq 41$$

$$6X + 5Y \leq 151$$

and

$$X \geq 0, \quad Y \geq 0$$

The new solution is $X = 13\frac{6}{7}$ and $Y = 13\frac{4}{7}$. The new level of profit is:

$$z = £2 \times 13\frac{6}{7} + £3 \times 13\frac{4}{7} = £68.43$$

The increased profit is $£68.43 - £67.14 = £1.29$.

This increased profit (subject to the small rounding error of £0.01) is the sum of the shadow prices ($£1.15 + £0.15$). It should be noted that the shadow prices calculated only apply while the constraints continue to work in the same way. If, for example, we continue to increase the supply of moulding material (because it can be obtained at a market price below the shadow price), other constraints may become active and change the value of the shadow price.

19.5 Computer-based solutions

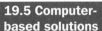

As we have seen, problems involving two variables can be solved graphically. We can consider 'special cases' (Section 19.3) by an examination of the feasible area and explore the effects of changing parameters. Shadow prices for each of the resources (constraints) can be determined by looking at the increase in the objective function (z) when the constraint totals are increased by one – marginal analysis. We think you will agree that this is a lot of work using graphs and equations. Even if we were prepared to do all this work, the graphical method would not help us to solve problems with three or more variables. Computer-based solutions must offer a quicker and better way of getting results, but what they cannot do is to formulate the problem. That is up to you.

For the most complex of problems you would probably use a specialist package, but Excel offers a built-in function which can deal with the more basic linear programming problems.

Let's again consider the following maximization problem:
Maximize

$$z = 2X + 3Y$$

subject to

$$X + 2Y \leq 40$$

$$6X + 5Y \leq 150$$

and

$$X \geq 0, \quad Y \geq 0$$

which we can solve using Excel. Other solution software is available but Excel offers easier access than most.

As a first step make sure that **Solver** is one of the loaded Add-in's on the computer that you are using. You can do this by opening Excel and clicking on the Tools menu. If the word Solver does not appear then it will be necessary to load the Add-in. (You can do this by clicking on the Tools menu, then clicking on Add-Ins, and then putting a tick in the box next to Solver Add in. This may not be possible if you are running on a network; see your technician.)

It will be necessary to set up the spreadsheet with the various values and parameters in the cells. (It is also a good idea to label the cells in some way.) You are then ready to use the Solver. Using the maximization problem, we have set up the spreadsheet with the labels in columns A and B, the variables and constraints in column C and the limits in column D. This is shown in Figure 19.12.

As you can see, the cell references are put into the various parts of the window, and then you click **Solve**.

Computer print-outs will vary according to the package used, but will have a number of common features. In our example, we have referred to variables X and Y, but some packages will use X_1 and X_2 (and extend this notation to X_3 and beyond) unless the variables are labelled at data entry stage. Excel produces three additional worksheets when you solve a linear programming problem. The first of these is the Answer Report. For our problem, this is shown in Figure 19.13.

The solution of $X = 14.285714$ and $Y = 12.857142$ corresponds to our previous answer of $14\frac{2}{7}$ and $12\frac{6}{7}$. The screen also gives the value of the objective function.

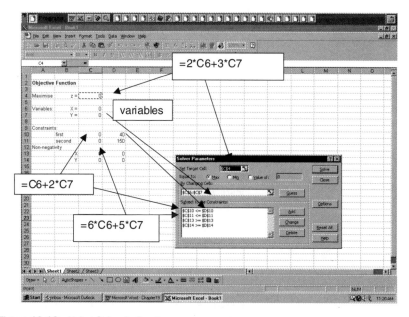

Figure 19.12 Using Solver in Excel.

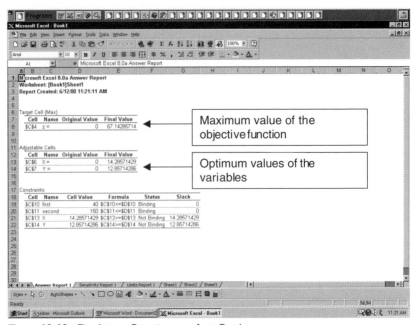

Figure 19.13   The Answer Report screen from Excel.

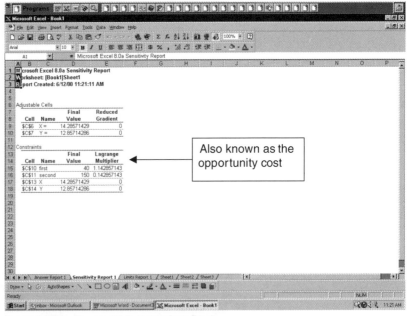

Figure 19.14   The sensitivity report from Excel.

Having found the optimal solution, we are interested in how sensitive this is to changes. To find out further information, we can look at the next screen, see Figure 19.14.

This screen tells us if there are any constraints that are not restricting us. In this case, the final value on both the first and second constraint is equal to the amount we had available, and therefore both constraints, in this case labour and materials, are active.

The formal version of this has S_1 and S_2 as slack variables, and referring to the two constraints. Essentially the slack variables allow us to make the two inequalities:

$$X + 2Y \leq 40$$

$$6X + 5Y \leq 150$$

into equations:

$$X + 2Y + S_1 = 40$$

$$6X + 5Y + S_2 = 150$$

The slack variables show how much of each resource is not being used. You will see from the screenshot that the 'slack' in each case is 0 (we are using all available labour and materials) and the shadow prices are £1.14 and £0.14 (allowing for rounding errors when making comparisons). The opportunity cost, or shadow price, is the value of an extra unit of that resource. So for labour, if we can get extra labour at less than £1.14 per hour, then it will be worthwhile and will increase the overall profit, as we showed above. A similar logic applies to extra litres of materials.

Case study

The Ressembler Group has acquired a small business making picture frames. It is not looking for any synergy between this acquired business and existing businesses but believes better management practice, as adopted elsewhere, will improve the business performance of this framing business. More attention will be given to financial control and production detail. The business currently produces two types of picture frame: the 'Mendip' and the 'Cotswold'. The time available for some of the key manufacturing stages is limited. At present only 2000 minutes of assembly time, 1000 minutes of paint time and 2500 minutes of packaging time is available each week. The following table was produced to show the time required to produce one frame in the three manufacturing stages.

	Time (in minutes)	
	Mendip	Cotswold
Assembly	2	2
Paint	2	2
Packaging	8	2

In addition, it is known that market conditions will limit the sale of Mendip to 300 or less each week.

The Mendip has a variable cost of £1.20 per frame and a selling price of £4.80 per frame, while the Cotswold has a variable cost of £2.10 per frame and a selling price of £4.50 per frame.

This is a two-product (or variable) problem and could be solved using the graphical method with additional calculations providing the shadow price and sensitivity analysis. The formulation is as follows:

Maximize

$$z = 3.60X_1 + 2.40X_2$$

where X_1 and X_2 are the products Mendip and Cotswold, and £3.60 is the profit contribution from the Mendip (£4.80 − £1.20) and £2.40 is the profit contribution from the Cotswold (£4.50 − £2.10)
subject to:

$$2X_1 + 2X_2 \leq 2000$$

$$2X_1 + 2X_2 \leq 1000$$

$$8X_1 + 2X_2 \leq 2500$$

$$X_1 \leq 300$$

and

$$X_1 \geq 0, \quad X_2 \geq 0$$

Exercise

> Solve this problem using the graphical method.
>   What comment would you make on the first constraint, assembly time?

However, it is more convenient to refer to a computer solution as shown in Figure 19.15.

It can be seen that the optimum solution is to produce 250 Mendip and 250 Cotswold each week, and that the contribution to profit would be £1500 per week. The first constraint, assembly, has 'slack' of 1000 minutes each week (obvious if you look at constraints 1 and 2 in the formulation) and the fourth constraint 50 units. You should recall that the fourth constraint was a marketing constraint on the sale of Mendip and that potentially we could be selling 50 more Mendip.

It should be noted that where 'slack' is available, S_1 and S_2, the opportunity cost or shadow price is 0 and the amount you can add at this price without changing the solution is unlimited (infinity). If you have spare resources, you will not pay extra for additional units and any extra units will not change your solution. An extra unit of paint time can add £1.00 to profit contribution and an extra unit of packaging time can add £0.20 (rounded) to the profit contribution. The opportunity cost or shadow price will also show how much profit contribution will fall, if a resource is reduced by one unit.

Case study

The managers of the small business making picture frames need to consider a proposal to produce a new type of frame to be called the 'Dale'. The Dale is expected to have a variable cost of £2.00 per frame and a selling price of £5.20 per frame. The new frame will require 3 minutes of assembly time, 2 minutes of paint time and 3 minutes of packaging time.

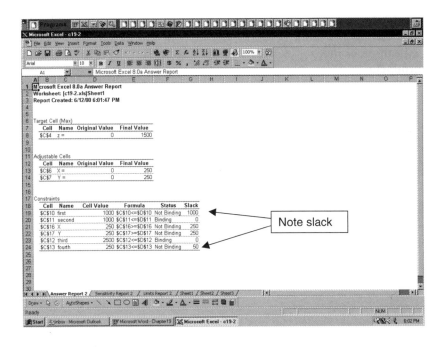

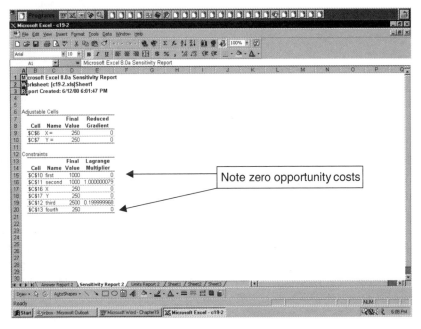

Figure 19.15 Solution screens for Ressembler problem.

There are two ways of approaching this problem; it can be reformulated and solved, or shadow prices can be examined.

19.5.1 Formulating and solving problems with three or more products

A problem involving three products is formulated in terms of X_1, X_2 and X_3. This notation can easily be extended to cover four or more products. The problem being considered by the Ressembler Group would be formulated as:

Maximize:

$$z = 3.60X_1 + 2.40X_2 + 3.20X_3$$

where X_3 is the number of Dale to be produced and £3.20 is the profit contribution (£5.20 − £2.00)
subject to

$$2X_1 + 2X_2 + 3X_3 \leq 2000$$

$$2X_1 + 2X_2 + 2X_3 \leq 1000$$

$$8X_1 + 2X_2 + 3X_3 \leq 2500$$

$$X_1 \leq 300$$

and

$$X_1 \geq 0, X_2 \geq 0, X_3 \geq 0$$

A typical computer-based solution is shown in Figure 19.16.

It can be seen that the solution is to produce 200 Mendip, 0 Cotswold and 300 Dale each week. It is not unusual to find a two-product formulation giving a one-product solution, or a three-product formulation giving a two-product solution, and so on. If one unit of the Cotswold, X_2, were brought into the solution, the total profit contribution would fall by £0.72 (rounded). It should be noted that constraints 1 and 4 both have 'slack' and therefore a zero opportunity cost or shadow price.

19.5.2 Using shadow price to evaluate new products

In the original two-product problem, the opportunity costs for assembly, painting and packaging were £0.00, £1.00 and £0.20 respectively. To produce one unit of Dale requires 3 minutes of assembly time, 2 minutes of paint time and 3 minutes of packaging time. The total opportunity cost of producing one Dale is therefore:

$$3 \times £0.00 + 2 \times £1.00 + 3 \times £0.20 = £2.60$$

To produce one unit of Dale will mean a reduction in the profit from other products of £2.60, but one unit of Dale will add a profit contribution of £3.20. This net gain of £0.60 per unit of Dale means that it is worth bringing Dale into production and reducing the other products. An examination of opportunity cost **does not provide the optimum solution**, but does provide a method of evaluating new products and indicating what type of new products are possible. In this case, the opportunity cost of assembly is 0, we have spare capacity, and it could be profitable to develop assembly-intensive new products. Alternatively, managers could consider ways of reducing the spare capacity in assembly and increasing the time available for painting and packaging.

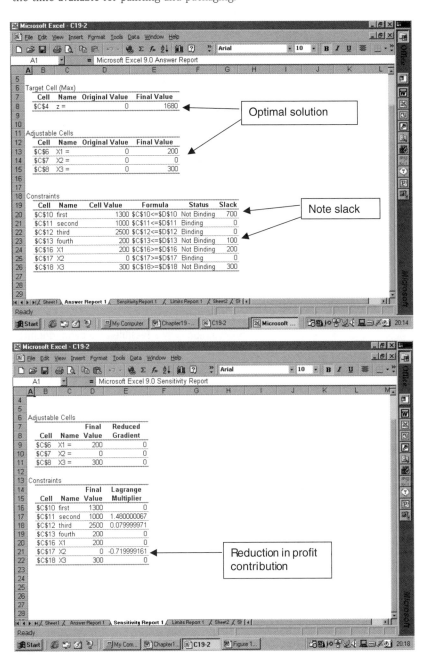

Figure 19.16 Solution to the three-variable problem.

19.6 Conclusions

Linear programming provides a way of formulating and solving a wide range of problems. If these problems are defined in terms of two variables, then the solution can be found using the graphical method. Of particular importance is the identification of the feasible area. In any given problem the feasible area may not exist or may not provide the manager with a satisfactory solution. In the longer term, the manager should **not only** seek the optimum solution but also seek to **improve on this solution** by the management of resources that constrain the solution.

In practice, problems are likely to have three or more variables, and computer-based solutions will have to be used. It is for this reason that writing out a problem as a set of equations and inequalities (formulation) is so important. Here we have only illustrated the solution of linear programming problem by the use of Excel; other, more specialized packages may give more detailed solutions. In addition to providing the optimum solution, typical print-outs will give the shadow prices (the opportunity cost of units of resource) and the sensitivity of the solution to changes in the parameters. A number of methods are available to solve the bigger linear programming problems, such as the **simplex method**, but these are beyond the scope of this book.

19.7 Problems

1 Show graphically the following inequalities:
 (a) $5X + 2Y \leq 80$
 (b) $X + Y \leq 30$
 (c) $X \leq 10$
 (d) $Y \geq 12$.

2 Show graphically a feasible area defined jointly by:
$$2Y + 5X \geq 140$$
$$X + Y \leq 40$$

3 (a) Maximize $z = 2x + 2y$

$$\text{subject to} \quad 3x + 6y \leq 300$$
$$4x + 2y \leq 160$$
$$y \leq 45$$
$$x \geq 0$$
$$y \geq 0$$

You should include in your answer the values of the optimum point and the corresponding value of z.
 (b) Determine the optimum solution if the objective function were changed to

$$z = x + 3y$$

 (c) Determine the optimum solution if the objective function were changed to

$$z = 5x + 10y$$

4 (a) Minimize $z = 2x + 3y$

$$\text{subject to} \quad 3x + y \geq 15$$
$$0.5x + y \geq 10$$
$$x + y \geq 13$$
$$x \geq 1$$
$$y \geq 4$$

(b) Identify the acting constraints (those that affect the solution).

(c) If the constraint $y \geq 4$ were changed to $y \geq 8$ how would this affect your solution?

5 A manufacturer of fitted kitchens produces two units, a base unit and a cabinet unit. The base unit requires 90 minutes in the production department and 30 minutes in the assembly department. The cabinet unit requires 30 minutes and 60 minutes respectively in these departments. Each day 21 hours are available in the production department and 18 hours are available in the assembly department. It has already been agreed that not more than 15 cabinet units are produced each day. If base units make a contribution to profit of £20 per unit and cabinet units £50 per unit, what product mix will maximize the contribution to profit and what is this maximum?

6 A confectionery company has decided to add two new lines to its product range: a chocolate fairy and a chocolate swan. It has been estimated that each case of chocolate fairies will cost £10.00 to produce and sell for £13.00, and that each case of chocolate swans will cost £11.40 to produce and sell for £15.00. Each case of chocolate fairies requires 30 oz of chocolate and each case of chocolate swans requires 50 oz. Only 450 oz of chocolate is available each day from existing facilities. It takes one hour of labour to produce either one case of chocolate fairies or chocolate swans and only 10 hours of labour are available each day.

(a) Formulate a linear programming model to maximize the daily profit from chocolate fairies and chocolate swans.

(b) Use the graphical linear programming method to solve the problem.

(c) Describe how the company could achieve an output of five cases of chocolate fairies and six cases of chocolate swans.

7 A company currently operates two types of van, the 'Loader' and the 'Superloader', which it uses for the transportation of pallets from its production plant to its distribution plant. The company fleet consists of four Loader vans and one Superloader which it needs to increase to meet increased demand. The Loader is capable of carrying 30 pallets a day and the Superloader 40 pallets a day with daily costs of £100 and £120 respectively. If demand estimates require the movement of 240 pallets each day, how should the fleet be increased (from the existing four Loaders and one Superloader) so as to minimize transportation costs?

8 A company has decided to produce a new cereal called 'Nuts and Bran' which contains only nuts and bran. It is to be sold in the standard size pack which must contain at least 375 g. To provide an 'acceptable' nutritional balance each pack should contain at least 200 g of bran. To satisfy the marketing manager, at least 20% of the cereal's weight should come from nuts. The production manager has advised you that nuts will cost 20 pence per 100 g

and that bran will cost 8 pence per 100 g.
(a) Formulate a linear programming model for the minimization of costs.
(b) Solve your linear programming model graphically.
(c) Determine the cost of producing each packet of cereal.

9 A production problem has been formulated as:
maximize:

$$z = 14X + 10Y$$

subject to

$$4X + 3Y \leq 240$$

$$2X + Y \leq 100$$

$$Y \leq 50$$

and

$$X \geq 0, \quad Y \geq 0$$

Explain the solution given by the following print-out.

		Final solution	

No.	Variable Names	Solution	Opportunity cost
1	X	+30.000000	0
2	Y	+40.000000	0
3	S_1	0	+3.0000000
4	S_2	0	+1.0000000
5	S_3	+10.000000	0

Maximized OBJ. = 820

	Sensitivity analysis for objective coefficients		
Variable	Min. $C(j)$	Original	Max. $C(j)$
X	+13.3333	+14.0000	+20.0000
Y	+7.00000	+10.0000	+10.5000

	Sensitivity analysis for RHS		
Constraint	Min. B(i)	Original	Max. B(i)
1	+200.000	+240.000	+250.000
2	+95.0000	+100.000	+120.000
3	+40.0000	+50.0000	+Infinity

10 A three-product problem has been formulated as:

maximize

$$z = 60X_1 + 70X_2 + 75X_3$$

subject to

$$3X_1 + 6X_2 + 4X_3 \leq 2400$$

$$5X_1 + 6X_2 + 7X_3 \leq 3600$$

$$3X_1 + 4X_2 + 5X_3 \leq 2600$$

and

$$X \geq 0, \quad Y \geq 0$$

Explain the solution given by the following print-out:

		Final solution	

Variable No.	Names	Solution	Opportunity cost
1	X_1	+500.00000	0
2	X_2	+150.00000	0
3	X_3	0	+.83333331
4	S_1	0	+3.3333333
5	S_2	+200.00000	0
6	S_3	0	+12.5000000

Maximized OBJ. = 40 500; Iteration = 4; Elapsed CPU second = .0625

	Sensitivity analysis for objective coefficients		

Variable	Min. $C(j)$	Original	Max. $C(j)$
X_1	+59.2857	+60.0000	+70.0000
X_2	+60.0000	+70.0000	+120.000
X_3	−Infinity	+75.0000	+75.8333

	Sensitivity analysis for RHS		

Constraint	Min. $B(i)$	Original	Max. $B(i)$
1	+1950.00	+2400.00	+3000.00
2	+3400.00	+3600.00	+Infinity
3	+1600.00	+2600.00	+2800.00

11 A company has just been awarded a contract to transport 594 000 litres of a liquid fertilizer each month and needs to make a decision on vehicle purchase. Two vehicles are known to be suitable, the Regular (denoted by X_1) and the Econ Tanker (denoted by X_2). The Regular has an estimated monthly operating cost of £800 and a capacity of 1980 litres. The Econ Tanker has a monthly operating cost of £1200 and a capacity 1500 litres. The number of deliveries made each month will depend on the vehicle, as they differ in speed and size. It has been estimated that each Regular will be able to make 20 deliveries each month and that each Econ Tanker will be able to make 36 deliveries each month. You have been allocated a budget of £500 000 for vehicle purchase and been told not to exceed this. The cost of a Regular is £25 000 and the cost of a Econ Tanker is £50 000. In addition, the company has made a requirement that at least five of the vehicles purchased are Econ Tankers, to meet existing contract commitments.

(a) Explain the solution, given in the following print-out.

	Final solution		
No.	Variable Names	Solution	Opportunity cost
1	X_1	+8.1818180	0
2	X_2	+5.0000000	0
3	S_1	+45454.547	0
4	S_2	0	+.02020202
5			
6	S_3	0	+109.09091
7			

Minimized OBJ. = 12545.46

(b) The company can now consider a third vehicle for purchase, the Jumbo Tanker. The Jumbo Tanker has an estimated monthly operating cost of £1500, has a capacity of 2800 litres and is able to make an estimated 16 deliveries each month. The cost of a Jumbo is £60 000. Should the company consider the purchase of the Jumbo?

12 A number of spreadsheets, Excel for example, can be used to solve linear programming problems. If possible, use a spreadsheet you are familiar with, to solve some of the problems in this section.

20 Networks

Any large project will involve the completion of a number of smaller jobs or tasks. Some of these tasks can be started straight away, some need to await the completion of other tasks and some tasks can be done in parallel. A network is a way of illustrating the various tasks and showing the relationship between them. A network can be used to show clearly the tasks or activities that need to be completed on time to keep the project on time and also those tasks that can be delayed without affecting the project time.

The technique of drawing up networks was developed during World War II and in the late 1950s in both the UK and the USA. In the UK it was developed by the Central Electricity Generating Board where its application reduced the overhaul time at a power station to 32% of the previous average. The US Navy independently developed the Programme Evaluation and Review Technique (PERT), while the DuPont company developed the critical path method, said to have saved the company $1 million in one year. All of these techniques are similar and have found applications in the building industry, in accountancy, and in the study of organizations, as well as their original uses.

Networks provide a planned approach to project management. To be effective, networks require a clear definition of all the tasks that make up the project and pertinent time estimates. If the project manager cannot clarify the necessary tasks and the resource requirements, then no matter how sophisticated the network, it will not compensate for these shortcomings. A number of claims have been made about the benefits of project management techniques but others have argued that in part, the benefits are due to managers having to know and clarify the tasks rather than the diagram which follows (which may by then be self-evident).

The objectives of network analysis are to locate the activities that must be kept to time, manage activities to make the most effective use of resources and look for ways of reducing the total project time. For any but the smallest projects, this analysis is likely to be done using a computer package, but your understanding of the output will only develop if you have some experience of the basic steps of analysis.

Objectives

After working through this chapter, you should be able to:

- construct a network diagram
- determine the earliest and latest start times
- identify the critical path
- explain the use of and calculate 'float' values
- construct Gantt charts
- suggest ways to manage the reduction of project time
- allow for uncertainty
- use and understand the printout from Microsoft Project.

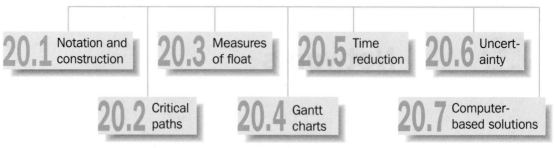

Figure 20.1   Structure of Chapter 20.

20.1 Notation and construction

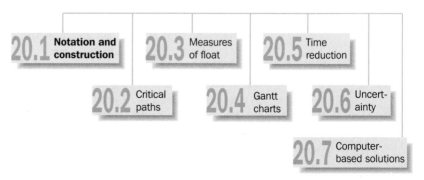

To construct and analyse networks you need to be familiar with the notation and language. An **activity** is a task or job that requires time and resources, such as counting the number of defective items, constructing a sampling frame or writing a report. An activity is represented by an unbroken arrow:

It should be noted that the method of network construction presented in this book is 'activity on the arrow' (the alternative is referred to as 'activity on the node').

An event or **node** is a point in time when an activity starts or finishes, for example, start counting the number of defective items or complete writing the report. An event or node is represented by a circle:

A **dummy activity** is used to maintain the logic of the network and does not require time or resources. A dummy activity is represented by a broken line with an arrow:

A **network** is a combination of activities and nodes which together show how the overall project can be managed.

Case study

The Ressembler Group is looking at the possible test launch of a new type of picture frame called 'Dale'. The main activities have been identified and times estimated as shown in Table 20.1.

Table 20.1 Activities necessary for test launch

Activity label	Description	Preceding activities	Duration (days)
A	Decide test market area	—	1
B	Agree marketing strategy	—	2
C	Agree production specification	—	3
D	Decide brand name	B	1
E	Prepare advertising plan	A	2
F	Agree advertising package	E	3
G	Design packaging	D	2
H	Production of test batch	C	5
I	Package and distribute	G, H	10
J	Monitor media support	F, D	3

The speedy construction of a network is a matter of practice and experience. A number of computer packages will construct the network diagram, but activities and precedence need to be fully specified. Network construction is an iterative process and several attempts may be needed to achieve a correct representation. Every network starts and finishes with a node; for good practice, you should avoid arrows that cross and arrows that point backwards.

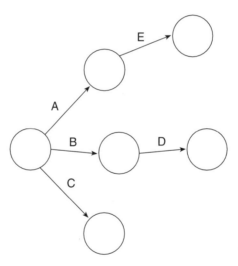

Figure 20.2

In Figure 20.2, we see the beginnings of the network for the project outlined in Table 20.1. Activities A, B and C can all begin immediately, since none have prerequisites. Once activity B is complete then activity D can begin. Each arrow is labelled, indicating that the method of construction is activity on the arrow.

The completed network is shown as Figure 20.3. The use of the dummy variable to maintain the logic should be noted; activity G follows only activity D whereas activity J follows both activity D and F.

Two situations when a dummy activity is likely to be required are shown in Figure 20.4.

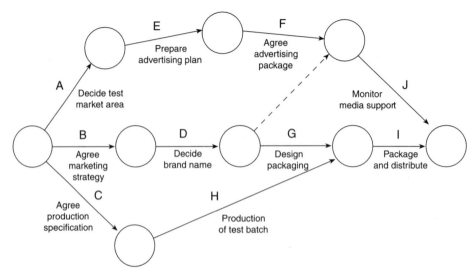

Figure 20.3   Completed network diagram.

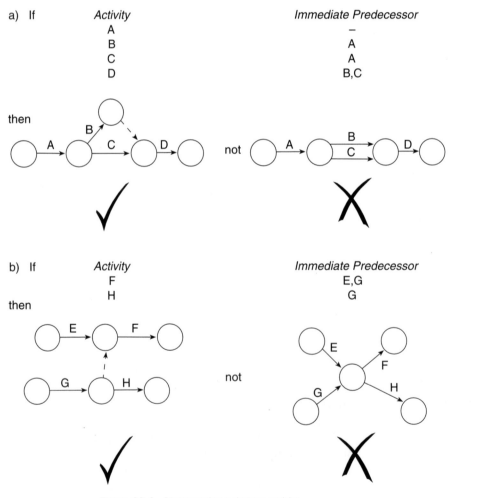

Figure 20.4   Diagramming a dummy activity.

Exercise

Check that you can use the information given in Table 20.2 to draw the network shown in Figure 20.5.

Table 20.2

Activity	Duration (days)	Preceding activities
A	10	–
B	3	A
C	4	A
D	4	A
E	2	B
F	1	B
G	2	C
H	3	D
I	2	E, F
J	2	I
K	3	G, H

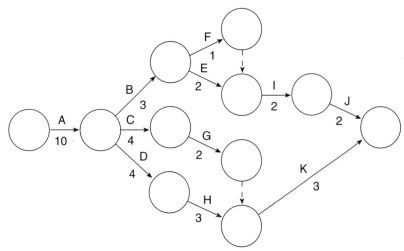

Figure 20.5

20.2 The critical path

The critical path is defined by those activities that must be completed on time for the project to be completed on time. To find the critical path we need to determine the earliest and latest times that an activity can begin and end. Each node is divided into three, as shown in Figure 20.6.

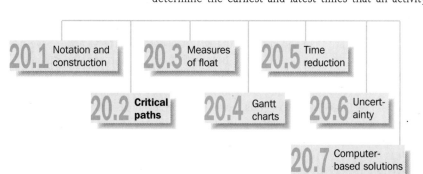

20.1 Notation and construction

20.2 **Critical paths**

20.3 Measures of float

20.4 Gantt charts

20.5 Time reduction

20.6 Uncertainty

20.7 Computer-based solutions

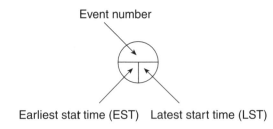

Figure 20.6

20.2.1 The earliest start time

The earliest start time (EST) is the earliest time that an activity could begin assuming all the preceding activities are completed as soon as possible. The convention is always to start with 0. As a rule of thumb the earliest start times are found by moving forward (from left to right) through the network adding activity times. If there are two or more routes leading to a node, the largest value is used.

20.2.2 The latest start times

The latest start time (LST) is the latest time an activity can begin without causing a delay in the overall duration of the project. It is necessary to use the overall duration given in the finish node and work backwards. As a rule of thumb the latest start times are found by moving backwards (from right to left) from the finish node, subtracting activity times. If two or more activities lead backwards to a node then the smallest value is used.

20.2.3 The critical path

In general, the critical path will pass through all those nodes where the EST is equal to the LST and this provides an easy way of scanning the network for the critical path. However, in some circumstances (see Section 20.3, Figure 20.9) this approach may suggest that a non-critical activity is critical. To be sure, you need to check that a measure called **total float is equal to 0** (no spare time).

20.2.4 Total float

The total float for an activity is the difference between the maximum time available for that activity and the duration of that activity. The total float for an activity with a start node of i and a finish node of j is given as:

$$\text{total float} = \text{LST for } j - \text{EST for } i - \text{duration of activity}$$

The earliest and latest start times for the project described in Table 20.1 are given in Figure 20.7.

The calculation of time begins with a 0 EST in node 1. We would then add a 1, 2 and 3 to 0 to get the EST at nodes 2, 3 and 4. The activity times for E and D, 2 and 1, are then added to get the EST at nodes 5 and 6. At node 7, we need to consider the cumulative times from activity F and through the dummy. To ensure the inclusion of a route using a dummy, the dummy can be given the value 0. In this case, from node 5 to 7 we have 3 + 3 and from node 6 to 7 we have 3 + 0; the largest value is 6 and therefore becomes the EST at node 7. At node 8 we need to consider 3 + 2 (node 6 to 8) and 3 + 5 (node 4 to 8); the largest value is 8 and becomes the EST at node 8. Finally the largest sum at node 9 is 8 +10 and 18 becomes the EST. It should be noted that even at this stage we can identify 18 days as the duration of the project. To find the LST's we work backwards using the duration of the project, 18, as the LST at the finish node.

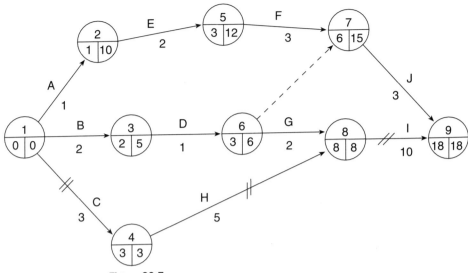

Figure 20.7

Subtraction of activity times J and I gives the LST of 15 at node 7, and 8 at node 8. At node 6 we need to consider moving backwards from node 7 (15 − 0) and from node 8 (8 − 2); the smallest value is 6 and this becomes the LST. The process continues until all the time measures are calculated. The critical path is formed by the activities C, H and I and is shown by the // symbol. In this case, the nodes where the EST = LST define the critical path. (It is clear by observation that activities C, H and I have 0 total float and all other activities have some total float.)

Example

Using the information given in Table 20.2 and the network given as Figure 20.5 check the times and critical path shown in Figure 20.8.

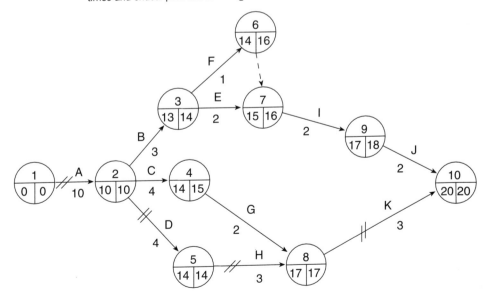

Figure 20.8

The EST's and LST's again make the critical path obvious: A, D, H, K. However, suppose that activities C and G were combined into a new activity L taking six days. This part of the network is shown in Figure 20.9.

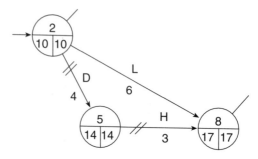

Figure 20.9

The EST's and LST's are equal on nodes 2 and 8 but activity L is not a critical activity. The total float for activity L is

$$= \text{LST for } j(\text{node 8}) - \text{EST for } i(\text{node 2}) - \text{duration of activity L}$$
$$= 17 - 10 - 6$$
$$= 1 \text{ day}$$

The total float is 0 on activities D and H and they remain critical.

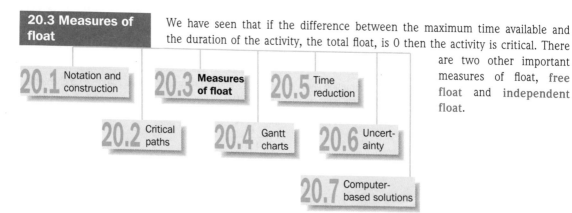

20.3 Measures of float

We have seen that if the difference between the maximum time available and the duration of the activity, the total float, is 0 then the activity is critical. There are two other important measures of float, free float and independent float.

20.3.1 Free float

Free float is the time that an activity could be delayed without affecting any of the activities that follow

$$\text{Free float} = \text{EST for } j - \text{EST for } i - \text{duration of activity}$$

However, free float does assume that previous activities run to time.

20.3.2 Independent float

The independent float gives the time that an activity could be delayed if all the previous activities are completed as late as possible and all the following activities are to start as early as possible.

$$\text{Independent float} = \text{EST for } j - \text{LST for } i - \text{duration of activity}$$

The determination of total, free and independent float is illustrated in Figure 20.10.

Consider the following activity M:

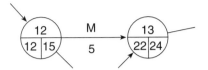

The 'measures of float' can be represented:

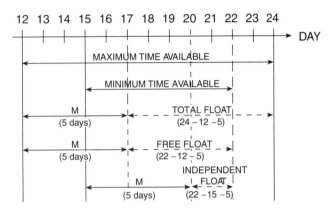

It can be seen that total float is seven days, free float is five days and independent float is two days.

Figure 20.10

20.4 Gantt charts and managing resources

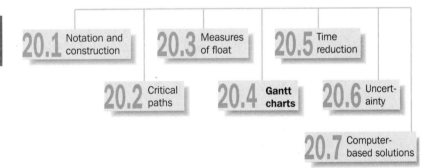

Case study

Consider again the possible test launch of a new type of picture frame by the Ressembler group, detailed in Table 20.1. A network showing the earliest and latest start times, and the critical path, is given as Figure 20.7. The Ressembler Group want to know how the project can be managed over (calendar) time and the administrative support required. Suppose that all the activities require the support of one member of administrative staff, except activity G which requires the support of two members of administrative staff. Suppose also that only three members of administrative staff, at most, are available for this kind of work.

A Gantt chart showing the timing of activities and a bar chart showing the overall administrative support required is given as Figure 20.11.

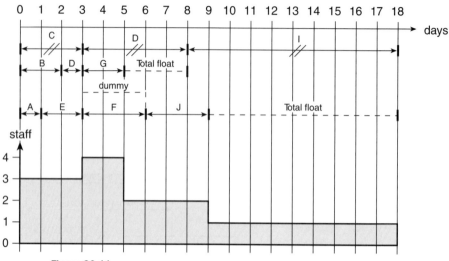

Figure 20.11

It should be noted that the critical path C, D and I defines the duration of the project and that other activities are drawn as parts of other pathways. Total float is usually given for the pathways as shown. The timing of each of the activities can be easily seen with this form of representation. The bar chart gives the total administrative support required by the activities directly above. For example, four administrative staff are required on days 4 and 5 (one for activity D, two for activity G and one for activity F). To more effectively manage the activities we can consider ways of delaying activities with total float. It can be seen in Figure 20.12 how we can reduce the total staff requirement on days 4 and 5 from four staff to three staff by delaying the start of activity F by 2 days.

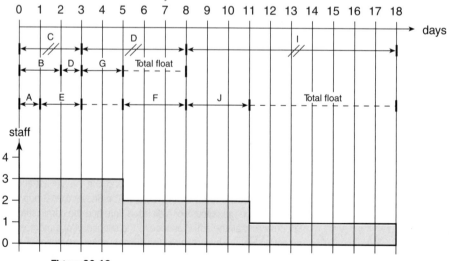

Figure 20.12

It should also be noted that we could choose to delay activity F by two days and avoid a clash with activity G because, for example, both activities required the same personnel or machinery.

Clearly network problems can be more complex than this, but we have tried to illustrate the principles involved. Essentially the Gantt chart allows the manager to see how the activities can proceed against calendar time and the bar chart can be used to show the resources required.

20.5 Project time reduction

One important aspect of project management is looking at ways of reducing the overall project time at an acceptable cost. Where an activity involves some chemical process it may be impossible to reduce the time taken successfully, but in most other activities the duration can be reduced at some cost. Time reductions may be achieved by using a different machine, adopting a new method of working, allocating extra personnel to the task or buying-in a part or a service. The minimum possible duration for an activity is known as the crash duration. Considerable care must be taken when reducing the time of activities on the network to make sure that the activity time is not reduced by so much that it is no longer critical. New critical paths will often arise as this time reduction exercise continues.

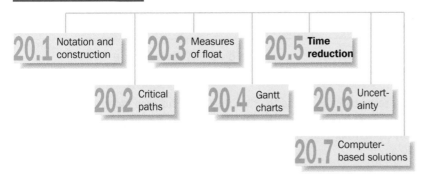

The project given in Table 20.3 will have a critical path consisting of activities A and D, a project time of 18 days and a cost of £580 if the normal activity times are used – as shown in Figure 20.13.

Table 20.3

Activity	Duration (days)	Preceding activities	Cost (£)	Crash duration (days)	Crash cost (£)
A	8	–	100	6	200
B	4	–	150	2	350
C	2	A	50	1	90
D	10	A	100	5	400
E	5	B	100	1	200
F	3	C, E	80	1	100

Since cost is likely to be of prime importance in deciding whether or not to take advantage of any feasible time reductions, the first step is to calculate the **cost increase per time period saved** for each activity. This is known as the **slope** for each activity. For activity A, this would be:

$$\frac{\text{Increase in cost}}{\text{Decrease in time}} = \frac{100}{2} = 50$$

The slopes for each activity are shown in Table 20.4.

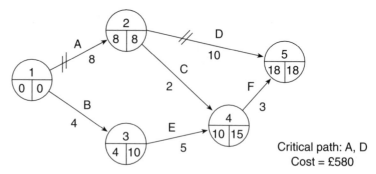

Critical path: A, D
Cost = £580

Figure 20.13

Table 20.4

Activity	A	B	C	D	E	F
Slope	50	100	40	60	25	10

A second step is to find the **free float time** for each non-critical activity. This is the difference between the earliest time that the activity can finish and the earliest starting time of the next activity. Free float times are shown in Table 20.5.

Table 20.5

Activity	ESTj	ESTi	Duration	Free float
B	4	0	4	0
C	10	8	2	0
E	10	4	5	1
F	18	10	3	5

To reduce the project time, select that activity on the critical path with the lowest slope (here A) and note the difference between its normal duration and its crash duration (here, 8–6=2). Look for the smallest (non-zero) free float time (here 1 for activity E), select the minimum of these two numbers and reduce the chosen activity by this amount (here A now has a duration of 7). Costs will increase by the time reduction multiplied by the slope (1 x 50). It is now necessary to reconstruct the network as shown in Figure 20.14.

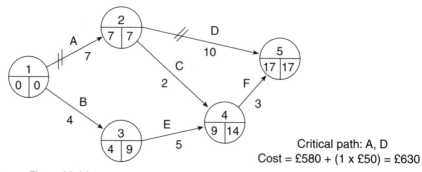

Critical path: A, D
Cost = £580 + (1 x £50) = £630

Figure 20.14

The procedure is now repeated and the new free float times are shown in Table 20.6.

Table 20.6

Activity	ESTj	ESTi	Duration	Free float
B	4	0	4	0
C	9	7	2	0
E	9	4	5	0
F	17	9	3	5

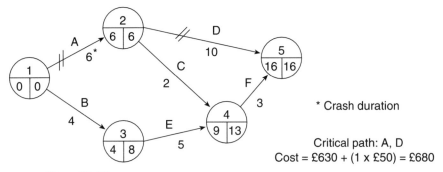

* Crash duration

Critical path: A, D
Cost = £630 + (1 x £50) = £680

Figure 20.15

The activity on the critical path with the lowest slope is still A, but it can only be reduced by one further time period. If this is done, we have the situation shown in Figure 20.15. Any further reduction in the project time must involve activity D, since activity A is now at its crash duration time. If we reduce activity D to six days (i.e. 10 − 4), we have the situation shown in Figure 20.16.

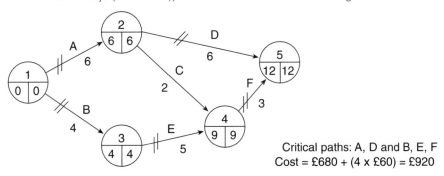

Critical paths: A, D and B, E, F
Cost = £680 + (4 x £60) = £920

Figure 20.16

There are now two critical paths through the network and thus for any further reduction in the project time it will be necessary to reduce both of these by the same amount. On the original critical path, only activity D can be reduced, and only by 1 time period at a cost of 60. For the second critical path, the activity with the lowest slope is F, at a cost of 10. If this is done, we have the situation in Figure 20.17.

Since both activities on the original critical path are now at their **crash durations**, it will not be possible to reduce the total project time further.

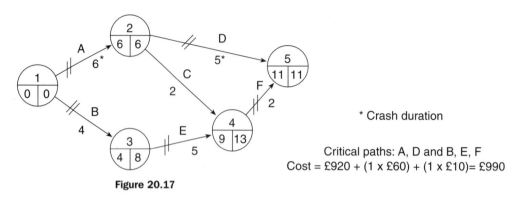

* Crash duration

Critical paths: A, D and B, E, F
Cost = £920 + (1 x £60) + (1 x £10)= £990

Figure 20.17

There are a number of variations on this type of cost reduction problem and again we have just tried to illustrate the general principles.

20.6 Uncertainty

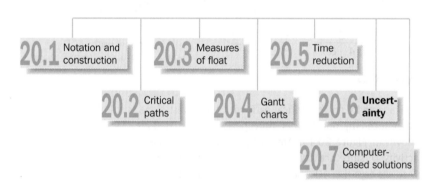

All the calculations in this chapter, so far, have assumed a certainty about the time required to complete an activity or activities. In practice, there will always be some uncertainty about the times taken by future activities. It is known that many activities are dependent upon the weather, and the element of chance is to be expected, e.g. when building a wall or completing a long sea journey. Other types of activity are also subject to uncertainty. How certain can we be that new software will work or be understood first time? How certain can we be that the response rate will allow interviewing to be completed in an exact time? To manage this uncertainty, we work with three estimates of time: **optimistic, most likely and pessimistic**. Activity time is usually assumed to follow the beta distribution (which need not concern us too much here). The mean and the variance for each activity given the three time estimates and the assumption of the beta distribution is:

$$\text{Mean} = \mu = \frac{O + 4M + P}{6}$$

$$\text{Variance} = \sigma^2 = \left(\frac{O - P}{6}\right)^2$$

where O is the optimistic time, M is the most likely time and P is the pessimistic time.

However, when managing a project, we are particularly interested in the overall duration, which is the sum of individual activity times. The sum of activity times (which individually follow the beta distribution) follows the normal distribution - an important statistical result. If we now consider the critical path

$$\text{the mean} = \text{the sum of activity means}$$
$$\text{the variance} = \text{the sum of activity variances}$$

An example is given in Table 20.7 where the critical path is given by the activities A, D, H and K.

Table 20.7

Activity	Optimistic (O)	Most likely (M)	Pessimistic (P)	Mean	Variance
A	8	10	14	10.333	1.000
D	3	4	6	4.167	0.250
H	2	3	6	3.333	0.444
K	1	3	7	3.333	1.000
				21.166	2.694

The mean of the critical path is $\mu = 21.666$ and the variance $\sigma^2 = 2.694$. The standard deviation is $\sigma = 1.641$.

We are now able to produce a **confidence interval** (and make other probability statements). The 95% confidence interval for the total project time (activities A, D, H and K)

$$= \mu \pm 1.96\sigma$$
$$= 21.666 \pm 1.96 \times 1.641$$
$$= 21.666 \pm 3.216$$

This is a simplified example where a single critical path A, D, H and K has been considered. However, in practice, as a more complex project proceeds, some activities will take longer than expected and others a shorter time. The critical path can therefore shift as the project progresses, and needs to be kept under review. The identification of the critical path and the calculations based on the means are expected outcomes and subject to chance.

20.7 Computer-based solutions to network problems

As with other quantitative methods, the techniques of network analysis have been made available in computer software. While early applications did little more than take the tedium out of calculating the duration of the critical path and task start and finish dates the latest versions are capable of supporting planning and management of the largest of engineering and construction projects. This section will look at some of the functions offered by one of the most commonly available pieces of project management software, **Microsoft Project for Windows**. There are many other pieces of software

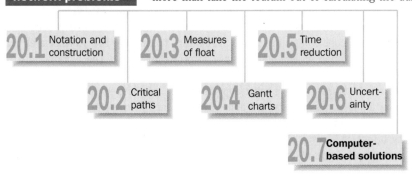

20.1 Notation and construction

20.2 Critical paths

20.3 Measures of float

20.4 Gantt charts

20.5 Time reduction

20.6 Uncertainty

20.7 Computer-based solutions

which have similar or more extensive facilities and cost a great deal more to purchase but Project for Windows (PfW) provides a good start in examining the basic features of such software.

20.7.1 Getting started

Assuming that the software has been installed on the computer and is ready for use the first thing to do is get it running by using the mouse to click the PfW icon. The opening screen will look something like that shown in Figure 20.18.

Figure 20.18   The opening screen for Microsoft Project.

As with most Microsoft software, tutorials and introductory options are available and it would benefit the new user to take the time to investigate these sources of information before checking the 'Do not show again' box. What follows assumes that the user has some understanding of the preceding text on networks and has the skills to use keyboard and mouse to enter data into the appropriate areas of the software.

A mouse click on 'Start a New Project' closes the Welcome inset and gives a clear view of the Task Table and Gantt Chart Screen, Figure 20.19.

This screen has the standard features of most Microsoft Windows products – a menu bar, icons for printing, etc., scroll bars for moving the viewing area but with the addition of functions specific to project planning. The working area of the screen is split vertically into a **spreadsheet** like table on the left and **calendar** on the right. Most of what appears can be customized, although there are some limits as to what can be done. The columns of the left-hand spreadsheet area are already titled from a preset selection. Double clicking on a column heading brings up a range of alternative column headings such as planned cost, %completed, etc. Some of these headings are linked to built in calculations such as %completed, some to user-entered data on which the calculations are based, e.g. duration and others are just information fields which have no effect on the operation of the software. Double clicking on the calendar headings allows the resolu-

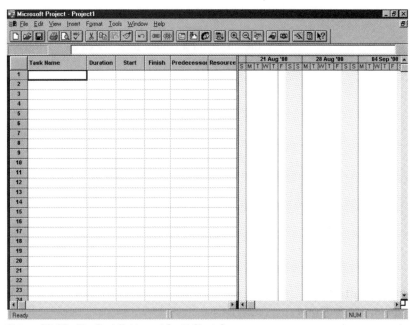

Figure 20.19   The Task Table and Gantt Chart Screen.

tion of the calendar to be changed to show hours, days, weeks, months or quarters with appropriate subdivisions.

The **calendar bar** is also a first clue that this software is intended for use in the **real world**. The dates shown on the calendar bar are the current and coming days, weeks or months, the date having been picked up from the computer's date function. Saturday and Sunday are shown with grey vertical strips to show that the software currently considers these to be non-working days.

20.7.2 Entering a project

This chapter commenced by looking at The Ressembler Group's project to launch a new type of picture frame. The data for this project is as shown in Table 20.8. with the modification that the unique task identifiers have been changed to numbers rather than letters of the alphabet to suit the needs of Project for Windows.

Table 20.8   Ressembler's new product launch

Activity label	Activity number	Description	Preceding activities	Duration (days)	Resources
A	1	Decide test market area	-	1	1 Staff
B	2	Agree marketing strategy	-	2	1 Staff
C	3	Agree production specification	-	3	1 Staff
D	4	Decide brand name	B	1	1 Staff
E	5	Prepare advertising plan	A	2	1 Staff
F	6	Agree advertising package	E	3	1 Staff
G	7	Design packaging	D	2	2 Staff
H	8	Production of test batch	C	5	1 Staff
I	9	Package and distribute	G, H	10	1 Staff
J	10	Monitor media support	F, D	3	1 Staff

Before making any entries into the software it is worth checking the default settings. Selecting the Tools Menu and then Options from that menu results in the display of a tabbed inset which allows various software settings to be changed. Selection of the Schedule Tab provides access to settings which govern the way in which tasks are scheduled. This tab includes a field titled 'Default Duration Type' which should show 'Fixed Duration'. If 'Resource Driven' is showing clicking on the down arrow next to the field box will allow the setting to be altered to 'Fixed Duration'. Tasks which are resource driven will have their duration changed if additional staff are assigned to them. As we will later be assigning two staff to carry out the task of 'designing the packaging' (Activity 7) and in our example **the duration is fixed at 2 days** we do not want the software to inadvertently reduce the duration to 1 day. This feature will be discussed further when we look at resource levelling.

While the 'Options' inset is still open click the 'Calculation' tab and confirm that 'Automatic' is selected. Clicking OK closes the inset and allows us to progress to entering tasks, precedence and durations.

20.7.3 Entering tasks, precedence and durations

Figure 20.20 shows the Task Table and Gantt Chart View with the 10 task descriptions entered. All that has been keyed in are the 10 task descriptions in the Task Names column. Everything else has been created by the software. Entering 'Decide test market area' in row 1 automatically gives that task the unique identifier of '1'. As with other table structures you can drag the contents of rows to new locations but the significance of the row number as a unique task identifier means that the logic of the network is also changed so you need to be careful when using this feature. If you choose to insert rows between existing entries all existing row references will be properly changed to reflect the new identifiers.

Continuing with Figure 20.20, the software has assumed a **default duration of 1 day** and that all tasks can start immediately. The start and finish dates shown

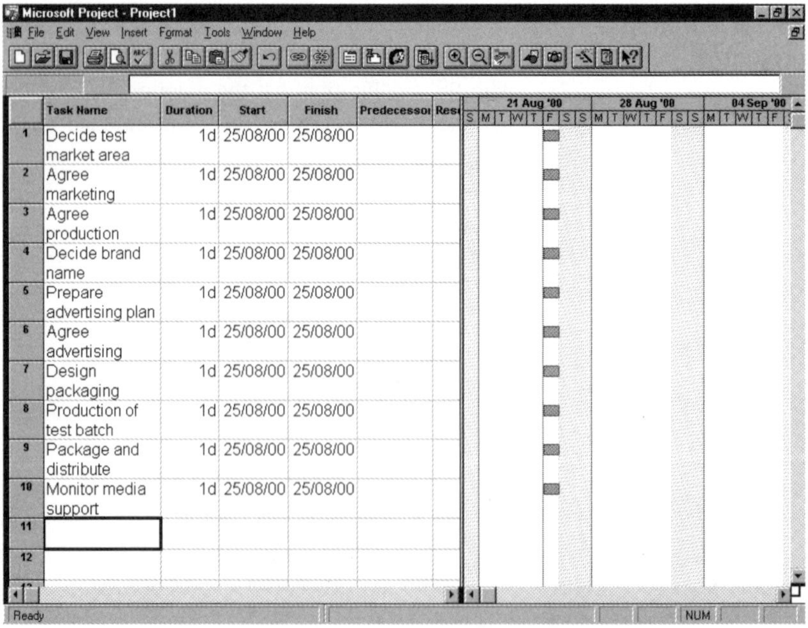

Figure 20.20   The Gantt Chart View with Task Descriptions Entered.

are the results of a network calculation based on those assumptions. On the right-hand side of the screen blue activity bars have appeared for each task synchronized to the assumed start date – this display is a Gantt chart.

A few manual changes have also been made to the screen display. The print size has been increased to make it more readable and the vertical split has been dragged to the right to accommodate the larger print. Dragging the centre split left or right allows the user to see more or less of the Gantt chart. Each side of the split has its own scroll controls at the bottom so you can scroll the table and the Gantt chart to see whichever columns or dates you wish. The vertical scroll bar works on both sides of the split at the same time so that the task names stay in line with their related Gantt chart bars.

Figure 20.21 shows the effect of entering the predecessors – the logic of the network.

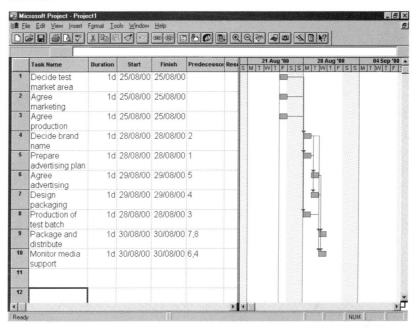

Figure 20.21   The Gantt Chart View with Predecessors Entered.

Tasks 1–3 have no predecessors and the predecessor column is left blank. Task 4 is dependent on task 2, so 2 is entered in the predecessor column. When the entry is complete the return key must be pressed to move on to the next cell or the tick checked on the line above the column titles. Mouse clicking on the next line merely adds that row number to the dependency - this can be annoying if you are used to selecting cells in a spreadsheet by clicking on them. Tasks 9 and 10 do have multiple predecessors. Once the first dependency has been keyed in, any others can be added by mouse clicking the appropriate row. The entry is ended for that cell by pressing return.

As the network logic is keyed in the appropriate links will appear on the Gantt chart and the task bars will move to reflect the Earliest Start Date permitted by preceding task durations and the logic. Figure 20.21 illustrates this, and also shows that the software is recognizing the weekend as non-working time and not scheduling work for Saturday and Sunday.

The next set of information to be entered is the duration of tasks

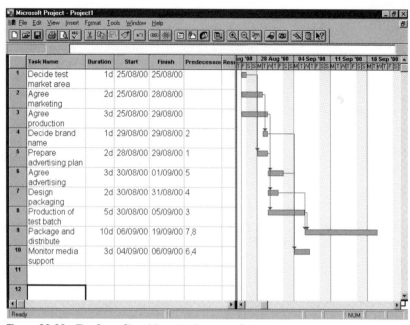

Figure 20.22 The Gantt Chart View with Durations Entered.

Figure 20.22 shows the results of entering the specified task durations.

The first task has been given a duration of 1d, 1 day, and as the present start day of the project is Friday it finishes before the weekend. The second task has a duration of 2 days but the Gantt chart bar is 4 days in length to allow for the task spanning the weekend. Allowing a task to be split by a non-working period normally presents no problems but occasionally can be difficult. If, for example, this task of Agreeing Marketing, involved marketing consultants travelling to visit the company we might prefer the 2 days to be sequential so we do not have to pay excessive travelling costs or hotel bills for the consultants staying over the weekend. In this example we could either change the project start date to the Monday or we could move this task to start on Monday. Moving the task can be achieved by clicking on the task bar, holding the right mouse button down and moving the task so that it starts on Monday. The size of the bar and the dates in the table will automatically change to reflect the change. There is however a hidden consequence of moving the bars or changing the calculated dates. Double clicking on row 1 will result in the display of the 'Task Information' inset. Clicking the 'Advanced' tab shows a selection of fields related to the task one of which is identified as 'Type'. This should be showing the default setting of 'as soon as possible'. This means that any change of logic, addition of new tasks, resource levelling, etc. will always have the effect of scheduling this task at it's earliest possible start date. In contrast double clicking on a task which has been manually moved from its calculated position and then accessing the 'Advanced' tab will show that type has changed to 'Start no earlier than' or some other constrained type. Any subsequent rescheduling will now treat the manually positioned task as having a fixed earliest start date.

Entering the task descriptions, their precedence and their durations has resulted in the software presenting a project plan in the form of a Gantt chart and a table of task start and finish dates. These indicate that the project commences on 25 August and that the last task is completed on the 19 September; a total of 26 calendar days. In this period there are four weekends each with two non-working days giving a total of 8 days lost to the project. The number of working days spent on the project is 26 minus 8, a total of 18 working days. This is the same as shown in Figure 20.11, the resource graph for the project drawn to carry out manual levelling.

20.7.4 The PERT chart view

The project can be viewed as a network diagram by selecting PERT chart from the View Menu. Figure 20.23 shows the screen displaying the PERT chart for the Ressembler project.

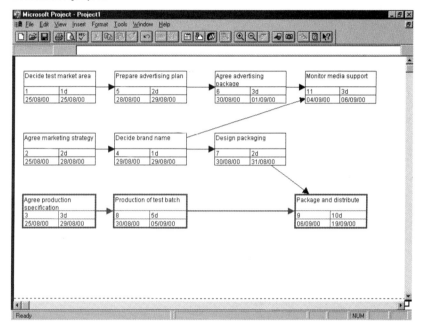

Figure 20.23 The PERT Chart View.

What is displayed is actually a **precedence diagram** not a PERT chart. A feature of PERT is the use of optimistic, expected and pessimistic durations for tasks, allowing a probability calculation to be made for the project finish date. As PfW itself will only allow one value for task duration it cannot produce a PERT chart.

The precedence diagram provided by this software, and most other software, differs significantly in appearance from the activity-on-arrow (AOA) diagrams used to manually draw a network but performs exactly the same function. The boxes represent the tasks and show their description, unique identifier, duration, start date and finish date. What is displayed in the task box can be changed using Box Styles from the Format Menu. The arrows serve only to show the linking logic between tasks but allow linkages not available with AOA diagrams. The AOA diagrams drawn in the preceding text are all based on finish–start links, the preceding task must be finished before its successor can start. However, consider a project such as painting a large bridge. The primer coat may be ready for top coat after a day of drying, but it could take a week to completely primer coat the structure. The best logical link between these activities would be 'start to start with a 1-day lag', i.e. whenever primer coating starts, top coating follows a day later. Double clicking on an arrow in the PERT view brings up an inset that allows the linkage type to be changed and lags or leads to be entered.

The lower three activity boxes in Figure 20.23 have a slightly heavier border and are red on screen to indicate that they are the critical path; 3, 8, 9.

The screen as shown has been rearranged from the way in which it was first presented by the computer by dragging the boxes into positions on the screen which give the clearest representation of the network. As computers have no sense of the aesthetics of diagrams, the PERT chart first presented can be very messy and difficult for the human eye to sort out. Re-arranging the diagram has no effect on the logic of the network. Tasks can however be added or removed and links changed in the PERT view if the user wishes to do so and is happy working in this graphical environment. Such changes are reflected in changes to the data in the Gantt chart and task table.

20.7.5 Adding resources

Returning to the Gantt Chart View, sliding the vertical screen split to the left reveals more columns, including one headed 'Resource Names'. If this column is not present insert a new column and select the 'Resource Name' field to head it. Keying in 'Staff' in the Resource Name cell of the first task automatically creates a resource named 'Staff'. Resources can be individuals (Fred Bloggs), groups of workers (painters) or physical equipment (crane, test bed, etc.). The example project uses only one resource type 'staff' for all tasks so the entry in the first cell can be copied and pasted to all other Resource cells. Task 7 however uses two staff where the other tasks only require one. Double clicking on row 7 brings up the 'Task Information' inset as shown in Figure 20.24.

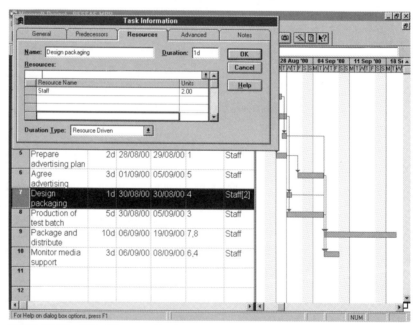

Figure 20.24   Adding Resource Usage.

Selecting the Resources Tab displays several fields and a Resource Table for this specific task. Staff is shown in the first row of the inset table as we have already entered this in the column – other resources could be added in the other rows so that the task employs multiple resources. Selecting the '1.00' in the units column next to staff and keying in '2' increases the requirement of staff for this task to the set level. In this screen shot 'Duration Type' is set as Resource Driven not Fixed Duration and the consequent halving of task duration can be seen in row 7. Changing to Fixed Duration returns the task duration to 2 days. The number of units allocated is shown in brackets in the Resource Names column.

While the software has automatically created a resource named 'staff' it knows nothing about this resource and will only have given it the default values. Figure 20.25 shows the Resource Sheet selected from the View menu.

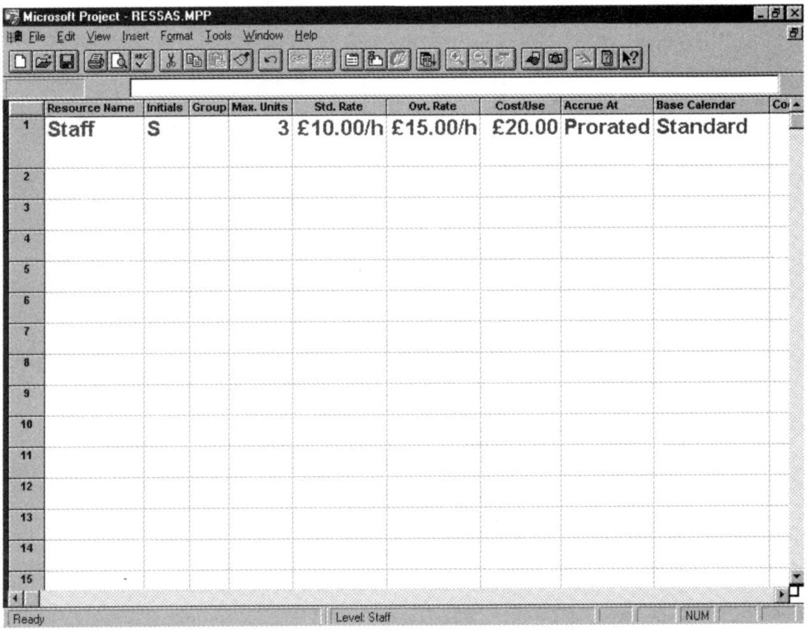

Figure 20.25   The Resource Table.

The entry on the resource sheet shown in Figure 20.25 has been amended to show max. units as 3, the number of staff available for this project, their standard rate as £10.00 per hour, overtime rate as £15.00 per hour and a cost/use of £20.00. Pro-rated indicates that costs will accrue on an hourly basis so that if 2 hours of actual work are reported then £20 will be added to actual costs. The resource calendar is shown as standard, which means that this resource is taken to be available during all normal working days and hours. An individual calendar can be created for each resource which reflects availability for longer working hours or weekend working and shows any annual holiday periods.

In the real world a company such as Ressemblers would probably have already created a full resource table for the year showing all its staff, their working hours and holidays. As PfW can handle multiple projects this table would be available to all projects being undertaken by the company and scheduling would reflect use of resources on other projects. Such complexity is beyond the intended scope of this text.

20.7.6 Resource levelling

Having entered detailed values for the 'staff' resource it is now sensible to look at the use of that resource. Figure 20.26 shows the view obtained by selecting 'Resource Graph' from the View menu.

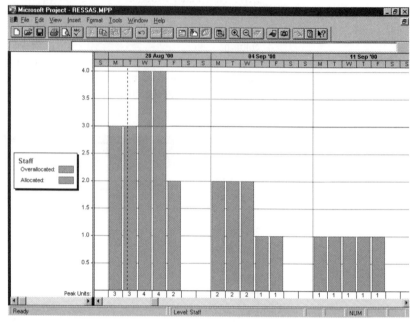

Figure 20.26 The Staff Resource Graph.

The graph shows the level of resource and the times at which it is required by the current plan. At present four staff are required for 2 days in the second week of the project. This exceeds the total number of staff available and this **excess demand** is shown in red on the screen. In fact, the staff resource was shown in red in the Resource Sheet which also indicated the existence of an overload.

Several remedies are possible. The total number of staff available could be increased, the working calendar could be modified to allow weekend or overtime working or the tasks could be adjusted from their earliest start dates to make best use of the existing resource. In extreme cases the logic of the project could also be reviewed.

The Tools menu allows the user to both carry out resource levelling, i.e. make best use of existing resources, and remove its effects. The later is important since if the results of resource levelling are an unacceptable increase in project duration it may be necessary to go back to the unlevelled schedule, implement some of the other remedies for overload and then Resource Level again. Figure 20.27 shows the screen with the Resource Levelling inset displayed.

Clicking on 'Level Now' invokes the levelling function. This is best left as a manual choice as 'Automatic' can have some unanticipated effects during the planning stage of a project. Figure 20.27 shows the data after it has been levelled, the inset has been recalled for the sake of the illustrative figure.

The task table shows that task 6 'Agree Advertising' has been moved later by 2 days to remove the overload. As a consequence this activity now spans a weekend. Activity 10 has also moved to a later start time as it is dependent on activity 6.

Figure 20.28 shows the revised graph for the staff resource after levelling with resource usage held within the maximum of three staff.

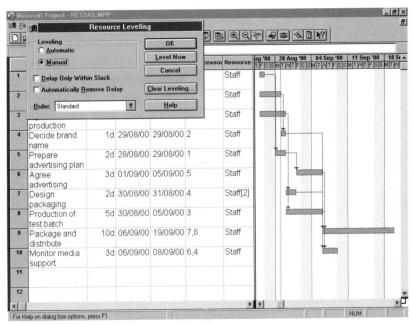

Figure 20.27   The Resource Levelling Function.

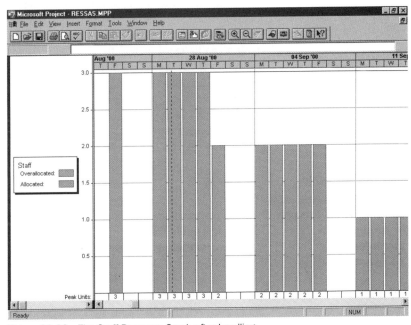

Figure 20.28   The Staff Resource Graph after Levelling.

20.7.7 Project statistics and reports

Figure 20.29 shows the project summary for the stage in planning currently reached. This inset is obtained from the File Menu, Summary Info., Statistics. More detailed reports can be produced but as printouts rather than screen displays.

The statistics inset gives a current (planned) duration of 18 (working) days for the project, a start date of 25 August and finish date of 19 September. Based on

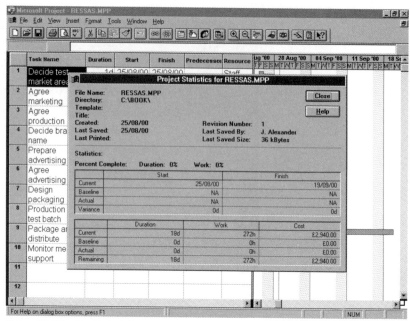

Figure 20.29   The Project Summary.

the costs entered for using staff the overall cost of the project is £2940 for 272 hours of work.

When planning is judged to be complete and accepted by the participants the project can be saved with a 'Baseline' which effectively freezes the plan. During execution of the project problems encountered may require some **replanning** of remaining activities which will change the 'current' plan. Having a frozen baseline always allows actual performance to be compared with the original as well as the current plan.

As the work progresses returns on actual work done can be entered at regular intervals so that the project can be effectively managed.

20.7.8 Concluding comments on the use of project management software

The coverage of PfW as an example of project management software is limited by the space available in this book. The software is rich in features to support realistic planning and monitoring of real-world projects. Learning to use such software for the occasional small project is probably not worth the effort but doing so for larger projects and situations where an organization manages its workload as a series of projects would be of great benefit.

In such larger projects uncertainty needs to be taken into account and can be dealt with by add-on software such as @Risk for Project. This piece of software allows probability distributions to be associated with both task durations and costs. The add-on uses Microsoft Excel to run multiple Monte Carlo trials with the data and then feeds back the results to PfW so that the most likely project durations and cost are displayed. The effects on the plan of uncertainty can then be evaluated and manual interventions made to address uncertainty and minimize cost and duration.

Computer-based solution of network problems is now a large and mature subsection of quantitative methods which allows extensive manual and automatic manipulation of the basic data using experienced qualitative judgement to arrive at manageable project plans.

20.8 Conclusions

The use of network diagrams and Gantt charts is accepted as a useful way of managing complex projects. To be effective, the activities need to be clearly defined, the relationships between activities established and accurate time estimates obtained. A sophisticated network diagram cannot be expected to make good other shortcomings.

A project can be monitored against a network diagram and problems quickly identified. A manager will need to ensure that all critical activities are completed on time and that non-critical paths do not become critical because of delays.

20.9 Problems

1 Draw a network diagram given the following information:

Activity	Preceding activity
A	–
B	A
C	A
D	C
E	C
F	B
G	D, E
H	F, G

2 Draw a network diagram given the following information:

Activity	Preceding activity
A	–
B	A
C	A
D	C
E	C
F	–
G	F
H	G
I	G
J	D, E
K	H, I
L	J, K
M	B, L

3 Determine the earliest start time and latest start time for each activity, identify the critical path and determine the project duration for the following network.

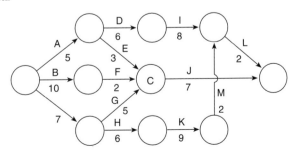

4 Determine the earliest start time and latest start time for each activity, identify the critical path and determine the project duration for the following network.

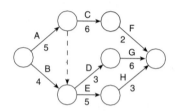

5 Construct the network, identify the critical path and determine the project duration for the following project:

Activity	Duration	Preceding activities
A	5	–
B	14	–
C	8	A
D	7	A
E	4	B, C
F	10	B, C
G	16	B, C
H	8	D, E
I	9	F
J	8	G
K	12	H, I

6 Construct the network, identify the critical path and determine the project duration for the following project:

Activity	Duration	Preceding activities
A	4	–
B	3	–
C	2	–
D	3	–
E	4	–
F	5	–
G	6	A
H	6	A
I	5	B
J	4	B
K	6	C
L	7	C
M	7	C
N	12	D
O	14	D
P	7	E
Q	8	F
R	4	H, I, K
S	6	J, L, N
T	8	J, L, N
U	5	J, L, N
V	6	M, P, Q
W	7	G, R, S
X	7	O, U, V

7 A company is planning to replace a computer system and has identified the following activities.

Activity	Description	Preceding activity	Duration (hours)
A	Shut down existing computer	–	2
B	Disconnect wiring	A	5
C	Remove existing computer	B	10
D	Remove existing wiring	B	20
E	Install new wiring	D, C	5
F	Test new wiring	E	3
G	Move in new computer	C	12
H	Connect new wiring	F, G	4
I	Install operating system	H	6
J	Check memory	I	2
K	Check operating system	I	3
L	Check peripherals	J, K	4

(a) Construct a network diagram for this project.

(b) Determine the critical path and show this on your diagram.

(c) What is the overall duration of the project?

(d) Explain what would happen if independently any of the following were to happen:

   (i) activity B took 7 hours

   (ii) activity G took 18 hours

   (iii) activity C took 21 hours

   (iv) activities A and B could be done at the same time.

8 A company has developed a new product and has identified the activities necessary for the product promotion campaign as follows.

Activity	Description	Preceding activity	Duration (weeks)	Support staff
A	Approval of training budget	–	1	1
B	Training of service people	A	8	2
C	Training of sales people	A	4	2
D	Sales promotion to distributors	C	4	1
E	Distribution to distributors	D	2	1
F	Distribution to retailers	E	4	1
G	Advertising brief	A	2	2
H	Advertising contract	G	1	1
I	Illustrations and text	H	4	1
J	Printing	I	4	2
K	Product launch	B, F, J	1	3

(a) Construct a network diagram for this project.

(b) Determine the critical path and show this on your diagram.

(c) Construct a Gantt chart and show the support staff requirement in an appropriate way.

(d) If the number of support staff is restricted to five, suggest how this project could most effectively be managed.

9 A person decides to move house and immediately starts to look for another property (this takes 40 working days). She also decides to get three valuations on her own property; the first takes two working days, the second, four working days and the third, three working days. When all of the valuations are available she takes one further day to decide on the price to ask for her own property. The process of finding a buyer takes 40 working days. When she has found a house, she can apply for a mortgage (this takes 10 days); have two structural surveys completed, the first taking 10 days and the second 8 days; and instruct a solicitor which takes 35 days. When the surveys have been done on her new property and her old house, the removal firm can be booked. (NB. Assume that her buyer's times for surveys, solicitors, mortgages, etc. are the same as her own.) When both solicitors have finished their work, and the removals are booked, she can finally move house, and this takes one day.

(a) Find the minimum total time to complete the move.

(b) What would be the critical activities in the move?

(c) What would be the effect of finding her new house after only 25 days?

10 Construct a network for the project outlined below and calculate the free float times on the non-critical activities and the critical path time. Find the cost and shortest possible duration for the whole project.

Activity	Preceding activities	Duration	Cost	Crash duration	Crash cost
A	–	4	10	2	60
B	–	6	20	3	110
C	A	5	15	4	50
D	B, C	4	25	3	70
E	B	3	15	1	55
F	E	5	25	1	65
G	D, F	10	20	4	50

11 (a) Construct a network for the project outlined below and calculate the free float time and slope for each activity.

(b) Identify the critical path and find the duration of the project using normal duration for each activity.

(c) Find the cost of reducing the duration for the whole project to 54 days.

(d) If there is a penalty of £500 per day over the contract time of 59 days and a bonus of £200 per day for each day less than the contract time, what will be the duration and cost of the project?

Activity	Preceding activities	Duration	Cost (£)	Crash duration	Crash cost (£)
A	—	5	200	4	300
B	A	7	500	3	1000
C	A	6	800	4	1400
D	—	6	500	5	700
E	D	6	700	3	850
F	D	8	900	5	1050
G	D	9	1000	5	1240
H	—	8	1000	4	1320
I	H	7	600	4	900
J	H	7	800	6	1000
K	—	5	1000	4	1200

(continues overleaf)

L	K	9	500	5	700
M	K	10	1200	8	1240
N	B	8	600	4	760
O	N, Q, S	14	1500	10	1780
P	R, U,	15	2000	10	2500
Q	C, E, Y	10	2000	8	2400
R	C, E, Y	15	1500	7	1900
S	F, I	20	3000	15	3750
T	V, W	10	2000	7	3200
U	G, J, L	14	1800	9	2250
V	G, J, L	22	5000	13	7700
W	M	18	4000	10	5280
X	O, P, T	11	3000	9	4000
Y	H	3	300	2	350

12 Construct a network for the project outlined below, and calculate the free float times on the non-critical activities, and the critical path time. Find the cost of the normal duration time. Find the shortest possible completion time, and its cost.

Activity	Preceding activities	Duration	Cost	Crash duration	Crash cost
A	—	5	100	4	200
B	—	4	120	2	160
C	A	10	400	4	1000
D	B	7	300	3	700
E	B	11	200	10	250
F	C	8	400	6	800
G	D, E, F	4	300	4	300

21 Modelling stock control and queues

This chapter considers two families of models, stock control and queues, which may or may not describe the problems faced by particular managers. These models are of interest because they illustrate that many **common problems** already have solutions and they demonstrate good modelling practice. Both of these situations should be relatively familiar to you, since we have all had to wait in queues often guessing which one of several to join, for example, in a supermarket. Similarly, we all keep a stock or store of something, and need to decide when to replenish stocks, maybe waiting until we discover, for example, that there is no coffee left, or maybe buying when we see the jar half empty. Even if managers are not directly involved in managing stock or queues, they are likely to work with cost constraints and they are likely to be concerned with performance measures. Once the models are understood, a manager can consider how the model or modelling principles can be adapted to meet their needs or the needs of the business.

Objectives

After working through this chapter, you should be able to:

- understand and use the 'basic' stock control model
- adapt the stock control model to allow for quantity discounts, lead time and probabilistic demand
- understand and use the 'basic' queuing model
- adapt the basic queuing model to allow for cost and multiple channels.

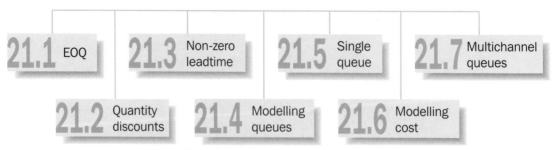

Figure 21.1 Structure of Chapter 21.

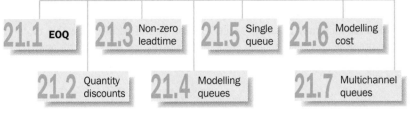

21.1 Introduction to the economic order quantity model

The economic order quantity (EOQ) model, or the economic batch quantity (EBQ) model as it is often called, is one mathematical representation of the costs involved in stock control. It is one of the simplest mathematical stock control models, but has been found to provide reasonable solutions to a number of practical problems.

The derivation of this model involves two simplifying assumptions (which we relax later):

- there is no uncertainty – demand is assumed to be known and constant
- the lead time is zero – the lead time is the time between placing an order and receiving the goods.

The economic order quantity model is developed from two types of cost associated with stock control: **ordering** costs and **holding** costs. Ordering costs are those costs incurred each time an order is placed. They can involve administrative work, telephone calls, postage, travel, or a combination of two or more of these. Holding costs are the costs of keeping an item in stock. These can include the cost of capital, handling, storage, insurance, taxes, depreciation, deterioration and obsolescence.

Example

A retailer, A to Z Limited, has a constant demand for 300 items each year. The cost of each item to the retailer is £20. The cost of ordering, handling and delivering is £18 per order regardless of the size of the order. The cost of holding items in stock amounts to 15% of the value of the stock held. If the lead time is zero, determine the order quantity that minimizes total inventory cost.

If Q is the quantity ordered on each occasion and D is the constant annual demand then the stock level as a function of time appears as in Figure 21.2.

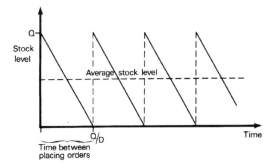

Figure 21.2 Stock level as a function of time.

We can start with an order of Q and since the lead time is zero the inventory rises from 0 units to Q units. Thereafter, the stock level falls at the constant rate of D units per year. The time taken for the stock level to reach 0 is Q/D years. If we consider the annual demand of 300 items per year as given in our example, the stock level would fall to 0 in 2 years if we placed orders for 600 and would fall to 0 in 3 months (1/4 of a year) if we placed orders for 75 each time.

It follows that the number of **orders per year** is D/Q. If orders of 75 units are placed to meet an annual demand of 300 then four deliveries are required each year. It also follows that the **average stock level** is $Q/2$ units, since the stock level varies uniformly between 0 and Q units. On this basis we are able to determine ordering costs and holding costs for this particular stock control model:

$$\text{Total ordering costs (TOC)} = \text{cost per order} \times \text{number of orders}$$

$$= 18 \times \frac{D}{Q}$$

$$\text{Total holding cost (THC)} = \text{cost of holding per item}$$
$$\times \text{average number of items in stock}$$

$$= (15\% \text{ of } \pounds 20) \times \frac{Q}{2}$$

$$= 3 \times \frac{Q}{2}$$

We can note that actual holding cost will vary directly with the stock level from $(15\% \text{ of } \pounds 20) \times Q$ to 0 and that the above is an average. We can also note that it is typical for holding costs to be expressed as a percentage of value.

In seeking a stock control policy we need to consider total variable cost (TVC) and total annual cost (TAC), where

$$\text{total variable cost} = \text{total ordering costs} + \text{total holding costs}$$

and

$$\text{total annual cost} = \text{total ordering costs} + \text{total holding costs}$$
$$+ \text{purchase cost (PC)}$$

We can determine an order quantity that will minimize the cost of stock control in our example by 'trial and error' as shown in Figure 21.3.

This type of analysis is ideal for spreadsheet analysis, with ordering cost and holding cost as parameters.

In this example both total variable cost and total annual cost are minimized if we place orders for 60 items, five times each year.

The structure of the table is as follows:

1 Demand is uniform at 300 items each year. It can be satisfied by one order of 300 each year, two orders of 150 every 6 months, three orders of 100 every 4 months and so on.

2 The cost of ordering remains fixed at £18 per order regardless of the size of an order. The total annual ordering cost is the ordering cost (C_0) multiplied by the ordering frequency. As the size of the order increases, the total annual order cost decreases.

3 Total annual holding cost is the cost of keeping an item in stock (often expressed as a percentage of value) multiplied by the average stock level. At the time an order is received, holding cost will be high but will decrease to zero as the stock level falls to zero. Total annual holding cost is derived from the averaging process. As the size of the order increases the total annual holding cost increases.

4 Total variable cost is the sum of one cost that decreases with order quantity (TOC) and one cost that increases with order quantity (THC). These cost functions are shown in Figure 21.4.

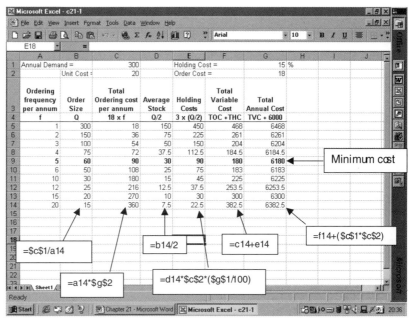

Figure 21.3 A simple costing model.

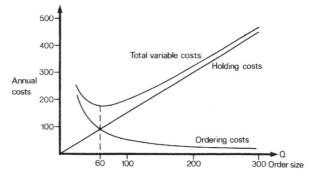

Figure 21.4

5 Total annual cost includes the annual purchase cost which in this case is £6000 (£20 × 300). The order quantity that minimizes the total variable cost need not minimize the more important total annual cost. It is the total annual cost of the inventory that the retailer will need to meet and it is this cost which can be affected by such factors as price discounts.

The order quantity that minimizes the total variable cost could have been determined by substitution into the following equation:

$$Q = \sqrt{\left(\frac{2 \times C_0 \times D}{C_H}\right)}$$

where Q is the economic order quantity, D is the annual demand, C_0 is the ordering cost and C_H is the holding cost per item per annum.

By substitution we are able to obtain the solution found by trial and error

$$Q = \sqrt{\left(\frac{2 \times 18 \times 300}{3.00} \right)} = 60$$

The derivation of this formula is given in Section 21.10.

Example

A company uses components at the rate of 500 a month which are bought at the cost of £1.20 each from the supplier. It costs £20 each time to place an order, regardless of the quantity ordered.

The total holding cost is made up of the capital cost of 10% per annum of the value of stock plus 3p per item per annum for insurance plus 6p per item per annum for storage plus 3p per item for deterioration.

If the lead time is zero, determine the number of components the company should order, the frequency of ordering and the total annual cost of the inventory.

$$\text{Annual demand} = 500 \times 12 = 6000$$
$$\text{Ordering cost} = £20$$
$$\text{Holding cost} = £1.20 \times 0.10 + £0.03 + £0.06 + £0.03 = £0.24$$

By substitution

$$Q = \sqrt{\left(\frac{2 \times 20 \times 6000}{0.24} \right)} = 1000$$

To minimize total variable cost, and in this case also total annual cost, we would place an order for 1000 components when the stock level is zero. The number of orders (or ordering frequency) is

$$\frac{D}{Q} = \frac{6000}{1000} = 6$$

We would therefore expect to place orders every 2 months.

$$\text{TAC} = \text{TOC} + \text{THC} + \text{PC}$$

$$= C_0 \times \frac{D}{Q} + C_H \times \frac{Q}{2} + \text{price} \times \text{quantity}$$

$$= £20 \times \frac{6000}{1000} + £0.24 \times \frac{1000}{2} + £1.20 \times 6000$$

$$= £120 + £120 + £7200$$

$$= £7440$$

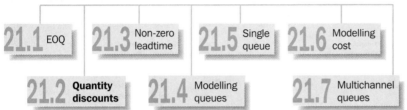

21.2 Quantity discounts

In practice it is common for a supplier to offer discounts on items purchased in larger quantities. This reduces the purchase cost. The ordering cost is also reduced since fewer orders need to be placed each year. However, holding costs are increased with the larger average stock level. The economic order quantity formula cannot be used directly since unit cost and hence purchase cost is no longer fixed.

Example

Suppose the retailer, A to Z Limited (see Section 21.1), is offered a discount in price of $2\frac{1}{2}\%$ on purchases of 100 or more. If the lead time remains zero, determine the order quantity that minimizes total inventory cost.

The total annual cost function, as shown in the last column of Figure 21.3, now becomes discontinuous at the point where the discount is effective. Not only is the purchase cost reduced, but also the holding cost changes, as it is expressed as a percentage of value (15% of the value of the stock level).

If $Q = 100$, the minimum order quantity to qualify for the discount, then:

$$\text{TOC} = £18 \times \frac{D}{Q} = £18 \times \frac{300}{100} = £54$$

purchase price (with $2\frac{1}{2}\%$ discount) $= £20 \times 0.975 = £19.50$

holding cost per item (15% of value) $= £19.50 \times 0.15 = £2.925$

$$\text{total holding cost} = £2.925 \times \frac{Q}{2} = £2.925 \times \frac{100}{2} = £146.25$$

total variable cost $= £54 + £146.25 = £200.25$

total annual cost $=$ total variable cost $+$ purchase cost

$$= £200.25 + £19.50 \times 300 = £6050.25$$

The discount on price of $2\frac{1}{2}\%$ will give the retailer (refer to Figure 21.3) an annual saving of $£6204 - £6050.25 = £153.75$ if orders of 100 are placed.

The effects on costs of the discount are shown in Figure 21.5.

If this were being modelled on a spreadsheet, only the holding cost column and percentage cost component of Figure 21.3 would need to be changed.

By inspection of total annual cost we can see that a minimum is achieved with an order size of 100. This function is shown in Figure 21.6.

The formula for the economic order quantity can be used within a price range to obtain a local minimum (up to 100 in this example) but needs to be used with caution.

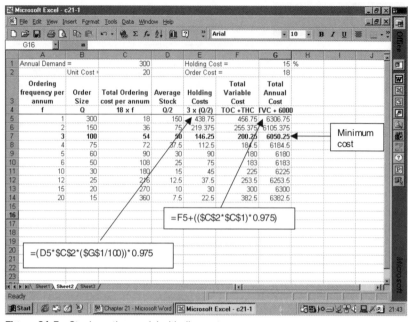

Figure 21.5 Stock costing model with discounts.

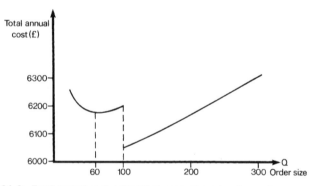

Figure 21.6 Total annual cost, taking inot account a price discount.

21.3 Non-zero lead time

One of the assumptions of the basic economic order quantity model (see Section 21.1) was a lead time of zero. The model can be made more realistic if we allow a time lag between placing an order and receiving it. The calculation of the economic batch quantity does not change; it is the time at which the order is placed that changes. The reorder point is the lowest level to which stock is allowed to fall (generally above 0).

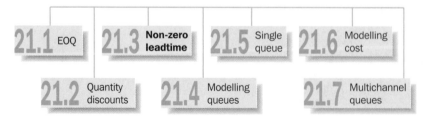

21.3.1 Non-zero lead time and constant demand

Suppose the retailer, A to Z Limited, retains a constant demand of 300 items each year but has a lead time of 1 month between placing an order and receiving the goods. The monthly demand for the items will be

$$300 \times \frac{1}{12} = 25$$

When the stock level falls to 25, the retailer will need to place the order (60 items in this case) to be able to meet the demand in one month's time. Stock levels are shown in Figure 21.7.

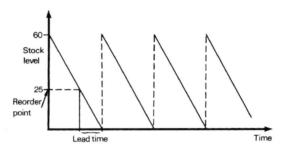

Figure 21.7   Stock levels with non-zero lead time.

21.3.2 Non-zero lead time and probabilistic demand

If demand is probabilistic, as shown in Figure 21.8, a reorder level cannot be specified that will **guarantee** the existing stock level reaching zero at the time the new order is received.

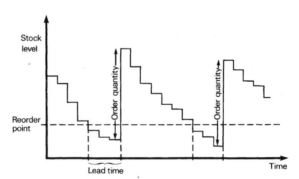

Figure 21.8   Stock levels with non-zero lead time and probabilistic demand.

To determine the reorder point we need to know the lead time (assumed constant), the risk a retailer or producer is willing to take of not being able to meet demand and the probabilistic demand function. The number of items held in stock when an order is placed will depend on the risk the stockholder is prepared to take in not being able to meet demand during the lead time. The probability distribution of demand is often described by the **Poisson** or **Normal** distributions.

Example

Suppose the retailer, A to Z Limited, now finds that the lead time is one month and that demand for items follows a Poisson distribution with a mean of 25 items per month. What should the reorder level be if the retailer is willing to take a risk of 5% of not being able to meet demand?

To determine the reorder level we need to refer to cumulative Poisson probabilities as shown in Table 21.1.

If the reorder level were 32, then

$$P(\text{not being able to meet demand}) = P(33 \text{ or more items in one month})$$

$$= 0.0715 \text{ or } 7.15\%$$

If the reorder level were 33, then

$$P(\text{not being able to meet demand}) = P(34 \text{ or more items in one month})$$

$$= 0.0498 \text{ or } 4.98\%$$

Given that the risk the retailer is willing to take of not being able to meet demand is at most 5%, the reorder level would be set at 33 items. Once stock falls to this level a new order is placed.

Table 21.1 Cumulative Poisson probabilities for $\lambda = 25$

r	$P(r \text{ or more})$
32	0.1001
33	0.0715
34	0.0498
35	0.0338

Example

You have been asked to develop a stock control policy for a company which sells office copiers.

The cost to the company of each machine is £250 and the cost of placing an order is £50. The cost of holding stock has been estimated to be 10% per annum of the value of the stock held. If the company runs out of stock then the demand is met by special delivery, but the company is willing to take only a 5% risk of this service being required. From past records it has been found that demand is normally distributed with a mean of 12 machines per week and a standard deviation of two machines.

If there is an interval of a week between the time of placing an order and receiving it, advise the company on how many machines it should order at one time and what its reorder level should be.

By substitution, we are able to calculate the economic order quantity.

$$Q = \sqrt{\left(\frac{2 \times C_0 \times D}{C_H} \right)}$$

Let

$$C_0 = £50$$
$$C_H = £250 \times 0.10 = £25$$
$$D = 12 \times 52 = 624$$

We have assumed that demand can exist for 52 weeks each year although production is not likely to take place for all 52 weeks:

$$Q = \sqrt{\left(\frac{2 \times 50 \times 624}{25} \right)} = 50$$

To minimize costs, assuming no quantity discounts, orders for 50 office copiers would be made at any one time. Given a lead time of a week, we need now to calculate a reorder level such that the probability of not being able to meet demand is no more than 5%.

Consider Figure 21.9. With reference to tables for the Normal distribution, a z-value of 1.645 will exclude 5% in the extreme right-hand tail area. By substitution, we are able to determine x:

$$z = \frac{x - \mu}{\sigma}$$

$$1.645 = \frac{x - 12}{2}$$

$$x = 12 + 1.645 \times 2 = 15.29$$

To ensure that the probability of not being able to meet demand is no more than 5%, a reorder level of 16 would need to be specified.

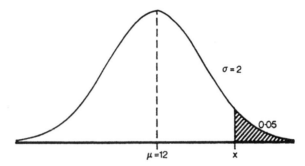

Figure 21.9 Weekly demand for office typewriters.

It should be noted that any increase in the reorder level will increase the average stock level and as a result will increase holding costs. As we include more factors, the model becomes more and more complex. The inclusion of uncertainty, for example, makes the stock control model more realistic in most circumstances. As we have seen, the economic order quantity can describe some stock control situations and can be developed to describe a number of others.

21.4 Introduction to modelling queues

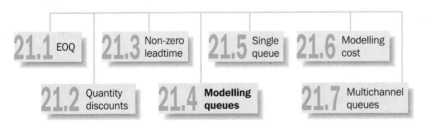

Queues in one form or another are part of our daily lives. We may queue in the morning to buy a newspaper, or as a car driver we may queue to enter a multistory car park. Not all queues involve people so directly. Queues may consist of jobs waiting to be processed on a computer, or aeroplanes waiting to land, or components in batches waiting to be machined.

A queue forms when a customer cannot get immediate service. To the customer the queue may represent a lack of quality in the service. To the business, eliminating queues may increase costs for little benefit. If, for example, an employee of an engineering company has to wait at a central store for a component, there is the opportunity cost associated with the possibility of lost production. However, to decrease waiting times at the central store, more service staff may need to be employed and the service arrangements changed at increased cost. Like other aspects of business, queues need to be managed and the relative costs balanced. If the service cost is high and the service has a high level of demand, e.g. dentists, queues may be lengthy. In other cases, where the service cost may be relatively cheap and lost custom expensive, queues may be short, e.g. supermarket checkouts and garages (you serve your own petrol).

Queues can take many forms and we can mathematically model only a few of these. Customers may form a single queue or separate queues at multiple service points. Customers may arrive singly, e.g. telephone calls through a switchboard, or in batches, e.g. passengers leaving a train. Customers may be served in the order of their arrival, e.g. football supporters, or on some other basis, e.g. oil tankers may be given priority in the entry to a port. The way a queue is managed or served is known as the queue discipline. In general, the length of any queue will depend on the rate of customer arrival, the time taken to serve the customer, the number of service points and the queue discipline. It is worth noting the difference between the system and the queue. The queue refers to the customers waiting to be served, whereas the system refers to the customers waiting to be served plus the customer or customers being served.

Simple queuing can be reasonably described mathematically but as we attempt to make the queuing model more realistic, the **mathematics quickly become complex**. In this chapter, the mathematics of the single-channel and multi-channel queues are presented. However, as the queuing situation becomes more complex, **simulation techniques** (see Chapter 22) offer a useful and practical alternative.

21.5 A model for a single queue

The derivation of a model for a single-server queue involves a number of simplifying assumptions.

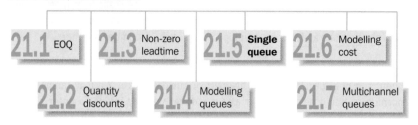

1 There is a single queue with a single service point.
2 The queue discipline is 'first come first served'.
3 The arrivals are random and can be described by the Poisson distribution. Given the integer nature of arrivals, only a discrete probability distribution would have the necessary characteristics. The Poisson distribution is particularly suitable for modelling random arrivals in intervals of time (see Section 9.3).
4 Service times follow a negative exponential distribution (**service times are random**). The properties of the negative exponential distribution need not concern us here. It is a continuous distribution and in many ways similar to the Poisson; it is defined by one parameter, the mean.
5 There is **no limit** to the number of customers waiting in the queue.
6 All customers **wait long enough to be served** (no reneging).

To understand the characteristics of a queue, we need to determine the average arrival rate, λ, and the average service rate, μ. The ratio of the average arrival rate to the average service rate, ρ, provides a useful measure of how busy the queue is, and is known as the **traffic intensity** or **utilization factor**.

$$\rho = \frac{\lambda}{\mu} = \frac{\text{average arrival rate}}{\text{average service rate}}$$

If, for example, the average arrival rate is 18 customers per hour and the average service rate is 60 customers per hour, the traffic intensity is given by

$$\rho = \frac{18}{60} = 0.3$$

If the average arrival rate is greater than the average service rate, $\lambda > \mu$, then customers will be entering the queue more frequently than they leave it, and the queue will get longer and longer. An important **assumption** of queuing theory is that $\rho < 1$ in which case the queue can achieve a steady-state or equilibrium.

The operating characteristics of the single queue model are as follows:

1 The probability of having to queue is given by the traffic intensity:

$$\rho = \frac{\lambda}{\mu}$$

Clearly, as the average service rate increases relative to the average arrival rate, the probability of having to queue decreases.

2 The probability that the customer is served immediately, only one in the system, is given by

$$P_0 = 1 - \rho$$

3 The probability of n customers in the system is
$$P_n = (1 - \rho)\rho^n$$

4 The average number of customers in the system is

$$L_s = \frac{\rho}{1 - \rho} = \frac{\lambda}{\mu - \lambda}$$

5 The average number of customers in the queue is

$$L_q = \frac{\rho^2}{1 - \rho} = \frac{\lambda^2}{\mu(\mu - \lambda)}$$

6 The average time spent in the system is

$$T_s = \frac{1}{\mu(1 - \rho)} = \frac{1}{\mu - \lambda}$$

7 The average time spent in the queue is

$$T_q = \frac{\rho}{\mu(1 - \rho)} = \frac{\lambda}{\mu(\mu - \lambda)}$$

Examples

1 Given the average arrival rate of 18 customers per hour and the average service rate of 60 customers per hour, determine the probability of having to queue, the probability of not having to queue and the probability of one, two and three customers in the system.

The probability of having to queue is equal to the trac intensity

$$\rho = \frac{18}{60} = 0.3$$

The probability of not having to queue is

$$P_0 = 1 - \rho = 1 - 0.3 = 0.7$$

The probabilities of there being one, two and three customers in the system is given by

$$P_1 = (1 - 0.3) \times 0.3 = 0.21$$

$$P_2 = (1 - 0.3) \times 0.3^2 = 0.063$$

$$P_3 = (1 - 0.3) \times 0.3^3 = 0.0189$$

With such a favourable service rate relative to the arrival rate, the chance of having to join a queue of any significant length is quite small.

2 Customers arrive at a small Post Oce with a single service point at an average rate of 12 per hour. Determine the average number of customers in the queue and the average time in the queue if service takes four minutes on average. If service times can be reduced to an average of three minutes, what would the impact be?
Given that $\lambda = 12$ and $\mu = 60/4 = 15$ then

$$L_q = \frac{\lambda^2}{\mu(\mu - \lambda)} = \frac{12^2}{15(15 - 12)} = 3.2$$

and

$$T_q = \frac{\lambda}{\mu(\mu - \lambda)} = \frac{12}{15(15 - 12)} = 0.27 \text{ hours} = 16.2 \text{ mins}$$

If average service time can be reduced to three minutes, then $\lambda = 12$ and $\mu = 60/3 = 20$, and

$$L_q = \frac{\lambda^2}{\mu(\mu - \lambda)} = \frac{12^2}{20(20 - 12)} = 0.9$$

and

$$T_q = \frac{\lambda}{\mu(\mu - \lambda)} = \frac{12}{20(20 - 12)} = 0.075 \text{ hours} = 4.5 \text{ mins}$$

The 25% reduction in average service time has made a considerable difference to the queue characteristics.

21.6 Queues – modelling cost

Whether a queue involves people or items there are associated costs. In the business situation it may be difficult to identify all the costs, e.g. the opportunity cost of an employee waiting in a queue (there may be benefits if these employees in the queue discuss ways of improving the quality of their work), but it is important to achieve a balance between those costs

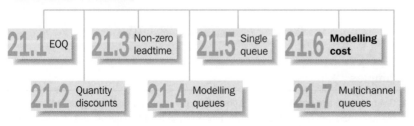

21.1 EOQ | 21.3 Non-zero leadtime | 21.5 Single queue | 21.6 **Modelling cost**

21.2 Quantity discounts | 21.4 Modelling queues | 21.7 Multichannel queues

increasing and those decreasing with queue length. Costs are generally of two types: service costs and queuing costs. Service costs include the labour necessary to provide the service and related equipment costs. Queuing cost is the opportunity cost of customers waiting and could include, for example, the lost contribution to profit incurred while an employee is waiting in a queue. The two following examples illustrate the possible methods of solution.

Examples

1 A spare parts department is manned by a receptionist with little technical knowledge of the product concerned. The receptionist is paid £3.50 per hour but can only deal with 12 requests per hour because of the need to refer to other staff for technical advice. A person with more technical knowledge would be paid £6.50 per hour but could deal with 15 requests per hour. On average, 10 requests per hour are made. If the opportunity cost of staff making the requests to the spare parts department is valued at £5 per hour, determine whether a change in staffing is worthwhile.

  If a receptionist with more technical knowledge is employed, the traffic density will fall from

$$\rho = \frac{10}{12} = 0.83$$

to

$$\rho = \frac{10}{15} = 0.67$$

If the receptionist has little technical knowledge, the average number of requests (or average number in the system) is

$$L_s = \frac{\rho}{1-\rho} = \frac{0.83}{1-0.83} = \frac{0.83}{0.17} = 4.88$$

and the average cost per hour = 4.88 × £5 + 3.50 = £27.90.

  If the receptionist has more technical knowledge, the average number of requests is

$$L_s = \frac{\rho}{1-\rho} = \frac{0.67}{1-0.67} = \frac{0.67}{0.33} = 2.03$$

and the average cost per hour = 2.03 × £5 + £6.50 = £16.65.

  In this case there are cost benefits in paying the higher hourly rate to have a receptionist with more technical knowledge.

2 Photocopying is done in batches. On average, six employees per hour need to use the facility and their time is valued at £3.00 per hour. A decision needs to be made on whether to rent a type A or type B photocopying machine. The type A machine can complete an average job in five minutes and has a rental charge of £16 per hour, whereas the type B machine will take eight minutes but has a rental charge of £12 per hour. Which machine is most cost effective?

  For the type A machine, the service rate is 60/5 = 12 per hour and the arrival rate is six per hour.

  The average time lost per worker per hour (average time spent in the system) is

$$T_s = \frac{1}{\mu - \lambda} = \frac{1}{12 - 6} = \frac{1}{6}$$

The cost per hour = 6 × 1/6 × £3 + £16 = £19.

For the type B machine, the service rate is $60/8 = 7.5$ per hour and the arrival rate is six per hour.

The average time lost per worker per hour is

$$T_s = \frac{1}{\mu - \lambda} = \frac{1}{7.5 - 6} = \frac{1}{1.5} = \frac{2}{3}$$

The cost per hour $= 6 \times 2/3 \times £3 + £12 = £24$.

In this case type A machine is most cost effective.

21.7 Modelling multi-channel queues

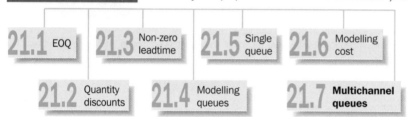

It is possible to reduce the average length of a queue and the average waiting times by increasing the number of service points. In this model a number of service points, S, are available and customers join a single queue. Whenever a service point becomes free, the next customer in line takes the place. This system is operated by a number of banks, building societies and car exhaust replacement centres. If the service rate at each channel is μ, then the traffic intensity is given by

$$\rho = \frac{\lambda}{\mu \times S}$$

To achieve a steady state, the traffic intensity, ρ, must be less than 1 or, in other words, the total arrival rate must be less than the total service rate, λ.

The derivation of the model for a multi-channel queue involves the following simplifying assumptions:

1 There is a **single** queue with S **identical** service points.
2 The queue discipline is '**first come first served**'.
3 Arrivals follow a **Poisson** distribution.
4 Service times follow the **negative exponential** distribution.
5 There is **no limit** to the number of customers waiting in the queue.
6 **All customers wait** long enough to be served.

The operating characteristics of the multi-channel queue model are as follows:

1 The probability of there being no one in the system is given by

$$P_0 = \frac{1}{\displaystyle\sum_{i=0}^{s-1} \frac{(\lambda/\mu)^i}{i!} + \frac{(\lambda/\mu)^S \times \mu}{(S-1)! \times (S \times \mu - \lambda)}}$$

2 The probability of there being n customers in the system is

$$P_n = \frac{(\lambda/\mu)^n}{S! \times S^{n-s}} \times P_0 \quad \text{for} \quad n > S$$

and

$$P_n = \frac{(\lambda/\mu)^n}{n!} \times P_0 \quad \text{for} \quad 0 \leq n \leq S$$

3 The average number of customers waiting for service is

$$L_q = \frac{(\lambda/\mu)^S \times \lambda \times \mu}{(S-1)! \times (S \times \mu - \lambda)^2} \times P_0$$

4 The average number of customers in the system is

$$L_s = L_q + \lambda/\mu$$

5 The average time spent in the queue is

$$T_q = \frac{L_q}{\lambda}$$

6 The average time spent in the system is

$$T_s = T_q + 1/\mu$$

Example

A fast food outlet, McBurger, has customers arriving at the rate of 50 per hour. There are three identical service points and each can handle 20 customers per hour. Determine the operating characteristics of the queue.

The probability of there being no one is the system in given by

$$P_0 = \frac{1}{\displaystyle\sum_{i=0}^{2} \frac{(50/20)^i}{i!} + \frac{(50/20)^3 \times 20}{(3-1)! \times (3 \times 20 - 50)}} = 0.0449$$

The probabilities of 1, 2, 3, 4 or 5 customers in the system are

$$P_1 = 50/20 \times 0.0449 = 0.1123$$

$$P_2 = \frac{(50/20)^2}{2} \times 0.0449 = 0.1403$$

$$P_3 = \frac{(50/20)^3}{6} \times 0.0449 = 0.1169$$

$$P_4 = \frac{(50/20)^4}{6 \times 3} \times 0.0449 = 0.0974$$

$$P_5 = \frac{(50/20)^5}{6 \times 9} \times 0.0449 = 0.0812$$

The average number of customers waiting for service is

$$L_q = \frac{(50/20)^3 \times 50 \times 20}{2! \times (3 \times 20 - 50)^2} \times 0.0449 = 3.51$$

The average number of customers in the system is

$$L_s = 3.51 + 2.5 = 6.01$$

The average time spent in the queue is

$$T_q = \frac{3.51}{50} = 0.0702 \text{ hours} = 4.212 \text{ minutes}$$

The average time spent in the system is

$$T_s = 0.0702 + \frac{1}{20} = 0.1202 \text{ hours} = 7.212 \text{ minutes}$$

Whether the queue characteristics are acceptable will depend on the business context. In terms of understanding the options available, the exercise could be repeated using four or five or more service points (let $S = 4$, or 5 or more). To make the problem more realistic, costs could also be considered. However, as we increase the complexity of the model, we increase the complexity of the mathematics and a point can be reached where the mathematics cannot be extended usefully.

21.8 Conclusions

In this chapter, we have developed mathematical models to describe stock control and the formation of queues. As we attempt to make these models more realistic, e.g. by including cost or more uncertainty, the mathematical description becomes more complex. These models have found useful business applications but it is important to recognize their limitations. It is always worth checking, for example, how meaningful the assumptions are. If demand tends to cluster in a predictable way, for example, this has major implications for how we model the business situation.

Some computer software is available for stock control and queuing models. This software is based on the type of mathematics presented in this chapter and essentially allows quicker computation. These models can also be developed in a spreadsheet format. However, merely analysing a model, even a very good model, does not necessarily provide the basis for the best business decision. A model tends to describe the business situation in one limited way. The economic order quantity model used in stock control, for example, is now seen as rather dated and management interest has turned to techniques like materials requirement planning (MRP) and just-in-time (JIT).

21.9 Problems

1 The demand for brackets of a particular type is 3000 boxes per year. Each order, regardless of the size of order, incurs a cost of £4. The cost of holding a box of brackets for a year is reckoned to be 60p. Determine the economic order quantity and the frequency of ordering.

2 A company needs to import programming devices for numerically controlled machines it produces and services. The necessary arrangements for import include considerable paperwork and managerial time which together have been estimated to cost the company £175 per order. Each item is bought at a cost of £50 and is not subject to discount. The cost of holding each item is made up of a storage charge of £2 and a capital charge of 12% per annum of the value of stock. Demand is for 700 programming devices per year.

(a) Determine the economic order quantity and the frequency of ordering.

(b) Determine the total cost of the ordering policy.

(c) If the lead time is one month and demand can be assumed constant, what should the reorder level be?

3   A company requires 28 125 components of a particular type each year. The ordering cost is £24 and the holding cost 10% of the value of stock held. The cost of a component is reduced from £6 to £5.80 if orders for 2000 or more are placed and reduced to £5.70 if orders for 3000 or more are placed. Determine the ordering policy that minimizes the total inventory cost.

4   Demand for a product follows a Normal distribution with a mean of 15 per week and standard deviation of four per week. If the lead time is 1 week, to what level should the reorder level be set to ensure that the probability of not meeting demand is no more than:
   (a)   5%
   (b)   1%?

5   If demand for a product follows a Poisson distribution with a mean of two items per week, to what level should the reorder level be set if:
   (a)   the probability of not meeting demand is to be no higher than 5% and the lead time is 1 week;
   (b)   the probability of not meeting demand is to be no higher than 1% and the lead time is 2 weeks?

6   Tourists arrive randomly at an Information Centre at an average rate of 24 per hour. There is only one receptionist and each enquiry takes 2 minutes on average. Determine
   (a)   the probability of queuing
   (b)   the probability of not having to queue
   (c)   the probabilities of there being one, two or three customers in the queue
   (d)   the average number of customers in the system
   (e)   the average number of customers in the queue
   (f)   the average time spent in the system, and
   (g)   the average time spent in the queue.

7   Products arrive for inspection at the rate of 18 per hour. What difference would it make to:
   (a)   the average number of products in the system
   (b)   the average number of products in the queue
   (c)   the average time in the system, and
   (d)   the average time in the queue.
   If the service rate could be increased from 24 to 30 per hour?

8   A maintenance engineer is able to complete five jobs each day on average, at a cost to the company of £110 per day. The number of completed jobs could be increased to seven each day if the engineer was given some unskilled assistance, at a total cost to the company of £150 per day. On average there are 4 new jobs each day. It has been estimated that the cost to the company of a job not done (an opportunity cost) is £200 per day. Should the engineer be given some unskilled assistance?

9   Lorries carrying animal feed can be unloaded manually or automatically. On average, eight lorries arrive each hour. The manual system can unload lorries at the rate of 11 per hour whereas the automatic system can deal with 13 per hour. The hourly cost of the manual system is £80 and the automatic system £95. If the cost of keeping a lorry waiting is £12 per hour, which

system is most cost effective?

10 A fast-food outlet, McBurger, has increased the number of service points from three to four. Customers still arrive at the rate of 50 per hour and each service point can handle, on average, 20 customers per hour. Determine the characteristics of the queue for the four-service-point system and compare with the three-service-point system.

11 Cars approach a three-tunnel system at the rate of 30 per hour. Each tunnel can accept 18 cars per hour. Determine the characteristics of the queue.

21.10 Appendix – proof of EBQ

This appendix need only concern those interested in the proof of the economic order quantity formula:

$$TVC = THC + TOC$$

$$= C_H \times \frac{Q}{2} + C_0 \times \frac{D}{Q}$$

We can differentiate this function with respect to Q (see Chapter 25) to obtain

$$\frac{d(TVC)}{dQ} = \frac{C_H}{2} - C_0 \frac{D}{Q^2}$$

To find a turning point we set this function for gradient equal to zero:

$$\frac{C_H}{2} - C_0 \frac{D}{Q^2} = 0$$

$$Q^2 = \frac{2 \times C_0 \times D}{C_H}$$

$$Q = \pm \sqrt{\left(\frac{2 \times C_0 \times D}{C_H} \right)}$$

To identify a maximum or a minimum we can differentiate a second time to obtain

$$\frac{d^2(TVC)}{dQ^2} = \frac{2 \times C_0 \times D}{Q^3}$$

This second derivative is positive (and thus identifies a turning point which is a minimum) when Q is positive. Hence the order quantity that minimizes total variable cost is

$$Q = +\sqrt{\left(\frac{2 \times C_0 \times D}{C_H} \right)}$$

22 Simulation

Mathematical models can be classified as being of one or two types: analytical or simulation. *Analytical models* are those solved by mathematical techniques, e.g. differentiation. We have seen in earlier chapters how aspects of business can be described using equations and how a solution can be obtained using appropriate techniques, e.g. breakeven analysis or linear programming. The model can be relatively simple, e.g. a linear equation relating total cost to output ($c = a + bx$) or fairly complex, e.g. the characteristics of multi-channel queues. The application of analytical models is limited by the necessary assumptions, e.g. linear constraints or arrivals following a Poisson distribution and the complexity of the mathematics. We have seen how intractable the mathematics can become as we attempt to make the model more realistic in business terms. When developing a model for queuing, we did not allow for the possibility of customers leaving very long queues, for example.

An alternative to solution by mathematical manipulation is simulation. *Simulation models* may also be specified in terms of equations and distributions but a solution is sought through experimentation rather than derivation. In a business application (there are many others), a simulation model attempts to imitate the reality of the business system in the same way that a flight simulator attempts to imitate aircraft flight. The flight simulator allows the experience of flight to be repeated any number of times, under varying conditions, and the consequences studied. A business simulation allows the business situation to be studied under a range of conditions. Managers, for example, may take the decisions in a simulation exercise and in a matter of hours be given the experience of running a business over a number of years. Simulation can be very quick and can effectively address the 'what if' type of question while at the same time incorporating other factors such as competitors reactions or step functions.

Simulation is included so that a manager can see the full range of quantitative methods that are available. It is likely that managers with the Ressembler Group (see Case 6) would have seen role play used with the training of sales staff and accepted the usefulness of such simulation but may not have considered the use of simulation to model a number of other business situations. This chapter is

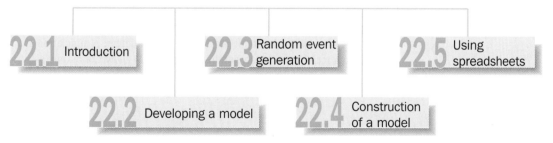

Figure 22.1 Structure of Chapter 22.

intended to introduce the principles of simulation and encourage the manager to consider the use of this approach.

Objectives

After working through this chapter, you should be able to:

- understand the construction of simulation models
- understand and apply random number generation
- develop simple simulation models by hand
- develop simple simulation models using spreadsheets
- understand the characteristics required of good simulation models.

22.1 An introduction to simulation models

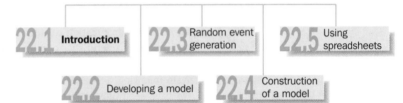

Typically, a simulation model will attempt to describe a business system by a number of equations. These equations are characterized by four types of variable.

1 Input variables are determined outside of the model; they are exogenous, and are subject to change for a particular simulation. It is the input variables that create the business situation; they give the model circumstances, e.g. increasing or decreasing demand. An input variable can be controllable, e.g. the manager may decide the reorder level, or uncontrollable, e.g. demand may be probabilistic and follow a known distribution.

2 Parameters (fixed variables) are input variables given a constant value for a particular simulation exercise. If, for example, a variable cost was allowed to increase during the simulation it would be regarded as an input variable; however, if its value were kept constant it would be a parameter.

3 A status variable, generally not included in the equations, gives definition to the simulation model. Status variables describe the state of the system. If, for example, the pattern of demand varies according to the month of the year, the status variable would specify the month. If a set of equations apply to the shoe industry and not the car components industry then this again is a matter of status.

4 The output variables provide the results of interest, e.g. the economic batch quantity or the number of service points that minimize queuing costs.

As the input variables are allowed to change, perhaps following a known distribution, the consequences on the output variables can be studied. The usefulness of any particular simulation model will depend on how well the equations relate the output variables to the input variables and parameters. A simulation may be 'run' through many times, even hundreds of times, to allow the output pattern to emerge.

22.2 Developing a simple simulation model

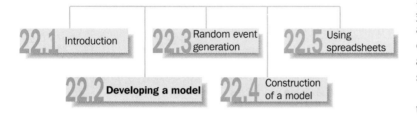

22.1 Introduction

22.3 Random event generation

22.5 Using spreadsheets

22.2 Developing a model

22.4 Construction of a model

Suppose you were invited to play a game of chance. If a fair coin shows a head you win £1 and if it shows a tail you lose £2. In this case the input variable represents a head or tail, the value that can be won or lost a parameter, the property of the coin (P(head) = $\frac{1}{2}$) a status variable and the actual amount won or lost each round the output variable. The system is represented in Figure 22.2.

To understand the characteristics of the system we would need to 'run' the model a number of times and observe the outcomes. To simulate the fairness of play, the input values are based on random numbers. The following sequence has been obtained from a random number generator:

$$6 \quad 7 \quad 8 \quad 8 \quad 9 \quad 9 \quad 9 \quad 2 \quad 5 \quad 9 \quad 3 \quad 0$$

To obtain a sequence of heads and tails, an even number (including zero as even) is taken to represent a head, and an odd number a tail. This rule can be written as a **look-up table** – see Table 22.1.

Table 22.1

0, 2, 4, 6, 8	Head
1, 3, 5, 7, 9	Tail

This will now give the sequence:

$$H \quad T \quad H \quad H \quad T \quad T \quad T \quad H \quad T \quad T \quad T \quad H$$

and the outcomes are then

$$£1 \quad -£2 \quad £1 \quad £1 \quad -£2 \quad -£2 \quad -£2 \quad £1 \quad -£2 \quad -£2 \quad -£2 \quad £1$$

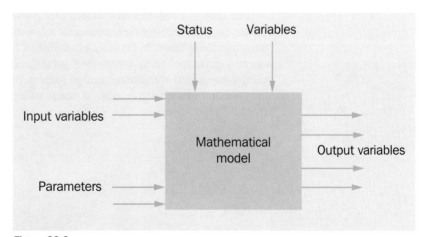

Figure 22.2

The average win in this short sequence is $-£9/12 = -£0.75$ (the total of the previous line divided by the number of events simulated).

We have not achieved the expected result of $-£0.50$ (see Section 8.5 on expected values) which is only likely to emerge with any stability over a longer run of plays, although even this is not certain, (see the comments in Chapter 8 on the frequency definition of probability).

Exercise

> Repeat the above simulation with a set of 20, 30 and 50 random numbers and compare the average win values.

One major advantage of simulation is that we can consider changes to the system that can make the mathematics far more difficult but yet require only modest adaptation to our simulation method.

Suppose the rules of the 'coin' game are changed to your advantage. If you get two heads in a row you win an extra £2, and if you get three heads in a row you win an extra £3 and so on.

Suppose the next set of random numbers give

$$6 \quad 1 \quad 5 \quad 0 \quad 0 \quad 4 \quad 3 \quad 9 \quad 8 \quad 2 \quad 1 \quad 8$$

the sequence of heads and tails would be

$$H \quad T \quad T \quad H \quad H \quad H \quad T \quad T \quad H \quad H \quad T \quad H$$

and the outcome

$$£1 \quad -£2 \quad -£2 \quad £1 \quad £1 \quad £1(+£3) \quad -£2 \quad -£2 \quad £1 \quad £1(+£2) \quad -£2 \quad £1$$

the average win in this case is $£2/12 = £0.17$.

To be sure of a reliable result, the number of trials would need to be increased.

Exercise

> Outline how you would solve the above problem mathematically.

Simulation is a step-by-step approach that gives results by iteration. The procedure can be effectively described using a flowchart and is particularly suitable for computer solution. In practice, all simulation is computer based. Procedures can be programmed using a 'high-level' language such as FORTRAN or BASIC or using a specialized simulation language such as GPSS (General Purpose Simulation System). In a limited number of cases, a simulation exercise can be undertaken on a spreadsheet.

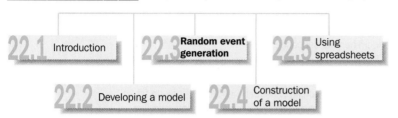

22.3 Random event generation

Using random numbers is an attempt to remove the chance of bias from any experimentation. In the previous example, we wanted to simulate the outcomes from a fair coin and used random even and odd numbers to do this. A sequence of alternating heads and tails, e.g. H T H T H T H is possible but should be no more likely than any other sequence. Typically, random number tables will give clusters of digits, e.g.

$$14 \quad 86 \quad 50 \quad 97 \quad 03$$

or

$$46905 \quad 41003 \quad 84341 \quad 28752$$

How you use such numbers clearly depends on the type of input variable you need to simulate.

If you are using random number tables, and always start from the top left-hand corner, the input values will be predictable and therefore not random. To ensure a random sequence from random number tables, you should always cross-off the used numbers. You should also check any random number generator to ensure a different sequence is generated for each simulation.

Random numbers can be used to generate all sorts of random events.

Example

If we know that in a given population, 20% of people are smokers, then given the following sequence of random numbers, we can simulate the sample selection of smokers.

Random Number sequence:

$$31338 \quad 28729 \quad 02095 \quad 07429$$

The first step will be to construct a look-up table. Since we know 20% of the population are smokers we need to assign two out of every 10 digits to represent smokers. This can give the following table:

Non-smokers	0–7
Smokers	8, 9

Note that we have been able to use the random numbers **singly** here since the original percentages were multiples of 10.

The result of this simulation would be:

NS, NS, NS, NS, S, NS, S, NS, NS, S, NS, NS, NS, S, NS, NS, NS, NS, NS, S

Example

It is estimated that 5% of a production output has one fault, and 1% has more than one fault. The rest are OK. Use the following sequence of random numbers to simulate a sampling procedure.

Random number sequence:

$$80272 \quad 64398 \quad 88249 \quad 06792$$

Here, because the given percentages are parts of 100, we need to assign the random numbers in **pairs**, and can construct the following look-up table:

No faults	0–93
One fault	94,95,96,97,98
More faults	99

The result of this simulation would be:

No. of Faults: 0, 0, 0, 0, 1, 0, 0, 0, 0, 0

Again, the digits could be assigned differently but in the same proportions. The following three examples show the generation of random events from the Binomial, Poisson and Normal distributions. In each case the following sequence of random numbers is used (to demonstrate method):

4　6　7　2　3　6　4　8　8　6　4　1　1　0　6　6　9　9　1　9

Example

In a market research test, groups of five adults are selected and their views on beverages sought. If 60% of adults prefer tea to coffee, simulate the possible groupings in terms of this preference.

This distribution is Binomial with $n = 5$. To use the available tables (Appendix A), we will need to let $p = 0.4$ (tables only give values for p up to 0.5) and consider the number of individuals that prefer coffee to tea. The results are shown in Table 22.2.

Using this probability information, we are able to assign sets of four random digits to each of the possible outcomes as shown in Table 22.3.

Using the random numbers given, the selected groups of five would have the following number who prefer tea: 3, 3, 4, 2, 5. This set of values could then be input to the simulation model.

Table 22.2

No. who prefer coffee (r)	Cumulative prob. (r or more)	Exact prob (r)	No. who prefer tea
0	1.0000	0.0778	5
1	0.9222	0.2592	4
2	0.6630	0.3456	3
3	0.3174	0.2304	2
4	0.0870	0.0768	1
5	0.0102	0.0102	0

Table 22.3

No. who prefer tea	Probability 'weight'	Allocation of random numbers
0	102	0000 to 0101
1	768	0102 to 0869
2	2304	0869 to 3137
3	3456	3174 to 6629
4	2592	6630 to 9221
5	778	9222 to 9999

Even from these figures the variation in the input data is apparent. It is possible to use the simulation model to consider the effect of more control on input variation, or in the case of this example, managing demand more carefully.

Example

The demand for a particular component follows a Poisson distribution with a mean of two per day. Generate input data for a simulation model of inventory costs.

Given the Poisson parameter of $\lambda = 2$ (see Appendix B), we can allocate random numbers as shown in Table 22.4.

Taking the random digits in sets of four, the number of components required over five simulated days is 2, 1, 4, 0, 6.

Table 22.4

Component demand	Cumulative probability	Exact probability	Allocation of random numbers
0	1.0000	0.1353	0000 to 1352
1	0.8647	0.2707	1353 to 4059
2	0.5940	0.2707	4060 to 6766
3	0.3233	0.1804	6767 to 8570
4	0.1429	0.0902	8571 to 9472
5	0.0527	0.0361	9473 to 9833
6	0.0166	0.0155	9834 to 9988
7 or more	0.0011	0.0011	9989 to 9999

Example

The time taken to complete a task follows a Normal distribution with a mean of 30 minutes and standard deviation of four minutes. Simulate events from this distribution.

The Normal distribution is continuous, and we need to consider the probability of an event being within a given interval. The determination of probability and the allocation of random numbers is shown in Table 22.5.

In this case, taking the random digits in sets of five, the time intervals are less than 22, 22 but less than 26 minutes, 26 but less than 30 minutes, 30 but less than 34 minutes, 34 but less than 38 minutes and 38 or more minutes. If the simulation model requires a more precise time, then the size of the interval would need to be reduced, e.g. 29 but less than 30 minutes.

Table 22.5

Interval	z (lower boundary)	Probability	Allocation of random numbers
<22		0.02275	00000 to 02274
22 < 26	−2	0.13595	02275 to 15869
26 < 30	−1	0.3413	15870 to 49999
30 < 34	0	0.3413	50000 to 84129
34 < 38	1	0.13595	84130 to 97724
38 or more	2	0.02275	97725 to 99999

As stated earlier, Tables 22.3–22.5 are often referred to as 'look-up' tables, since they allow you to look up the effect on the model of a particular random number.

Random numbers generated by computer may not be strictly random, since they are often produced using some form of algorithm. This is unlikely to be a problem for the small-scale simulations we are discussing here, but you can perform a check by generating a long sequence of random numbers and testing to see that the distribution of each digit (0–9) is uniform.

22.4 The construction of a simulation model

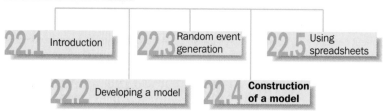

The generation of random event values for the input variables is only one aspect of simulation. The models developed for a business system need to make business sense. A model can only be as good as its specification, and the relationships between variables need to be investigated and understood. Building the model is, in many ways, the easiest part. The most difficult bit is understanding how the situation actually functions and the range of possible outcomes at each stage of the process. Suppose, for example, that a simulation model was developed to manage demand for an assembled product but did not take account of the need to supply replacement parts. The model may meet the requirements of production planning but not the needs of the purchasing department. The purchasing department would require an estimate that included components for assembly and replacement.

Simulation modelling should demonstrate an understanding of the business system. It involves:

1. the formulation of the problem
2. problem analysis
3. model development
4. implementation
5. the questioning of results; and
6. the questioning of the model.

Any of these steps may be repeated a number of times in the light of experience. To understand the business problem a description will be required from those concerned – the problem owners. See Figure 22.3.

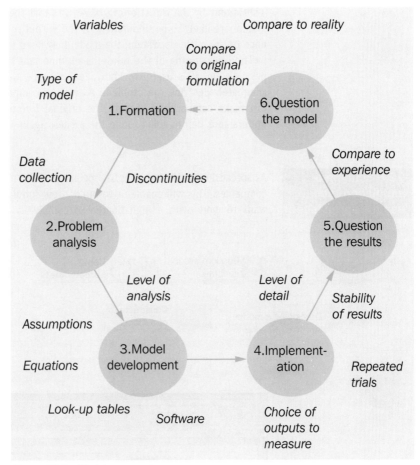

Figure 22.3 Model construction.

Formulation of the problem requires an agreement on what the variables will be and to what extent they should be controllable. It also needs to be recognized that simulation is only one method of problem solution and there are likely to be others. Simulation may model the existing stock control system, for example, but should the problem owners consider alternative systems of stock management such as 'just-in-time'? Having agreed the type of model and the variables, the system will need further investigation and observation.

Problem analysis will involve the collection of data and other information on the variables included.

Model development involves the description of the business system by a set of equations. It may be necessary to make some simplifying assumptions, e.g. that the relationships are linear, but the equations must reasonably relate the output variables to the input variables and parameters.

The implementation of a simulation model is likely to involve the repetition of a set of calculations using a computer. It is only when some stability emerges in the pattern of output that the number of repetitions is likely to be sufficient. We can then apply statistical method to these output results, e.g. calculate averages. It is not for the problem owner, i.e. the manager, just to accept the output results; rather he/she should question their meaning.

It is only by the **questioning of results** that the implications for the business system are likely to be understood. If the output results do not match the experience of the business system, the model may need to be improved.

The **questioning of the model** is an important step in understanding the business system. It is useful to know, for example, that linear relationships do not adequately describe the problem. As well as comparing the model to reality, we can make comparisons back to the original formulation, and build in any new understands before going round the process again.

22.5 Building a simulation model using a spreadsheet

As spreadsheets develop, they are given more and more functions by the software companies. This will enable us to build a few simulation models where we do not want to work with a high degree of complexity, but where the mathematics might prove complicated. One particularly useful feature in this context is the ability of spreadsheet cells to generate random numbers.

Load the spreadsheet COINS from the website and look at the structure used (see Figure 22.4).

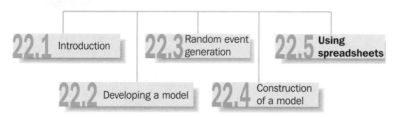

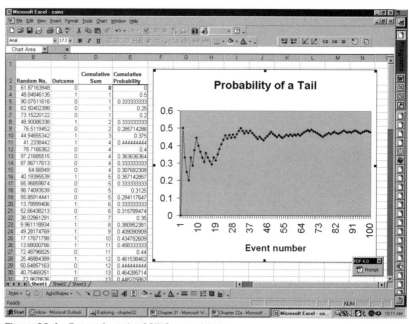

Figure 22.4 Extract from the COINS spreadsheet.

The first column is a simple record of the number of events, while the second column uses the inbuilt function RAND() in Excel and multiplies the result by 100 so that we get a sequence of random numbers from 0 to 100. The third column uses a logic statement to determine whether we have a Head (0) or a Tail (1). Between them, these three columns represent the basic simulation, and they can be extended for as many rows (events) as you wish. The Cumulative Sum column keeps a record of the number of Tails so far and the final column is a running calculation of the probability of a Tail (number so far divided by the number of events simulated).

It is often useful to graph the output variable in order to see if it is converging to a particular figure. In this case, you would graph the fifth column, and as you should expect in this case, it converges to a value of 0.5.

Exercise

> Run the COINS simulation 20 times and make a note of the final Probability of a Tail. (Press F9 to run the simulation once) What do these results show?

As a second example, look at the simple queue spreadsheet on the website (see Figure 22.5). In this case, the average amount of time spent in the queue is not as obvious as the probability used in the last example. (It could be calculated using the formulae in Chapter 21.)

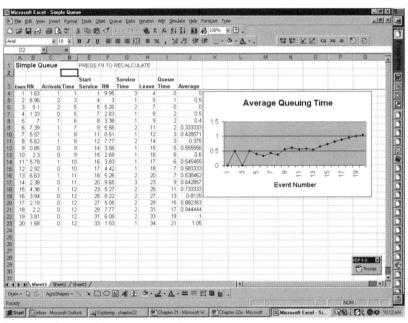

Figure 22.5 Extract from the Simple Queue spreadsheet.

Here we have chosen to measure time in the queue as our output variable, although others would be possible. The look-up tables used in construction this simulation are shown below.

Table 22.6

Random number	Inter-arrival time
<4	1
4 < 8	2
8+	3

Table 22.7

Random number	Service time
<5	1
5 < 9	2
9 +	3

As you can see from the graph on the spreadsheet, the average time in the queue is very low given these parameters.

Exercise

> Run the Simple Queue simulation 10 times and make a note of the average queue time after 20 events in each case (press F9 to run once). Are there any conclusions that you can draw from this?
>   Edit cells C4 and/or G4 to change the parameters of the model and note the effects on the average time in the queue.

22.6 Conclusions

Simulation provides a useful alternative to other methods of problem solving. Equations that can be solved with relative ease for simple business situations can become increasingly difficult as we attempt to take account of the realities of business. In practice, all business simulation is likely to be computer based. It is not the calculations that require the investment of time but rather the development of an adequate mathematical model.

Simulation does not provide all of the answers, but it does allow the repeated running of a situation to check for stability and various 'what if …' possibilities. The graphical interface of some specialist packages makes it easy to explain and use with managers, and thus help them in making decisions.

22.7 Problems

1 Why use simulation to solve business problems when most equations can be solved using other techniques provided a number of simplifying assumptions are made?

2 Why are random numbers important in simulation?

3 Lead time for a particular component can be 1, 2 or 3 weeks with respective probabilities of 0.3, 0.2 and 0.5. Using random numbers generate a sequence of 20 lead times.

4 Use random numbers to take ten random samples from:
   (a) a Binomial distribution where $n = 5$ and $p = 0.45$
   (b) a Poisson distribution with a mean of six, and
   (c) a Normal distribution with a mean of 70 and standard deviation of five.

5 The number of cars arriving each hour has the following probability distribution:

Arrivals	0	1	2	3	4	5
Probability	0.12	0.24	0.33	0.14	0.12	0.05

The service times vary according to the needs of the customer. There is a 40% chance that two cars are serviced in one hour and a 60% chance that three cars are serviced. Simulate a 35-hour working week and assess the maximum queue length.

6 Use the COINS spreadsheet (use sheet 2) and note the probability of a tail at 10, 100, 500, 1000, and 5000 events. Run the simulation 20 times and record the probability at each point each time. Find the average for each point. What can you imply about both probability and simulation from your results?

7 Use the SIMPLE QUEUE spreadsheet and extend the sequence to 500 events (do this by copying the last row down the spreadsheet). Record the average time in the queue at 10, 50, 100 and 500 events. Run the simulation30 times, again recording the average time at these points. Now calculate the average answer at each point. What can you deduce from these results?

Part 6   Conclusions

Part 6 has presented a range of models that have been applied to business problems. The communication of ideas and the justification of solutions is particularly important in business, and models provide an effective description of how a problem is perceived and a means for examining the consequences of any actions that might be taken. Once a model has been developed it can be compared to the reality of the workplace. A model may allow us to look at the original problem in different ways, ask different questions and consider different data. The whole process of problem-solving can be transformed, if we have new ways of exploring the original problem and more creative ways of developing solutions.

The models may find direct application, if they happen to 'fit' a current problem. Perhaps of more importance is the **competence to model**. Each of the models considered illustrates good modelling practice and the approaches can be adapted to solve other problems. Managers will be interested in measures of performance now and in the future, they will need to make choices subject to constraints, they will need to manage change over time, they will need to understand complex systems and they should benefit from a good understanding of the approaches to modelling.

Part 7 Mathematical background

Part 7 differs significantly from the other parts of the book, since it deals with a certainty which can exist in mathematical relationships but which does not arise within statistical relationships. It is also different in the way in which it deals with the topics. Because of the 'theoretical' nature of this subject matter we have decided not to include a case study for this section, but you will find that we draw a large number of our examples from economics and economic theory.

You may already have read through and used *Improve Your Maths* at the start of this book, and it is the intention of this part to build on that basic foundation. Mathematics offers a concise, exact method of describing a situation together with a set of analysis tools which are well proven and well used. Such analysis allows different perspectives to be explored and new relationships to be determined. These can then be tested in the practical context by the use of the statistical techniques which have been introduced in the earlier Parts of this book. Subjects such as microeconomic theory can be developed almost wholly using mathematics, and if your course uses mathematics extensively in this way, then this part will give you the background that you need.

23 Mathematical relationships

The explicit use of mathematics within business courses has changed over recent years from being a strong, separate part of the course in quantitative methods, to more of a background subject, referred to when needed by other techniques, but which students are often assumed to bring with them. This is not always true, and reference to *Improve Your Maths* could be helpful. This chapter aims to build on the sort of basic knowledge you will have developed and illustrate the ways in which a mathematical model may help to understand business situations.

The use of mathematics provides a precise way of describing a range of situations. Whether we are looking at production possibilities or constructing an economic model of some kind, mathematics will effectively communicate our ideas and solutions. Reference to x or y should not take us back to the mysteries of school algebra but should give us a systematic way of modelling and solving some problems. Most things we measure will take a range of values, for example, people's height, and mathematically they are described by variables. A variable z, for example, could measure temperature throughout the day, the working speed of a drilling machine or a share price. The important point here is that the vari-

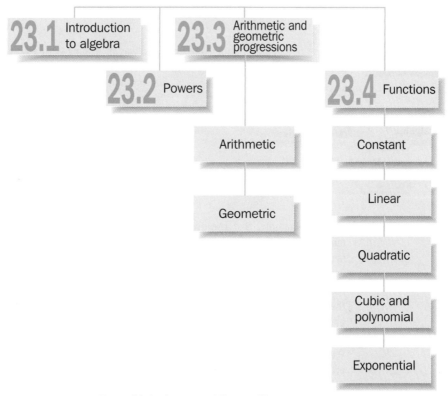

Figure 23.1 Structure of Chapter 23.

able z is what we want it to be. The value that z actually takes will depend on a number of other factors. Temperature throughout the day will depend on location; the working speed of a drilling machine on the materials being used; and share price on economic forecasts.

Suppose you wish to calculate your weekly pay. If you work 40 hours each week and you are paid £10 per hour, then your weekly pay will be $40 \times £10 = £400$, which is a specific result. However, if the number of hours you work each week and the hourly rate of pay can both vary, you will need to go through the same calculation each week to determine your weekly pay. To describe this process, you can use mathematical notation. If the letter h represents the number of hours worked and another letter p represents the hourly rate of pay in pounds per hour, then the general expression hp describes the calculation of weekly pay. If $h = 33$ and $p = 11$ then weekly pay will equal £363. It should be noted at this point that the symbol for multiplication is used only when necessary; hp is used in preference to $h \times p$ although they mean exactly the same. To show the calculation of the £363 we would need to write 33 hours $\times £11$ per hour.

Objectives

After working through this chapter, you should be able to:

- use linear functions for breakeven analysis
- state the format of mathematical progressions
- solve quadratic functions for roots
- find equilibria points for simple economic situations.

23.1 Introduction to algebra

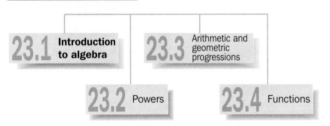

The use of algebra provides one powerful way of dealing with a range of business problems. As we shall see, a single answer is rarely the result of one correct method, or likely to provide a complete solution to a business-related problem. Consider **breakeven analysis** as an example. Problems are often specified in such a way that a single value of output can be found where the total cost of production is equal to total revenue, i.e. no profit is being made. This single figure may well be useful. A manager could be told, for example, that the division will need to break even within the next six months. This single figure will not, however, tell the manager the level of output necessary to achieve an acceptable level of profits by the end of the year. Even where a single figure is sufficient, it will still be the result of a number of simplifying assumptions. If, for example, we relax the assumption that price remains the same regardless of output the modelling of this business problem will become more complex.

Suppose a company is making a single product and selling it for £390. Whether the company is successful or not will depend on both the control of production costs and the ability to sell. If we recognize only two types of cost, fixed cost and variable cost, we are already making a simplifying assumption. Suppose also that fixed costs, those that do not vary directly with the level of output (e.g. rent, lighting, administration), equal £6500 and variable costs, those costs that do vary directly with the level of output (e.g. direct labour, materials), equal £340 per unit.

The cost and revenue position of this company could be described or modelled using a spreadsheet format. Output levels could be shown in the first column (A) of the spreadsheet. The corresponding levels of revenue and costs could then be calculated in further columns (assuming that all output is sold). Finally a profit or loss figure could be determined as shown in Table 23.1.

Table 23.1 Format of spreadsheet to determine breakeven point for single-product company

	A Output	B Revenue	C Fixed cost	D Variable cost	E Total cost	F Profit/loss
1	100	39 000	6500	34 000	40 500	−1500
2	110	42 900	6500	37 400	43 900	−1000
3	120	46 800	6500	40 800	47 300	−500
4	130	50 700	6500	44 200	50 700	0
5	140	54 600	6500	47 600	54 100	500
6	150	58 500	6500	51 000	57 500	1000

Algebra provides an ideal method to describe the steps taken to construct this spreadsheet. If we use x to represent output and p to represent price then revenue r is equal to px. Finally, total cost c is made up of two components: a fixed cost, say a, and a variable cost which is the product of variable cost per unit, say b, and the level of output. To summarize:

$$r = px$$

and

$$c = a + bx$$

Profit, usually written as π (it would be confusing to use p for price and profit), is the difference between revenue and total cost:

$$\pi = r - c$$
$$= px - (a + bx)$$

It should be noted here that the brackets signify that all the enclosed terms should be taken together. The expression for profit can be simplified by collecting the x terms together:

$$\pi = px - bx - a$$

The minus sign before the bracket was an instruction to subtract each of the terms included within the bracket. As x is now common to the first two terms we can take x outside new brackets to obtain:

$$\pi = (p - b)x - a$$

or using the numbers from the example above:

$$\pi = (390 - 340)x - 6500$$
$$= 50x - 6500$$

These few steps of algebra have two important consequences in this example. Firstly, we can see that profit has a fairly simple relationship to output, x. If we would like to know the profit corresponding to an output level of 200 units, we merely let $x = 200$ to find that profit is equal to $50 \times 200 - 6500$ or £3500. Secondly, algebra allows us to develop new ideas or concepts. The difference shown between price and variable cost per unit $(p - b)$, or £50 in this example, is known as **contribution to profit**. Each unit sold represents a gain of £50 for the company; whether a loss or a profit is being made depends on whether the fixed costs have been covered or not.

Consider now the breakeven position. If profit is equal to 0, then

$$0 = 50x - 6500$$

To obtain an expression with x on one side and numbers on the other, we can subtract $50x$ from both sides:

$$-50x = -6500$$

To determine the value of x, divide both sides by -50:

$$x = \frac{-6500}{-50} = 130$$

You need to remember here that if you divide a minus by a minus the answer is a plus.

The purpose of the above example is not to show the detail of breakeven analysis but rather the power of algebra. If we consider the above division, the following interpretation is possible: the breakeven level is the number of £50 gains needed to cover the fixed cost of £6500. You are likely to meet the concept of contribution to profit in studies of both accountancy and marketing.

In reality, few companies deal with a single product. Suppose a company produces two products, x and y, selling for £390 and £365 and with variable cost per unit of £340 and £305 respectively. Fixed costs are £13 700. A spreadsheet model could be developed as shown in Table 23.2 (again assuming all output is sold).

The spreadsheet reveals that a two-product company can break even in more than one way. The notation used so far can be extended to cover companies producing two or more products. Using x and y subscripts on price and variable costs, for a two-product company we could write

Table 23.2 Format of spreadsheet to determine breakeven points for a two-product company

A	B	C	D	E	F	G
Output X	Output Y	Revenue	Fixed cost	Variable cost	Total cost	Profit/loss
120	110	86 950	13 700	74 350	88 050	−1100
130	110	90 850	13 700	77 750	91 450	−600
140	110	94 750	13 700	81 150	94 850	−100
142	110	95 530	13 700	81 830	95 530	0
120	120	90 600	13 700	77 400	91 100	−500
130	120	94 500	13 700	80 800	94 500	0
140	120	98 400	13 700	84 200	97 900	500

$$r = p_x x + p_y y \qquad c = a + b_x x + b_y y$$

Profit is the difference between revenue and total costs:

$$\pi = (p_x x + p_y y) - (a + b_x x + b_y y)$$
$$= (p_x - b_x)x + (p_y - b_y)y - a$$

It can be seen in this case that profit is made up of the contribution to profit from two products less fixed cost. Substituting the numbers from the example we have

$$\pi = (390 - 340)x + (365 - 305)y - 13\,700$$
$$= 50x + 60y - 13\,700$$

You may recall that one equation with one variable or indeed two independent equations with two variables can be uniquely solved. As we have shown, one equation with two variables does not have a unique solution but rather a range of possible solutions. In a business context, a range of solutions may be preferable as these offer management a choice.

23.2 Powers

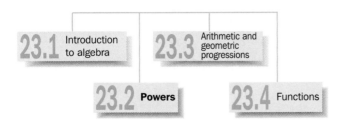

When multiplying the same quantity together several times, it will be much more convenient to use powers rather than to write that quantity down each time, thus:

$$a \times a = a^2$$
$$a \times a \times a = a^3$$
$$a \times a \times a \times a = a^4, \text{ etc.}$$
(where a is any number)

Once we begin to manipulate quantities raised to powers, we can develop rules as follows:

$$a^3 \times a^2 = (a \times a \times a) \times (a \times a) = a^5 = a^{3+2}$$

i.e. for multiplication, we *add* the powers.

$$a^3 / a^2 = \frac{a \times a \times a}{a \times a} = a^1 = a^{3-2}$$

i.e. for division, we *subtract* the powers. Note also that a on its own can be written as a^1.

Another example:

$$a^4 / a^4 = \frac{a \times a \times a \times a}{a \times a \times a \times a} = a^{4-4} = a^0 = 1$$

Using the rule for division, we can see that any number (apart from 0) divided by itself will give a result to the power zero. In terms of the number system, any number raised to the power zero will be equal to 1.

If we have a^3 / a^6, this is:

$$\frac{a \times a \times a}{a \times a \times a \times a \times a \times a} = \frac{1}{a \times a \times a} = a^{3-6} = a^{-3} = \frac{1}{a^3}$$

i.e. a negative power means that we take the *reciprocal* value.

Since we add powers when multiplying, then:

$$a^{1/2} \times a^{1/2} = a^1 = a$$

and the number which multiplied by itself gives *a* must be the square root of *a*, thus:

$$a^{1/2} = \sqrt{a}$$
$$a^{1/3} = \sqrt[3]{a} \text{ the cube root of } a$$
$$a^{3/2} = \sqrt{a^3} \text{ the square root of } a \text{ cubed, and so on}$$

If we are faced by brackets which are raised to powers, we treat them in the same ways; so that

$$x(a + b) = ax + bx$$
$$(a + b)^2 = (a + b)(a + b)$$
$$= a(a + b) + b(a + b)$$
$$= a^2 + ab + ba + b^2$$

and since

$$ab = ba = a^2 + 2ab + b^2$$
$$(a - b)^2 = a^2 - 2ab + b^2$$

(Note: $a^3 + a^2$ cannot be simplified by expressing it as *a* to a single power.)

23.3 Arithmetic and geometric progressions

In much of our work we are not just looking at a single number but rather a range of numbers. In developing a solution to a business problem, perhaps using a spreadsheet model, we are likely to generate a list of figures rather than a single figure. A **sequence** of numbers is just an ordered list, for example:

18, 23, 28, 33, 38, 43

and

9, 36, 144, 576, 2304, 9216

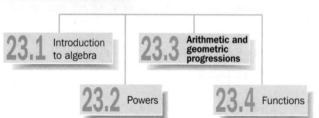

If the sequence can be produced using a particular rule or law, it is referred to as a **series** or a **progression**. It is particularly useful if we can recognize a pattern in a list. In the first example, 5 is being added each time and in the second example, the previous term is being multiplied by 4. As we shall see, the first sequence is an arithmetic progression and the second sequence is a geometric progression.

23.3.1 Arithmetic progression

A sequence is said to be an arithmetic progression if the difference between the terms remains the same, e.g. +5. This difference is called the **common differ-**

ence and is usually denoted by d. If a is the first term, then the successive terms of an arithmetic progression are given by

$$a, a + d, a + 2d, a + 3d, \ldots$$

The nth term is given by

$$t_n = a + (n - 1)d$$

Example

Given the sequence

18, 23, 28, 33, 38, 43

find the value of the 12th term.

Solution: Since $a = 18$ and $d = 5$

$$t_{12} = 18 + 11 \times 5 = 73$$

The sum of an arithmetic progression with n terms is given by

$$S_n = \frac{n}{2}[2a + (n - 1)d]$$

A proof of this formula is given in section 23.7 for the interested reader.

Example

Given the same sequence

18, 23, 28, 33, 38, 43

what is the sum of this sequence:

(i) with the 6 terms shown; and
(ii) continued to 12 terms?

Solution: By substitution:

(i) $S_6 = 6/2(2 \times 18 + 5 \times 5) = 3(36 + 25) = 183$.
(ii) $S_{12} = 12/2(2 \times 18 + 11 \times 5) = 6(36 + 55) = 546$.

Example

You have been offered a new contract. The starting salary is £12 000 per annum and there is an incremental increase of £800 at the end of each year. What will your salary be at the end of year 6 and what will your total earnings be over this period?

We could, of course, develop a spreadsheet solution to this problem as shown in Table 23.3.

As an alternative, we could substitute values as necessary:

$$t_6 = 12\,000 + 5 \times 800 = 16\,000$$

$$S_6 = 6/2(2 \times 12\,000 + 5 \times 800) = 84\,000$$

Table 23.3 Format of spreadsheet to show incremental salary increases

A Year	B Salary
1	12 000
2	12 800
3	13 600
4	14 400
5	15 200
6	16 000
Total	84 000

23.3.2 Geometric progression

A sequence is said to be a geometric progression if the ratio between the terms remains the same, e.g. 4. This constant ratio, r, is called the common ratio. The successive terms of a geometric progression are given by

$$a, ar, ar^2, ar^3, ar^4, \ldots$$

The nth term is given by

$$t_n = ar^{n-1}$$

Example

Given the sequence

$$9, 36, 144, 576, 2304, 9216$$

find the value of the 12th term.
 Solution: Since $a = 9$ and $r = 4$,

$$t_{12} = 9 \times (4)^{11} = 9 \times 4\,194\,304 = 37\,748\,736$$

It should be noted at this point that geometric progressions have a tendency to grow very quickly when the value of r is greater than 1. This observation has major implications if human populations or the debtors of a company can be modelled using such a progression.
 The sum of a geometric progression with n terms is given by

$$S_n = \frac{a(r^n - 1)}{r - 1}$$

A proof of this formula is given in section 23.8 for the interested reader.

Example

Given the same sequence

$$9, 36, 144, 576, 2304, 9216$$

what is the sum of this sequence:

(i) with the six terms shown; and

(ii) continued to 12 terms?

Solution: By substitution,

(i) $S_6 = \dfrac{9(4^6 - 1)}{4 - 1} = 9 \times \dfrac{4095}{3} = 12\,285$

(ii) $S_{12} = \dfrac{9(4^{12} - 1)}{4 - 1} = 9 \times \dfrac{16\,777\,215}{3} = 50\,331\,645$

Example

You have just been offered a different contract. The starting salary is still £12 000 per annum but the increase will be 6% of the previous year's salary rather than a fixed rate. What will your salary be at the end of year 6 and what will your total earnings be over this period?

We could again develop a spreadsheet solution as shown in Table 23.4.

Table 23.4 Format of spreadsheet to show percentage salary increases

A Year	B Salary
1	12 000.00
2	12 720.00
3	13 483.20
4	14 294.19
5	15 149.72
6	16 058.71
Total	83 703.82

Again, by the substitution of values, using $r = 1.06$ (to give a 6% increase per annum)

$$t_6 = 12\,000(1.06)^6 = 16\,058.71$$

and

$$S_6 = 12\,000\,\frac{(1.06^6 - 1)}{1.06 - 1} = 83\,703.82$$

23.4 Functions

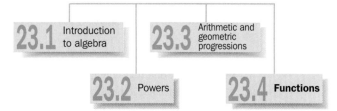

23.4.1 Constant functions

Here, no matter what value x takes, the value of y remains the same, as shown in Figure 23.2. The line representing $y = k$ passes through the y-axis at a value k and goes off, at least in theory, to infinity in both directions. Constant functions of this type will appear in linear programming problems.

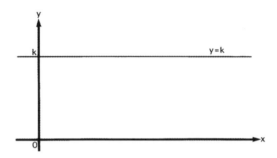

Figure 23.2

If the values of x for which $y = k$ are limited to a particular group (known as the **domain** of the function) then we may use the following symbols:

$$< x \text{ means less than } x;$$
$$> x \text{ means greater than } x;$$
$$\leq x \text{ means less than or equal to } x; \text{ and}$$
$$\geq x \text{ means greater than or equal to } x$$

Now if the constant function only applies between $x = 0$ and $x = 10$, then this will be as illustrated in Figure 23.3 and is written as:

$$y = k \text{ for } 0 \leq x \leq 10$$
$$= 0 \text{ elsewhere}$$

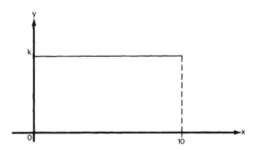

Figure 23.3

We could use this idea of the domain of a function to represent graphically a book price with discounts for quantity purchase, as in Figure 23.4.

23.4.2 Linear functions

A linear function is one that will give a straight line when we draw the graph (at least in two dimensions; it has a similar, but more complex meaning in three, four or more dimensions). This function occurs frequently when we try to apply quantitative techniques to business-related problems, and even where it does not

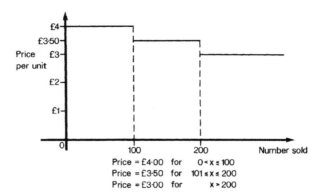

Price = £4·00 for   0 < x ≤ 100
Price = £3·50 for   101 ≤ x ≤ 200
Price = £3·00 for   x > 200

Figure 23.4

apply exactly, it may well form a close enough approximation for the use to which we are putting it.

If the value of y is always equal to the value of x then we shall obtain a graph as shown in Figure 23.5. This line will pass through the origin and will be at an angle of $45°$ to the x-axis, provided that both axes use the same scale. This function was used when we considered Lorenz curves. We may change the

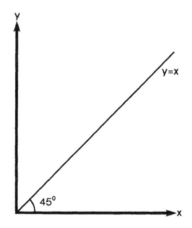

Figure 23.5

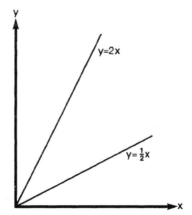

Figure 23.6

angle or slope of the line by multiplying the value of x by some constant, called a **coefficient**. Two examples are given in Figure 23.6. These functions still go through the origin; to make a function go through some other point on the y-axis, we add or subtract a constant. The point where the function crosses the y-axis is called the **intercept**. Two examples are given in Figure 23.7.

The general format of a linear function is $y = a + bx$, where a and b are constants. To draw the graph of a linear function we need to know two points which satisfy the function. If we take the function

$$y = 100 - 2x$$

then we know that the intercept on the y-axis is 100 since this is the value of y if $x = 0$. If we substitute some other, convenient, value for x, we can find another point through which the graph of the function passes. Taking $x = 10$, we have:

$$y = 100 - 2 \times 10 = 100 - 20 = 80$$

so the two points are (0, 100) and (10, 80). Marking these points on a pair of axes, we may join them up with a ruler to obtain a graph of the function (see Figure 23.8).

Fixed costs, such as £6500 and £13 700 in our breakeven examples, do not change with output level and would be represented by a horizontal straight line.

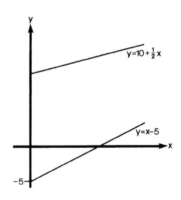

Figure 23.7

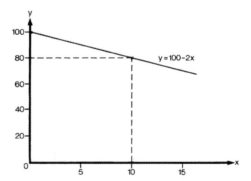

Figure 23.8

Exercise

> Given the following revenue and total cost functions, show graphically where they cross, and thus where breakeven occurs. Develop the spreadsheet model of these functions to confirm your answer.
>
> $$r = 390x$$
> $$c = 6500 + 340x$$
>
> where x is equal to production and sales.

To illustrate a technique which we used in linear programming we will solve a similar problem. A two-product company has a profit function as follows:

$$\pi = 50x + 60y - 13\,700$$

and we wish to show the breakeven position graphically ($\pi = 0$). We first let $x = 0$ and find the value of y and then let $y = 0$ and find the value of x. When $x = 0$, $y = 228$ (rounded down). When $y = 0$, $x = 274$. We can then plot a breakeven line, as in Figure 23.9, showing the various product combinations of output that allow the company to break even. On the same graph, we can also see how the company can make a loss or a profit.

We may use two linear functions to illustrate the market situation in economics, allowing one to represent the various quantities demanded over a range of prices, and the other to represent supply conditions. Where these two functions cross is known as the **equilibrium point** (point E in Figure 23.10), since at this price level the quantity demanded by the consumers is exactly equal to the amount that the suppliers are willing to produce, and thus the market is cleared. Note that we follow the tradition of the economics texts and place quantity on the x-axis and price on the y-axis. Figure 23.10 might illustrate a situation in which the **demand function** is:

$$P = 100 - 4Q$$

and the **supply function** is:

$$P = 6Q$$

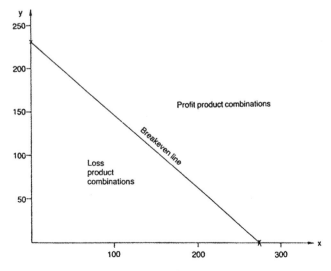

Figure 23.9

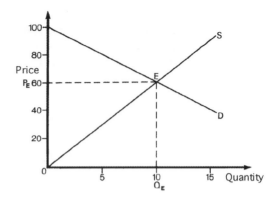

Figure 23.10

Since we know that the price (P_E) will be the same on both functions at the equilibrium point we can manipulate the functions to find the numerical values of P_E and Q_E:

$$\text{(Demand)} \qquad P = P \qquad \text{(Supply)}$$
$$100 - 4Q = 6Q$$
$$100 = 10Q$$
$$10 = Q$$

If $Q = 10$, then

$$P = 100 - 4Q = 100 - 40 = 60$$

thus $P_E = 60$ and $Q_E = 10$.

This system could also be used to solve pairs of linear functions which are both true at some point; these are known as **simultaneous equations** (a more general version of the demand and supply relationship above). Taking each equation in turn we may construct a graph of that function; where the two lines cross is the solution to the pair of simultaneous equations, i.e. the values of x and y for which they are both true. If

$$5x + 2y = 34 \qquad \text{and} \qquad x + 3y = 25$$

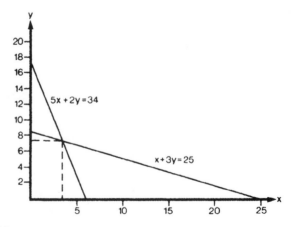

Figure 23.11

then reading from the graph in Figure 23.11, we find that $x = 4$ and $y = 7$ is the point of intersection of the two linear functions.

This system will work well with simple equations, but even then is somewhat time-consuming: there is a simpler method for solving simultaneous equations, which does not involve the use of graphs.

Looking at two examples will illustrate this point.

Examples

1 $5x + 2y = 34$

$x + 3y = 25$

If we multiply each item in the second equation by 5 we have

$5x + 15y = 125$

and since both equations are true at the same time we may subtract one equation from the other (here, the first from the new second)

$$5x + 15y = 125$$
$$5x + 2y = 34$$
$$13y = 91$$

Therefore,			$y = 7$

Having found the value of y, we can now substitute this into either of the original equations to find the value of x:

$5x + 2(7) = 34$

$5x + 14 = 34$

$5x = 20$

$x = 4$

2 $6x + 5y = 27$
$7x - 4y = 2$

Multiply the first equation by 4 and the second by 5

$$24x + 20y = 108$$

$$35x - 20y = 10$$

Add together	$59x = 118$

Therefore,			$x = 2$

substitute	$6 \times 2 + 5y = 27$

$12 + 5y = 27$

$5y = 15$

$y = 3$

and thus the solution is $x = 2$, $y = 3$

There is yet another way of solving simultaneous equations using matrices which we will illustrate in the next chapter.

Returning now to graphs of linear functions, we may use the method developed above to find the equation of a linear function that passes through two particular points.

Examples

1 If a linear function goes through the points $x = 2$, $y = 5$ and $x = 3$, $y = 7$ we may substitute these values into the general formula for a linear function, $y = a + bx$, to form a pair of simultaneous equations:

for (2, 5) $5 = a + 2b$

for (3, 7) $7 = a + 3b$

Subtracting the first equation from the second gives

$2 = b$

and substituting back into the first equation gives

$$5 = a + 2 \times 2$$
$$5 = a + 4$$

Therefore, $a = 1$

Now substituting the values of a and b back into the general formula, gives

$$y = 1 + 2x$$

2 A linear function goes through (5, 40) and (25, 20), thus

$$40 = a + \;\; 5b$$
$$20 = a + 25b$$
$$\overline{20 = \;\;\;\; -20b}$$

Therefore $b = -1$

$$40 = a - 5$$

Therefore $a = 45$ and thus $y = 45 - x$

An alternative method for finding the equation is to label the points as (x_1, y_1) and (x_2, y_2) and then substitute into

$$\frac{y_1 - y}{y_2 - y_1} = \frac{x_1 - x}{x_2 - x_1}$$

Taking the last example, we have:

$$\frac{40 - y}{20 - 40} = \frac{5 - x}{25 - 5}$$
$$\frac{40 - y}{-20} = \frac{5 - x}{20}$$

Multiplying both sides by 20 gives:

$$-(40 - y) = (5 - x)$$
$$-40 + y = 5 - x$$
$$45 - x = y$$

23.4.3 Quadratic functions

A quadratic function has the general equation

$$y = ax^2 + bx + c$$

and once the values of a, b and c are given we have a specific function. This function will produce a curve with one bend, or change of direction. (It is usually said to have one **turning point**.) If the value assigned to the coefficient of x^2, a, is

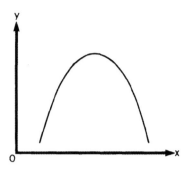

Figure 23.12

negative, then the shape in Figure 23.12 will be produced, while if *a* is positive, the shape in Figure 23.13 will result.

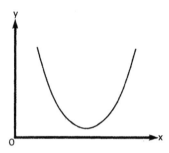

Figure 23.13

To construct the graph of a function it is necessary to produce a table of values, plot the pairs of values, and join them together.

Example

Consider the function:

$$y = x^2 - 2x + 1$$

Here we decide on a range of *x* values and write out each part of the function across the top of the table. We then work out each column for each value of *x*. Finally we add across each row to give the *y*-value on the right-hand side.

x	x^2	$-2x$	$+1$	y
−2	4	+4	+1	9
−1	1	+2	+1	4
0	0	0	+1	1
1	1	−2	+1	0
2	4	−4	+1	1
3	9	−6	+1	4

It is not always obvious at this stage for which range of values of *x* we should calculate the equivalent values of *y*; but we usually want to show where the curve changes direction and you will notice from the table above that the change in *y* decreases as we approach the change in direction and then increases again, so

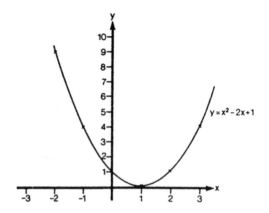

Figure 23.14

that if the changes in y are increasing as x increases, you should try lower values of x (Figure 23.14). Producing such a table and graph is easily done using a spreadsheet.

Exercise

Construct a table and graph of the function $y = -x^2 - 2x + 1$.

We are often interested in finding where a quadratic function crosses the x axis; the values of x when $y = 0$. These points are known as the roots of the quadratic equation, e.g. the values of x that make $ax^2 + bx + c = 0$. These roots may be found in several ways.

We can graph the function and refer directly to the graph. It can be seen in Figure 23.15, for example, that the roots are -1 and 6. To check the correctness of this result, we can substitute these values back into the original equation and obtain a y value of 0.

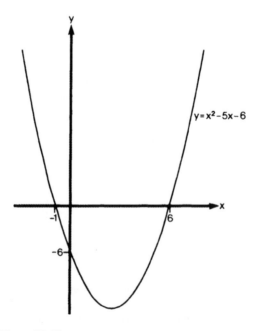

Figure 23.15

A second method is to recognize that an equation such as $x^2 - 5x - 6 = 0$ can often be written as the product of two factors, say $(x + p)$ and $(x + q)$ where p and q are two real numbers:

$$x^2 - 5x - 6 = (x + p)(x + q) = 0$$

In this case it can be shown that $p = -6$ and $q = 1$. So

$$(x - 6)(x + 1) = 0$$

Now for this to be true, either the first factor equals zero or the second factor equals zero, hence

$$x - 6 = 0, \quad \text{i.e. } x = 6$$

or

$$x + 1 = 0, \quad \text{i.e. } x = -1$$

and so the roots are -1 and $+6$. The easiest functions to break down in this way are those with just x^2 as the first term. The two numbers in the factors when multiplied together must give the third term, or constant; and the numbers when multiplied by the other x and added (allowing for signs) must give the coefficient of x (here $-6x + 1x = -5x$).

Example

If

$$x - 5x + 4 = 0$$

then

$$(x - 4)(x - 1) = 0$$

and roots are $+1$ and $+4$.

The third method of finding roots is to use a formula. If $ax^2 + bx + c = 0$ then the roots are at

$$x = \frac{-b \pm \sqrt{(b^2 - 4ac)}}{2a}$$

and we substitute the appropriate values of a, b and c to find these roots. A proof of this formula is given in section 23.9 for the interested reader.

Example

$$x^2 - 5x + 4 = 0$$

then $a = 1$, $b = -5$, $c = +4$ so the roots are at

$$x = \frac{-(-5) \pm \sqrt{[(-5)^2 - 4(1)(4)]}}{2(1)}$$

$$= \frac{5 \pm \sqrt{[25 - 16]}}{2}$$

$$= \frac{5 \pm \sqrt{9}}{2}$$

$$= \frac{5 + 3}{2} \quad \text{or} \quad \frac{5 - 3}{2}$$

$$= \frac{8}{2} \quad \text{or} \quad \frac{2}{2}$$

$$= 4 \quad \text{or} \quad 1$$

Example

$$2x^2 - 4x - 10 = 0$$

then $a = 2$, $b = -4$, $c = -10$ so the roots are at

$$x = \frac{4 \pm \sqrt{(16 + 80)}}{4}$$

$$= \frac{4 \pm \sqrt{96}}{4}$$

$$= \frac{4 \pm 9.8}{4}$$

$$= \frac{13.8}{4} \quad \text{or} \quad \frac{-5.8}{4}$$

$$= 3.45 \quad \text{or} \quad -1.45$$

This method will always give the roots, but beware of *negative* values for the expression under the square root sign $(b^2 - 4ac)$. If this is negative then the function is said to have **imaginary roots**, since in normal circumstances we cannot take the square root of a negative number. At this level, these imaginary roots need not concern us.

Quadratic functions are often used to represent cost equations, such as marginal cost or average cost, and sometimes profit functions. When profit is represented by a quadratic function, then we can use the idea of roots either to find the range of output for which any profit is made, or we can specify a profit level and find the range of output for which at least this profit is made.

Example

If profit $= -x^2 + 8x + 1$ where x represents output, then if the specified profit level is 8, we have:

$$-x^2 + 8x + 1 = 8$$
$$-x^2 + 8x - 7 = 0$$
$$(x - 7)(1 - x) = 0$$
$$x = 7 \quad \text{or} \quad 1$$

This is illustrated in Figure 23.16.

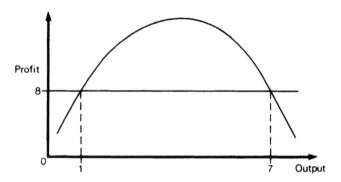

Figure 23.16

We have seen how two pairs of points are required to determine the equation of a straight line. To find a quadratic function, three pairs of points are required. One method of doing this is shown below.

Example

If a quadratic function goes through the points

$$x = 1, \quad y = 7$$
$$x = 4, \quad y = 4$$
$$x = 5, \quad y = 7$$

then we may take the general equation $y = ax^2 + bx + c$ and substitute:

$$(1, 7) \quad 7 = a(1)^2 + b(1) + c = a + b + c \tag{23.1}$$
$$(4, 4) \quad 4 = a(4)^2 + b(4) + c = 16a + 4b + c \tag{23.2}$$
$$(5, 7) \quad 7 = a(5)^2 + b(5) + c = 25a + 5b + c \tag{23.3}$$

Rearranging the first equation 23.1 gives

$$c = 7 - a - b$$

and substituting for c into equation 23.2 gives

$$4 = 16a + 4b + (7 - a - b)$$
$$4 = 15a + 3b + 7$$
$$-3 = 15a + 3b \tag{23.4}$$

Substituting again for c but into equation 23.3 gives

$$7 = 25a + 5b + (7 - a - b)$$
$$7 = 24a + 4b + 7$$
$$0 = 24a + 4b \qquad\qquad 23.5$$

We now have two simultaneous equations (23.4 and 23.5) in two unknowns (a and b).

Multiplying equation 23.4 by 4 and equation 23.5 by 3 to create equal coefficients of b, gives:

$$-12 = 60a + 12b$$
$$0 = 72a + 12b$$

which, if one is subtracted from the other, gives

$$12 = 12a$$

Therefore, $a = 1$.

From equation 23.5, we have:

$$0 = 24 + 4b$$

Therefore,

$$b = -6$$

and from equation 23.1

$$c = 7 - a - b$$
$$= 7 - 1 + 6 = 12$$

and thus the quadratic function is:

$$y = x^2 - 6x + 12$$

23.4.4 Cubic and polynomial functions

A cubic function has the general equation:

$$y = ax^3 + bx^2 + cx + d$$

and will have two turning points when $b^2 > 4ac$. They are often used to represent total cost functions in economics (Figures 23.17 and 23.18). We could go on extending the range of functions by adding a term in x^4, and then one in x^5 and so on. The general name for functions of this type is **polynomials**. For most business and economic purposes we do not need to go beyond cubic functions, but for some areas, the idea that a sufficiently complex polynomial will model any situation will appear.

23.4.5 Exponential functions

Within mathematics, those values that arise in a wide variety of situations and a broad spectrum of applications (and often run to a vast number of decimal places) tend to be given a special letter. One example most people will have met is π (pi). Exponential functions make use of another such number, e (Euler's constant) which is a little over 2.7. Raising this number to a power gives the graph in Figure 23.19, or, if the power is negative, the result is as Figure 23.20.

The former often appears in growth situations, and was of use in considering money and interest. The latter may be incorporated into models of failure rates, market sizes and many probability situations.

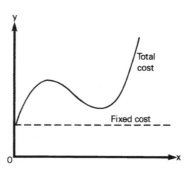

Figure 23.17

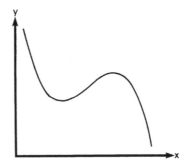

Figure 23.18

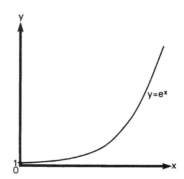

Figure 23.19

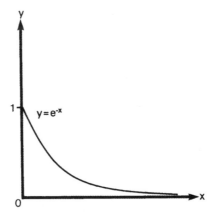

Figure 23.20

23.5 Conclusions

Mathematical relationships have proved useful in describing, analysing and solving business-type problems. They have been particularly used in the development of economic theory, where their exactness and certainty have sometimes led to extra insights into relationships. Even if reality is not as exact as a mathematical relationship implies, the shapes of graphs and functions provide a clear way of thinking about certain problems.

23.6 Problems

For each of the following, find x in terms of y:

1 $4x + 3y = 2x + 21y$

2 $2x - 4y = 3(x + y) + 5y$

3 $x + y + 3x = 4y + 2x + 7y - 2x$

4 $3x + 5(x - y) = 2(2x + 3y) + y$

5 $x^2(x^2 + 4x + 3) - 3y(4y + 10) - x^4 - 4(x^3 - 10) = 3x^2 - 2y(6y - 10)$

Simplify each of the following expressions:

6 $a^2 \times a^3$

7 $a \times a^2 / a^3$

8 $(a + b)a^2 / b$

9 $(a^{1/2} \times a^2 \times a^{3/2}) / a^2$

10 $a^2(a^5 + a^8 - a^{10}) / a^4$

11 $(1/2)^3$

12 $6x(2 + 3x) - 2(9x^2 - 5x) - 484 = 0$

Expand the brackets and simplify the following expressions:

13 $a(2a + b)$

14 $(a + b)^2 - ab$

15 $(a + 2b)^2$

16 $(a + b)(a - b)$

17 $(a + b)^3 - (a - b)^3$

18 $(3x + 6y)(4x - 2y)$

For each of the following arithmetic progressions, find (a) the 10th term and (b) the sum given that each progression has 12 terms:

19 5, 8, 11, 14, 17, ...

20 112, 126, 140, 154, ...

21 57, 49, 41, 33, 25, ...

22 If the third term and the sixth term of an arithmetic progression are 11 and 23 respectively, determine the ninth term.

For each of the following geometric progressions, find (a) the 7th term and (b) the sum given that the progression has 10 terms:

23 4, 12, 36, 108, ...

24 6, −12, 24, −48, ...

25 4, 3.2, 2.56, 2.048, ...

26 If the third term and the fifth term of a geometric progression are 225 and 5625 respectively, determine the sixth term.

Construct a graph and mark the following points upon it:

27 (a) (1, 1)
 (b) (2, 3)
 (c) (3, 2)
 (d) (−1, 4)
 (e) (2, −3)
 (f) (−2, −1)

28 Shade the area on the graph where x is above 0 but below 4 and y is above 0 but below 2.

29 Construct graphs of the following functions:
 (a) $y = 0.5x$ $0 < x < 20$
 (b) $y = 2 + x$ $0 < x < 15$
 (c) $y = 25 - 2x$ $0 < x < 15$
 (d) $y = 2x + 4$ $3 < x < 6$
 (e) $y = 3$
 (f) $x = 4 - 0.5y$ $0 < x < 5$
 (g) $y = x^2 - 5x + 6$ $-2 < x < 6$
 (h) $y = x^3 - 2x^2 + x - 2$ $-2 < x < 6$

30 Solve the following simultaneous equations:
 (a) $4x + 2y = 11$
   $3x + 4y = 9$
 (b) $2x - 3y = 20$
   $x + 5y = 23$
 (c) $6x + 2y = 20$
   $2x - 3y = 14$
 (d) $4x + 5y = 22.75$
   $7x + 6y = 31.15$

31 Find the roots of the following functions:
 (a) $x^2 - 3x + 2 = 0$
 (b) $x^2 - 2x - 8 = 0$
 (c) $x^2 + 13x + 12 = 0$
 (d) $2x^2 - 6x - 40 = 0$
 (e) $3x^2 - 15x + 8 = 0$
 (f) $x^3 - 6x^2 - x + 6 = 0$ (Hint, use a graph)
 (g) $-2x^2 + 5x + 40 = 0$

32 A demand function is known to be linear $[P = f(Q)]$, and to pass through the following points:

$$Q = 10, P = 50; \quad Q = 20, P = 30$$

Find the equation of the demand function.

33 A supply function is known to be linear $[P = f(Q)]$, and to pass through the following points:

$$Q = 15, P = 10; \quad Q = 40, P = 35$$

Find the equation of the supply function.

34 Use your answers to questions 32 and 33 to find the point of equilibrium for a market having the respective demand and supply functions.

35 A demand function is known to be linear $[P = f(Q)]$ and to pass through the following points:

$$Q = 5, P = 25; \quad Q = 9, P = 7$$

An associated supply function is also linear $[P = f(Q)]$ and passes through:

$$Q = 5, P = 10; \quad Q = 25, P = 12$$

Find the equations of each function, and hence find the point of equilibrium. If the demand function now shifts to the right, to pass through the points:

$$Q = 10, P = 25; \quad Q = 14, P = 7$$

find the new equilibrium position.

36 A market has been analysed and the following points estimated on the demand and supply curves:

Output demand	Price
1	1902
5	1550
10	1200

Output supplied	Price
1	9
5	145
10	540

(a) Determine the equations of the demand and supply functions, assuming that in both cases price $= f$(output), and that the function is quadratic. (NB. Neither function applies above an output of 20.)

(b) Determine the equilibrium price and quantity in this market.

23.7 Proof of AP

Proof that

$$S_n = \frac{n}{2}[2a + (n - 1)d]$$

The sum of an arithmetic progression with n terms may be written

$$S_n = a + (a + d) + (a + 2d) + \cdots + [a + (n - 1)d]$$

This expression may also be written as

$$S_n = [a + (n - 1)d] + [a + (n - 1)d - d]$$
$$+ [a + (n - 1)d - 2d] + \cdots + (a + d) + a$$

by reversing the order of the terms on the right-hand side of the equation. Adding the two equations gives

$$2S_n = [2a + (n - 1)d] + [a + d + a + (n - 1)d - d]$$
$$+ \cdots + [a + (n - 1)d + a]$$
$$= [2a + (n - 1)d] + [2a + (n - 1)d] + \cdots$$

but as there are n terms

$$2S_n = n[2a + (n - 1)d]$$

and thus,

$$S_n = \frac{n}{2}[2a + (n - 1)d]$$

23.8 Proof of GP

Proof that

$$S_n = \frac{a(r^n - 1)}{r - 1}$$

The sum of a geometric progression with n terms may be written:

$$S_n = a + ar + ar^2 + ar^3 + \cdots + ar^{n-1}$$

multiplying both sides by r gives

$$rS_n = ar + ar^2 + ar^3 + ar^4 + \cdots + ar^n$$

Subtracting the first equation from the second gives

$$rS_n - S_n = ar^n - a$$
$$S_n(r - 1) = a(r^n - 1)$$

So,

$$S_n = \frac{a(r^n - 1)}{r - 1}$$

23.9 Proof of quadratic formula

Proof that

$$x = \frac{-b \pm \sqrt{(b^2 - 4ac)}}{2a}$$

The form of a quadratic equation is

$$ax^2 + bx + c = 0$$

Dividing by a and rearranging gives

$$x^2 + \frac{b}{a}x = -\frac{c}{a}$$

Adding $\dfrac{b^2}{4a^2}$ to both sides gives

$$x^2 + \frac{b}{a}x + \frac{b^2}{4a^2} = \frac{b^2}{4a^2} - \frac{c}{a}$$

This can be written

$$\left(x + \frac{b}{2a}\right)^2 = \frac{b^2 - 4ac}{4a^2}$$

Taking the square root of each side gives

$$x + \frac{b}{2a} = \pm\frac{\sqrt{(b^2 - 4ac)}}{2a}$$

Hence

$$x = \frac{-b \pm \sqrt{(b^2 - 4ac)}}{2a}$$

24 Matrices

The algebra used in Chapter 23 allows the specification and solution of a range of business problems and provides the framework for the theoretical development of disciplines such as economics. We have seen single equations solved in terms of x and simultaneous equations solved in terms of x and y. However, we must be able to deal effectively with more complex problems. Matrix notation provides a way of describing these more complex problems and the rules of matrices provide a way of manipulating these problems. It must be remembered that matrix algebra only provides a convenient notation and some new methods of solution. Matrices will not solve problems that are not amenable to solution by other methods. As with the application of all quantitative methods, problem specification is the important first step.

Matrices are particularly useful in solving sets of simultaneous equations. The compact form of notation is consistent with the use of arrays in computer programming, and many computer packages will offer the facility of matrix manipulation, e.g. MINITAB. To illustrate the application of matrix algebra, an input/

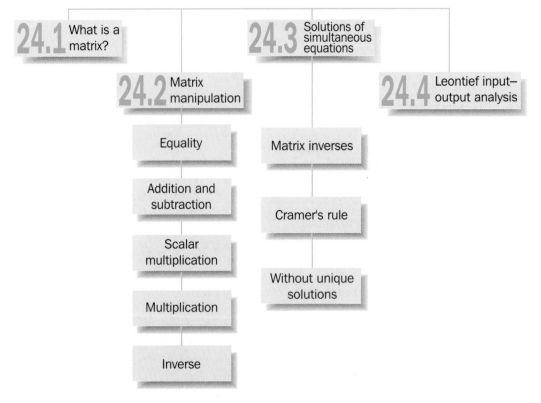

Figure 24.1 Structure of Chapter 24.

output model is developed in section 24.4 where it is shown how output from one industry is the input to another. Clearly, as the number of industries considered increases, the complexity of modelling increases. Matrices are also used extensively in multiple regression (Chapter 17), and the methods of matrix algebra in more advanced work on linear programming (Chapter 19). Further applications include probability modelling, such as Markov Chains (Chapter 8).

Objectives

After working through this chapter, you should be able to:

- define a matrix
- add and subtract matrices
- multiply matrices
- invert square matrices
- solve simultaneous equations using matrices.

24.1 What is a matrix?

A matrix is a rectangular array of numbers arranged in rows and columns and is characterized by its size (or order), written as (no. of rows × no. of columns). The whole matrix is usually referred to by a capital letter, whilst individual numbers, or elements, within the matrix are referred to by lower case letters, usually with a suffix to identify in which row and in which column they appear. Note that a matrix does not have a numerical value; it is merely a convenient way of representing an array of numbers. If

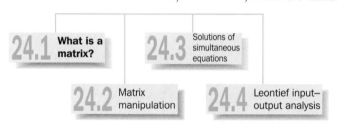

$$A = \begin{bmatrix} 4 & 8 & 17 & 12 \\ 21 & 3 & 19 & 17 \\ 10 & 21 & 4 & 2 \end{bmatrix}$$

then the order of matrix A is (3×4) and the element a_{13} is the 17 since it is in the first row and the third column.

It is often convenient to use the double subscript notation where a_{ij} denotes the element located in the ith row and jth column of matrix A and b_{ij} denotes the element located in the ith row and jth column of matrix B.

If a matrix has only one row, then it is known as a row vector; if it has only one column, then it is a column vector, e.g.

$$B = \begin{bmatrix} 4 & 8 & 7 \end{bmatrix} \qquad C = \begin{bmatrix} 10 \\ 12 \\ 28 \\ 49 \\ 102 \end{bmatrix}$$

The order of matrix B is (1×3) and the order of matrix C is (5×1).

Example

A company, Comfy Chairs Ltd, produces three types of chair, the Classic, the Victorian and the Modern in its existing factory. Weekly output of the Classic, the Victorian and the Modern types in mahogany are 30, 60 and 80 respectively, and in teak, 20, 30 and 40 respectively.

This information can easily be represented by a matrix, say **D**, where columns refer to chair type and the rows to the wood used.

$$D = \begin{bmatrix} 30 & 60 & 80 \\ 20 & 30 & 40 \end{bmatrix}$$

24.2 Matrix manipulation

We will find that many sets of data and many situations can be described in terms of rows and columns. In this section we consider the rules governing matrix algebra that will allow us to manipulate these rows and columns.

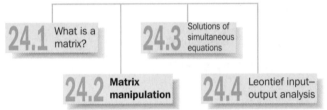

24.1 What is a matrix?

24.3 Solutions of simultaneous equations

24.2 Matrix manipulation

24.4 Leontief input–output analysis

24.2.1 Equality of matrices

The concept of equality is fundamental to all algebra. In matrix algebra, two matrices **A** and **B** are equal only if they are of the same size *and* their corresponding elements are equal, $a_{ij} = b_{ij}$ for *all* values of i and j.

Example

Matrix **E** is only equal to matrix **D** if matrix **E** is of the same size, (2×3), and each element of matrix **E** is equal to each element of matrix **D** (above). If

$$E = \begin{bmatrix} a & b & c \\ d & e & f \end{bmatrix}$$

then $a = 30$, $b = 60$, $c = 80$, $d = 20$, $e = 30$ and $f = 40$.

24.2.2 Addition and subtraction of matrices

To add or subtract matrices they must be of the same order; they are then said to be **conformable for addition**. When this is true, for addition the corresponding elements in each matrix are added together and for subtraction, each of the second elements is subtracted from the corresponding elements in the first. If **A** and **B** are two matrices of the same size, then $A + B = [a_{ij} + b_{ij}]$ and $A - B = [a_{ij} - b_{ij}]$ for all values of i and j. If

$$A = \begin{bmatrix} 10 & 15 \\ 20 & 14 \end{bmatrix} \qquad B = \begin{bmatrix} 21 & 13 \\ 12 & 17 \end{bmatrix}$$

then

$$A + B = \begin{bmatrix} 10 & 15 \\ 20 & 14 \end{bmatrix} + \begin{bmatrix} 21 & 13 \\ 12 & 17 \end{bmatrix}$$

$$= \begin{bmatrix} (10+21) & (15+13) \\ (20+12) & (14+17) \end{bmatrix} = \begin{bmatrix} 31 & 28 \\ 32 & 31 \end{bmatrix}$$

and

$$\mathbf{A} - \mathbf{B} = \begin{bmatrix} 10 & 15 \\ 20 & 14 \end{bmatrix} - \begin{bmatrix} 21 & 13 \\ 12 & 17 \end{bmatrix}$$

$$= \begin{bmatrix} (10-21) & (15-13) \\ (20-12) & (14-17) \end{bmatrix} = \begin{bmatrix} -11 & 2 \\ 8 & -3 \end{bmatrix}$$

If we have:

$$\begin{bmatrix} 10 & 20 \\ 50 & 5 \end{bmatrix} - \begin{bmatrix} 10 & 20 \\ 50 & 5 \end{bmatrix} = \begin{bmatrix} 0 & 0 \\ 0 & 0 \end{bmatrix}$$

then the result is a zero matrix (which performs the same function as zero in ordinary arithmetic).

Addition of matrices is said to be commutative (sequence makes no difference) since $\mathbf{A} + \mathbf{B} = \mathbf{B} + \mathbf{A}$.

Example

The company, Comfy Chairs Ltd, have presented proposed weekly output from a new factory in matrix **E** shown below.

$$\mathbf{E} = \begin{bmatrix} 30 & 30 & 0 \\ 18 & 20 & 0 \end{bmatrix}$$

Again, columns represent chair types and rows represent wood used. What would the combined weekly output be from the existing and new factories?

$$\mathbf{D} + \mathbf{E} = \begin{bmatrix} 30 & 60 & 80 \\ 20 & 30 & 40 \end{bmatrix} + \begin{bmatrix} 30 & 30 & 0 \\ 18 & 12 & 0 \end{bmatrix} = \begin{bmatrix} 60 & 90 & 80 \\ 38 & 42 & 40 \end{bmatrix}$$

24.2.3 Scalar multiplication of a matrix

A matrix may be multiplied by a single number or scalar. To do this we multiply each element of the matrix by the scalar, e.g.

$$5 \times \begin{bmatrix} 4 & 8 & 3 \\ 17 & 2 & 12 \end{bmatrix} = \begin{bmatrix} 20 & 40 & 15 \\ 85 & 10 & 60 \end{bmatrix}$$

In terms of matrix notation, if **A** is the matrix and c is the scalar then $c\mathbf{A} = [ca_{ij}]$ for all values of i and j.

Example

Suppose Comfy Chairs Ltd plan to increase output across the range from their existing factory by 10%. To increase output across the range by 10%, we need to multiply each element by 1.1.

$$1.1 \times \mathbf{D} = \begin{bmatrix} 1.1 \times 30 & 1.1 \times 60 & 1.1 \times 80 \\ 1.1 \times 20 & 1.1 \times 30 & 1.1 \times 40 \end{bmatrix}$$

$$= \begin{bmatrix} 33 & 66 & 88 \\ 22 & 33 & 44 \end{bmatrix}$$

It is often convenient to reverse this argument, and take a common factor out of the matrix (cf. taking a common factor out of a bracket), to make further calculations easier, e.g.

$$\begin{bmatrix} 10 & 170 & 100 \\ 20 & 90 & 95 \\ 140 & 30 & 50 \end{bmatrix} = 10 \times \begin{bmatrix} 1 & 17 & 10 \\ 2 & 9 & 9.5 \\ 14 & 3 & 5 \end{bmatrix}$$

When the *same* matrix is to be multiplied by a series of scalars, we have:

$$a\mathbf{A} + b\mathbf{A} + c\mathbf{A} + d\mathbf{A} = (a + b + c + d)\mathbf{A}$$

24.2.4 Multiplication of matrices

When two matrices are to be multiplied together it is first necessary to check that the multiplication is possible; the matrices must be conformable for multiplication. This condition is satisfied if the number of columns in the first matrix is equal to the number of rows in the second matrix. The outcome of multiplication will be a matrix with the same number of rows as the first matrix and the same number of columns as the second matrix. If matrix **A** is of order $(a \times b)$ and matrix **B** is of order $(c \times d)$, then for multiplication to be possible, b must equal c and the new matrix produced by the product **AB** will be of order $(a \times d)$.

Thus a (2×3) matrix multiplied by a (3×2) matrix will give a (2×2) result; whereas a (3×2) matrix multiplied by a (2×3) matrix will give a (3×3) result. It is not possible to multiply a matrix of order (4×8) by a matrix of order (5×3). Note that even though the resultant matrix may be of the same order, multiplication is **not commutative** since **AB** does not necessarily equal **BA**.

The process of multiplication involves using a particular row from the first matrix and a particular column from the second matrix; placing the result as a single element in the result matrix, e.g.

$$\mathbf{A} = \begin{bmatrix} 1 & 4 & 7 \\ 2 & 5 & 8 \\ 3 & 6 & 9 \end{bmatrix} \qquad \mathbf{B} = \begin{bmatrix} 10 & 13 \\ 11 & 14 \\ 12 & 15 \end{bmatrix} \qquad \mathbf{A} \times \mathbf{B} = \mathbf{C}$$

To find $\mathbf{A} \times \mathbf{B}$, we will work out each element separately. For example, taking the first row of **A** and the first column of **B** gives the element c_{11} in the first row and column of **C**, i.e.

$$[1 \quad 4 \quad 7] \begin{bmatrix} 10 \\ 11 \\ 12 \end{bmatrix} = (1 \times 10) + (4 \times 11) + (7 \times 12) = 138 \\ = c_{11}$$

Note that we have gone along the row of the first matrix and down the column of the second matrix, multiplying the corresponding elements.

To find the *second* element in the first row of **C**, we take the first row of the matrix **A**, and the *second* column of the matrix **B**.

$$[1 \quad 4 \quad 7] \begin{bmatrix} 13 \\ 14 \\ 15 \end{bmatrix} = (1 \times 13) + (4 \times 14) + (7 \times 15) = 174 \\ = c_{12}$$

This process continues, using the second row from **A**, and then the third row. In general, the mth row of **A** by the nth column of **B** gives the element c_{mn} in the mth row and nth column of **C**.

Exercise

> Calculate the remaining elements of the matrix **C**
>
> Answer:
>
> $$\mathbf{C} = \begin{bmatrix} 138 & 174 \\ 171 & 216 \\ 204 & 258 \end{bmatrix}$$

Note that it is not possible to find $\mathbf{B} \times \mathbf{A}$ because the number of columns in $\mathbf{B}$ is not the same as the number of rows in $\mathbf{A}$.

Example

$$\begin{bmatrix} 1 & 3 \\ 2 & 5 \end{bmatrix}\begin{bmatrix} 3 & 2 \\ 5 & 7 \end{bmatrix} = \begin{bmatrix} 18 & 23 \\ 31 & 39 \end{bmatrix}$$

$$\begin{bmatrix} 3 & 2 \\ 5 & 7 \end{bmatrix}\begin{bmatrix} 1 & 3 \\ 2 & 5 \end{bmatrix} = \begin{bmatrix} 7 & 19 \\ 19 & 50 \end{bmatrix}$$

This confirms that $\mathbf{A} \times \mathbf{B} \neq \mathbf{B} \times \mathbf{A}$. Since the order in which the matrices are multiplied together is crucial to the product obtained, it is useful to have a terminology to show this. In the product $\mathbf{A} \times \mathbf{B}$, $\mathbf{A}$ is said to be **postmultiplied** by $\mathbf{B}$, whilst $\mathbf{B}$ is said to be **premultiplied** by $\mathbf{A}$.

We have seen, above, the zero matrix which performs the same function in matrix algebra as a zero in arithmetic. We also need a matrix which will perform the function that unity (one) plays in arithmetic (e.g. $5 \times 1 = 5$).

For any matrix $\mathbf{A}$, what matrix has no effect when $\mathbf{A}$ is postmultiplied by it? i.e.

$$\mathbf{A} \times ? = \mathbf{A}$$

The matrix which performs this function is known as the **identity matrix** and consists of 1s on the diagonal from top left (a_{11}) to bottom right (a_{nn}) and 0s for all other elements. Note that all identity matrices are *square* and that the 1s are said to be on the leading or **principal diagonal**.

Thus:

$$\begin{bmatrix} 5 & 3 \\ 2 & 1 \end{bmatrix}\begin{bmatrix} 1 & 0 \\ 0 & 1 \end{bmatrix} = \begin{bmatrix} 5 & 3 \\ 2 & 1 \end{bmatrix}$$

or

$$\begin{bmatrix} 6 & 8 & 10 \\ 12 & 20 & 14 \\ 2 & 75 & 3 \end{bmatrix}\begin{bmatrix} 1 & 0 & 0 \\ 0 & 1 & 0 \\ 0 & 0 & 1 \end{bmatrix} = \begin{bmatrix} 6 & 8 & 10 \\ 12 & 20 & 14 \\ 2 & 75 & 3 \end{bmatrix}$$

If $\mathbf{I}$ is used for the identity matrix, we have:

$$\mathbf{AI} = \mathbf{IA} = \mathbf{A}$$

Example

Suppose Comfy Chairs Ltd. decide not to implement an overall increase in output across the range by 10% from their existing factory but rather increase the Classic by 5%, the Victorian by 10% and the Modern by 15% in both mahogany and teak.

This can be done by multiplying matrix **D** by a new matrix **F**:

$$\mathbf{DF} = \begin{bmatrix} 30 & 60 & 80 \\ 20 & 30 & 40 \end{bmatrix} \begin{bmatrix} 1.05 & 0 & 0 \\ 0 & 1.10 & 0 \\ 0 & 0 & 1.15 \end{bmatrix} = \begin{bmatrix} 31.5 & 66 & 92 \\ 21 & 33 & 46 \end{bmatrix}$$

Example

Suppose we want to know the total number of chairs produced in mahogany or teak from the existing factory before any increase.

We can do this by multiplying matrix **D** by a column vector of 1s:

$$\begin{bmatrix} 30 & 60 & 80 \\ 20 & 30 & 40 \end{bmatrix} \begin{bmatrix} 1 \\ 1 \\ 1 \end{bmatrix} = \begin{bmatrix} 170 \\ 90 \end{bmatrix}$$

24.2.5 Inverse of a matrix

As in arithmetic, division of one matrix by a second matrix is equivalent to multiplying the first by the **inverse** of the second, bearing in mind that the operation is not commutative. To transfer a matrix from one side of an equation to the other, both sides are multiplied by the inverse of the matrix. The inverse of a matrix **A** is denoted by $\mathbf{A}^{-1}$. A matrix multiplied in either order by the inverse of itself will always give the identity matrix, i.e.

$$\mathbf{AA}^{-1} = \mathbf{A}^{-1}\mathbf{A} = \mathbf{I}$$

Suppose we have:

$$\mathbf{A} \times \mathbf{B} = \mathbf{C}$$

and we want to find **B**. If we *premultiply* both sides by $\mathbf{A}^{-1}$:

$$\mathbf{A}^{-1}\mathbf{AB} = \mathbf{A}^{-1}\mathbf{C}$$

$$\mathbf{IB} = \mathbf{A}^{-1}\mathbf{C}$$

$$\mathbf{B} = \mathbf{A}^{-1}\mathbf{C}$$

To find **A**, we can postmultiply both sides by $\mathbf{B}^{-1}$:

$$\mathbf{ABB}^{-1} = \mathbf{CB}^{-1}$$

$$\mathbf{AI} = \mathbf{CB}^{-1}$$

$$\mathbf{A} = \mathbf{CB}^{-1}$$

Finding the inverse of a matrix is *only* possible if the matrix is *square*, i.e. has the same number of rows as columns, and we shall initially look at the special case of a (2×2) matrix and then suggest a method for larger matrices.

The inverse of a (2×2) *matrix*

If

$$\mathbf{A} = \begin{bmatrix} a_{11} & a_{12} \\ a_{21} & a_{22} \end{bmatrix}$$

then

$$\mathbf{A}^{-1} = \frac{1}{(a_{11}a_{22}) - (a_{12}a_{21})} \begin{bmatrix} a_{22} & -a_{12} \\ -a_{21} & a_{11} \end{bmatrix}$$

Putting this into words, the fraction in front of the matrix is one over the product of the two elements on the principal diagonal minus the product of the other two elements. Within the matrix the two elements on the principal diagonal change places and the other two elements change sign, e.g. if

$$\mathbf{A} = \begin{bmatrix} 1 & 2 \\ 3 & 4 \end{bmatrix}$$

then

$$\mathbf{A}^{-1} = \frac{1}{(1 \times 4) - (2 \times 3)} \begin{bmatrix} 4 & -2 \\ -3 & 1 \end{bmatrix}$$

$$= -\frac{1}{2} \begin{bmatrix} 4 & -2 \\ -3 & 1 \end{bmatrix}$$

$$= \begin{bmatrix} -2 & 1 \\ 1.5 & -0.5 \end{bmatrix}$$

To check the answer we can find $\mathbf{AA}^{-1}$:

$$\begin{bmatrix} 1 & 2 \\ 3 & 4 \end{bmatrix} \begin{bmatrix} -2 & 1 \\ 1.5 & -0.5 \end{bmatrix} = \begin{bmatrix} 1 & 0 \\ 0 & 1 \end{bmatrix}$$

The denominator of the fraction we calculated above is called the determinant of the matrix, and if this is 0 then the matrix is said to be singular, and does not have an inverse.

Larger matrices

To find the inverse for a matrix larger than (2×2) is a somewhat more complex procedure and may be done by using the method of cofactors or by row operations. We shall give an example of how to find the inverse for a (3×3) matrix. (For larger matrices it is probably advisable to use a computer program.)

To find an inverse using the cofactors, we must first find the determinant of the matrix, denoted by $|\mathbf{A}|$. If

$$\mathbf{A} = \begin{bmatrix} a_{11} & a_{12} & a_{13} \\ a_{21} & a_{22} & a_{23} \\ a_{31} & a_{32} & a_{33} \end{bmatrix}$$

then

$$|\mathbf{A}| = a_{11} \begin{vmatrix} a_{22} & a_{23} \\ a_{32} & a_{33} \end{vmatrix} - a_{21} \begin{vmatrix} a_{12} & a_{13} \\ a_{32} & a_{33} \end{vmatrix} + a_{31} \begin{vmatrix} a_{12} & a_{13} \\ a_{22} & a_{23} \end{vmatrix}$$

$$= a_{11}(a_{22}a_{33} - a_{23}a_{32}) - a_{21}(a_{12}a_{33} - a_{13}a_{32})$$

$$+ a_{31}(a_{12}a_{23} - a_{13}a_{22})$$

A cofactor of an element consists of the determinant of those elements which are not in the same row and not in the same column as that element. Thus the cofactor of a_{11} is:

$$\begin{vmatrix} a_{22} & a_{23} \\ a_{32} & a_{33} \end{vmatrix}$$

and this can be evaluated as $(a_{22}a_{33}) - (a_{23}a_{32})$. To form the inverse matrix, each element of the initial matrix is replaced by its *signed* cofactor. (Note the signs of the cofactors replacing a_{12}, a_{21}, a_{23} and a_{32}.)

$$\begin{bmatrix} \begin{vmatrix} a_{22} & a_{23} \\ a_{23} & a_{33} \end{vmatrix} & -\begin{vmatrix} a_{21} & a_{23} \\ a_{31} & a_{33} \end{vmatrix} & \begin{vmatrix} a_{21} & a_{22} \\ a_{31} & a_{32} \end{vmatrix} \\ -\begin{vmatrix} a_{12} & a_{13} \\ a_{32} & a_{33} \end{vmatrix} & \begin{vmatrix} a_{11} & a_{13} \\ a_{31} & a_{33} \end{vmatrix} & -\begin{vmatrix} a_{11} & a_{12} \\ a_{31} & a_{32} \end{vmatrix} \\ \begin{vmatrix} a_{12} & a_{13} \\ a_{22} & a_{23} \end{vmatrix} & -\begin{vmatrix} a_{11} & a_{13} \\ a_{21} & a_{23} \end{vmatrix} & \begin{vmatrix} a_{11} & a_{12} \\ a_{21} & a_{22} \end{vmatrix} \end{bmatrix}$$

After each element of this new matrix has been evaluated, it is transposed, that is, each column is written as a row, so that a_{12} becomes a_{21}. The resultant matrix is then multiplied by one over the determinant. This is likely to become considerably clearer as we work through an example.

Example

$$A = \begin{bmatrix} 1 & 2 & 3 \\ 1 & 3 & 5 \\ 1 & 5 & 12 \end{bmatrix}$$

then

$$|A| = 1 \times (3 \times 12 - 5 \times 5) - 1 \times (2 \times 12 - 3 \times 5) + 1 \times (2 \times 5 - 3 \times 3)$$
$$= 11 - 9 + 1 = 3$$

Replacing elements by cofactors, we have:

$$\begin{bmatrix} \begin{vmatrix} 3 & 5 \\ 5 & 12 \end{vmatrix} & -\begin{vmatrix} 1 & 5 \\ 1 & 12 \end{vmatrix} & \begin{vmatrix} 1 & 3 \\ 1 & 5 \end{vmatrix} \\ -\begin{vmatrix} 2 & 3 \\ 5 & 12 \end{vmatrix} & \begin{vmatrix} 1 & 3 \\ 1 & 12 \end{vmatrix} & -\begin{vmatrix} 1 & 2 \\ 1 & 5 \end{vmatrix} \\ \begin{vmatrix} 2 & 3 \\ 3 & 5 \end{vmatrix} & -\begin{vmatrix} 1 & 3 \\ 1 & 5 \end{vmatrix} & \begin{vmatrix} 1 & 2 \\ 1 & 3 \end{vmatrix} \end{bmatrix}$$

$$= \begin{bmatrix} 11 & -7 & 2 \\ -9 & 9 & -3 \\ 1 & -2 & 1 \end{bmatrix}$$

Transposing this, we have:

$$\begin{bmatrix} 11 & -9 & 1 \\ -7 & 9 & -2 \\ 2 & -3 & 1 \end{bmatrix}$$

This is called the **adjunct matrix**, and A^{-1} is this matrix multiplied by the reciprocal of the determinant.

$$A^{-1} = \frac{1}{3} \begin{bmatrix} 11 & -9 & 1 \\ -7 & 9 & -2 \\ 2 & -3 & 1 \end{bmatrix}$$

Exercise

> Check that $\mathbf{AA}^{-1} = \mathbf{I}$.

To find the inverse using row operations, we create a partitioned matrix, by putting an identity matrix alongside the original matrix:

$$\left[\begin{array}{ccc|ccc} 1 & 2 & 3 & 1 & 0 & 0 \\ 1 & 3 & 5 & 0 & 1 & 0 \\ 1 & 5 & 12 & 0 & 0 & 1 \end{array}\right]$$

Our objective now is to multiply and divide each row, or add and subtract rows until the matrix on the *left* of the partition is an identity. At that point, whatever is to the *right* of the partition will be the inverse of the original matrix. We already have a 1 at a_{11}, so to change the 1 at a_{21} to 0 we may subtract row 1 from row 2. (Note that we subtract corresponding elements for the *whole* row.)

$$\left[\begin{array}{ccc|ccc} 1 & 2 & 3 & 1 & 0 & 0 \\ 0 & 1 & 2 & -1 & 1 & 0 \\ 1 & 5 & 12 & 0 & 0 & 1 \end{array}\right]$$

To alter the 1 to a 0 at a_{31}, we again subtract row 1 from row 3.

$$\left[\begin{array}{ccc|ccc} 1 & 2 & 3 & 1 & 0 & 0 \\ 0 & 1 & 2 & -1 & 1 & 0 \\ 0 & 3 & 9 & -1 & 0 & 1 \end{array}\right]$$

To alter the 2 to a 0 at a_{12}, we can subtract *two* times row 2 from row 1.

$$\left[\begin{array}{ccc|ccc} 1 & 0 & -1 & 3 & -2 & 0 \\ 0 & 1 & 2 & -1 & 1 & 0 \\ 0 & 3 & 9 & -1 & 0 & 1 \end{array}\right]$$

To alter the 3 to a 0 at a_{32}, subtract *three* times row 2 from row 3.

$$\left[\begin{array}{ccc|ccc} 1 & 0 & -1 & 3 & -2 & 0 \\ 0 & 1 & 2 & -1 & 1 & 0 \\ 0 & 0 & 3 & 2 & -3 & 1 \end{array}\right]$$

To reduce the 3 to a 1 at a_{33}, divide row 3 by 3.

$$\left[\begin{array}{ccc|ccc} 1 & 0 & -1 & 3 & -2 & 0 \\ 0 & 1 & 2 & -1 & 1 & 0 \\ 0 & 0 & 1 & 2/3 & -1 & 1/3 \end{array}\right]$$

To alter the -1 to a 0 at a_{13}, add row 3 to row 1.

$$\left[\begin{array}{ccc|ccc} 1 & 0 & 0 & 11/3 & -3 & 1/3 \\ 0 & 1 & 2 & -1 & 1 & 0 \\ 0 & 0 & 1 & 2/3 & -1 & 1/3 \end{array}\right]$$

To alter the 2 to a 0 at a_{23}, subtract *two* times row 3 from row 2.

$$\left[\begin{array}{ccc|ccc} 1 & 0 & 0 & 11/3 & -3 & 1/3 \\ 0 & 1 & 0 & -7/3 & 3 & -2/3 \\ 0 & 0 & 1 & 2/3 & -1 & 1/3 \end{array}\right]$$

As you will see, this is the same answer that was achieved by using the cofactors method. Note also that $(\mathbf{AB})^{-1} = \mathbf{B}^{-1}\mathbf{A}^{-1}$ as you can easily prove.

24.3 Solutions of simultaneous equations

Matrices provide a methodology to manage sets of equations. Most of the equations encountered at this level will give unique solutions, e.g. $x = 5$, but this is not always the case. The single equation $x + 2y = 10$ does not have a unique solution but a number of solutions, e.g. $x = 0$ and $y = 5$ *or* $x = 2$ and $y = 4$ *or* $x = -10$ and $y = 10$ etc. In general, we need the same number of independent equations as variables to obtain unique solutions.

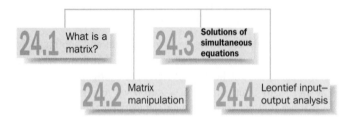

24.3.1 Solving equations using matrix inverses

Consider the equations:

$$7x + 4y = 80 \qquad 5x + 3y = 58$$

This pair of equations may be written in matrix notation as:

$$\begin{bmatrix} 7 & 4 \\ 5 & 3 \end{bmatrix} \begin{bmatrix} x \\ y \end{bmatrix} = \begin{bmatrix} 80 \\ 58 \end{bmatrix}$$

or

$$\mathbf{Ax} = \mathbf{b}$$

If we premultiply by $\mathbf{A}^{-1}$ we have

$$\mathbf{A}^{-1}\mathbf{Ax} = \mathbf{A}^{-1}\mathbf{b}$$
$$\mathbf{Ix} = \mathbf{A}^{-1}\mathbf{b}$$
$$\mathbf{x} = \mathbf{A}^{-1}\mathbf{b}$$

The result will hold for *any* set of simultaneous equations where there are as many equations as unknowns; and thus if we premultiply the vector $\mathbf{b}$ by the inverse of the matrix $\mathbf{A}$, we shall be able to find the values of the unknowns that satisfy the equations. Here:

$$\mathbf{A}^{-1} = \frac{1}{21 - 20} \begin{bmatrix} 3 & -4 \\ -5 & 7 \end{bmatrix}$$

Therefore,

$$\begin{bmatrix} x \\ y \end{bmatrix} = \begin{bmatrix} 3 & -4 \\ -5 & 7 \end{bmatrix} \begin{bmatrix} 80 \\ 58 \end{bmatrix}$$
$$= \begin{bmatrix} 240 - 232 \\ -400 + 406 \end{bmatrix}$$
$$= \begin{bmatrix} 8 \\ 6 \end{bmatrix}$$

Thus $x = 8$ and $y = 6$.

If we have three equations:

$$4x_1 + 3x_2 + x_3 = 8$$
$$2x_1 + x_2 + 4x_3 = -4$$
$$3x_1 \quad\quad + x_3 = 1$$

then it can be shown that the inverse of the **A** matrix is:

$$\mathbf{A}^{-1} = \frac{1}{31}\begin{bmatrix} 1 & -3 & 11 \\ 10 & 1 & -14 \\ -3 & 9 & -2 \end{bmatrix}$$

and that the solution to the equations is:

$$\begin{bmatrix} x_1 \\ x_2 \\ x_3 \end{bmatrix} = \frac{1}{31}\begin{bmatrix} 1 & -3 & 11 \\ 10 & 1 & -14 \\ -3 & 9 & 1 \end{bmatrix}\begin{bmatrix} 8 \\ -4 \\ 1 \end{bmatrix}$$

$$= \frac{1}{31}\begin{bmatrix} 31 \\ 62 \\ -62 \end{bmatrix}$$

$$= \begin{bmatrix} 1 \\ 2 \\ -2 \end{bmatrix}$$

thus $x_1 = 1$, $x_2 = 2$ and $x_3 = -2$.

24.3.2 Solving equations using Cramer's rule

Consider the following set of equations:

$$a_1 x + b_1 y + c_1 z = k_1$$
$$a_2 x + b_2 y + c_2 z = k_2$$
$$a_3 x + b_3 y + c_3 z = k_3$$

We need to find the determinant of the matrix of coefficients, **A**, and the determinant of this matrix with each column replaced in turn by the column vector of constant terms.

The determinant of matrix **A** may be written

$$|\mathbf{A}| = \begin{vmatrix} a_1 & b_1 & c_1 \\ a_2 & b_2 & c_2 \\ a_3 & b_3 & c_3 \end{vmatrix}$$

and the determinants of the matrix with column replacement

$$|\mathbf{A}_1| = \begin{vmatrix} k_1 & b_1 & c_1 \\ k_2 & b_2 & c_2 \\ k_3 & b_3 & c_3 \end{vmatrix} \quad |\mathbf{A}_2| = \begin{vmatrix} a_1 & k_1 & c_1 \\ a_2 & k_2 & c_2 \\ a_3 & k_3 & c_3 \end{vmatrix} \quad |\mathbf{A}_3| = \begin{vmatrix} a_1 & b_1 & k_1 \\ a_2 & b_2 & k_2 \\ a_3 & b_3 & k_3 \end{vmatrix}$$

If $|\mathbf{A}| \neq 0$, the unique solution is given by:

$$x = \frac{|\mathbf{A}_1|}{|\mathbf{A}|} \quad\quad y = \frac{|\mathbf{A}_2|}{|\mathbf{A}|} \quad\quad z = \frac{|\mathbf{A}_3|}{|\mathbf{A}|}$$

Example

Given

$$\begin{bmatrix} 4 & 3 & 1 \\ 2 & 1 & 4 \\ 3 & 0 & 1 \end{bmatrix} \begin{bmatrix} x_1 \\ x_2 \\ x_3 \end{bmatrix} = \begin{bmatrix} 8 \\ -4 \\ 1 \end{bmatrix}$$

$$|\mathbf{A}| = \begin{vmatrix} 4 & 3 & 1 \\ 2 & 1 & 4 \\ 3 & 0 & 1 \end{vmatrix} = 31 \qquad |\mathbf{A}_1| = \begin{vmatrix} 8 & 3 & 1 \\ -4 & 1 & 4 \\ 1 & 0 & 1 \end{vmatrix} = 31$$

$$|\mathbf{A}_2| = \begin{vmatrix} 4 & 8 & 1 \\ 2 & -4 & 4 \\ 3 & 1 & 1 \end{vmatrix} = 62 \qquad |\mathbf{A}_3| = \begin{vmatrix} 4 & 3 & 8 \\ 2 & 1 & -4 \\ 3 & 0 & 1 \end{vmatrix} = -62$$

Therefore

$$x_1 = \frac{|\mathbf{A}_1|}{|\mathbf{A}|} = \frac{31}{31} = 1$$

$$x_2 = \frac{|\mathbf{A}_2|}{|\mathbf{A}|} = \frac{62}{31} = 2$$

$$x_3 = \frac{|\mathbf{A}_3|}{|\mathbf{A}|} = \frac{-62}{31} = -2$$

24.3.3 Equations without unique solutions

Not all sets of simultaneous equations will give unique solutions like $x_1 = 1$, $x_2 = 2$ and $x_3 = -2$. Consider the following.

$$2x + y - z = 9$$

$$x - y + 2z = 5$$

$$4x + 2y - 2z = 18$$

This can be written:

$$\begin{bmatrix} 2 & 1 & -1 \\ 1 & -1 & 2 \\ 4 & 2 & -2 \end{bmatrix} \begin{bmatrix} x \\ y \\ z \end{bmatrix} = \begin{bmatrix} 9 \\ 5 \\ 18 \end{bmatrix}$$

Using the notation $\mathbf{Ax} = \mathbf{b}$, we could attempt to find the inverse of $\mathbf{A}$ by the partitioned matrix method.

$$\left[\begin{array}{ccc|ccc} 2 & 1 & -1 & 1 & 0 & 0 \\ 1 & -1 & 2 & 0 & 1 & 0 \\ 4 & 2 & -2 & 0 & 0 & 1 \end{array} \right]$$

Divide row 1 by 2, so that $a_{11} = 1$

$$\left[\begin{array}{ccc|ccc} 1 & 1/2 & -1/2 & 1/2 & 0 & 0 \\ 1 & -1 & 2 & 0 & 1 & 0 \\ 4 & 2 & -2 & 0 & 0 & 1 \end{array} \right]$$

Take row 1 from row 2, so that $a_{21} = 0$

$$\left[\begin{array}{ccc|ccc} 1 & 1/2 & -1/2 & 1/2 & 0 & 0 \\ 0 & -3/2 & 5/2 & -1/2 & 1 & 0 \\ 4 & 2 & -2 & 0 & 0 & 1 \end{array}\right]$$

Take 4 times row 1 from row 3, so that $a_{31} = 0$

$$\left[\begin{array}{ccc|ccc} 1 & 1/2 & -1/2 & 1/2 & 0 & 0 \\ 0 & -3/2 & 5/2 & -1/2 & 1 & 0 \\ 0 & 0 & 0 & -2 & 0 & 1 \end{array}\right]$$

Divide row 2 by 5 and add to row 1, so that $a_{13} = 0$

$$\left[\begin{array}{ccc|ccc} 1 & 1/5 & 0 & 2/5 & 1/5 & 0 \\ 0 & -3/2 & 5/2 & -1/2 & 1 & 0 \\ 0 & 0 & 0 & -2 & 0 & 1 \end{array}\right]$$

Multiply row 2 by $-2/3$, so that $a_{22} = 1$

$$\left[\begin{array}{ccc|ccc} 1 & 1/5 & 0 & 2/5 & 1/5 & 0 \\ 0 & 1 & -5/3 & -1/3 & -2/3 & 0 \\ 0 & 0 & 0 & -2 & 0 & 1 \end{array}\right]$$

We cannot obtain an identity matrix on the left hand side. In this case we cannot obtain a unique solution, because the equations are not independent. Looking at the original set of equations, it can be seen that the third equation was double the first. We only have two independent equations and three variables.

If we now consider only the first two independent equations, and take twice equation 1 from equation 2 we can eliminate z:

$$5x + y = 23$$

At this stage we can identify combinations of x and y that satisfy this condition, but not unique solutions.

24.4 Leontief input–output analysis

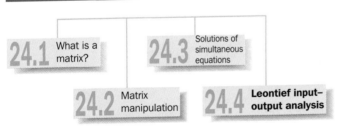

This input–output model was first developed by Leontief in the 1940s. It recognized that industries or economic sectors do not produce in isolation but rely on each other. The output from one sector may be the input to the same sector, or to another productive sector, or may be consumed as final demand. Final demand refers to the non-productive sector, including households and exports. The agricultural sector, for example, will retain some of its output as seed or breeding stock, transfer the majority of its output to the food processing industries with the remainder of its output going directly for home consumption or exports.

In a complex economy, the segregation may be fairly arbitrary since many companies produce a range of products, that may fall into several sectors, but in the UK and the rest of Europe, the segregation is still attempted by using a Standard Industrial Classification of about 40 sectors. We will not attempt to model a complete system of 40 sectors (!) but will analyse a hypothetical economy with three productive sectors. The same approach, however, could be applied to a larger system.

If the three productive sectors have outputs x_1, x_2 and x_3, part of this output will go to other productive sectors and part to final demand. An important feature of this model is that for each sector, total output is equal to total input. An example is given in Table 24.1.

Table 24.1

Outputs from:	sector X_1	Inputs to: sector X_2	sector X_3	Final demand	Total output
X_1	20	10	60	110	200
X_2	40	10	50	150	250
X_3	10	60	30	400	500
Other inputs	130	170	360		
	200	250	500		

The first three columns show the inputs to sectors X_1, X_2 and X_3. It can be seen in the first column that sector X_1 uses 20 units of its own output, 40 units from sector X_2 and 10 units from sector X_3. The remaining 130 units are referred to as **primary inputs** and include items such as labour and raw materials. The fourth column gives final demand which is the difference between total output and output used by other productive sectors. A **control economy** would attempt to manage these final demand levels through the planning of industry or sectoral output. Clearly if high proportions of output are destined for other industries, the amount available for final demand will be reduced.

To produce 200 units from sector X_1, we need inputs of 20, 40 and 10 from sectors X_1, X_2 and X_3 respectively. Expressed as proportions we get 0.10, 0.20 and 0.05 which are known as the **input–output coefficients** or **technical coefficients**. Another important feature of the model is that these input–output coefficients remain *constant* regardless of changes in final demand. These input–output coefficients are placed into a matrix **A**, and using matrix notation,

$$a_{11} = 0.10 \qquad a_{21} = 0.20 \qquad a_{31} = 0.05$$

In general a_{ij} is the proportion of input from X_i in the output of X_j. The matrix **A** is known as the **input–output matrix** and can be written:

$$\mathbf{A} = \begin{bmatrix} a_{11} & a_{12} & a_{13} \\ a_{21} & a_{22} & a_{23} \\ a_{31} & a_{32} & a_{33} \end{bmatrix} = \begin{bmatrix} 0.10 & 0.04 & 0.12 \\ 0.20 & 0.04 & 0.10 \\ 0.05 & 0.24 & 0.06 \end{bmatrix}$$

The input–output system shown in Table 24.1 can be represented by matrices:

$$\begin{bmatrix} 0.10 & 0.04 & 0.12 \\ 0.20 & 0.04 & 0.10 \\ 0.05 & 0.24 & 0.06 \end{bmatrix} \begin{bmatrix} 200 \\ 250 \\ 500 \end{bmatrix} + \begin{bmatrix} 110 \\ 150 \\ 400 \end{bmatrix} = \begin{bmatrix} 200 \\ 250 \\ 500 \end{bmatrix}$$

In summary:

$$\mathbf{AX} + \mathbf{D} = \mathbf{X}$$

where **X** is the output matrix and **D** is the demand matrix. This equation is known as the input–output equation. To find outputs to meet any projected final demands we must solve the above equation in terms of **X**. Using matrix algebra:

$$\mathbf{X} - \mathbf{AX} = \mathbf{D}$$

$$(\mathbf{I} - \mathbf{A})\mathbf{X} = \mathbf{D}$$

$$(\mathbf{I} - \mathbf{A})^{-1}(\mathbf{I} - \mathbf{A})\mathbf{X} = (\mathbf{I} - \mathbf{A})^{-1}\mathbf{D}$$

$$\mathbf{X} = (\mathbf{I} - \mathbf{A})^{-1}\mathbf{D}$$

Taking matrix **A** away from the identity matrix **I** gives

$$(\mathbf{I} - \mathbf{A}) = \begin{bmatrix} 0.90 & -0.04 & -0.12 \\ -0.20 & 0.96 & -0.10 \\ -0.05 & -0.24 & 0.94 \end{bmatrix}$$

The inverse of $(\mathbf{I} - \mathbf{A})$ is

$$\frac{10}{77\,132} \begin{bmatrix} 8784 & 664 & 1192 \\ 1930 & 8400 & 1140 \\ 960 & 2180 & 8560 \end{bmatrix}$$

Exercise

Using the figures from the example show that
$$\mathbf{X} = (\mathbf{I} - \mathbf{A})^{-1}\mathbf{D}$$

Having developed the input–output model, we are now told that the final demand levels required from X_1, X_2 and X_3 are 77 132, 231 396 and 385 660 respectively. To achieve these levels we can solve the input–output equation in terms of **X**, to determine industry outputs.

$$\begin{bmatrix} X_1 \\ X_2 \\ X_3 \end{bmatrix} = \frac{10}{77\,132} \begin{bmatrix} 8784 & 664 & 1192 \\ 1930 & 8400 & 1140 \\ 960 & 2180 & 8560 \end{bmatrix} \begin{bmatrix} 77\,132 \\ 231\,396 \\ 385\,660 \end{bmatrix}$$

$$= \begin{bmatrix} 167\,360 \\ 328\,300 \\ 503\,000 \end{bmatrix}$$

The outputs required from sectors X_1, X_2 and X_3 are 167 360, 328 300 and 503 000.

The equation system can be used to show the relationships between industries within an economy and the effect on final demand of various output and input changes. It was used in the UK in the early 1970s to look at the effects of a reduction of oil supplies on various industrial sectors. However, the difficulty with this system is not only its size for a complex economy but the implicit assumption that the input–output coefficients remain constant. This is unlikely to remain true as technical progress and innovation change the ways in which some industries operate. To take this into account, a new matrix of input–output coefficients must be derived every few years and this can be a costly and time-consuming process.

The measurement of inputs/outputs can also present problems. It is generally assumed that we can express all inputs and outputs in the same units, say millions of pounds. Clearly the outputs from some sectors are more tangible than others. In a model of the economy, how do we measure the output from the public sector?

24.5 Conclusions

As we have seen, matrices provide a method for solving a range of problems. Once we begin to work with more than two equations, the algebra can become quite tedious. In practice, we would look for a computer-based method of solution for the larger problems. In terms of specifying a problem, matrices provide a convenient form of notation.

24.6 Problems

$$A = [4 \quad 2 \quad 1] \qquad B = \begin{bmatrix} 6 \\ 8 \\ 10 \end{bmatrix} \qquad C = \begin{bmatrix} 2 & 1 \\ 4 & 5 \end{bmatrix}$$

$$D = \begin{bmatrix} 5 & 7 \\ 8 & 10 \end{bmatrix} \qquad E = \begin{bmatrix} 4 & 12 & 8 \\ 7 & 2 & 5 \\ 9 & 1 & 3 \end{bmatrix} \qquad F = \begin{bmatrix} 1 & 2 & 10 \\ 1 & 0 & 2 \\ 1 & 7 & 1 \end{bmatrix}$$

$$G = \frac{1}{58} \begin{bmatrix} -14 & 68 & 4 \\ 1 & -9 & 8 \\ 7 & -5 & -2 \end{bmatrix} \qquad H = [10 \quad 5 \quad 8]$$

Using the matrices given above, find:

1 $C + D$

2 $4A$

3 AB

4 CD

5 DC

6 BC

7 $E + F$

8 AE

9 EF

10 FB

11 C^{-1}

12 E^{-1}

13 $(AB)^{-1}$

14 BA

15 FG

16 GF

17 HF

18 BH

19 HB

20 BHF

Solve the following sets of simultaneous equations using matrix algebra:

21 $4x + 2y = 11$

   $3x + 4y = 9$

22 $6x + 3y = 18$

   $2x - 3y = 14$

23 $5x - 3y = 26$

   $2x + 2y = 4$

24 $10x + 10y = 6$

   $3x + 7y = 11$

25 $4a + 3b + c = 15$

   $2a + 5b + 7c = 47$

   $5a + 6b - 2c = 7$

26 $10a + 3b + 4c = 8$

   $20a - 5b + 2c = 12$

   $25a - b + 5c = 16$

Using Cramer's rule, solve

27 $4x + 2y - z = 57$

   $x - y + 2z = -12$

   $2x + 2y + z = 31$

28 $2a + b + c = 7$

   $b + 3c = 9$

   $4a - b = 1$

Using a computer-based method, solve

29 $a + 2b + c + 2d = 5$

   $2a + b + c - d = 1$

   $3a + b + 3c + d = 10$

   $a - b - c + 2d = 4$

30 $4x_1 + 2x_2 - x_3 + x_4 = 29$

   $x_1 + x_2 + x_3 - x_4 = 7$

   $2x_1 + 3x_2 + x_4 = 20$

   $3x_1 + x_2 - x_3 + 2x_4 = 22$

Using the notation developed in the previous section, analyse the following input–output tables:

31

Outputs from	Inputs to:		Final demand	Total output
	A	B		
A	10	40	50	100
B	30	20	50	100
Other	60	40		
	100	100		200

(a) Identify **A**, **X**, **D** and verify that the relationship holds.
(b) If final demand changes to 220 for A and 170 for B, find the level of output required from each sector.
(c) Reconstruct the table above given these new final demands.

32

Outputs from	Inputs to:		Final demand	Total output
	A	B		
A	5	85	10	100
B	20	120	360	500
Other	75	295		
	100	100		600

(a) Verify that the relationship between **A**, **X** and **D** holds for this table.
(b) Find the outputs necessary from each sector if the final demand changes to 50 for A and 500 for B.

33

Outputs from	Inputs to:			Final demand	Total output
	A	B	C		
A	20	10	30	40	100
B	20	40	60	80	200
C	20	20	120	140	300
Other	40	130	90		
	100	200	300		600

(a) Find the matrix of technical coefficients.
(b) Find $(\mathbf{I} - \mathbf{A})$.

(c) Show that the inverse of $(\mathbf{I} - \mathbf{A})$ is:

$$\frac{10}{342}\begin{bmatrix} 46 & 4 & 9 \\ 16 & 46 & 18 \\ 18 & 9 & 63 \end{bmatrix}$$

(d) Confirm that $\mathbf{X} = (\mathbf{I} - \mathbf{A})^{-1}\mathbf{D}$.

(e) If final demand now changes to 342 for A, 684 for B and 1026 for C, find the levels of output necessary for each sector.

34

Outputs from	Inputs to:			Final demand	Total output
	A	B	C		
A	10	10	10	170	200
B	10	20	20	350	400
C	100	300	70	530	1000
Other	80	70	900		
	200	400	1000		1600

Using the information given above, find the levels of output necessary from each sector to support final demand figures of 1000 for A, 1000 for B and 400 for C.

25   **Use of calculus**

We have seen in Chapter 23 how a range of business situations and economic relationships can be described mathematically. Using equations, we can determine the breakeven point of a company or the price where market demand is equal to market supply. However, we are not always just interested in a particular level of output, or price, or point in time but often how *change* is taking place. We may be interested in the rate at which profits are increasing or decreasing, or the effects of a unit change in price, or sales growth over time.

Calculus is about the measurement of change. It provides the methodology to calculate the rate of change and indeed, whether the rate of change is increasing or decreasing. Calculus also provides a notation to describe change. Given that y is a function of x $(y = f(x))$, the change in y resulting from a change in x is written dy/dx.

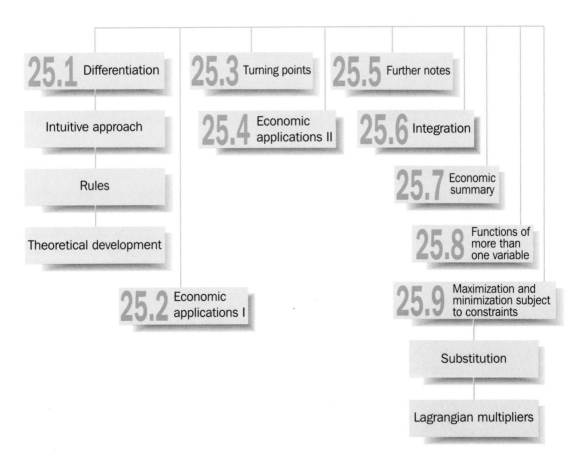

Figure 25.1   Structure of Chapter 25.

Objectives

After working through this chapter, you should be able to:

- define differentiation
- differentiate simple functions of one variable
- apply this differentiation to economic models
- find maxima and minima of one-variable functions
- differentiate more complex one-variable functions
- differentiate functions of more than one variable
- find maxima and minima subject to constraints
- integrate simple functions.

25.1 Differentiation

In this section we consider an intuitive development of differentiation (25.1.1), the rules of differentiation (25.1.2) and finally, the theoretical development of differentiation (25.1.3). Many courses at the first-year level will not expect you

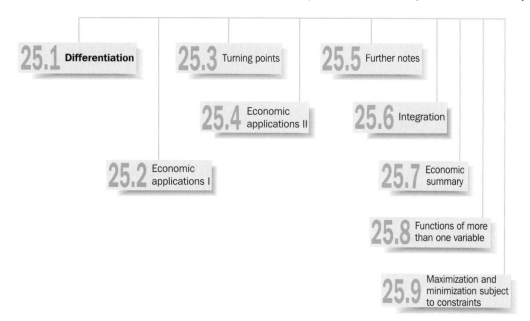

to be familiar with the theoretical aspects of the subject and you can leave section 25.1.3 without a loss of continuity. Indeed, if you are new to the subject, you can choose to leave section 25.1.3 on your first reading of this chapter and come back to it at a later stage.

25.1.1
Differentiation: an
intuitive approach

Differentiation measures the change in y resulting from a change in x. If we were looking at a graph plot of y against x, we would think of this change in terms of gradient; and this is helpful to some extent. Suppose we were interested in the cost of car hire. The total cost in pounds, y, could be plotted against the number of miles, x. The gradient would give the cost of each extra mile.

If the total cost was a fixed charge of £200, then $y = 200$. No x term is included as the cost does not depend on the number of miles travelled. The change in y resulting from a change in x is 0, as an extra mile makes no difference to the cost. In this case, $dy/dx = 0$.

If the cost of car hire had two components, a fixed charge of £150 and a mileage charge of 20p per mile, then $y = 150 + 0.20x$. Each extra mile would cost 0.20 and $dy/dx = 0.20$. In the case of a linear function, the gradient is constant (the increase in y resulting from a unit increase in x) and the change measured by the differential dy/dx is equal to the gradient.

The cost of car hire could be expressed in a more complex form:

$$y = 150 + 0.30x - 0.0005x^2 \qquad \text{for } 0 \leq x \leq 500$$
$$y = 150 + 0.04x \qquad \text{for } x > 500$$

There are two major issues to consider here: continuity and the meaning of change. Firstly, for differentiation to be valid the function must be **continuous** within the relevant range, i.e. no gaps or jumps. In this case, we are dealing with two functions, both continuous within the ranges given; differentiation is valid except at the boundary points of $x = 0$ and $x = 500$ where the functions are discontinuous. As we will see from the rules of differentiation given in section 25.1.2:

$$dy/dx = 0.30 - 0.001x \qquad \text{for } 0 < x < 500$$
$$dy/dx = 0.04 \qquad \text{for } x > 500$$

The rate of change is not constant and will depend on the value of x

$$\text{e.g. when } x = 100, \ dy/dx = 0.2, \text{ and}$$
$$\text{when } x = 200, \ dy/dx = 0.1$$

But does dy/dx measure gradient? Now the y values corresponding to $x = 100$ and $x = 200$ are 175 $(150 + 0.30 \times 100 - 0.0005 \times 100^2)$ and 190 $(150 + 0.03 \times 200 - 0.0005 \times 200^2)$ respectively. This increase of 15 units in y achieved over an increase of 100 units in x gives an average increase of 0.15 units in y for each unit increase in x, over this range. Differentiation does not give the average increase over a range of x values but rather a measure of change occurring at a point on the curve. Differentiation can be thought of as giving the gradient of a tangent to a curve (see section 25.1.3).

Before moving on to consider the rules of differentiation, let us again consider the equation $y = 150 + 0.30x - 0.0005x^2$. We are often presented with such equations and given little explanation, but it is worth pausing to examine the structure of this equation. It can be rewritten $y = 150 + (0.30 - 0.0005x)x$. The three components of cost are therefore a fixed cost, plus a charge per mile times the number of miles. The charge per mile, $(0.30 - 0.0005x)$, depends on the number of miles, x, and as x increases, the charge made decreases. You will also find this form of equation when you look at revenue functions where price per unit falls as sales increase. It is always worth asking the question 'What does this equation mean?'.

**25.1.2
Differentiation:
the rules**

To obtain a value for dy/dx we can apply a number of rules:

1   Given a function of the form $y = k$ where k is a constant,

$$\frac{dy}{dx} = 0$$

A constant, by its nature, does not change (depend on x) and therefore has zero rate of change.

Example

If $y = 23$, then $dy/dx = 0$.

2   Given a function of the form $y = ax$ where a is a constant,

$$\frac{dy}{dx} = a$$

For each unit increase in x there is a constant a unit increase in y – the rate of change.

Example

If $y = 4.5x$, then $dy/dx = 4.5$.

3   Given a function of the form $y = ax + k$

$$\frac{dy}{dx} = a$$

When we differentiate this more general linear function, the rate of change is still a constant given by the x coefficient, because for a function consisting of several parts or terms, we differentiate each part separately and then put the results together.

Example

If $y = 32.8 + 0.96x$, then $dy/dx = 0.96$.

4   Given a function of the form $y = ax^n$

$$\frac{dy}{dx} = nax^{n-1}$$

Example

If $y = 6x^3$, then $a = 6$ and $n = 3$

$$\frac{dy}{dx} = 3 \times 6x^2 = 18x^2$$

Example

If $y = 10x^2$, then $dy/dx = 2 \times 10x^1 = 20x$ since $x^1 = x$.

Example

If $y = 6x^3 - 4x^2 + 10x - 50 + 3x^{-2}$, then

$$\frac{dy}{dx} = 3 \times 6x^2 - 2 \times 4x^1 + 10x^0 + 0 + (-2) \times 3x^{-3}$$

$$= 18x^2 - 8x + 10 - 6x^{-3}$$

NB. $x^0 = 1$.

To complete this section, we need to consider certain other functions which have special **derivatives**. A derivative is the result of differentiation and is also called the differential coefficient.

Example

If $y = e^x$ (the exponential function described in section 23.4.5) then $dy/dx = e^x$.

Example

If $y = \log_e x$ then $dy/dx = 1/x$.

**25.1.3
Differentiation:
the theoretical
development**

Calculus, as discussed, is about the measurement of change. However, the rate of change can be interpreted in two ways: the average rate of change and the instantaneous rate of change. If you consider a car travelling from Birmingham to Manchester, then its **average speed** (i.e. the rate of change of distance in relation to time) may be 50 miles per hour, but its speed at any particular moment may be much less (for example 5 m.p.h. in heavy traffic) or rather more (for example 70 m.p.h. on the motorway). The driver, looking at the speedometer at a particular moment, can tell the **instantaneous speed**. Now it has been argued that there can be no such thing as instantaneous speed, since speed is to do with the distance travelled, but try convincing someone who has been driving and met a wall (which was presumably stationary) that they could not be travelling at a speed at the instant the vehicle and the wall met!

Calculus, then, is about the **instantaneous rate of change** and *not* the average rate of change.

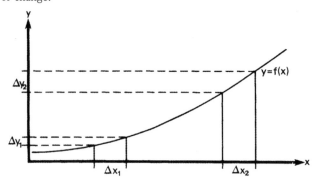

Figure 25.2

If we consider the graph of a function (as in Figure 25.2) we can see that there is an increasing change in y ($\Delta y_1 < \Delta y_2$) for equal changes in x ($\Delta x_1 = \Delta x_2$) as we move to the right. The symbol Δ (Greek capital delta) indicates an interval in the value of a variable, e.g. Δx represents the difference between two values of x. Thus there are different rates of change in y for different values of x. To find the average rate of change between two values of x, we would divide the change in y by the change in x. Here,

$$\frac{\Delta y_1}{\Delta x_1} \text{ is less than } \frac{\Delta y_2}{\Delta x_2}$$

If a small section of the graph of the function is magnified (as in Figure 25.3), we see that the ratio $\Delta y/\Delta x$ not only measures the average rate of change of y between two values of x, but also the gradient or slope of a straight line (known as a **chord**) drawn between two points on the graph. As we reduce the change in x the straight line gets shorter and shorter and hence its slope or rate of change gets nearer and nearer to the slope of the curve at a single point; but if we have no change in x, then there is no change in y, although the function is still changing (cf. the car mentioned above). Now if the straight line between the two points can represent the average slope of the graph between the points, is there a straight line at one point whose slope is that of the graph at that point? The answer is yes!

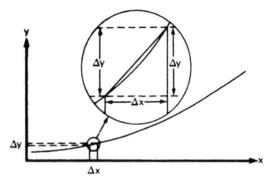

Figure 25.3

Figure 25.4 shows a straight line which touches the curve at just one point (known as a **tangent**) and the slope of this line will be the slope of the curve at the point where the two touch. Since there is only one point now in question, the slope of the curve at that point measures not the average but the instantaneous rate of change.

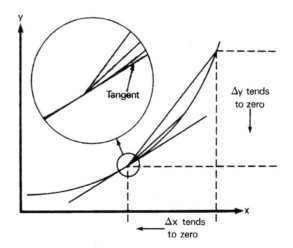

Figure 25.4

Formally, the measurement of the slope of a curve at a point is valid if the function is **continuous**, i.e. there are no gaps or jumps in the function; thus if $y = f(x)$ then both x and y are continuous variables in the relevant range or domain. The slope of the tangent at the point where it touches the function is defined as the limit of the ratio $\Delta y/\Delta x$ as Δx tends to zero.

A limit is a number to which the ratio gets nearer and nearer, as the interval Δx gets smaller. For example the ratio

$$\frac{x + \Delta x}{x}$$

will get closer and closer to 1 as the interval Δx gets smaller. Thus the limit of this expression will be 1.

Formally, the change in x is thought of as a distance, h, so that the interval Δx is from x to $x + h$. Since $y = f(x)$, the interval Δy extends from $f(x)$ to $f(x + h)$ (see Figure 25.5).

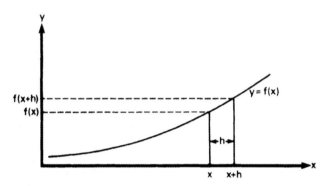

Figure 25.5

We saw above that an average slope of a function was

$$\frac{\Delta y}{\Delta x} = \frac{y_2 - y_1}{x_2 - x_1}$$

or, in Figure 25.5

$$\frac{f(x + h) - f(x)}{(x + h) - x} = \frac{f(x + h) - f(x)}{h}$$

and it is this ratio, as h gets smaller and smaller, that gives the instantaneous rate of change of $f(x)$ at a single value of x.

The limit of $\Delta y / \Delta x$ as $h \to 0$ is denoted by dy/dx or $f'(x)$. Here d signifies an infinitesimal change in a variable (cf. Δ which signifies finite change).

Example

If $f(x) = x^2$, then $f(x + h) = (x + h)^2 = x^2 + 2xh + h^2$ so the ratio $\Delta y / \Delta x$ is given by:

$$\frac{\Delta y}{\Delta x} = \frac{(x^2 + 2xh + h^2) - x^2}{h}$$

$$= \frac{2xh + h^2}{h}$$

$$= 2x + h$$

and as $h \to 0$, this gets closer and closer to $2x$. Thus, for $y = x^2$,

$$\frac{dy}{dx} = 2x$$

So the rate of change of the function $y = x^2$ at $x = 2$, for example, is $2x = 2 \times 2 = 4$.

Note that the rate of change of a function is often referred to as the **gradient** of the function.

25.2 Economic applications I

Within economics, there are several functions which are related to each other as function to derivative. For instance, the total cost (TC) represents the cost of producing a particular amount of the product; the marginal cost (MC) is the cost

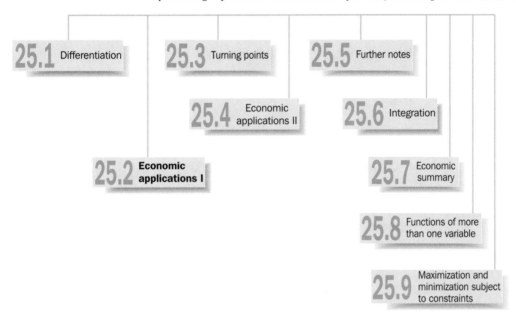

of producing one extra unit and so the marginal cost of a given output is the rate at which total cost is changing at that output. Thus, if we have a total cost function and differentiate it, we will find that the result is a marginal cost function. This relationship will also hold for revenue functions.

Thus if $y = TC$ then

$$\frac{dy}{dx} = MC$$

and if $y = TR$ then

$$\frac{dy}{dx} = MR$$

where TR is total revenue and MR is marginal revenue. For example, if

$$TC = 40 + 10x + 2x^2 + x^3$$

then

$$MC = 10 + 4x + 3x^2$$

(per unit change in output when output $= x$). For example, if

$$TR = 4x$$

then

$$MR = 4$$

(per unit change in sales when sales $= x$).

We may also use the idea of averaging to find the average cost of, or revenue from, a given output. If total revenue is £100 from an output of 5, then the average revenue (AR) will be $100/5 = £20$ per unit. So if

$$TR = 100x - 10x^2$$

$$AR = \frac{1}{x}(100x - 10x^2)$$

$$= 100 - 10x$$

(per unit when x units are sold). Similarly if

$$AC = x^2 - 10x + 38$$

then

$$TC = x(x^2 - 10x + 38)$$
$$= x^3 - 10x^2 - 38x$$

If we begin with a cubic total cost function and find the marginal cost function, which will be quadratic, and the average cost function, which will also be quadratic, then graph these, we will have the typical economic diagram, as in Figure 25.6.

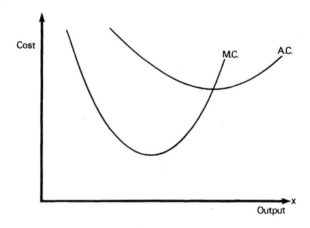

Figure 25.6

With a linear demand curve (i.e. an AR function) we will obtain the relationship in Figure 25.7, where the marginal revenue function will also be linear, but have a slope twice that of the average revenue function. Since if

$$AR = a + bx$$

then

$$TR = ax + bx^2$$

and

$$MR = a + 2bx$$

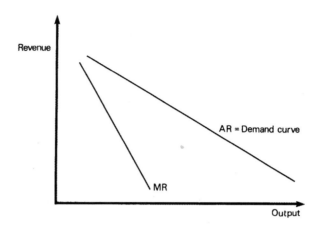

Figure 25.7

Elasticity of demand measures the responsiveness of the quantity sold to various factors: initially, the most useful of these is price elasticity (E_D), which is defined as:

$$E_D = \frac{dq}{dp} \times \frac{p}{q}$$

Example

If the demand curve is given by

$$p = 100 - 5q$$

then

$$\frac{dp}{dq} = -5$$

and since

$$\frac{dq}{dp} = \frac{1}{dp/dq}$$

then

$$\frac{dq}{dp} = -\frac{1}{5}$$

At a quantity of 10, the price is:

$$p = 100 - 5(10) = 50$$

So, elasticity is:

$$E_D = \frac{1}{-5} \times \frac{50}{10} = -1$$

At a quantity of 8, the price is 60 and elasticity is

$$E_D = \frac{1}{-5} \times \frac{60}{8} = -1.5$$

So for a linear demand function, there is a *constant* slope but a *changing* elasticity.

25.3 Turning points

In section 23.4.3 we considered quadratic functions and stated that the sign of the coefficient of x^2 determined the shape of the function. With functions that

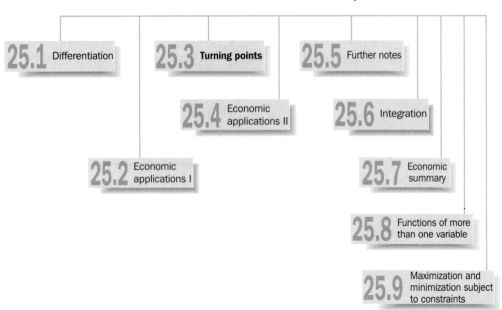

include higher powers of x the graph will have several changes of direction. We can use the method of differentiation developed above to locate both *where* these changes of direction occur and which way the change affects the function. This is the method of locating the **maximum** and **minimum** values of a function. By a 'maximum' value we do not necessarily mean one that is greater than all other values (a **global** maximum), it could simply be greater than all neighbouring values only (a **local** maximum). Similar definitions apply to minimum values. For the economic applications at this stage, this distinction between local and global maxima is not of great importance.

If you consider the function represented in Figure 25.8 you can see that the slopes of the function at points A, B and C are quite different. At point A the function is increasing and thus the slope is positive. At point C the function is

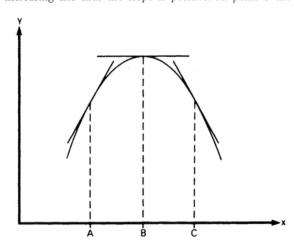

Figure 25.8

decreasing, and thus the slope is negative. However, at point B, it is just changing direction and is neither going up nor going down; therefore the slope is zero. Now the slope of a function can be found by differentiation.

In Figure 25.9, the first graph represents a function which has a maximum at the point A and a minimum at the point B, and the second graph shows dy/dx against x for this function. This shows that to the left of the point A there is a positive slope, but that slope is decreasing. At A there is zero slope, and so $dy/dx = 0$. After A the original function is decreasing and so dy/dx is negative, but when the point B is reached, where the original function begins to go up again, there will be zero slope, and dy/dx will again be 0.

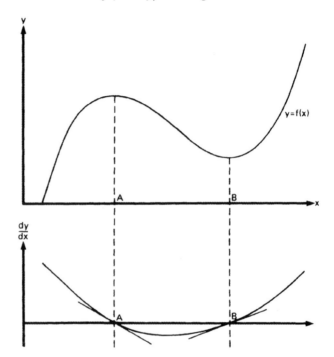

Figure 25.9

To the right of point B, the original function is increasing, and hence the second graph is in the positive region.

From this we see that at both **turning points** the value of $dy/dx = 0$.

There is more information that we can gain from Figure 25.9. Looking at the second graph we find that at the point A, the value of dy/dx is decreasing as x increases, and thus has a negative slope, while at the point B, the value of dy/dx is increasing, and thus has a positive slope. Thus, if we can find the slope of the dy/dx function, we can distinguish between maximum points like A and minimum points like B.

In effect, this means differentiating the original function a second time, and we denote this second differential by a slightly different symbol:

$$\frac{d^2y}{dx^2}$$

The process of differentiating is still the same, only it is now applied to the function dy/dx.

The rules for distinguishing a maximum from a minimum value are thus:

for a maximum, $\quad \dfrac{dy}{dx} = 0 \quad$ and $\quad \dfrac{d^2y}{dx^2} < 0$

for a minimum, $\quad \dfrac{dy}{dx} = 0 \quad$ and $\quad \dfrac{d^2y}{dx^2} > 0$

Examples

1 If

$$y = 2x^2 - 8x + 50$$

then

$$\frac{dy}{dx} = 4x - 8 = 0$$

therefore $4x = 8$ and hence

$$x = \frac{8}{4} = +2$$

so the turning point is at $x = 2$.

$$\frac{d^2y}{dx^2} = 4$$

which is *positive*, and so the turning point is a minimum.

The value of y is $2(2)^2 - 8(2) + 50 = 8 - 16 + 50 = 42$.

Thus the function $y = 2x^2 - 8x + 5$ has a minimum point at $(2, 42)$.

2 $y = \frac{1}{3}x^3 - 4x^2 + 15x + 10$

then

$$\frac{dy}{dx} = x^2 - 8x + 15 = 0$$

factorizing gives $(x - 3)(x - 5) = 0$ so $x = 3$ and 5 and there is a turning point for each of these values.

$$\frac{d^2y}{dx^2} = 2x - 8$$

Now at $x = 3$, $2x - 8 = 2(3) - 8 = -2$ (negative $\Rightarrow$ maximum) and at $x = 5$, $2x - 8 = 2(5) - 8 = +2$ positive $\Rightarrow$ minimum).

Also, at $x = 3$,

$$y = \frac{1}{3}(3)^3 - 4(3)^2 + 15(3) + 10 = 28$$

and, at $x = 5$,

$$y = \frac{1}{3}(5)^3 - 4(5)^2 + 15(5) + 10 = 26.67$$

Thus the function

$$y = \frac{1}{3}x^3 - 4x^2 + 15x + 10$$

has a maximum at $(3, 28)$ and a minimum at $(5, 26.67)$.

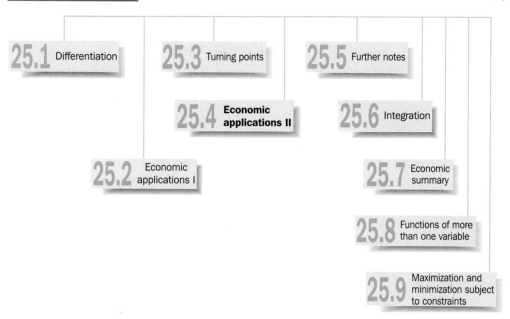

25.4 Economic applications II

The economist is often interested in finding the maximum value of certain relationships for profit, sales revenue, welfare, etc., and minimum values, particularly

25.1 Differentiation

25.3 Turning points

25.5 Further notes

25.4 **Economic applications II**

25.6 Integration

25.2 Economic applications I

25.7 Economic summary

25.8 Functions of more than one variable

25.9 Maximization and minimization subject to constraints

for cost functions. The methods developed above will allow us to specify where these turning points occur.

Examples

1 If a firm has

$$TR = 40x - 8x^2$$

and $$TC = 8 + 16x - x^2$$

where $x =$ thousands of units of product, then its profit function (π) will be the difference between its total revenue and total cost:

$$\pi = TR - TC$$
$$= (40x - 8x^2) - (8 + 16x - x^2)$$
$$= 40x - 8x^2 - 8 - 16x + x^2$$
$$= -8 + 24x - 7x^2$$

If we wish to find where maximum profit occurs, we differentiate to give

$$\frac{d\pi}{dx} = 24 - 14x$$

and find that $d\pi/dx$ is zero when $x = 24/14 = 1.714$, and $d^2\pi/dx^2 = -14$, which is negative, thus representing a maximum.

So the firm will achieve maximum profit at an output of $x = 1.714$, and since x is measured in thousands, this is 1714 units. Profit here is 12.571. If the firm wishes to maximize its sales revenue (i.e. TR), we have:

$$TR = 40x - 8x^2$$

$$\frac{dTR}{dx} = 40 - 16x$$

and when $dTR/dx = 0$, $x = 40/16 = 2.5$ and $d^2TR/dx^2 = -16$, which is negative, thus representing a maximum.

So, maximum sales revenue will be at 2500 units, but profit here would only be 8.25.

2 If a firm's total cost function is

$$TC = 200 + 10x - 6x^2 + x^3$$

then the 200 represents fixed cost, since it does not vary with the level of output. If we remove this fixed cost, we will be left with total variable cost (TVC) with

$$TVC = 10x - 6x^2 + x^3$$

and average variable cost (AVC) can be found by dividing by x:

$$AVC = 10 - 6x + x^2$$

We want to know the output to give minimum AVC, and to find this, we differentiate:

$$\frac{dAVC}{dx} = -6 + 2x = 0$$

therefore $x = 3$ and

$$\frac{d^2AVC}{dx^2} = +2$$

positive, thus minimum.

So, minimum average variable cost is at an output of 3.

25.5 Further notes

1 If the function to be differentiated is the product of two functions, then there is a method of differentiating without having to multiply the two functions out, e.g.

$$y = (2x^2 + 10x + 5)(6x^3 + 12x^2)$$

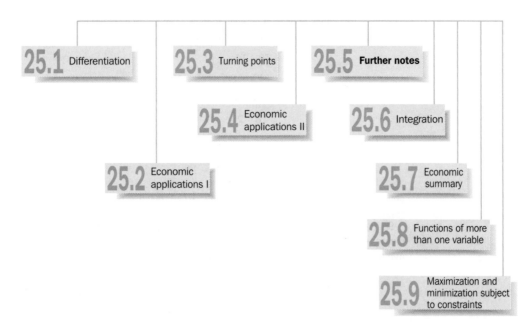

25.1 Differentiation

25.3 Turning points

25.5 **Further notes**

25.4 Economic applications II

25.6 Integration

25.2 Economic applications I

25.7 Economic summary

25.8 Functions of more than one variable

25.9 Maximization and minimization subject to constraints

let

$$u = 2x^2 + 10x + 5$$

and

$$v = 6x^3 + 12x^2$$

so if $y = uv$ then, in general,

$$\frac{dy}{dx} = v \times \frac{du}{dx} + u \times \frac{dv}{dx}$$

Now if $u = 2x^2 + 10x + 5$ then

$$\frac{du}{dx} = 4x + 10$$

and if $v = 6x^3 + 12x^2$ then

$$\frac{dv}{dx} = 18x^2 + 24x$$

and

$$\begin{aligned}
\frac{dy}{dx} &= (6x^3 + 12x^2)(4x + 10) + (2x^2 + 10x + 5)(18x^2 + 24x) \\
&= 24x^4 + 60x^3 + 48x^3 + 120x^2 + 36x^4 + 48x^3 + 180x^3 \\
&\quad + 240x^2 + 90x^2 + 120x \\
&= 60x^4 + 336x^3 + 450x^2 + 120x
\end{aligned}$$

2   If the function to be differentiated consists of one function divided by another, then the following method is appropriate.

$$y = \frac{(10x^2 + 6x + 5)}{(12x^3 + 15x^2)}$$

so let

$$u = 10x^2 + 6x + 5$$

and

$$v = 12x^3 + 15x^2$$

Then if $y = u/v$, then, in general,

$$\frac{dy}{dx} = \frac{v \times \dfrac{du}{dx} - u \times \dfrac{dv}{dx}}{v^2}$$

If $u = 10x^2 + 6x + 5$ then

$$\frac{du}{dx} = 20x + 6$$

and if $v = 12x^3 + 15x^2$ then

$$\frac{dv}{dx} = 36x^2 + 30x$$

$$\frac{dy}{dx} = \frac{(12x^3 + 15x^2)(20x + 6) - (10x^2 + 6x + 5)(36x^2 + 30x)}{(12x^3 + 15x^2)^2}$$

$$= \frac{240x^4 + 300x^3 + 72x^3 + 90x^2}{} \\ \frac{-(360x^4 + 300x^3 + 216x^3 + 180x^2 + 180x^2 + 150x)}{144x^6 + 360x^5 + 225x^4}$$

$$= \frac{-120x^4 - 144x^3 - 270x^2 - 150x}{144x^6 + 360x^5 + 225x^4}$$

$$= \frac{-6x(20x^3 + 24x^2 + 45x + 25)}{x^4(144x^2 + 360x + 225)}$$

Note that it is not always necessary to carry out all of the multiplications and simplifications can help if made earlier in the calculation.

3   If the function to be differentiated is a function of another function, then the following method is appropriate. If

$$y = (10x^2 + 6x)^3$$

let $u = 10x^2 + 6x$ thus

$$\frac{du}{dx} = 20x + 6$$

and $y = u^3$ thus

$$\frac{dy}{du} = 3u^2$$

and, in general,

$$\frac{dy}{dx} = \frac{dy}{du} \times \frac{du}{dx}$$
$$= 3u^2(20x + 6)$$
$$= 3(10x^2 + 6x)^2(20x + 6)$$
$$= 3(100x^4 + 120x^3 + 36x^2)(20x + 6)$$

This could be further simplified.

4   If the function is exponential then if $y = e^x$

$$\frac{dy}{dx} = e^x$$

If $y = e^{ax}$ then

$$\frac{dy}{dx} = a\,e^{ax}$$

e.g.

$$y = e^{2x} \qquad \text{and} \qquad \frac{dy}{dx} = 2e^{2x}$$

If $y = e^{(x^2 + 6)}$ let $u = x^2 + 6$. Then

$$y = e^u \qquad \text{and} \qquad \frac{du}{dx} = 2x$$

so

$$\frac{dy}{dx} = \frac{dy}{du} \times \frac{du}{dx} = e^u \times 2x = 2x\,e^{(x^2 + 6)}$$

5   If $y = \log_e x$ then

$$\frac{dy}{dx} = \frac{1}{x}$$

If

$$y = \log_e(x^3 + 2x + 3)$$

let

$$u = x^3 + 2x + 3$$

then

$$y = \log_e(u) \qquad \text{and} \qquad \frac{du}{dx} = 3x^2 + 2$$

so

$$\frac{dy}{dx} = \frac{dy}{du} \times \frac{du}{dx}$$
$$= \frac{1}{u}(3x^2 + 2)$$
$$= \frac{3x^2 + 2}{x^3 + 2x + 3}$$

25.6 Integration

Differentiation has allowed us to find an expression for the rate of change of a function; what we need now is some method of reversing the process, i.e. obtaining the original function when the rate of change is known. This process is integration

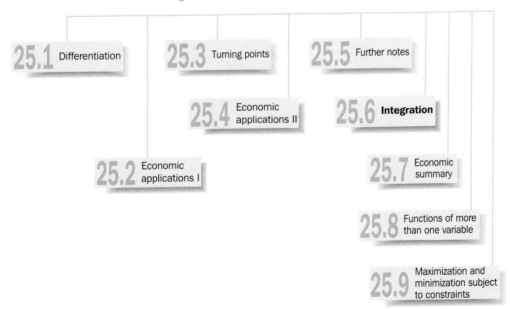

If $y = x^2$, we know that the rate of change is $2x$; thus if we integrate $2x$ we must get x^2. However, if $y = x^2 + 10$, the rate of change is still $2x$, and integrating will give us x^2 and *not* $x^2 + 10$. This is because the constant in the initial expression has a zero rate of change, and therefore disappears when it is differentiated. When we integrate, then, we should add a constant c to the expression we obtain, and we will need some further information if we are to find the specific value of this constant. For example, if

$$\frac{dy}{dx} = 2x$$

integrating gives

$$y = x^2 + c$$

If we also know that $y = 10$ when $x = 0$ we have

$$10 = 0^2 + c$$

therefore $c = 10$ and hence $y = x^2 + 10$.

The symbol used for integration is an 'old style' S, i.e. $\int$, and it is usual to put dx after the expression to show that we are integrating with respect to x. Thus

$$\int (2x)\, dx = x^2 + c$$

As with differentiation, there is a general formula for integration. If

$$y = ax^n$$

where a is a constant then

$$\int y\, dx = \frac{ax^{n+1}}{n+1} + c$$

for all values of n except $n = -1$. In that special case we have

$$\int \frac{1}{x}\, dx = \log_e x + c$$

Reverting to the more usual case, if $y = 15x^4$, then $a = 15$ and $n = 4$ so

$$\int y\, dx = \frac{15x^{4+1}}{4+1} + c = \frac{15x^5}{5} + c$$

$$= 3x^5 + c$$

When there are several terms in the function we must treat each separately, e.g.

$$
\begin{array}{lllll}
y & = 10x^3 & + 6x^2 & - 4x & + 10 \\
 & (a = 10 & (a = 6 & (a = -4 & (a = 10 \\
 & n = 3) & n = 2) & n = 1) & n = 0)
\end{array}
$$

$$
\begin{aligned}
\int y\, dx &= \frac{10x^{3+1}}{3+1} + \frac{6x^{2+1}}{2+1} - \frac{4x^{1+1}}{1+1} + 10x + c \\
&= \frac{10x^4}{4} + \frac{6x^3}{3} - \frac{4x^2}{2} + 10x + c \\
&= 2.5x^4 + 2x^3 - 2x^2 + 10x + c
\end{aligned}
$$

Note that integrating the constant, 10, gives $10x$.

Integration can be viewed as a summation process; for example, if you sum all of the marginal costs (the cost of producing one more) up to some point, then you will obtain total cost. This idea can also be used to find a sum of areas bounded by a curve between two points (known as definite integration), provided the function is positive. For example,

$$\int_2^5 (10x + 5)\,dx$$

means find the integral of $10x + 5$, i.e. find the sum of all products $(10x + 5)\,dx$, between the values $x = 2$ and $x = 5$.

Integrating gives the indefinite integral $5x^2 + 5x + c$, but the definite integral is usually written as

$$[5x^2 + 5x]_2^5$$

omitting the constant, because in the evaluation (below) it cancels itself out.

Now we evaluate this at $x = 5$, and at $x = 2$ and subtract the second from the first:

$$(5x^2 + 5x)_{(x=5)} - (5x^2 + 5x)_{(x=2)}$$
$$= (125 + 25) - (20 + 10)$$
$$= \quad 150 \quad - \quad 30$$
$$= 120$$

This problem is illustrated in Figure 25.10, the shaded area being the 120 found above.

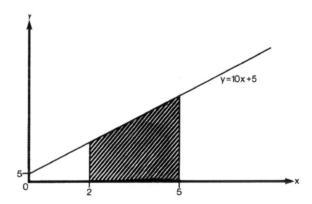

Figure 25.10

Within economics, the process of integration will allow us to go from marginal functions to total functions. Thus

$$\int MR\,dx = TR + c$$

$$\int MC\,dx = TC + c'$$

But with TR, there is rarely any revenue if output is zero, so in general the constant will also be zero. With TC there is a cost to the firm even if no production takes place, and so c' will be non-zero and positive. It will represent fixed cost.

If the marginal cost function for a firm has been identified as:

$$MC = 2x^2 - 8x + 10$$

and fixed costs are known to be 100, then the total cost function is given by:

$$TC = \int MC\,dx$$

$$= \int (2x^2 - 8x + 10)\,dx$$

$$= \tfrac{2}{3}x^3 - 4x^2 + 10x + c$$

$TC = 100$ when $x = 0$, therefore $c = 100$, and thus the total cost function is:

$$TC = \tfrac{2}{3}x^3 - 4x^2 + 10x + 100$$

25.7 Economic summary

Figure 25.11 summarizes the application of calculus to simple economic models, usually of one firm or one market.

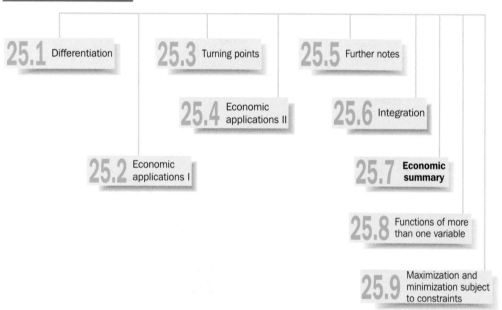

25.1 Differentiation

25.3 Turning points

25.5 Further notes

25.4 Economic applications II

25.6 Integration

25.2 Economic applications I

25.7 Economic summary

25.8 Functions of more than one variable

25.9 Maximization and minimization subject to constraints

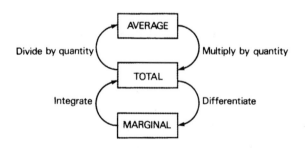

Figure 25.11

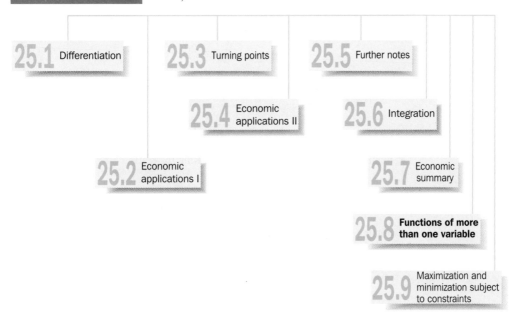

25.8 Functions of more than one variable

The simple economic models discussed so far embody functions of only one variable. As we move on to consider more complex economic and business situations, we find that there are *several* factors which will affect the outcome. Where

the exact relationship is unknown, it must be estimated by statistical means (see Chapter 17), but for many examples in economics an exact function can be specified. For a duopoly (two sellers), the sales of one company's product may be a function of their own price, the price charged by the competitor and the respective amounts spent on advertising. In production theory, it is the *combination* of both capital and labour which determines the amount produced and not just one factor of production; while in welfare economics we consider the combination of goods which is available to the consumer or community. For all of these situations, while we are interested in the final outcome of particular decisions, we are also interested in the individual contribution of each factor, and also in ways of making optimal decisions on each factor in order to optimize the final outcome.

Consider a function such that y is determined by x and z; for example:

$$y = 10x^2 + 6xz + 15z^2$$

We can find the rate of change of y with respect to x if we *temporarily* hold the value of z constant when differentiating. (To distinguish this from differentiation where $y = f(x)$ we now use a new symbol $\delta y/\delta x$.)

Differentiating, we have:

$$\frac{\delta y}{\delta x} = 20x + 6z$$

Note that $10x^2$ was treated as before, giving $20x$, $6xz$ was treated as $(6z)x$, giving $6z$ and $15z^2$ was treated as a constant.

We can also find the rate of change of y with respect to z by temporarily holding x constant. Thus

$$\frac{\delta y}{\delta z} = 6x + 30z$$

The process described above is known as **partial differentiation** since we are finding a rate of change while part of the function is held constant. As with the functions of single variables, each of the functions we have obtained may be differentiated again, but note the number of outcomes:

$$\frac{\delta^2 y}{\delta x^2} = 20$$

i.e. holding z constant again and differentiating $20x + 6z$.

$$\frac{\delta^2 y}{\delta z^2} = 30$$

i.e. holding x constant again and differentiating $6x + 30z$.

$$\frac{\delta^2 y}{\delta x\, \delta z} = 6$$

i.e. holding x constant and differentiating $20x + 6z$.

$$\frac{\delta^2 y}{\delta z\, \delta x} = 6$$

i.e. holding z constant and differentiating $6x + 30z$.
   Note, however, that

$$\frac{\delta^2 y}{\delta x\, \delta z} = \frac{\delta^2 y}{\delta z\, \delta x}$$

in general.
   We may now use these results to find the maxima and minima for a particular function. For a **maximum**:

$$\frac{\delta y}{\delta x} = 0 \qquad \text{and} \qquad \frac{\delta y}{\delta z} = 0$$

$$\frac{\delta^2 y}{\delta x^2} < 0 \qquad \text{and} \qquad \frac{\delta^2 y}{\delta z^2} < 0$$

and

$$\left(\frac{\delta^2 y}{\delta x^2} \right) \left(\frac{\delta^2 y}{\delta z^2} \right) \geq \left(\frac{\delta^2 y}{\delta x\, \delta z} \right)^2$$

For a **minimum**:

$$\frac{\delta y}{\delta x} = 0 \qquad \text{and} \qquad \frac{\delta y}{\delta z} = 0$$

$$\frac{\delta^2 y}{\delta x^2} > 0 \qquad \text{and} \qquad \frac{\delta^2 y}{\delta z^2} > 0$$

and

$$\left(\frac{\delta^2 y}{\delta x^2} \right) \left(\frac{\delta^2 y}{\delta z^2} \right) \geq \left(\frac{\delta^2 y}{\delta x\, \delta z} \right)^2$$

Examples

1 If $y = 5x^2 + 4xz - 60x - 40z + 4z^2 + 1000$ then

$$\frac{\delta y}{\delta x} = 10x + 4z - 60 = 0 \qquad \frac{\delta y}{\delta z} = 4x - 40 + 8z = 0$$

Rearranging gives two simultaneous equations:

$$10x + 4z = 60 \qquad 4x + 8z = 40$$

Multiplying the first by 2 gives

$$20x + 8z = 120$$

Subtracting the second gives

$$16x = +80$$

Therefore $x = 5$ and substituting gives $z = 2.5$

$$\frac{\delta^2 y}{\delta x^2} = 10, \qquad \frac{\delta^2 y}{\delta z^2} = 8$$

both positive, and

$$\frac{\delta^2 y}{\delta x \, \delta z} = 4, \qquad \left(\frac{\delta^2 y}{\delta x^2}\right)\left(\frac{\delta^2 y}{\delta z^2}\right) = 10 \times 8 = 80$$

which is greater than

$$\left(\frac{\delta^2 y}{\delta x \, \delta z}\right)^2 = 4^2 = 16$$

Thus the function has a minimum when $x = 5$ and $z = 2.5$ (the y value will be 800).

2 Consider a monopolist with two products, A and B, who wishes to maximize total profit. The two demand functions are:

$$P_A = 80 - q_A \qquad P_B = 50 - 2q_B$$

and the total cost function is

$$TC = 100 + 8q_A + 6q_B + 14q_A^2 + 4q_B^2 + 4q_A q_B$$

The total revenue function will consist of two parts:

$$TR_A = 80q_A - q_A^2 \qquad \text{and} \qquad TR_B = 50q_B - 2q_B^2$$

thus the total profit function is

$$\pi = TR_A + TR_B - TC$$

$$= -100 + 72q_A + 44q_B - 15q_A^2 - 6q_B^2 - 4q_A q_B$$

$$\frac{\delta \pi}{\delta q_A} = 72 - 30q_A - 4q_B = 0$$

$$\frac{\delta \pi}{\delta q_B} = 44 - 12q_B - 4q_A = 0$$

Rearranging these equations gives two simultaneous equations:

$$30q_A + 4q_b = 72$$
$$4q_A + 12q_B = 44$$

Multiplying the first by 3 gives

$$90q_A + 12q_B = 216$$

Then subtracting the second gives

$$86q_A = 172$$

Therefore $q_A = 2$ and substituting gives $q_B = 3$.

$$\frac{\delta^2 \pi}{\delta q_A^2} = -30, \qquad \frac{\delta^2 \pi}{\delta q_B^2} = -12$$

both negative, and

$$\frac{\delta^2 \pi}{\delta q_A \, \delta q_B} = -4$$

but $(-30)(-12) > (-4)^2$.

Thus maximum profit is where the quantity of A sold is two, and the quantity of B sold is three. By substitution into the profit function, we find that the level of profit will be 38. The prices charged for the two products are found from the demand functions; the price of A is 78 and the price of B is 44.

25.9 Maximization and minimization subject to constraints

Many economics problems have limitations upon the solution that may be obtained: production may be limited by the available supply of raw materials or by the capacity of the current factory; cost may have to be kept to a minimum subject to certain, predetermined, output levels; or a consumer's utility maximized subject to the budget. In each of these cases, if we simply maximize or

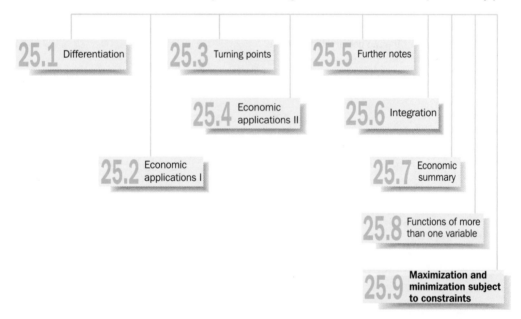

minimize the function it is unlikely that this will also satisfy the constraint and thus we need to take the constraint into consideration during the optimization process. For simple functions is may be possible to substitute the constraint into the function, but the more general method is that of Lagrangian multipliers.

25.9.1 Substitution

For functions of two variables, we may be able to substitute the constraint into the function. Consider the following example:

$$z = 10 + 3x - 4x^2 + 10y - 2y^2 + xy$$

to be maximized subject to the constraint that $x = 2y$.

Substituting, we have:

$$z = 10 + 6y - 16y^2 + 10y - 2y^2 + 2y^2$$

$$= 10 + 16y - 16y^2$$

$$\frac{\mathrm{d}z}{\mathrm{d}y} = 16 - 32y = 0 \text{ therefore } y = \tfrac{1}{2} \text{ and hence } x = 1$$

$$\frac{\mathrm{d}^2z}{\mathrm{d}y^2} = -32 \text{ therefore it is a maximum.}$$

Exercise

Find the values of x and y to maximize the function without the constraint. (Answer: $x = 22/31$; $y = 83/31$.)

25.9.2 Lagrangian multipliers

For more complex functions the method above will become tedious and thus we now place the constraint as part of the function we are to optimize. Again, we will work through an example. Suppose

$$z = 10x - 4x^2 + 20y - y^2 + 4xy$$

subject to $2x + 4y = 100$. The first step is to rearrange the constraint as follows:

$$100 - 2x - 4y = 0$$

We the multiply it by a new unknown value, λ (lambda), which is the Lagrangian multiplier:

$$\lambda(100 - 2x - 4y) = 0$$

and this function is added to the function z to give a new function z^*. Therefore

$$z^* = 10x - 4x^2 + 20y - y^2 + 4xy + \lambda(100 - 2x - 4y)$$

This will not change the value of z since, as we have just seen, the term added is equal to 0, and it can be shown that if we find the optimum point or points of the function z^* then these will be the optimum points for the original function z subject to the constraint.

Partially differentiating z^*, we have:

$$\frac{\delta z^*}{\delta x} = 10 - 8x + 4y - 2\lambda = 0$$

$$\frac{\delta z^*}{\delta y} = 20 - 2y + 4x - 4\lambda = 0$$

and

$$\frac{\delta z^*}{\delta \lambda} = 100 - 2x - 4y = 0$$

(thus satisfying the constraint).

As we have seen in Chapter 24, there are several ways of solving three simultaneous equations and these will not be repeated here: solving the three equations gives $x = 10$, $y = 20$ and $\lambda = 5$ (although strictly we do not need to know the value of λ).

Further differentiation shows that $x = 10$, $y = 20$ is a maximum for the function z subject to the constraint. (Note that although

$$\left(\frac{\delta^2 z}{\delta x^2}\right)\left(\frac{\delta^2 z}{\delta y^2}\right)$$

equals

$$\left(\frac{\delta^2 z}{\delta y\,\delta x}\right)^2$$

and is not 'greater than', there is sufficient evidence to show a maximum.)

Example

Consider a Cobb–Douglas production function

$$Q = 10L^{1/2}K^{1/2}$$

subject to the constraint that $4L + 10K = 100$, where L is the amount of labour and K is the amount of capital.

Rearranging the constraint, we have

$$100 - 4L - 10K = 0 \quad \text{and} \quad \lambda(100 - 4L - 10K) = 0$$

We thus wish to maximize the function

$$Q^* = 10L^{1/2}K^{1/2} + \lambda(100 - 4L - 10K)$$

$$\frac{\delta Q^*}{\delta L} = 5L^{-1/2}K^{1/2} - 4\lambda = 0 \quad \text{so} \quad \lambda = \frac{5K^{1/2}}{4L^{1/2}}$$

$$\frac{\delta Q^*}{\delta K} = 5L^{1/2}K^{-1/2} - 10\lambda = 0 \quad \text{so} \quad \lambda = \frac{5L^{1/2}}{10K^{1/2}}$$

Thus

$$\frac{5K^{1/2}}{4L^{1/2}} = \frac{5L^{1/2}}{10K^{1/2}}$$

$$50K = 20L$$

$$L = 2.5K$$

But

$$\frac{\delta Q^*}{\delta \lambda} = 100 - 4L - 10K = 0$$

$$100 - 10K - 10K = 0$$

$$\text{so } K = 5 \text{ and hence } L = 12.5$$

Further partial differentiation shows this to be a maximum, and thus production (Q) is maximized subject to the constraint when $L = 12.5$ and $K = 5$. Note that the function Q itself, when unconstrained, has no maximum value.

Exercise

> Check that the constraint is satisfied and find the value of Q.
> (Answer: $Q = 79.06$.)

25.10 Conclusions

Calculus was developed during the seventeenth century from problems experienced by physical scientists, like Galileo, trying to describe the world. It arose from the need to give an adequate mathematical account of changes in motion. Although the more immediate applications of calculus were found in the physical sciences, calculus has played an important part in the development of other disciplines, particularly economics. Whenever we need to consider the effects of change, calculus provides a language and a set of techniques to manage the mathematical descriptions.

25.11 Problems

Differentiate each of the following functions:

1 $y = 12$

2 $y = 10 + 2x$

3 $y = 3x + 7$

4 $y = 2x^5$

5 $y = 14x^3 - 3x^4$

6 $y = 17x^2 + 14x + 10$

7 $x = 7y + 2y^2 - 4y^3$

8 $p = 2q^2 - 10 + 4q^{-1}$

9 $y = e^x$

10 $r = 2s + 4s^2 + s^{1/2}$

11 $y = 20x^7 + 4x^6 - 3x^5 - 7x^4 + 4x^3 - 11x^2 + 2x + 57 - x^{-1} + 9x^{-2}$

12 $y = \dfrac{1}{x^2} - \dfrac{2}{x^3} - \dfrac{4}{3x^4}$

13 If average revenue for a company is given by:

$$AR = 100 - 2x \ (x \text{ is quantity})$$

find the marginal revenue function and the price when $MR = 0$.
   Graph the AR, MR and TR functions.

14 A firm's total costs are given by the following function:

$$TC = \tfrac{1}{3}x^3 - 5x^2 + 30x \ (x \text{ is quantity})$$

Find the average cost and marginal cost functions. Graph the three functions.

15 Using the information in the previous two questions, find the quantity levels where:
   (a) $AR = AC$; and
   (b) $MR = MC$.

16 A firm's average cost function is given as:

$$AC = x^2 - 3x + 25 \ (x = \text{quantity})$$

and its total revenue function as:

$$TR = 60x - 2x^2$$

Find the quantity levels where:
(a)   MC = MR; and
(b)   AR = AC

17 Given the following demand function:

$$P = 40 - 2x$$

find the price elasticity of demand at quantity levels of:
(a)   8;
(b)   10;
(c)   12.

Construct a graph of the average revenue and marginal revenue functions. What may you determine from this graph and your previous calculations?

18 An oligopolist sells a product at £950 per unit of production, and estimates the average revenue function to consist of two linear segments.

Below ten units,   AR = 990 when output =   2

AR = 965 when output =   7

Above ten units,   AR = 840 when output = 12

AR = 565 when output = 17

(a) Find the equation of the average revenue function below ten units of output.
(b) Find the equation of the average revenue function above ten units of output.
(c) Find the equations of the *two* marginal revenue functions, assuming that total revenue is zero if output is zero.
(d) Sketch the average and marginal revenue functions.
(e) Find the price elasticity of demand for a price rise and for a price fall from the current position.
(f) Find the range of marginal cost figures that are consistent with profit maximization and the current output level. (NB. For profit maximization MC = MR.)

Find the maximum and/or minimum of each of the following functions, checking the second order conditions (i.e. d^2y/dx^2):

19 $y = x^2 - 10x + 25$

20 $y = -2x^2 + 40x + 1000$

21 $y = 3x^2 + 6x + 5$

22 $y = 2x + 1$

23 $y = 3x^2 - 10x + 4$

24 $y = \frac{1}{3}x^3 - 3x^2 + 5x + 10$

25 $a = -4b^2 + 2b + 10$

26 $y = 2x^3 - 12x^2 + 12x + 10$

27 $y = 25$

28 $y = 0.25x^4 - 2x^3 + 5.5x^2 - 6x + 40$

29 A company's profit function is given by:

$$\text{profit} = -2x^2 + 40x + 10 \text{ (where } x = \text{output)}$$

Determine the profit maximizing output level for the company, checking that the second order conditions are met.

30 A firm has the following total revenue and total cost functions:

$$TR = 100x - 2x^2$$

$$TC = \frac{1}{3}x^3 - 5x^2 + 30x \text{ (where } x = \text{output)}.$$

Find the output level to maximize profits, and the level of profit achieved at this output.

31 A firm has an average revenue of

$$AR = 60 - 2x$$

and an average cost function of

$$AC = x^2 - 3x + 25 \text{ (where } x = \text{output)}$$

Find the output level for maximum profit, and the level of profit. (NB. Check the second order conditions.)

32 Given the following average cost function, show that the marginal cost function cuts the average cost function at the latter's minimum point.

$$AC = 2x^2 - 4x + 100$$

Use the rules of differentiation outlined in section 25.5 to find the first differentials of the following functions:

33 $y = (x^2 + 6)(2x + 5)$

34 $y = (3x^2 + 4x + 10)(2x^2 + 6x + 5)$

35 $y = (4x^3 + 6x^2)^3$

36 $y = e^{x^2/2}$

37 $y = (4x + 6x^2)/(x^3 + 6x)$

38 $y = 2x/(3x^4 + 10)$

Integrate each of the following functions:

39 $y = 4x$

40 $y = 3x^2 + 4x + 10$

41 $y = 2x^3 + 6x^2 + 10x^{-2}$

42 $y = 25$

43 $y = x^2 + x + 5$

44 $y = x + 10$ between limits of $x = 2$ and $x = 4$.

45 $y = 3x^2 + 4x + 5$ between limits of $x = 0$ and $x = 5$.

46 $y = -2x^2 + 40x + 10$ between limits of $x = 0$ and $x = 10$.

47 A firm's market demand function is

$$AR = 150 - 2x$$

and its marginal cost function is

$$MC = \frac{1}{3}x^2 - 5x + 30$$

(a) Find the level of output to maximize profits.
(b) Find the level of profit and the price at this point.

(c) Find the price elasticity of demand at this point.

(d) Sketch the total revenue, total cost and profit functions.

(e) If a tax of 10 per unit is imposed on the firm, what effect will this have on the production level for maximum profit?

48 A company is faced by the following marginal cost and marginal revenue functions:

$$MC = 16 - 2Q$$
$$MR = 40 - 16Q$$

It is also known that fixed costs are 8 when production is zero. Find:

(a) the total cost function,

(b) the total revenue function,

(c) the output to give maximum sales revenue,

(d) the output to give maximum profit,

(e) the range of output for which profit is at least 1; and

(f) sketch the total cost and total revenue functions.

Find the first partial derivatives for each of the following:

49 $y = 4xz$

50 $y = 2x^3 + 4x^2z + 2xz^2 - 3z^3$

51 $y = 7x + 3x^2z - x^3z^2 - 2xz + 5xz^2 - 7xz^3 - 10$

52 $y = 10x^4 - 15z^3$

53 $q = 2p_1^2 - 3p_1 + 4p_1p_2 - 5p_2 + p_2^2$

54 $r = 2t(s^2 + s + 5) - 3s(2t^2 - 4t + 7)$

Find the maxima or minima of the following functions:

55 $y = 2x^2 - 2xz + 2.5z^2 - 2x - 11z + 20$

56 $y = 3x^2 + 3xz + 3z^2 - 21x - 33z + 100$

57 $y = -2x^2 - 2xz - 4z^2 + 40x + 90z - 150$

58 $z = 19x - 4x^2 + 16y - 2xy - 4y^2$

59 $y = x^2 + xz + z^2$

60 $y = 100 - 4x$

61 If a firm's total costs are related to its workforce and capital equipment by the function

$$TC = 10L^2 + 10K^2 - 25L - 50K - 5LK + 2000$$

where L = thousands of employees and K = thousands of pounds invested in capital equipment, find the combination of labour and capital to give minimum total cost. Find this cost and show that it is a minimum.

62 A monopolist has two products, X and Y, which have the following demand functions:

$$P_X = 26.2 - X$$
$$P_Y = 24 - 2Y$$

The total costs of the monopolist are given by:

$$TC = 5X^2 + XY + 3Y^2$$

Determine the amounts of X and Y the monopolist should produce to maximize profit, and the amount of profit produced.

63 A company is able to sell two products, X and Y, which have demand functions:

$$P_X = 52 - 2X$$
$$P_Y = 20 - 3Y$$

and has a total cost function

$$TC = 10 + 3X^2 + 2Y^2 + 2XY$$

(a) Determine the profit maximizing levels of output for X and Y and the level of this profit.

(c) If product Y were not produced, determine the production level of X to maximize profit and the level of this profit.

(c) If product X were not produced, determine the production level of Y to maximize profit and the level of this profit.

Maximize or minimize the following functions subject to their constraints, using either substitution or the method of Lagrangian multipliers.

64 $y = 10 + 20x + 6z - 4x^2 - 2z^2$ subject to $x = 3z$.

65 $y = 4x^2 - 6x + 7z + 3z^2 + 2xz$ subject to $x = 5z$.

66 $z = 10x - x^2 + 14y - 2y^2 + xy$ subject to $2x + 3y = 3100$.

67 $z = 6x - 3x^2 + 40y - 8y^2 + 5xy$ subject to $4x - 5y = 30$.

68 $z = 2xy$ subject to $x + y = 1$.

69 Maximize the production function

$$Q = 4L^{1/2}K^{1/2}$$

subject to the constraint that $3L + 5K = 200$, finding the values for L and K.

70 Find the values of L and K to maximize the production function

$$Q = 8L^{1/4}K^{3/4}$$

subject to the constraint $L + 8K = 1000$.

71 A firm's profit is given by the function

$$\pi = 600 - 3x^2 - 4x + 2xy - 5y^2 + 48y$$

where π denotes profit, x output and y advertising expenditure.

(a) Find the profit maximizing values of x and y and hence determine the maximum profit. Confirm that second order conditions are satisfied.

(b) If now the firm is subject to a budget constraint $2x + y = 5$ determine the new values of x and y which maximize profit.

(c) Without further calculation determine the effect of a constraint $2x + y = 8$.

Part 7 Conclusions

In Part 7 you have been looking at the precise subject area of mathematics, rather than the more uncertain areas of statistics and modelling. You should now be in a position to deal with the sort of mathematics used in economics courses. At first sight the mathematics may seem limited to this theoretical perspective, but it does provide an underpinning for all of statistics and for many of the modelling techniques that we have discussed earlier in the book. Later in your course, or in a post-graduate course, you may be expected to understand the theory behind such models, and Part 7 will have given you a useful background on which to build.

A facility with basic mathematical ideas can help you to develop new ways of looking at a relationship between variables, and even if the set of equations (model) that you build does not exactly match the particular business situation, it is likely to give you pointers which will help in deciding upon the direction in which to develop your ideas.

Appendices

A

Cumulative Binomial probabilities

		$p = 0.01$	0.05	0.10	0.20	0.30	0.40	0.45	0.50
$n = 5$	$r = 0$	1.0000	1.0000	1.0000	1.0000	1.0000	1.0000	1.0000	1.0000
	1	0.0490	0.2262	0.4095	0.6723	0.8319	0.9222	0.9497	0.9688
	2	0.0010	0.0226	0.0815	0.2627	0.4718	0.6630	0.7438	0.8125
	3		0.0012	0.0086	0.0579	0.1631	0.3174	0.4069	0.5000
	4			0.0005	0.0067	0.0308	0.0870	0.1312	0.1875
	5				0.0003	0.0024	0.0102	0.0185	0.0313
$n = 10$	$r = 0$	1.0000	1.0000	1.0000	1.0000	1.0000	1.0000	1.0000	1.0000
	1	0.0956	0.4013	0.6513	0.8926	0.9718	0.9940	0.9975	0.9990
	2	0.0043	0.0861	0.2639	0.6242	0.8507	0.9536	0.9767	0.9893
	3	0.0001	0.0115	0.0702	0.3222	0.6172	0.8327	0.9004	0.9453
	4		0.0010	0.0128	0.1209	0.3504	0.6177	0.7430	0.8281
	5		0.0001	0.0016	0.0328	0.1503	0.3669	0.4956	0.6230
	6			0.0001	0.0064	0.0473	0.1662	0.2616	0.3770
	7				0.0009	0.0106	0.0548	0.1020	0.1719
	8				0.0001	0.0016	0.0123	0.0274	0.0547
	9					0.0001	0.0017	0.0045	0.0107
	10						0.0001	0.0003	0.0010

where
p is the probability of a characteristic (e.g. a defective item),
n is the sample size and
r is the number with that characteristic.

B Cumulative Poisson probabilities

	$\lambda = 1.0$	2.0	3.0	4.0	5.0	6.0	7.0
$x = 0$	1.0000	1.0000	1.0000	1.0000	1.0000	1.0000	1.0000
1	0.6321	0.8647	0.9502	0.9817	0.9933	0.9975	0.9991
2	0.2642	0.5940	0.8009	0.9084	0.9596	0.9826	0.9927
3	0.0803	0.3233	0.5768	0.7619	0.8753	0.9380	0.9704
4	0.0190	0.1429	0.3528	0.5665	0.7350	0.8488	0.9182
5	0.0037	0.0527	0.1847	0.3712	0.5595	0.7149	0.8270
6	0.0006	0.0166	0.0839	0.2149	0.3840	0.5543	0.6993
7	0.0001	0.0011	0.0335	0.1107	0.2378	0.3937	0.5503
8		0.0002	0.0119	0.0511	0.1334	0.2560	0.4013
9			0.0038	0.0214	0.0681	0.1528	0.2709
10			0.0011	0.0081	0.0318	0.0839	0.1695
11			0.0003	0.0028	0.0137	0.0426	0.0985
12			0.0001	0.0009	0.0055	0.0201	0.0534
13				0.0003	0.0020	0.0088	0.0270
14				0.0001	0.0007	0.0036	0.0128
15					0.0002	0.0014	0.0057
16					0.0001	0.0005	0.0024
17						0.0002	0.0010
18						0.0001	0.0004
19							0.0001

where
λ is the average number of times a characteristic occurs and
x is the number of occurrences.

Areas in the right-hand tail of the Normal distribution

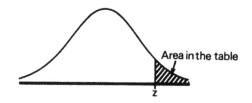

Area in the table

z	.00	.01	.02	.03	.04	.05	.06	.07	.08	.09
0.0	.5000	.4960	.4920	.4880	.4840	.4801	.4761	.4721	.4681	.4641
0.1	.4602	.4562	.4522	.4483	.4443	.4404	.4364	.4325	.4286	.4247
0.2	.4207	.4168	.4129	.4090	.4052	.4013	.3974	.3936	.3897	.3859
0.3	.3821	.3783	.3745	.3707	.3669	.3632	.3594	.3557	.3520	.3483
0.4	.3446	.3409	.3372	.3336	.3300	.3264	.3228	.3192	.3156	.3121
0.5	.3085	.3050	.3015	.2981	.2946	.2912	.2877	.2843	.2810	.2776
0.6	.2743	.2709	.2676	.2643	.2611	.2578	.2546	.2514	.2483	.2451
0.7	.2420	.2389	.2358	.2327	.2296	.2266	.2236	.2206	.2177	.2148
0.8	.2119	.2090	.2061	.2033	.2005	.1977	.1949	.1922	.1894	.1867
0.9	.1841	.1814	.1788	.1762	.1736	.1711	.1685	.1660	.1635	.1611
1.0	.1587	.1562	.1539	.1515	.1492	.1496	.1446	.1423	.1401	.1379
1.1	.1357	.1335	.1314	.1292	.1271	.1251	.1230	.1210	.1190	.1170
1.2	.1151	.1132	.1112	.1093	.1075	.1056	.1038	.1020	.1003	.0985
1.3	.0968	.0951	.0934	.0918	.0901	.0885	.0869	.0853	.0838	.0823
1.4	.0808	.0793	.0778	.0764	.0749	.0735	.0721	.0708	.0694	.0681
1.5	.0668	.0655	.0643	.0630	.0618	.0606	.0594	.0582	.0571	.0559
1.6	.0548	.0537	.0526	.0516	.0505	.0495	.0485	.0475	.0465	.0455
1.7	.0446	.0436	.0427	.0418	.0409	.0401	.0392	.0384	.0375	.0367
1.8	.0359	.0351	.0344	.0336	.0329	.0322	.0314	.0307	.0301	.0294
1.9	.0287	.0281	.0274	.0268	.0262	.0256	.0250	.0244	.0239	.0233
2.0	.02275	.02222	.02169	.02118	.02068	.02018	.01970	.01923	.01876	.01831
2.1	.01786	.01743	.01700	.01659	.01618	.01578	.01539	.01500	.01463	.01426
2.2	.01390	.01355	.01321	.01287	.01255	.01222	.01191	.01160	.01130	.01101
2.3	.01072	.01044	.01017	.00990	.00964	.00939	.00914	.00889	.00866	.00842
2.4	.00820	.00798	.00776	.00755	.00734	.00714	.00695	.00676	.00657	.00639
2.5	.00621	.00604	.00587	.00570	.00554	.00539	.00523	.00508	.00494	.00480
2.6	.00466	.00453	.00440	.00427	.00415	.00402	.00391	.00379	.00368	.00357
2.7	.00347	.00336	.00326	.00317	.00307	.00298	.00289	.00280	.00272	.00264
2.8	.00256	.00248	.00240	.00233	.00226	.00219	.00212	.00205	.00199	.00193
2.9	.00187	.00181	.00175	.00169	.00164	.00159	.00154	.00149	.00144	.00139

(Continued)

z	.00	.01	.02	.03	.04	.05	.06	.07	.08	.09
3.0	.00135									
3.1	.00097									
3.2	.00069									
3.3	.00048									
3.4	.00034									
3.5	.00023									
3.6	.00016									
3.7	.00011									
3.8	.00007									
3.9	.00005									
4.0	.00003									

D Student's *t* critical points

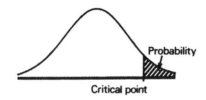

Probability

Critical point

Probability	0.10	0.05	0.025	0.01	0.005
$\nu = 1$	3.078	6.314	12.706	31.821	63.657
2	1.886	2.920	4.303	6.965	9.925
3	1.638	2.353	3.182	4.541	5.841
4	1.533	2.132	2.776	3.747	4.604
5	1.476	2.015	2.571	3.365	4.032
6	1.440	1.943	2.447	3.143	3.707
7	1.415	1.895	2.365	2.998	3.499
8	1.397	1.860	2.306	2.896	3.355
9	1.383	1.833	2.262	2.821	3.250
10	1.372	1.812	2.228	2.764	3.169
11	1.363	1.796	2.201	2.718	3.106
12	1.356	1.782	2.179	2.681	3.055
13	1.350	1.771	2.160	2.650	3.012
14	1.345	1.761	2.145	2.624	2.977
15	1.341	1.753	2.131	2.602	2.947
16	1.337	1.746	2.120	2.583	2.921
17	1.333	1.740	2.110	2.567	2.898
18	1.330	1.734	2.101	2.552	2.878
19	1.328	1.729	2.093	2.539	2.861
20	1.325	1.725	2.086	2.528	2.845
21	1.323	1.721	2.080	2.518	2.831
22	1.321	1.717	2.074	2.508	2.819
23	1.319	1.714	2.069	2.500	2.807
24	1.318	1.711	2.064	2.492	2.797
25	1.316	1.708	2.060	2.485	2.787
26	1.315	1.706	2.056	2.479	2.779
27	1.314	1.703	2.052	2.473	2.771
28	1.313	1.701	2.048	2.467	2.763
29	1.311	1.699	2.045	2.462	2.756
30	1.310	1.697	2.042	2.457	2.750
40	1.303	1.684	2.021	2.423	2.704
60	1.296	1.671	2.000	2.390	2.660
120	1.289	1.658	1.980	2.358	2.617
∞	1.282	1.645	1.960	2.326	2.576

where ν is the number of degrees of freedom.

E χ^2 critical values

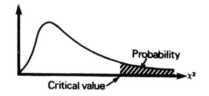

Probability	0.250	0.100	0.050	0.025	0.010	0.005	0.001
$\nu = 1$	1.32	2.71	3.84	5.02	6.63	7.88	10.8
2	2.77	4.61	5.99	7.38	9.21	10.6	13.8
3	4.11	6.25	7.81	9.35	11.3	12.8	16.3
4	5.39	7.78	9.49	11.1	13.3	14.9	18.5
5	6.63	9.24	11.1	12.8	15.1	16.7	20.5
6	7.84	10.6	12.6	14.4	16.8	18.5	22.5
7	9.04	12.0	14.1	16.0	18.5	20.3	24.3
8	10.2	13.4	15.5	17.5	20.3	22.0	26.1
9	11.4	14.7	16.9	19.0	21.7	23.6	27.9
10	12.5	16.0	18.3	20.5	23.2	25.2	29.6
11	13.7	17.3	19.7	21.9	24.7	26.8	31.3
12	14.8	18.5	21.0	23.3	26.2	28.3	32.9
13	16.0	19.8	22.4	24.7	27.7	29.8	34.5
14	17.1	21.1	23.7	26.1	29.1	31.3	36.1
15	18.2	22.3	25.0	27.5	30.6	32.8	37.7
16	19.4	23.5	26.3	28.8	32.0	34.3	39.3
17	20.5	24.8	27.6	30.2	33.4	35.7	40.8
18	21.6	26.0	28.9	31.5	34.8	37.2	42.3
19	22.7	27.2	30.1	32.9	36.2	38.6	43.8
20	23.8	28.4	31.4	34.2	37.6	40.0	45.3
21	24.9	29.6	32.7	35.5	38.9	41.4	46.8
22	26.0	30.8	33.9	36.8	40.3	42.8	48.3
23	27.1	32.0	35.2	38.1	41.6	44.2	49.7
24	28.2	33.2	36.4	39.4	43.0	45.6	51.2
25	29.3	34.4	37.7	40.6	44.3	46.9	52.6
26	30.4	35.6	38.9	41.9	45.6	48.3	54.1
27	31.5	36.7	40.1	43.2	47.0	49.6	55.5
28	32.6	37.9	41.3	44.5	48.3	51.0	56.9
29	33.7	39.1	42.6	45.7	49.6	52.3	58.3
30	34.8	40.3	43.8	47.0	50.9	53.7	59.7
40	45.6	51.8	55.8	59.3	63.7	66.8	73.4
50	56.3	63.2	67.5	71.4	76.2	79.5	86.7
60	67.0	74.4	79.1	83.3	88.4	92.0	99.6

(Continued)

Probability	0.250	0.100	0.050	0.025	0.010	0.005	0.001
70	77.6	85.5	90.5	95.0	100	104	112
80	88.1	96.6	102	107	112	116	125
90	98.6	108	113	118	124	128	137
100	109	118	124	130	136	140	149

where ν is the number of degrees of freedom.

F

Present value factors

Years	1%	2%	3%	4%	5%	6%	7%	8%	9%	10%
1	.9901	.9804	.9709	.9615	.9524	.9434	.9346	.9259	.9174	.9091
2	.9803	.9612	.9426	.9426	.9070	.8900	.8734	.8573	.8417	.8264
3	.9706	.9423	.9151	.8890	.8638	.8396	.8163	.7938	.7722	.7513
4	.9610	.9238	.8885	.8548	.8227	.7921	.7629	.7350	.7084	.6830
5	.9515	.9057	.8626	.8219	.7835	.7473	.7130	.6806	.6499	.6209
6	.9420	.8880	.8375	.7903	.7462	.7050	.6663	.6302	.5963	.5645
7	.9327	.8706	.8131	.7599	.7107	.6651	.6227	.5835	.5470	.5132
8	.9235	.8535	.7894	.7307	.6768	.6274	.5820	.5403	.5019	.4665
9	.9143	.8368	.7664	.7026	.6446	.5919	.5439	.5002	.4604	.4241
10	.9053	.8203	.7441	.6756	.6139	.5584	.5083	.4632	.4224	.3855
11	.8963	.8043	.7224	.6496	.5847	.5268	.4751	.4289	.3875	.3505
12	.8874	.7885	.7014	.6246	.5568	.4970	.4440	.3971	.3555	.3186
13	.8787	.7730	.6810	.6006	.5303	.4688	.4150	.3677	.3262	.2897
14	.8700	.7579	.6611	.5775	.5051	.4423	.3878	.3405	.2992	.2633
15	.8613	.7430	.6419	.5553	.4810	.4173	.3624	.3152	.2745	.2394
16	.8528	.7284	.6232	.5339	.4581	.3936	.3387	.2919	.2519	.2176
17	.8444	.7142	.6050	.5134	.4363	.3714	.3166	.2703	.2311	.1978
18	.8360	.7002	.5874	.4936	.4155	.3503	.2959	.2502	.2120	.1799
19	.8277	.6864	.5703	.4746	.3957	.3305	.2765	.2317	.1945	.1635
20	.8195	.6730	.5537	.4564	.3769	.3118	.2584	.2145	.1784	.1486
21	.8114	.6598	.5375	.4388	.3589	.2942	.2415	.1987	.1637	.1351
22	.8034	.6468	.5219	.4220	.3418	.2775	.2257	.1839	.1502	.1228
23	.7954	.6342	.5067	.4057	.3256	.2618	.2109	.1703	.1378	.1117
24	.7876	.6217	.4919	.3901	.3101	.2470	.1971	.1577	.1264	.1015
25	.7798	.6095	.4776	.3751	.2953	.2330	.1842	.1460	.1160	.0923

(Continued)

Years	11%	12%	13%	14%	15%	16%	17%	18%	19%	20%
1	.9009	.8929	.8850	.8772	.8696	.8621	.8547	.8475	.8403	.8333
2	.8116	.7972	.7831	.7695	.7561	.7432	.7305	.7182	.7062	.6944
3	.7312	.7118	.6931	.6750	.6575	.6407	.6244	.6086	.5934	.5787
4	.6587	.6355	.6133	.5921	.5718	.5523	.5337	.5158	.4987	.4823
5	.5935	.5674	.5428	.5194	.4972	.4761	.4561	.4371	.4190	.4019
6	.5346	.5066	.4803	.4556	.4323	.4104	.3898	.3704	.3521	.3349
7	.4817	.4523	.4251	.3996	.3759	.3538	.3332	.3139	.2959	.2791
8	.4339	.4039	.3762	.3506	.3269	.3050	.2848	.2660	.2487	.2326
9	.3909	.3606	.3329	.3075	.2843	.2630	.2434	.2255	.2090	.1938
10	.3522	.3220	.2946	.2697	.2472	.2267	.2080	.1911	.1756	.1615
11	.3173	.2875	.2607	.2366	.2149	.1954	.1778	.1619	.1476	.1346
12	.2858	.2567	.2307	.2076	.1869	.1685	.1520	.1372	.1240	.1122
13	.2575	.2292	.2042	.1821	.1625	.1452	.1299	.1163	.1042	.0935
14	.2320	.2046	.1807	.1597	.1413	.1252	.1110	.0985	.0876	.0779
15	.2090	.1827	.1599	.1401	.1229	.1079	.0949	.0835	.0736	.0649
16	.1883	.1631	.1415	.1229	.1069	.0930	.0811	.0708	.0618	.0541
17	.1696	.1456	.1252	.1078	.0929	.0802	.0693	.0600	.0520	.0451
18	.1528	.1300	.1108	.0946	.0808	.0691	.0592	.0508	.0437	.0376
19	.1377	.1161	.0981	.0826	.0703	.0596	.0506	.0431	.0367	.0313
20	.1240	.1037	.0868	.0728	.0611	.0514	.0433	.0365	.0308	.0261
21	.1117	.0926	.0768	.0638	.0531	.0443	.0370	.0309	.0259	.0217
22	.1007	.0826	.0680	.0560	.0462	.0382	.0316	.0262	.0218	.0181
23	.0907	.0738	.0601	.0491	.0402	.0329	.0270	.0222	.0183	.0151
24	.0817	.0659	.0532	.0431	.0349	.0284	.0231	.0188	.0154	.0126
25	.0736	.0588	.0471	.0378	.0304	.0245	.0197	.0160	.0129	.0105

G Mann–Whitney test statistic

n_1	p	$n_2=2$	3	4	5	6	7	8	9	10	11	12	13	14	15	16	17	18	19	20
2	.001	0	0	0	0	0	0	0	0	0	0	0	0	0	0	0	0	0	0	0
	.005	0	0	0	0	0	0	0	0	0	0	0	0	0	0	0	0	0	1	1
	.01	0	0	0	0	0	0	0	0	0	0	0	1	1	1	1	1	1	2	2
	.025	0	0	0	0	0	0	1	1	1	1	2	2	2	2	2	3	3	3	3
	.05	0	0	0	1	1	1	2	2	2	2	3	3	4	4	4	4	5	5	5
	.10	0	1	1	2	2	2	3	3	4	4	5	5	5	6	6	7	7	8	8
3	.001	0	0	0	0	0	0	0	0	0	0	0	0	0	0	0	1	1	1	1
	.005	0	0	0	0	0	0	0	1	1	1	2	2	2	3	3	3	3	4	4
	.01	0	0	0	0	0	1	1	2	2	2	3	3	3	4	4	5	5	5	6
	.025	0	0	0	1	2	2	3	3	4	4	5	5	6	6	7	7	8	8	9
	.05	0	1	1	2	3	3	4	5	5	6	6	7	8	8	9	10	10	11	12
	.10	1	2	2	3	4	5	6	6	7	8	9	10	11	11	12	13	14	15	16
4	.001	0	0	0	0	0	0	0	0	1	1	1	2	2	2	3	3	4	4	4
	.005	0	0	0	0	1	1	2	2	3	3	4	4	5	6	6	7	7	8	9
	.01	0	0	0	1	2	2	3	4	4	5	6	6	7	9	8	9	10	10	11
	.025	0	0	1	2	3	4	5	5	6	7	8	9	10	11	12	12	13	14	15
	.05	0	1	2	3	4	5	6	7	8	9	10	11	12	13	15	16	17	18	19
	.10	1	2	4	5	6	7	8	10	11	12	13	14	16	17	18	19	21	22	23
5	.001	0	0	0	0	0	0	1	2	2	3	3	4	4	5	6	6	7	8	8
	.005	0	0	0	1	2	2	3	4	5	6	7	8	8	9	10	11	12	13	14
	.01	0	0	1	2	3	4	5	6	7	8	9	10	11	12	13	14	15	16	17
	.025	0	1	2	3	4	6	7	8	9	10	12	13	14	15	16	18	19	20	21
	.05	1	2	3	5	6	7	9	10	12	13	14	16	17	19	20	21	23	24	26
	.10	2	3	5	6	8	9	11	13	14	16	18	19	21	23	24	26	28	29	31
6	.001	0	0	0	0	0	0	2	3	4	5	5	6	7	8	9	10	11	12	13
	.005	0	0	1	2	3	4	5	6	7	8	10	11	12	13	14	16	17	18	19
	.01	0	0	2	3	4	5	7	8	9	10	12	13	14	16	17	19	20	21	23
	.025	0	2	3	4	6	7	9	11	12	14	15	17	18	20	22	23	25	26	28
	.05	1	3	4	6	8	9	11	13	15	17	18	20	22	24	26	27	29	31	33
	.10	2	4	6	8	10	12	14	16	18	20	22	24	26	28	30	32	35	37	39

Source: Conover W. J. (1971), Practical Nonparametric Statistics, New York: Wiley, 384–8.

(Continued)

n_1	p	$n_2=2$	3	4	5	6	7	8	9	10	11	12	13	14	15	16	17	18	19	20
7	.001	0	0	0	0	1	2	3	4	6	7	8	9	10	11	12	14	15	16	17
	.005	0	0	1	2	4	5	7	8	10	11	13	14	16	17	19	20	22	23	25
	.01	0	1	2	4	5	7	8	10	12	13	15	17	18	20	22	24	25	27	29
	.025	0	2	4	6	7	9	11	13	15	17	19	21	23	25	27	29	31	33	35
	.05	1	3	5	7	9	12	14	16	18	20	22	25	27	29	31	34	36	38	40
	.10	2	5	7	9	12	14	17	19	22	24	27	29	32	34	37	39	42	44	47
8	.001	0	0	0	1	2	3	5	6	7	9	10	12	13	15	16	18	19	21	22
	.005	0	0	2	3	5	7	8	10	12	14	16	18	19	21	23	25	27	29	31
	.01	0	1	3	5	7	8	10	12	14	16	18	21	23	25	27	29	31	33	35
	.025	1	3	5	7	9	11	14	16	18	20	23	25	27	30	32	35	37	39	42
	.05	2	4	6	9	11	14	16	19	21	24	27	29	32	34	37	40	42	45	48
	.10	3	6	8	11	14	17	20	23	25	28	31	34	37	40	43	46	49	52	55
9	.001	0	0	0	2	3	4	6	8	9	11	13	15	16	18	20	22	24	26	27
	.005	0	1	2	4	6	8	10	12	14	17	19	21	23	25	28	30	32	34	37
	.01	0	2	4	6	8	10	12	15	17	19	22	24	27	29	32	34	37	39	41
	.025	1	3	5	8	11	13	16	18	21	24	27	29	32	35	38	40	43	46	49
	.05	2	5	7	10	13	16	19	22	25	28	31	34	37	40	43	46	49	52	55
	.10	3	6	10	13	16	19	23	26	29	32	36	39	42	46	49	53	56	59	63
10	.001	0	0	1	2	4	6	7	9	11	13	15	18	20	22	24	26	28	30	33
	.005	0	1	3	5	7	10	12	14	17	19	22	25	27	30	32	35	38	40	43
	.01	0	2	4	7	9	12	14	17	20	23	25	28	31	34	37	39	42	45	48
	.025	1	4	6	9	12	15	18	21	24	27	30	34	37	40	43	46	49	53	56
	.05	2	5	8	12	15	18	21	25	28	32	35	38	42	45	49	52	56	59	63
	.10	4	7	11	14	18	22	25	29	33	37	40	44	48	52	55	59	63	67	71
11	.001	0	0	1	3	5	7	9	11	13	16	18	21	23	25	28	30	33	35	38
	.005	0	1	3	6	8	11	14	17	19	22	25	28	31	34	37	40	43	46	49
	.01	0	2	5	8	10	13	16	19	23	26	29	32	35	38	42	45	48	51	54
	.025	1	4	7	10	14	17	20	24	27	31	34	38	41	45	48	52	56	59	63
	.05	2	6	9	13	17	20	24	28	32	35	39	43	47	51	55	58	62	66	70
	.10	4	8	12	16	20	24	28	32	37	41	45	49	53	58	62	66	70	74	79
12	.001	0	0	1	3	5	8	10	13	15	18	21	24	26	29	32	35	38	41	43
	.005	0	2	4	7	10	13	16	19	22	25	28	32	35	38	42	45	48	52	55
	.01	0	3	6	9	12	15	18	22	25	29	32	36	39	43	47	50	54	57	61
	.025	2	5	8	12	15	19	23	27	30	34	38	42	46	50	54	58	62	66	70
	.05	3	6	10	14	18	22	27	31	35	39	43	48	52	56	61	65	69	73	78
	.10	5	9	13	18	22	27	31	36	40	45	50	54	59	64	68	73	78	82	87
13	.001	0	0	2	4	6	9	12	15	18	21	24	27	30	33	36	39	43	46	49
	.005	0	2	4	8	11	14	18	21	25	28	32	35	39	43	46	50	54	58	61
	.01	1	3	6	10	13	17	21	24	28	32	36	40	44	48	52	56	60	64	68
	.025	2	5	9	13	17	21	25	29	34	38	42	46	51	55	60	64	68	73	77
	.05	3	7	11	16	20	25	29	34	38	43	48	52	57	62	66	71	76	81	85
	.10	5	10	14	19	24	29	34	39	44	49	54	59	64	69	75	80	85	90	95
14	.001	0	0	2	4	7	10	13	16	20	23	26	30	33	37	40	44	47	51	55
	.005	0	2	5	8	12	16	19	23	27	31	35	39	43	47	51	55	59	64	68
	.01	1	3	7	11	14	18	23	27	31	35	39	44	48	52	57	61	66	70	74
	.025	2	6	10	14	18	23	27	32	37	41	46	51	56	60	65	70	75	79	83
	.05	4	8	12	17	22	27	32	37	42	47	52	57	62	67	72	78	83	88	93
	.10	5	11	16	21	26	32	37	42	48	53	59	64	70	75	81	86	92	98	103

(Continued)

n_1	p	$n_2=2$	3	4	5	6	7	8	9	10	11	12	13	14	15	16	17	18	19	20
15	.001	0	0	2	5	8	11	15	18	22	25	29	33	37	41	44	48	52	56	60
	.005	0	3	6	9	13	17	21	25	30	34	38	43	47	52	56	61	65	70	74
	.01	1	4	8	12	16	20	25	29	34	38	43	48	52	57	62	67	71	76	81
	.025	2	6	11	15	20	25	30	35	40	45	50	55	60	65	71	76	81	86	91
	.05	4	8	13	19	24	29	34	40	45	51	56	62	67	73	78	84	89	95	101
	.10	6	11	17	23	28	34	40	46	52	58	64	69	75	81	87	93	99	105	111
16	.001	0	0	3	6	9	12	16	20	24	28	32	36	40	44	49	53	57	61	66
	.005	0	3	6	10	14	19	23	28	32	37	42	46	51	56	61	66	71	75	80
	.01	1	4	8	13	17	22	27	32	37	42	47	52	57	62	67	72	77	83	88
	.025	2	7	12	16	22	27	32	38	43	48	54	60	65	71	76	82	87	93	99
	.05	4	9	15	20	26	31	37	43	49	55	61	66	72	78	84	90	96	102	108
	.10	6	12	18	24	30	37	43	49	55	62	68	75	81	87	94	100	107	113	120
17	.001	0	1	3	6	10	14	18	22	26	30	35	39	44	48	53	58	62	67	71
	.005	0	3	7	11	16	20	25	30	35	40	45	50	55	61	66	71	76	82	87
	.01	1	5	9	14	19	24	29	34	39	45	50	56	61	67	72	78	83	89	94
	.025	3	7	12	18	23	29	35	40	46	52	58	64	70	76	82	88	94	100	106
	.05	4	10	16	21	27	34	40	46	52	58	65	71	78	84	90	97	103	110	116
	.10	7	13	19	26	32	39	46	53	59	66	73	80	86	93	100	107	114	121	128
18	.001	0	1	4	7	11	15	19	24	28	33	38	43	47	52	57	62	67	72	77
	.005	0	3	7	12	17	22	27	32	38	43	48	54	59	65	71	76	82	88	93
	.01	1	5	10	15	20	25	31	37	42	48	54	60	66	71	77	83	89	95	101
	.025	3	8	13	19	25	31	37	43	49	56	62	68	75	81	87	94	100	107	113
	.05	5	10	17	23	29	36	42	49	56	62	69	76	83	89	96	103	110	117	124
	.10	7	14	21	28	35	42	49	56	63	70	78	85	92	99	107	114	121	129	136
19	.001	0	1	4	8	12	16	21	26	30	35	41	46	51	56	61	67	72	78	83
	.005	1	4	8	13	18	23	29	34	40	46	52	58	64	70	75	82	88	94	100
	.01	2	5	10	16	21	27	33	39	45	51	57	64	70	76	83	89	95	102	108
	.025	3	8	14	20	26	33	39	46	53	59	66	73	79	86	93	100	107	114	120
	.05	5	11	18	24	31	38	45	52	59	66	73	81	88	95	102	110	117	124	131
	.10	8	15	22	29	37	44	52	59	67	74	82	90	98	105	113	121	129	136	144
20	.001	0	1	4	8	13	17	22	27	33	38	43	49	55	60	66	71	77	83	89
	.005	1	4	9	14	19	25	31	37	43	49	55	61	68	74	80	87	93	100	106
	.01	2	6	11	17	23	29	35	41	48	54	61	68	74	81	88	94	101	108	115
	.025	3	9	15	21	28	35	42	49	56	63	70	77	84	91	99	106	113	120	128
	.05	5	12	19	26	33	40	48	55	63	70	78	85	93	101	108	116	124	131	139
	.10	8	16	23	31	39	47	55	63	71	79	87	95	103	111	120	128	136	144	152

H Wilcoxon test statistic

Table of critical values of T in the Wilcoxon matched-pairs signed-ranks test

N	Level of significance, direction predicted		
	.025	.01	.005
	Level of significance, direction not predicted		
	.05	.02	.01
6	0	—	—
7	2	0	—
8	4	2	0
9	6	3	2
10	8	5	3
11	11	7	5
12	14	10	7
13	17	13	10
14	21	16	13
15	25	20	16
16	30	24	20
17	35	28	23
18	40	33	28
19	46	38	32
20	52	43	38
21	59	49	43
22	66	56	49
23	73	62	55
24	81	69	61
25	89	77	68

Source: S. Siegel, Nonparametric Statistics. McGraw-Hill Book Company, New York, 1956, table G.

Runs test

Source: Swed, Frieda S. and Eisenhart, C. (1943) Tables for testing randomness of grouping in a sequence of alternatives, Ann. Math. Statist., 14, 66–87

Lower critical values of r in the runs test

n_1 \ n_2	2	3	4	5	6	7	8	9	10	11	12	13	14	15	16	17	18	19	20
2											2	2	2	2	2	2	2	2	2
3			2	2	2	2	2	2	2	2	2	2	2	3	3	3	3	3	3
4		2	2	2	3	3	3	3	3	3	3	3	3	4	4	4	4	4	4
5		2	2	3	3	3	3	3	4	4	4	4	4	4	4	4	5	5	5
6	2	2	3	3	3	3	4	4	4	4	5	5	5	5	5	5	6	6
7	2	2	3	3	3	4	4	5	5	5	5	5	6	6	6	6	6	6
8	2	3	3	3	4	4	5	5	5	6	6	6	6	6	7	7	7	7
9	2	3	3	4	4	5	5	5	6	6	6	7	7	7	7	8	8	8
10	2	3	3	4	5	5	5	6	6	7	7	7	7	8	8	8	8	9
11	2	3	4	4	5	5	6	6	7	7	7	8	8	8	9	9	9	9
12	2	2	3	4	4	5	6	6	7	7	7	8	8	8	9	9	9	10	10
13	2	2	3	4	5	5	6	6	7	7	8	8	9	9	9	10	10	10	10
14	2	2	3	4	5	5	6	7	7	8	8	9	9	9	10	10	10	11	11
15	2	3	3	4	5	6	6	7	7	8	8	9	9	10	10	11	11	11	12
16	2	3	4	4	5	6	6	7	8	8	9	9	10	10	11	11	11	12	12
17	2	3	4	4	5	6	7	7	8	9	9	10	10	11	11	11	12	12	13
18	2	3	4	5	5	6	7	8	8	9	9	10	10	11	11	12	12	13	13
19	2	3	4	5	6	6	7	8	8	9	10	10	11	11	12	12	13	13	13
20	2	3	4	5	6	6	7	8	9	9	10	10	11	12	12	13	13	13	14

Note: For the one-sample runs test, any value of r that is equal to or smaller than that shown in the body of this table for given value of n_1 and n_2 is significant at the 0.05 level.

Upper critical values of r in the runs test

n_1 \ n_2	2	3	4	5	6	7	8	9	10	11	12	13	14	15	16	17	18	19	20
2																			
3																			
4				9	9														
5			9	10	10	11	11												
6			9	10	11	12	12	13	13	13	13								
7				11	12	13	13	14	14	14	14	15	15	15					
8				11	12	13	14	14	15	15	16	16	16	16	17	17	17	17	17
9					13	14	14	15	16	16	16	17	17	18	18	18	18	18	18
10					13	14	15	16	16	17	17	18	18	18	19	19	19	20	20
11					13	14	15	16	17	17	18	19	19	19	20	20	20	21	21
12					13	14	16	16	17	18	19	19	20	20	21	21	21	22	22
13						15	16	17	18	19	19	20	20	21	21	22	22	23	23
14						15	16	17	18	19	20	20	21	22	22	23	23	23	24
15						15	16	18	18	19	20	21	22	22	23	23	24	24	25
16							17	18	19	20	21	21	22	23	23	24	25	25	25
17							17	18	19	20	21	22	23	23	24	25	25	26	26
18							17	18	19	20	21	22	23	24	25	25	26	26	27
19							17	18	20	21	22	23	23	24	25	26	26	27	27
20							17	18	20	21	22	23	24	25	25	26	27	27	28

J

Durbin–Watson statistic

To test H_0: no positive serial correlation,

if $d < d_L$, reject H_0;
if $d > d_U$, accept H_0;
if $d_L < d < d_U$, the test is inconclusive.

To test H_0: no negative serial correlation, use $d = U - d$.

Level of significance $\alpha = 0.05$

n	$p = 2$		$p = 3$		$p = 4$		$p = 5$		$p = 6$	
	d_L	d_U	d_L	d_U	d_L	d_U	d_L	d_U	d_L	d_U
15	1.08	1.36	0.95	1.54	0.82	1.75	0.69	1.97	0.56	2.21
16	1.10	1.37	0.98	1.54	0.86	1.73	0.74	1.93	0.62	2.15
17	1.13	1.38	1.02	1.54	0.90	1.71	0.78	1.90	0.67	2.10
18	1.16	1.39	1.05	1.53	0.93	1.69	0.82	1.87	0.71	2.06
19	1.18	1.40	1.08	1.53	0.97	1.68	0.86	1.85	0.75	2.02
20	1.20	1.41	1.10	1.54	1.00	1.68	0.90	1.83	0.79	1.99
21	1.22	1.42	1.13	1.54	1.03	1.67	0.93	1.81	0.83	1.96
22	1.24	1.43	1.15	1.54	1.05	1.66	0.96	1.80	0.86	1.94
23	1.26	1.44	1.17	1.54	1.08	1.66	0.99	1.79	0.90	1.92
24	1.27	1.45	1.19	1.55	1.10	1.66	1.01	1.78	0.93	1.90
25	1.29	1.45	1.21	1.55	1.12	1.66	1.04	1.77	0.95	1.89
26	1.30	1.46	1.22	1.55	1.14	1.65	1.06	1.76	0.98	1.88
27	1.32	1.47	1.24	1.56	1.16	1.65	1.08	1.76	1.01	1.86
28	1.33	1.48	1.26	1.56	1.18	1.65	1.10	1.75	1.03	1.85
29	1.34	1.48	1.27	1.56	1.20	1.65	1.12	1.74	1.05	1.84
30	1.35	1.49	1.28	1.57	1.21	1.65	1.14	1.74	1.07	1.83
31	1.36	1.50	1.30	1.57	1.23	1.65	1.16	1.74	1.09	1.83
32	1.37	1.50	1.31	1.57	1.24	1.65	1.18	1.73	1.11	1.82
33	1.38	1.51	1.32	1.58	1.26	1.65	1.19	1.73	1.13	1.81
34	1.39	1.51	1.33	1.58	1.27	1.65	1.21	1.73	1.15	1.81
35	1.40	1.52	1.34	1.58	1.28	1.65	1.22	1.73	1.16	1.80
36	1.41	1.52	1.35	1.59	1.29	1.65	1.24	1.73	1.18	1.80
37	1.42	1.53	1.36	1.59	1.31	1.66	1.25	1.72	1.19	1.80
38	1.43	1.54	1.37	1.59	1.32	1.66	1.26	1.72	1.21	1.79
39	1.43	1.54	1.38	1.60	1.33	1.66	1.27 .	1.72	1.22	1.79
40	1.44	1.54	1.39	1.60	1.34	1.66	1.29	1.72	1.23	1.79
45	1.48	1.57	1.43	1.62	1.38	1.67	1.34	1.72	1.29	1.78
50	1.50	1.59	1.46	1.63	1.42	1.67	1.38	1.72	1.34	1.77
55	1.53	1.60	1.49	1.64	1.45	1.68	1.41	1.72	1.38	1.77
60	1.55	1.62	1.51	1.65	1.48	1.69	1.44	1.73	1.41	1.77

(Continued)

n	p = 2		p = 3		p = 4		p = 5		p = 6	
	d_L	d_U	d_L	d_U	d_L	d_U	d_L	d_U	d_L	d_U
65	1.57	1.63	1.54	1.66	1.50	1.70	1.47	1.73	1.44	1.77
70	1.58	1.64	1.55	1.67	1.52	1.70	1.49	1.74	1.46	1.77
75	1.60	1.64	1.57	1.68	1.54	1.71	1.51	1.74	1.49	1.77
80	1.61	1.66	1.59	1.69	1.56	1.72	1.53	1.74	1.51	1.77
85	1.62	1.67	1.60	1.70	1.57	1.72	1.55	1.75	1.52	1.77
90	1.63	1.68	1.61	1.70	1.59	1.73	1.57	1.75	1.54	1.78
95	1.64	1.69	1.62	1.71	1.60	1.73	1.58	1.75	1.56	1.78
100	1.65	1.69	1.63	1.72	1.61	1.74	1.59	1.76	1.57	1.78

Random number table

The table gives 2500 random digits, from 0 to 9, arranged for convenience in blocks of 5.

87024	74221	69721	44518	58804	04860	18127	16855	61558	15430
04852	03436	72753	99836	37513	91341	53517	92094	54386	44563
33592	45845	52015	72030	23071	92933	84219	39455	57792	14216
68121	53688	56812	34869	28573	51079	94677	23993	88241	97735
25062	10428	43930	69033	73395	83469	25990	12971	73728	03856
78183	44396	11064	92153	96293	00825	21079	78337	19739	13684
70209	23316	32828	00927	61841	64754	91125	01206	06691	50868
94342	91040	94035	02650	36284	91162	07950	36178	42536	49869
92503	29854	24116	61149	49266	82303	54924	58251	23928	20703
71646	57503	82416	22657	72359	30085	13037	39608	77439	49318
51809	70780	41544	27828	84321	07714	25865	97896	01924	62028
88504	21620	07292	71021	80929	45042	08703	45894	24521	49942
33186	49273	87542	41086	29615	81101	43707	87031	36101	15137
40068	35043	05280	62921	30122	65119	40512	26855	40842	83244
76401	68461	20711	12007	19209	28259	49820	76415	51534	63574
47014	93729	74235	47808	52473	03145	92563	05837	70023	33169
67147	48017	90741	53647	55007	36607	29360	83163	79024	26155
86987	62924	93157	70947	07336	49541	81386	26968	38311	99885
58973	47026	78574	08804	22960	32850	67944	92303	61216	72948
71635	86749	40369	94639	40731	54012	03972	98581	45604	34885
60971	54212	32596	03052	84150	36798	62635	26210	95685	87089
06599	60910	66315	96690	19039	39878	44688	65146	02482	73130
89960	27162	66264	71024	18708	77974	40473	87155	35834	03114
03930	56898	61900	44036	90012	17673	54167	82396	39468	49566
31338	28729	02095	07429	35718	86882	37513	51560	08872	33717
29782	33287	27400	42915	49914	68221	56088	06112	95481	30094
68493	88796	94771	89418	62045	40681	15941	05962	44378	64349
42534	31925	94158	90197	62874	53659	33433	48610	14698	54761
76126	41049	43363	52461	00552	93352	58497	16347	87145	73668
80434	73037	69008	36801	25520	14161	32300	04187	80668	07499
81301	39731	53857	19690	39998	49829	12399	70867	44498	17385
54521	42350	82908	51212	70208	39891	64871	67448	42988	32600
82530	22869	87276	06678	36873	61198	87748	07531	29592	39612
81338	64309	45798	42954	95565	02789	83017	82936	67117	17709
58264	60374	32610	17879	96900	68029	06993	84288	35401	56317
77023	46829	21332	77383	15547	29332	77698	89878	20489	71800
29750	59902	78110	59018	87548	10225	15774	70778	56086	08117
08288	38411	69886	64918	29055	87607	37452	38174	31431	46173
93908	94810	22057	94240	89918	16561	92716	66461	22337	64718
06341	25883	42574	80202	57287	95120	69332	19036	43326	98697
23240	94741	55622	79479	34606	51079	09476	10695	49618	63037
96370	19171	40441	05002	33165	28693	45027	73791	23047	32976
97050	16194	61095	26533	81738	77032	60551	31605	95212	81078
40833	12169	10712	78345	48236	45086	61654	94929	69169	70561
95676	13582	25664	60838	88071	50052	63188	50346	65618	17517
28030	14185	13226	99566	45483	10079	22945	23903	11695	10694
60202	32586	87466	83357	95516	31258	66309	40615	30572	60842
46530	48755	02308	79508	53422	50805	08896	06963	93922	99423
53151	95839	01745	46462	81463	28669	60179	17880	75875	34562
80272	64398	88249	06792	98424	66842	49129	98939	34173	49883

Answers

Chapter 1

1 Systems and solutions evolve to meet needs at a particular points in time. It is very common for a solution that did make sense (allocate drivers to an area) to become less appropriate as the environment changes. It is often said that what is most predictable is change. The problem is knowing what this change is going to be. In this case, the business is finding that it is working in a more complex environment; demand has become less predictable. The use of both quantitative and qualitative information will help the business. Quantitative information could be used to develop models of operational efficiency. It would be useful to be able to model demand by area and to examine whether a more considered response to demand would allow us to even-out the workload. It would also be useful to look at the points of delivery and the distances, and the journey times involved. It could be that revised timings and routes would lead to a more efficient allocation of vehicles and drivers. However, any change needs to be managed and qualitative information could be used to better understand the concerns of the drivers. It would be useful to know the range of problems encountered by the drivers and what they perceive as fair and unfair.

2 In this case, the problem could be easily stated as 'making a presentation in 2 weeks' time'. As stated the problem would be seen as clearly structured and a structured approach (an information search) could be used. The topic is clearly defined (employment trends) and the information is known to exist. However, the problem situation described is not about presentation content but is more about the group process of making a presentation. The problem could therefore be redefined as 'how can we, as a group, prepare the required presentation in the next 2 weeks'. It then becomes necessary to consider information about the group, like the time individuals are prepared to commit to groupwork and the skills of individuals to complete the necessary tasks.

3 To ensure that your research is feasible you would need to establish the resource requirements (typically budgets in business problems) and time constraints. If you are limited to a few hours and a notepad, you may interview one or two available individuals in depth and try to typify experience – present your findings as case-studies or case-histories. If you could take a larger sample, you would need to show how this was fairly selected and in what ways the sample was representative. Your approach would be seen as reliable if others could repeat your approach and achieve consistent results. Giving all students a fair chance of selection is one way of ensuring your approach is reliable. Validity can be more difficult to assess. It is tempting for students to report that they would have done better if staff sickness had not been a problem. The structure of any questionnaire is important (see Chapter 3). It is useful to consider factual detail first (e.g. how may hours of class time were missed), then response (e.g. what did you do during a cancelled class, how much course material did you miss) and finally the judgement as to what difference it made on a number of key aspects (e.g. did you receive sufficient guidance on the following topic areas ...?).

4 Data is a collection of the facts and figures we plan to work with. To have taken notes on the behaviour of customers as they enter a supermarket or to have made a made a record of what they bought is data. Once we organize data in a meaningful way it becomes information. To know that '50% of customers look at the magazine area first' or 'that 60% or more purchase milk during a visit' is information.

5 Queues begin to form when service (or throughput) is not immediate. In general, the length of any queue will depend on the arrival rate, the time taken to provide the service, the number of service points (or lanes) and the way the queue is managed (like 'first come first served'). A mathematical model would link together the various flow rates. A model would also allow us to develop measures of performance like the probability of having to queue, the average number in the queue and the average time in the queue. If we can model one kind of queue then we should be able to adapt this model to describe other queuing systems.

6 Fixed costs are those you would incur regardless of the level of use, such as the purchase cost and insurance. Variable costs are related to the level of use and could include electricity, telephone connection, paper and ink.

7 Total cost (Exone) $= 400 + 20x$, where $x =$ number of copies in thousands. By substitution of x values into the above equation, we can produce the table shown below:

x (000's)	Total cost (£'s)
10	600
20	800
30	1000
40	1200
50	1400

We can repeat the exercise for Fastrack:

$$\text{Total cost (Fastrack)} = 300 + 25x$$

and produce a similar table:

x (000's)	Total cost (£'s)
10	550
20	800
30	1050
40	1300
50	1550

When 20 000 copies are being produced the cost levels are the same for both machines. We can see that the machine with the lower fixed costs will tend to have the lower total cost at low levels of production (the answer will still depend on other costs) and that the machine with the lower variable costs (running costs) will tend to have the lower total cost at higher levels of production.

8 We use the term 'chance' when we are not sure about a future outcome. If we see two equally likely outcomes then we can talk about the chances being 50/50. The classic example used is tossing a fair coin where the chance of a head or tail is equal, and hence 50/50. As we progress with the topic, we are more likely to talk about a probability of 0.5 (or indeed 50%).

9 The benefits we would expect from computers (or more precisely computer software) would include the ability to:
- receive and send data
- search for data (using the Internet)
- calculate a range of statistics relatively easily
- produce a range of diagrams
- develop mathematical and other models
- produce documentation (e.g. questionnaires)
- include statistical work within text.

10 (a) A question like *'what do you dislike about your travel to work?'* is going to produce an open-ended response. Analysis will involve working with the statements produced; some of which we could quote directly in a report. We could look at the number of times a particular point was made and even refer to the frequency of some issues. We need to be careful not to assume that because a particular point was not mentioned by a particular respondent that it is not important to that respondent. Open-ended questions can improve our understanding of a particular problem but provide little, if any, quantitative data.

(b) Questions that require a *yes/no* type answer produce nominal (or categorical) data.

(c) Time is seen as continuous but the measurement may make it appear as discrete (to the nearest minute). Time has the properties of ratio measurement.

(d) As (c) above. Age is rounded down (last birthday).

(e) Rating scales produce an ordinal level of measurement.

11 (a) 234.56
(b) 234.6.

12 (a) £1 285 900
(b) £ 1286 000.

13 (a) Using the additive model we get the totals of 34, 39 and 34 for products A, B and C, respectively. On the basis of the results, we would see product B as the winner and be indifferent between A and C.

(b) Doubling the scores corresponding to 'market trends' would give the new totals of 37, 42 and 38. In this case, product B would stay the winner and product C would become second choice.

(c) The model could be improved by trying to identify a more complete range of factors, not just marketing (e.g. financial, operations). A more systematic approach could be used to evaluate the importance of certain factors. Attempts could be made, for example, to model future demand and model cash flows.

14 (a) By allocating the given scores to the ticks we get a total of 11 for Prototype a and 15 for Prototype B. The additive model would support the choice of Prototype B.

(b) By multiplying the scores for 'product life expectancy' by 3 and 'expected returns' by 2 we get the new totals of 20 and 19. In this case

Prototype A would be seen as the winner. However, we need to remember that the use of scoring models is not a precise science and that a difference of 1 would be seen as being of little importance.

15 The overall totals (multiplying weight by score in each case) are 104, 146, 104 and 66, respectively. Location B would be seen as the winner with A and C joint second choice (again indifference). In practice we would probably reject D as it has little to commend it and do further analysis on them others, e.g. a financial analysis. Scoring models can be used as a low cost screen, where options with few strengths are rejected while the others are further scrutinized.

Chapter 2

1 Data is essentially sets of numbers. Once we give numbers meaning they become data. We could, for example, look at the number of cars passing a roadside recorder. Once those recordings are translated into flowrate they become more informative. Further analysis will identify the numbers that exceed the local speed limit, which can further inform decision making.

2 Bias in a statistical sense, is the difference between the result we should get if our methodology works and the result we actually got. If we are undertaking a survey of chocolate or beer consumption and those that consume most choose not to participate or under-report their consumption, then a bias exists. There are many sources of bias. Bias can be a result of survey design. If those individuals selected (on average) are not representative of the population of interest then this is a source of bias. The construction and asking of questions can also be a source of bias. Stereotyped concepts, negatives, emotional phases, leading questions, presuming questions and hypothetical questions are all sources of bias.

3 Once information is collected we generalize to the population of interest. Unless we are clear about this population, we cannot ensure that the secondary or primary data is sufficiently representative. The defined population allows us to clearly state who is included or needs to be included and who is excluded. In any survey on driving we would need to clarify the term. A checklist of questions is often useful. In this case we would need to clarify whether we are only referring to those with a full licence, what vehicles are included, whether the driving include business and pleasure and a number of other considerations.

4 This question is concerned with the process of definition and as a class exercise you will see a number of possible definitions emerge. It is important to relate the definition to the objective of the survey, e.g. in a survey on the attitudes to smoking in the workplace we need to consider whose attitude and to what extent we need to ensure a fair representation of smokers and non-smokers. The definition can be geographic, e.g. a defined residential area or in terms of objects and items, e.g. a particular set of vehicles or components.

5 This question is a test of you and your local library. Most of the information required is available from government sources as listed in Table 2.1. You should try to find the *Guide to Official Statistics*.

6 This question is intended to give you some experience of using a search engine. With a little 'trial and error' you should soon be tapping into the pool of internet information. A shortage of information should not be a problem. You should consider issues like how to select information, how to follow leads, how the information has been presented and the validity of the information.

7 As we shall see (throughout this book), an effectively selected sample can give good results. If we were to take a number of samples of 1000 people from a large population and calculate the average age, we would see that this sample average would vary from sample to sample. However, *on average*, the sample average will give the result expected from the overall population. Some sample results will not be as close as we would like but most will be acceptably close. Even for sample sizes of 100 or 200 or 300 the sample results can get acceptably close to the result expected from the overall population.

Chapter 3

1 Reference to the Gallup, MORI and NOP websites will give you a good indication of the range and scale of survey work undertaken. Market research is now an important industry.

2 To ensure that all respondents had an equal chance of selection, we would need to check that our sampling frame (a listing of our population of interest) was complete and that we had a method of selection that gave an equal chance. In a simple situation we could assign each possible respondent a number and then pick numbers out of a hat. In most cases, a number would be assigned to each individual, usually a three- or four-digit number, and selection would be made by taking three- or four-digit numbers from random number tables or a random number generator.

3 To ensure a known chance of selection, we would again need a complete sampling frame. To double the chance of inclusion in the sample of part-time students we could simply give each part-time student two numbers. The problem with this is that overall you may still not get the balance sought between part-time and full-time students and you could select the same part-time students twice. In this case you could see the mode of study as strata and then randomly select the desired number from each strata. To avoid a bias in your results, the results would need to be weighted so that the calculations were adjusted to allow for the different chances.

4 The Electoral Register is a popular sampling frame because it offers a fairly complete listing of those entitled to vote (includes most of the adult population) and a fairly complete listing of households. It does cover the UK and it is updated.

5 Interest in postcodes for survey work is increasing. Essentially, a postcode identifies a cluster of households but not individual homes or individuals. Postcodes offer a cost-effective way of undertaking postal surveys and can be used to provide a random starting point for other survey designs.

6 (a) In this case we would need to clarify what we meant by clothes shops and whether we would include shops that sell clothes alongside other items. Generally reference to the *Yellow Pages* would be sufficient for this kind of survey.

   (b) We are unlikely to find a listing of those that eat chocolate regularly. Given that the percentage eating chocolate is high, we would work with a suitable listing of all people and exclude those that don't eat chocolate (e.g. with a filter question).

   (c) Again we would need to clarify who would be included as customers. Given that no direct listing of names would exist, we could examine flow rates through the week and sample from the flow. If the sampling was in proportion to the flow rate and included the element of random selection, we could ensure a representative sample.

(d) In this case we should be able to use the enrolment lists (if we have access). The process can become more complex if we also want to know the reasons why some students have chosen not to enrol, why some students have withdraw or failed (and no longer enrolled) and the views of past students.

7 A stratified sample is designed to ensure that particular groups or strata are correctly represented. The individuals within strata can be selected by a method of our choice including simple random sampling. Quota sampling is designed to ensure the right balance in terms of certain characteristics, such as age and gender, but final selection, critically, is still left to the interviewer.

8 In a random sample survey, an interviewer is given a list of names (or addresses) to work with and it is easy to identify the number of non-response and some of the reasons why. In a quota sample, typically no record of those that say no is kept. Even is we kept a count of those that said no, that would only include those discussed under the category 'those who refuse to co-operate'.

9 If separate surveys are taken over time, change can be a consequence of two factors: a genuine change in behaviour over time and a change in the sample. A panel eliminates the second factor but we still need to be aware of panel conditioning factors. Using a panel has the additional advantages of speed (your respondents are already located), cost (you only incur the cost of selection once) and the possibility of building-up a depth of information.

Questions 10–15 are particularly suitable for class discussion. It is not a matter of finding the right answer but rather exploring ideas and effectively applying material from the chapter.

Chapter 4

1 This question is concerned with the variety of representation used. It is important to distinguish between discrete (where essentially counts are made, e.g. the number of cars sold by model or the number of defective items) and continuous data (where 'ruler' type measurement is made, e.g. travel distance or time).

2 The diagram presented in question 2 needs to represent the 25 unit increase from year 1 to year 2 and the 25 unit increase from year 2 to year 3 (equal increases). A 'traditional' bar chart showing sales in a single dimension would do this without misrepresentation.

3 The data is discrete. A component bar chart, for example, could be constructed showing sales by country (with a key showing product type) or sales by product type (with a key showing country).

4 A bar chart can be drawn using the frequency distribution:

New orders (x) 0 1 2 3 4 5 6
Frequency (f) 3 3 4 5 6 3 1

5 (a) A spreadsheet or protractor is needed for this exercise. To draw you need to calculate the number of degrees (share of 360). Type X (198, 90, 72) and Type Y (270, 36, 54).

(b) Bar charts could be constructed using either absolute or percentage values.

6 The frequencies are: 1, 3, 7, 14, 8, 5, 2.

7 Starting at £4 and constructing intervals of £1, the frequencies are: 1, 2, 0, 4, 3, 4, 2, 1, 0, 1, 1, 1.

8 The cost of travel will need to be represented by a histogram. An upper boundary will need to be assumed for the last interval and the histogram will need to be scaled to reflect the differing interval widths.

Questions 9, 10 and 11. Histograms need to be drawn that are scaled to allow for unequal widths. Any assumptions made regarding the lower boundary of the first group or upper boundary of the last group needs to be clearly stated.

12 This question is intended to illustrate the type of results produced by a survey. Question 1 should be represented by a histogram (remember that number of years is often rounded down, like age last birthday) and the remainder by pie charts or bar charts.

13 You should try to produce graphs like this on a software package, such as Excel. It should be noted that sales have increased noticeably since the low point in year 3.

Questions 14 and 15 give typical time series data. A plot should be made against time and any pattern of seasonal variation noted.

16 Plotting the original data suggest percentage growth over time. Taking the log values produces a more linear plot (again suggesting percentage growth).

Questions 17 and 18 both require the figures as cumulative percentages (see Section 4.5.4).

Chapter 5

1 This question is concerned with selecting statistics that are appropriate for the measurement being considered. The mean would usefully describe the average travel time (a) whereas, the mode would be used to describe the most popular model of car. In many cases, we would look at two or more statistics. The mean and median would both give us useful information on the earnings of manual workers (c). We are also likely to want measures of difference.

2 This question is intended as a basis of discussion. Typically the arithmetic mean is the value calculated and is the value that most people would accept and understand. The statistics calculated depend both on the measurement achieved and the purpose of the analysis.

3 Any comparison of the median and mean tends to be concerned with the importance of skew within the distribution (see Figure 5.7). When considering an earnings distribution, it is useful to have both statistics as together they are more informative.

4 Mean = 2.84 orders
median = 3 orders
mode = 4 orders

5 Mean = 1.3611 faults
median = 1 fault
mode = 0 faults

6 Mean = 146.8 miles
median = 146 miles

mode = 135 miles
New mean = 162.22 miles
new median = 149.5 miles
new mode = 135 miles

7 Mean = £9.2365
median = £8.86

8 (Using mid-points 0.5, 3 and 7.5)
Local: mean = £2.40
         median = £1.90
Comm: mean = £3.11
         median = £2.70
Long: mean = £6.41
         median = £6.70

9 (Using mid-points 50, 125, 175, 300, 500)
Mean = £204.35
median = £161.92
mode = £147.62

10 (Using mid-points 0.5, 2.0, 6.5, 12.5, 17.5) Mean = 5.59 miles
median = 3.66 miles
mode = 0.94 miles

11 (Using mid-points –17.5, –12.5, –7.5, –2.5, 5, 15, 25) Mean = £0.21
median = –£1.41

12 Mean = 1.14 days
median = 1 day
mode = 0 day

13 Arithmetic mean = 7.26
Geometric mean = 6.5849

14 Mean = £108.26

15 Mean = £454.44
Total value = £818 000

Chapter 6

1 This question, like question 5.1, is concerned with selecting appropriate statistics for the measurement achieved. If the mean is being used as the measure of location then the standard deviation will be used as the measure of spread (a). It is important to consider both the measurement and the purpose.

2 The standard deviation is a measure of difference that is valid if ruler type measurement is achieved eg time and distance. Differences about the mean are used (squared, averaged and then the square-root taken). The standard deviation can be used for comparisons (to compare the spread of one set of data against another) or for more advanced statistic work (see Chapters 10 onwards).

3 One of the important characteristics of the standard deviation is that is uses all the data, whereas the range only uses the two most extreme observations. As we will see, the standard deviation also has nice statistical properties (defines the spread of the normal distribution).

4 Range = 6
quartile deviation = 1.5
standard deviation = 1.67

5 Range = 4
quartile deviation = 1
standard deviation = 1.3775

6 Range = 57
quartile deviation = 8
Standard deviation = 12.89
New range = 171
New quartile deviation = 13.25
New Standard deviation = 36.44

7 Mean = £9.2365
standard deviation = £2.825

8 (Using mid-points 0.5, 3 and 7.5)
Local: s.d. = 1.877
Comm: s.d. = 2.2143
Long: s.d. = 1.9296

9 (Using mid-points 50, 125, 175, 300, 500)
s.d. = 138.7852
range determined by assumed lower and upper boundaries
quartile deviation = $\frac{1}{2}(266.17 - 113.42) = 76.375$

10 First quartile = 1.5833
third quartile = 8.3446
quartile deviation = 3.38065
standard deviation = 4.9672

11 Mean = £0.21
standard deviation = £7.306

12 Range = 5 − 0 = 5
quartile deviation = 1
standard deviation = 1.2492

13 (a) Mean = £89.80
standard deviation = £48.7330
(b)   53.258%
(c)   median = £85.21
first quartile = £46.38
third quartile = £123.66
quartile deviation = £38.64

14 (a)   A £12 208.33
£1946.77
B £12 375
£1235.33
(b)   A coefficient of var = 15.95%
skew = 0.03
B coefficient of var = 9.98%
skew = −0.06

15 (a)   6.25 minutes
(b)   20.5 minutes
(c)   27.22 minutes

Chapter 7

1 (a) 80, 92, 96, 100, 104, 116
(b) 4.167%, 11.538%

2 (a) 52.63, 63.16, 84.21
(b) 247, 266, 285, 313.5

3 (a) 100, 175, 475
(b) 100, 168.75, 431.25
(c) 100, 168, 420.75
(d) 100, 167, 417
(e) 100, 167, 386
(f) 100, 97.8, 108.7
(g) 100, 97.4, 100.5

4 (a) 100, 105.9, 105.9
(b) 100, 117.86, 139.29
(c) 100, 122.5, 150
(d) 100, 121, 141.2

5 100, 110.4, 120.3

6 and 7 require the use of recently published figures. You should try to develop a spreadsheet model.

8 (a) 100, 102.2, 106.47,
112.59, 118.58, 133.59,
146.28, 153.49, 159.5,
167.78, 174.46, 182.48,
192.63, 200.78, 212.27,
228.83
(b) 100, 103.2, 107.5,
112.4, 123.4, 136.3,
145.1, 146.3, 150.5,
158.1, 166.64, 176.91,
191.46, 205.53, 223.4,
244.42

9 (a) 103.81, 111.3
(b) 108.34, 109.98
(c) 104.35, 113.9
(d) 108.92, 112.56
(e) 113.06, 125.27
(f) 103.81, 111.3,
108.34, 135.40,
104.35, 112.50,
108.92, 136.86,
113.06, 152.33
(g) 103.81, 111.9,
108.34, 109.98,
104.35, 116.25,
108.92, 114.25,
113.06, 127.85
(h) 103.81, 111.30,
108.34, 98.29,
104.35, 113.65,
108.92, 100.36,
113.06, 111.70

Chapter 8

1   0.25

2   0.25

3   0.125

4   0.375

5   0.64, 0.16, 0.032, 0.096

6   25

7   640, 160, 32, 96

8   (a) 1/18
    (b) 1/9
    (c) 1/6
    (d) 1/6

9   1/36

10 9/16

11 1/12

12 7/12

13 1/144

14 (a) 1/13
    (b) 1/4
    (c) 16/52
    (d) 1/52
    (e) 3/13
    (f) 1/2
    (g) 1/26
    (h) 3/26

15 (a) 0.159 76
    (b) 0.159 879

16 116/221

17 (a) 0.25
    (b) 0.5

18 $p(A) = 0.1; p(B) = 0.0333; p(C) = 0.025$

19 (a) 140
    (b) 180
    (c) 402
    (d) Expect 120, so yes

20 (i)   0.075
    (ii) 0.08
    (iii) 0.0018279 (assuming 50% of each colour)

21 (b) 0.9

22 (a) 0.555 55
    (b) 0.288 888 9
    (c) 0.4

23 0.0081

24 (a) 0.2
(b) 0.6
(c) 0.8
(d) 0.6
(e) 0.25

25 (a) 0.25
(b) 0.28

26 (a) 0.168
(b) 0.4667
(c) 0.5368

27 (a) 0.729
(b) 0.081
(c) 0.0081

28 £410

29 Project A: £5600
Project B: £4350

30 EMV(ad) = £6300
EMV(no ad) = £5750

31 Option (a)   £16 000
Option (b)   £29 000
Option (c)   −£3500

32 You need carefully to include all the choices in this decision tree. The advice is to undertake the survey. If the survey results are favourable, then manufacture; and if unfavourable, sell the design (if still possible).

33 Refer to Section 8.8.

34
$$\mathbf{P}^2 = \begin{bmatrix} 0.04 & 0.96 \\ 0 & 1 \end{bmatrix}$$

$$\mathbf{P}^4 = \begin{bmatrix} 0.0016 & 0.9984 \\ 0 & 1 \end{bmatrix}$$

$$\mathbf{P}^8 = \begin{bmatrix} 0.000\,002\,56 & 0.999\,997\,44 \\ 0 & 1 \end{bmatrix}$$

35 (a) [85   48   11   16]
(b) [75.5   44.3   11.4   28.8]
(c) [69.02   40.67   11.27   39.04]
(d) [64.419   37.52   10.829   47.232]

36 (a) [60   105   110   145   160   70]
(b) [1.62   7.695   20.63   66.245   265.83   287.98]
(c) £47 000
(d) £40 717.20

37 (a) [74.5   20   10.5]
(b) [70.15   14.9   19.95]
(c) [62.515   14.03   28.455]

(d) [54.9845   21.5605]

(e) [55.737 55   20.807 45]

(f) [55.662 245   20.882 755]

Chapter 9

1 (a) 3

(b) 120, 120

(c) 38 760, 1

(d) 45, 10, 1

(e) 635 013 559 600

2 0.000 000 000 006 299 078

3 $P(0) = 0.0256$, $P(1) = 0.1536$, $P(2) = 0.3456$, $P(3) = 0.3456$, $P(4) = 0.1296$

4 (a) 0.004 096

(b) 0.276 48

(c) 0.544 32

5 (a) 0.000 655 36

(b) 0.123 863 04

(c) 0.315 394 56

6 (i) 0.142 625

(ii) 0.020 341 9

7 0.659 002

8 $P(0) = 0.135 33$, $P(1) = 0.270 67$, $P(2) = 0.270 67$, $P(3) = 0.180 447$, $P(4) = 0.090 223 5$, $P(5) = 0.036 089 4$

9 $P(>4) = 0.0527$, $P(>5) = 0.0166$

10 $P(0) = 0.223 13$, $P(1) = 0.334 695$, $P(2) = 0.251 021$, $P(3) = 0.125 511$, $P(4) = 0.047 066 5$

11 $P(0) = 0.3679$, $P(1) = 0.3679$, $P(2) = 0.1839$, $P(3) = 0.0613$, $P(4) = 0.0153$, $P(5+) = 0.0037$

Days = 134.28, 134.28, 67.14, 22.38, 5.59, 1.35

Cost = £364 748 (depending upon assumptions)

12 $P(>3) = 0.566 527 6$, expected no. = 2719.332 48

Total cost = $4 \times 4800 \times £7.84 = £150 528$

13 Demand > supply for approximately 16 days.

14 (a) 0.0803

(b) 8083

(c) £1 204 520.96

(d) TR = £10 000 000

TC = £6 000 000; + £1 204 500 (claims)

expected profit = £2 795 479.04

15 (a) 0.327 68

(b) 0.000 32

(c) 0.737 28

Chapter 10

1 (a) 1
(b) −0.6
(c) −1.54
(d) 2.14
(e) 0

2 (a) 0.1587
(b) 0.002 19
(c) 0.977 25
(d) 0.5398
(e) 0.0446
(f) 0.3821
(g) 0.8023
(h) 0.995 06
(i) 0.044 82
(j) 0.1747
(k) 0.1337
(l) 0.045 56
(m) 0.9258
(n) 0.95

3 (a) 1.3
(b) −0.52
(c) 0.02
(d) −2.44
(e) 1.77
(f) ±1.96
(g) −1.645 and +1.645

4 (a) 3.01%
(b) 10.75%
(c) 93.39%
(d) £109.46
(e) £92.40

5 (a) 0.9973
(b) 0.993 79
(c) 0.3085

6 (a) 0.2451
(b) 0.3650
(c) 0.8495

7 (a) 0.135 335
(b) 0.1446

8 (a) 0.0537
(b) 0.259 45

9 (a) 0.834
(b) 0.9292
(c) 0.905 31

10 0.2643

11 (a) 0.8023
(b) cannot assume $p = 0.3$, so cannot estimate (sums give 0.8186)

12 0.0516

13 (a) 0.648
    (b) 0.123

14 (a) 0.2514
    (b) 0.0632

15 0.3156

16 (a) 0.002 48
    (b) 0.1867

17 (a) 0.3446
    (b) 0.014 26

18 0.1575

| Chapter 11 |

1 £1796 ± £138.59, £1796 ± £277.19, £1796 ± £554.37
2 £287 ± £13.73, £287 ± £18.05
3 £6.30 ± £0.41, £6.30 ± £0.29, £6.30 ± £0.21
4 (a) $\bar{x} = 146.8$, $s = 13.05$
      $146.8 ± 4.044$
    (b) 73
5 $\bar{x} = 1.14$, $s = 1.25$
    95%: $1.14 ± 0.155$
    99%: $1.14 ± 0.2040$
6 (Using mid-points 0.5, 15., 3.5, 7.5, 12.5) (a) 0.5295
                                               (b) 90
7 95%: $6\% ± 2.7\%$
    99%: $6\% ± 3.5\%$
8 (a) $20\% ± 2.5\%$
    (b) 6147
9 (a) 374.33
    (b) 384.16
10 $-5\,g ± 6.34\,g$
11 $-0.244 ± 0.223$
12 $11\% ± 7.1\%$
13 $4\% ± 5\%$
14 $1\% ± 7.7\%$
15 (a) $£5.40 ± £0.61$
    (b) 45.33
16 $15\% ± 12.122\%$
17 (a) $98\% ± 3.36\%$
18 $£1259 ± £102.78$
19 7.9 minutes ± 0.83 minutes
20 2.245 and 5.755

21 -1.099 and 3.499

22 14.8 and 26.198 ordered values; $142 \leq$ median ≤ 150

Chapter 12

2 See Section 11.10, question 1. Only when s.d. $= £2000$ would you accept the null hypothesis.

3 See section 11.10, question 2. Accept null hypothesis using 95% and 99%.

4 $H_0 = £850$, $H_1 < £850$, $z = -1.48$ cannot reject H_0

5 $z = 2.357$, reject H_0

6 $z = -12.39$, reject H_0

7 $z = 2.817$, reject H_0

8 $z = -13.622$, reject H_0

9 $z = 1.52$, cannot reject H_0

10 $z = -1.353$, cannot reject H_0

11 $z = -1.67$, reject H_0

12 $z = -0.2887$, cannot reject H_0

13 $z = -1.3066$, cannot reject H_0

14 $z = -2.449$, reject H_0

15 $z = -0.9405$, cannot reject H_0

16 $z = 2.225$, reject H_0

17 $t = -0.9035$, cannot reject H_0

18 $t = -2.871$, reject H_0

19 $t = 0.8017$, cannot reject H_0

20 $t = 0.0942$, cannot reject H_0

21 $t = 1.6978$, cannot reject H_0

22 (a) 29.46%
(b) 19.63%

Chapter 13

1 Chi-sq. $= 4.254$, cannot reject H_0

2 Chi-sq. $= 14.46$, reject H_0

3 Chi-sq. $= 140.555$, reject H_0

4 Chi-sq. $= 2.517$, cannot reject H_0

5 Chi-sq. $= 8.67$, cannot reject H_0

6 Chi-sq. $= 20.559$ reject H_0

7 Chi-sq. $= 4.4211$, cannot reject H_0

8 Chi-sq. $= 1.277$, cannot reject H_0

9 Chi-sq. $= 41.5244$, reject H_0 (last two groups combined)

10 Chi-sq. $= 50.7$, reject H_0 (last four groups combined)

11 $T = 52$, reject H_0

12 $z = 2.789$, reject H_0

13 $T = 39$, reject H_0

14 $T = 143$, reject H_0

15 Reject H_0

16 $z = -1.0665$, cannot reject H_0

17 Reject H_0

18 $z = -0.7653$, cannot reject H_0

19 Cannot reject H_0

20 Reject H_0

Chapter 14

1 (b) 101.75, 103.125, 104.125, 105.625, 107.625, 109.125, 110.125, 111.5, 113.5, 115.25, 116.5, 118.0

(c) Q1 = 8.0833, Q2 = -2.375, Q3 = -14.625, Q4 = 9.5

2 (b) 200.57, 202, 202.71, 204.14, 203.57, 206.43, 212.14, 225, 221.43, 219.29, 217.86, 220.71, 224.29, 227.14, 230, 229.29, 227.86, 226.43, 225, 222.86, 230, 231.43, 229.29, 226.43, 226.43, 222.14, 225, 226.43, 227.86

(c) M = -110.373, T = -93.409, W = -63.945, TH = -17.688, F = 35.734, S = 175.912, SU = 73.769

3 (b) 39.75, 44.875, 50.375, 55.875, 59.75, 62.625, 65.75, 69.0, 71.375, 72.5, 73.375, 73.875

(c) $Q_1 = -14.83$, $Q_2 = -5.25$, $Q_3 = 1.375$, $Q_4 = 18.67$

4 Trend $= 28.88 + 3.67647X$

$Q_1 = -13.365$, $Q_2 = -5.0425$, $Q_3 = 1.03$, $Q_4 = 17.355$

5 (b) Trend $= 11.33 + 1.3X$

(c) $Q_1 = 1.88$, $Q_2 = 0.89$, $Q_3 = 0.39$, $Q_4 = 0.83$

6 (a) Average $R = 0$

(b) Average $R = 1$

7 (a) Trend $= 12.83 + 0.769X$

(c) $Q_1 = -3.35$, $Q_2 = -1.12$, $Q_3 = 0.78$, $Q_4 = 3.68$

8 (b) Trend $= 59.595 - 1.3136X$

(c) 26.765

9 (b) Trend $= 95.32 + 1.136$ Time

(c) 38.67, 73, 109.67, 127.33, 136, 138.67, 139.67, 140, 138.67, 136, 130.67, 124.67, 119.33, 115.33, 111.67, 104.67, 94, 81

10 (a) $\alpha = 0.2$ 225, 225, 226, 226, 228.8, 232.04, 237.632, 246.1056, 258.8845, 271.1076, 272.8861, 266.3089, 257.0471, 251.6377, 248.9101, 246.9281, 247.7425, 249.194, 252.3552, 261.8842, 274.5073, 290.8059, 292.6447, 286.9157, 275.7326

(b) $\alpha = 0.4$ 225, 225, 227, 226.6, 231.96, 237.176, 246.3056, 259.7834, 279.87, 295.922, 289.5532, 269.7319, 249.8392, 241.9035, 240.3421,

239.8053, 244.2832, 248.5699, 255.1419, 273.0852, 293.8511, 318.7107, 311.2264, 292.3358, 267.8015

(c) $\alpha = 0.6$ 225, 225, 228, 226.8, 234.72, 240.888, 252.3552, 268.9421, 293.5768, 309.4307, 291.7723, 260.7089, 236.2836, 232.5134, 235.8054, 237.7221, 245.6889, 251.2755, 259.5102, 283.8041, 308.5216, 337.0087, 314.8035, 284.3214, 252.3286

11 $\alpha = 0.3$ MAD $= 776.73042$; $\alpha = 0.99$ MAD $= 452.14828$

12 (a) 3.9787

(b) 2.88016

(c) 2.27242

(d) 2.16142

Chapter 15

1 $r_s = 0.2$

2 $r_s = -0.524$

3 0.6

4 0.595

5 $r_s = -0.333$

6 $r_s = 0.5515$

8 $r = 0.989949$

9 $r = -0.99056$, $r^2 = 0.98121$

10 $r = 0.9745$, $r^2 = 0.9496$

11 $r = 0.70295$

12 $r = 0$

13 $r = 0.744$, $r^2 = 0.554$

14 $r^2 = 0.8177$

15 $r = 0.9284$

16 $r = 0.88713$, $r^2 = 0.787$

17 (b) 0.6898

18 $r = -0.9825$

19 $r = 0.9328$, $r^2 = 0.87$, using X and log Y gives $r = 0.999$, $r^2 = 0.998$

Chapter 16

1 $y = 0.6 + 2.8x$

2 $y = 34.5151 - 0.90606x$

3 $y = 2.537 + 2.045x$

4 $y = 2.24 + 0.815x$, $y_{(x=7)} = 7.94$

5 $y = 4$

6 $y = -15.9 + 5.21x$

7 (a) $y = 833.8028 + 65.7356x$

(b) 3003.08

(c) 4777.94

8 (a) ATC $= 56.5619 - 3.304$ production

(c) 28.48, 3.705, -9.5095

9 $y = 133.73 - 1.394x$

10 $y = -10.1714 + 0.8993x$, $\log y = 0.586 + 0.01695x$

11 Report score $= 30.615 + 0.5962$ Pre-test score

12 (b) Time $= 29.5895 - 1.05813$ Experience

(c) (i) 25.36, (ii) -33.89

Chapter 17

1 (a) $Y = 7.52 + 3.93X_2$

(b) $Y = 33.267 - 2.32X_3$

(c) $Y = 31.18 + 0.323X_2 - 2.135X_3$

(e) $R^2 = 0.9957$

2 $R^2 = 0.926$, adj. $R^2 = 0.897$

$Y = 134.12 - 0.005X_1 - 0.3798X_2 - 0.1691X_3 + 0.003432X_4$

Note multicollinearity

3 $R^2 = 0.975$, adj. $R^2 = 0.968$, problem is perfect correlation X_1 and X_2, need to delete one of them

$Y = -283.512 + 16.49X_2 + 7.08X_3$

4 (a) Amount spent $= -4.1 + 1.76$ Size $+ 0.986$ Income

5 Income $= 4.705 + 1.728$ Education $+ 1.133$ In post $+ 0.986$ Prev. jobs

Chapter 18

5 (a) £367.04

(b) £390.23

6 (a) £16 105.10

(b) £16 145.04

(c) £15 529.69

7 £641.09

8 £3519.61

9 £3485.10

10 £353.90

11 Project 1: £18 708.70

Project 2: £17 021.20

12 Large process £4 297 560

Two med processes £3 960 220

13 £1221.02

14 £1343.12

15 £12 646.93

16 £593.26

17 £996.26

18 27%

19 (a) 23.1%

(b) 21.6%

(c) 16.6%

20 1.94%

Chapter 19

3 (a) $x = 20, y = 40, z = 120$
(b) $x = 10, y = 45, z = 145$
(c) Multiple solution including $x = 20, y = 40$ and $x = 10, y = 45, z = 500$

4 (a) $x = 6, y = 7, z = 33$
(b) Acting constraints $0.5x + y \geq 10, x + y \geq 13$
(c) $x = 5, y = 8, z = 34$

5 6 base units, 15 cabinet units, profit contribution $= £870$

6 (b) $F = 2.5$ and $S = 7.5$
(c) Increase labour available to 11 hours

7 Add 2 more Superloaders

8 (b) $N = 75$ and $B = 300$
(c) 39p

Chapter 20

3 Critical path: C, H, K, M, L: time $= 26$

4 Critical path: A, D, G: time $= 14$

5 Critical path: B, F, I, K: time $= 45$

6 Critical path: D, N, S, W: time $= 28$

Chapter 21

1 200, 15

2 (a) 175, 4
(b) £36 400
(c) 58.33

3 (i) $Q = 1500$, TC $= £169\,650$
(ii) TC $= £164\,042.50$
(iii) TC $= £161\,392.50$

4 (a) 21.58
(b) 24.3052

5 (a) 5
(b) 6

6 (a) 0.8
(b) 0.2
(c) $P_1 = 0.16$
$P_2 = 0.128$
$P_3 = 0.1024$
(d) 4
(e) 3.2
(f) 0.167 hrs or 10 mins
(g) 0.133 hrs or 8 mins

7 (a) reduced by 1.5
(b) reduced by 1.35
(c) reduced by 5 mins
(d) reduced by 4.5 mins

8 Cost without ass. $= £910$/day

Cost with ass. = £416.64/day

9 Manual system £112.00/hr
Auto system £114.20

10 4 service point system:
$P_0 = 0.0737$
$P_1 = 0.1843$
$P_2 = 0.2303$
$P_3 = 0.1919$
$P_4 = 0.1200$
$P_5 = 0.0750$
$L_q = 0.53$
$L_s = 3.03$
$T_q = 0.636$ mins
$T_s = 3.636$ mins

11 $P_0 = 0.1727$
$P_1 = 0.2878$
$P_2 = 0.2399$
$P_3 = 0.1333$
$P_4 = 0.0740$
$P_5 = 0.0411$
$L_q = 0.3747$
$L_s = 2.0414$
$T_q = 0.7494$ mins
$T_s = 24.08$ mins

Chapter 23

1 $x = 9y$

2 $x = -12y$

3 $x = 2.5y$

4 $x = 3y$

5 $y = -4$

6 a^5

7 1

8 $a^2 + a^3/b$

9 a^2

10 $a^3 + a^6 - a^8$

11 1/8

12 $x = 22$

13 $2a^2 + ab$

14 $a^2 + ab + b^2$

15 $a^2 + 4ab + 4b^2$

16 $a^2 - b^2$

17 $6a^2b + 2b^3$

18 $12x^2 + 18xy - 12y^2$

19 (a) 32
(b) 258

20 (a) 238
(b) 2268

21 (a) -15
(b) 156

22 35

23 (a) 2916
(b) 118 096

24 (a) 384
(b) -2046

25 (a) 1.048 576
(b) 17.852 52

26 28 125

30 (a) $x = 2.6, y = 0.3$
(b) $x = 13, y = 2$
(c) $x = 4, y = -2$
(d) $x = 1.75, y = 3.15$

31 (a) $x = 1$ or 2
(b) $x = -2$ or 4
(c) $x = -1$ or -12
(d) $x = -3.217$ or 6.217
(e) $x = 0.607$ or 4.393
(f) $x = -1, +1$ or 6
(g) $x = -3.3935$ or 5.8935

32 $P = 70 - 2Q$

33 $P = -5 + Q$

34 $P = 20, Q = 25$

35 $P = 47.5 - 4.5Q$
$P = 9.5 + 0.1Q$
Equilibrium $Q = 8.26$
$P = 10.33$
New equilibrium $P = 10.815$
$Q = 13.152$
New demand $P = 70 - 4.5Q$

36 (a) demand: $P = 2Q^2 - 100Q + 2000$
supply: $P = 5Q^2 + 4Q$
(b) $Q = 13.765, P = 1002.45$

Chapter 24

1 $\begin{bmatrix} 7 & 8 \\ 12 & 15 \end{bmatrix}$

2 $\begin{bmatrix} 16 & 8 & 4 \end{bmatrix}$

3 50

4 $\begin{bmatrix} 18 & 24 \\ 60 & 78 \end{bmatrix}$

5 $\begin{bmatrix} 38 & 40 \\ 56 & 58 \end{bmatrix}$

6 impossible

7 $\begin{bmatrix} 5 & 14 & 18 \\ 8 & 2 & 7 \\ 10 & 8 & 4 \end{bmatrix}$

8 $\begin{bmatrix} 39 & 53 & 45 \end{bmatrix}$

9 $\begin{bmatrix} 24 & 64 & 72 \\ 14 & 49 & 79 \\ 13 & 39 & 95 \end{bmatrix}$

10 $\begin{bmatrix} 122 \\ 26 \\ 72 \end{bmatrix}$

11 $\frac{1}{6} \begin{bmatrix} 5 & -1 \\ -4 & 2 \end{bmatrix}$

12 $\frac{1}{204} \begin{bmatrix} 1 & -28 & 44 \\ 24 & -60 & 36 \\ -11 & 104 & -76 \end{bmatrix}$

13 1/50

14 $\begin{bmatrix} 24 & 12 & 6 \\ 32 & 16 & 8 \\ 40 & 20 & 10 \end{bmatrix}$

15 $\begin{bmatrix} 1 & 0 & 0 \\ 0 & 1 & 0 \\ 0 & 0 & 1 \end{bmatrix}$

16

$$\begin{bmatrix} 1 & 0 & 0 \\ 0 & 1 & 0 \\ 0 & 0 & 1 \end{bmatrix}$$

17 $[23 \quad 76 \quad 118]$

18

$$\begin{bmatrix} 60 & 30 & 48 \\ 80 & 40 & 64 \\ 100 & 50 & 80 \end{bmatrix}$$

19 180

20

$$\begin{bmatrix} 138 & 456 & 708 \\ 184 & 608 & 944 \\ 230 & 760 & 1180 \end{bmatrix}$$

21 $x = 2.6, y = 0.3$

22 $x = 4, y = -2$

23 $x = 4, y = -2$

24 $x = -1.7, y = 2.3$

25 $a = 1, b = 2, c = 5$

26 $a = 0.1, b = -1, c = 2.5$

27 $x = 8, y = 10, z = -5$

28 $a = 1, b = 3, c = 2$

29 $a = 1, b = -1, c = 2, d = 2$

30 $x_1 = 5.727\,27, x_2 = 2.454\,55, x_3 = 0, x_4 = 1.181\,82$

31 $X_1 = 406.67, X_2 = 365$

32 $A = 178.779, B = 704.942$

33 (e) $A = 810, B = 1620, C = 2250$

34 $X_1 = 1103.49, X_2 = 1151.81, X_3 = 1952.26$

Chapter 25

1 0

2 2

3 3

4 $10x^4$

5 $42x^2 - 12x^3$

6 $34x + 14$

7 $7 + 4y - 12y^2$

8 $4q - 4q^{-2}$

9 e^x

10 $2 + 8s + 0.5s^{-1/2}$

11 $f'(x) = 140x^6 + 24x^5 - 15x^4 - 28x^3 + 12x^2 - 22x + 2 + x^{-2} - 18x^{-3}$

12 $-2x^{-3} + 6x^{-4} + (16/3)x^{-5}$

13 MR $= 100 - 4x$; $P = 50$

14 AC $= (1/3)x^2 - 5x + 30$

MC $= x^2 - 10x + 30$

15 (a) $x = 19.67399$

(b) $x = 11.888$

16 MC $=$ MR at $x = 3.765$

AC $=$ AR at $x = 6.437$

17 (a) $E_D = -1.5$

(b) $E_D = -1.0$

(c) $E_D = -0.667$

18 (a) $P = 1000 - 5Q$

(b) $P = 1500 - 55Q$

(c) MR $= 1000 - 10Q$, MR $= 1500 - 110Q$

(e) $E_D = -19$, $E_D = -1.727$

(f) $400 <$ MC < 900

19 Min. at $x = 5$, $y = 0$

20 Max. at $x = 10$, $y = 1200$

21 Min. at $x = -1$, $y = 2$

22 Linear, so no max. or min.

23 Min. at $x = 1.6667$, $y = -4.3333$

24 Max. at $x = 1$, $y = 12.33$

Min. at $x = 5$, $y = 1.67$

25 Max. at $b = 0.25$, $a = 10.25$

26 Max. at $x = 0.5858$, $y = 13.3137$

Min. at $x = 3.414$, $y = -9.3137$

27 Neither, a constant cannot have max. or min.

28 Min. at $x = 1$, $y = 37.75$

Max. at $x = 2$, $y = 38$

Min. at $x = 3$, $y = 37.75$

29 Output $(x) = 10$: Profit $= 210$

30 Max. profit at $x = 11.888$ when profit $= £696.1116$

31 Output $(x) = 3.765$: Profit $= 92.58$

32 Min. AC at $x = 1$, when AC $=$ MC $= 98$

33 $f'(x) = 6x^2 + 10x + 12$

34 $(2x^2 + 6x + 5)(6x + 4) + (3x^2 + 4x + 10)(4x + 6)$

35 $f'(x) = 3(4x^3 + 6x^2)^2(12x^2 + 12x)$

$= 576x^8 + 2304x^7 + 3024x^6 + 1296x^5$

36 $x e^{x^2/2}$

37 $f'(x) = [(x^3 + 10)(4 + 12x) - (4x + 6x^2)(3x^2 + 6)]/[(x^3 + 6x)^2]$

$= [-6x^4 - 8x^3 + 36x^2]/[x^6 + 12x^4 + 36x^2]$

38 $[20 - 18x^4]/(9x^8 + 60x^4 + 100)$

39 $2x^2 + c$

40 $x^3 + 2x^2 + 10x + c$

41 $0.5x^4 + 2x^3 - 10x^{-1} + c$

42 $25x + c$

43 $\dfrac{x^3}{3} + \dfrac{x^2}{2} + 5x + c$

44 26

45 200

46 1433.33

47 (a) $x = 20.53$

(b) 108.93

(c) $E_D = -2.65$

(e) Production level decreases.

48 (a) TC $= 16Q - Q^2 + 8$

(b) TR $= 40Q - 8Q^2$

(c) $Q = 2.5$

(d) $Q = 1.71429$

(e) 3 or 0.428 57

49 $\delta y/\delta x = 4z$, $\delta y/\delta z = 4x$

50 $\delta y/\delta x = 6x^2 + 8xz + 2^2$, $\delta y/\delta z = 4x^2 + 4xz - 9z^2$

51 $\delta y/\delta x = 7 + 6xz - 3x^2z^2 - 2z + 5z^2$,
$\delta y/\delta z = 3x^2 - 2x^3z - 2x + 10xz - 21xz^2$

52 $\delta y/\delta x = 40x^3$, $\delta y/\delta z = -45z^2$

53 $\delta q/\delta p_1 = 4p_1 - 3 + 4p_2$, $\delta q/\delta p_2 = 4p_1 - 5 + 2p_2$

54 $\delta r/\delta t = 2s^2 + 2s + 10 - 12ts + 12s = 2s^2 + 14s - 12ts + 10$,
$\delta r/\delta s = 4ts + 2t - 6t^2 + 12t - 21 = -6t^2 + 14t + 14ts - 21$

55 $z = 3, x = 2, y = 1.5$

56 $z = 5, x = 1, y = 7$

57 $z = 10, x = 5, y = 400$

58 $y = 1.5, x = 2, z = 31$

59 $z = 0, x = 0, y = 0$

60 Linear so no max. or min.

61 $L = 2, K = 3$, TC $= 1900$

62 $X = 2, Y = 2.2, \Pi = 52.6$

63 (a) $Y = 1, X = 5, \Pi = 130$

(b) $Y = 0, X = 5.2, \Pi = 125.2$

(c) $X = 0, Y = 2, \Pi = 10$

64 $z = 0.868, x = 2.605$

65 $z = 0.101\,77, x = 0.508\,85$

66 $y = 539.0435, x = 741.4348, z = -716\,234.7$

67 $y = 2.233$, $x = 10.2913$, $z = -91.652\,71$

68 $y = 0.5$, $x = 0.5$, $z = 0.5$

69 $L = 33.333\,333$, $K = 20$, $Q = 103.2796$

70 $K = 119.403$, $L = 44.776$, $Q = 747.5016$

71 (a) $x = 1$, $y = 5$, $\Pi = 718$

 (b) $y = 4.629\,29$, $x = 0.185\,185$, $\Pi = 715.9252$

Index